AF888305

METHODS IN MOLECULAR BIOLOGY

Series Editor
John M. Walker
School of Life and Medical Sciences
University of Hertfordshire
Hatfield, Hertfordshire, AL10 9AB, UK

For further volumes:
http://www.springer.com/series/7651

The Nuclear Envelope

Methods and Protocols

Edited by

Sue Shackleton

Department of Biochemistry, University of Leicester, Leicester, UK

Philippe Collas

Institute of Basic Medical Sciences, University of Oslo, Oslo, Norway

Eric C. Schirmer

Wellcome Trust Centre for Cell Biology, University of Edinburgh, Edinburgh, UK

Humana Press

Editors
Sue Shackleton
Department of Biochemistry
University of Leicester
Leicester, UK

Philippe Collas
Institute of Basic Medical Sciences
University of Oslo
Oslo, Norway

Eric C. Schirmer
Wellcome Trust Centre for Cell Biology
University of Edinburgh
Edinburgh, UK

ISSN 1064-3745 ISSN 1940-6029 (electronic)
Methods in Molecular Biology
ISBN 978-1-4939-3528-4 ISBN 978-1-4939-3530-7 (eBook)
DOI 10.1007/978-1-4939-3530-7

Library of Congress Control Number: 2016937417

© Springer Science+Business Media New York 2016
This work is subject to copyright. All rights are reserved by the Publisher, whether the whole or part of the material is concerned, specifically the rights of translation, reprinting, reuse of illustrations, recitation, broadcasting, reproduction on microfilms or in any other physical way, and transmission or information storage and retrieval, electronic adaptation, computer software, or by similar or dissimilar methodology now known or hereafter developed.
The use of general descriptive names, registered names, trademarks, service marks, etc. in this publication does not imply, even in the absence of a specific statement, that such names are exempt from the relevant protective laws and regulations and therefore free for general use.
The publisher, the authors and the editors are safe to assume that the advice and information in this book are believed to be true and accurate at the date of publication. Neither the publisher nor the authors or the editors give a warranty, express or implied, with respect to the material contained herein or for any errors or omissions that may have been made.

Cover illustration: Super-resolution light microscopy reconstruction of a nucleus from a cultured human dermal fibroblast accumulating immature lamin A, labelled to show prelamin A (green), mature lamin A/C (orange) and DNA (blue). For further information, see Chapter 10. Image credit of Marek Drozdz and David Vaux.

Printed on acid-free paper

This Humana Press imprint is published by Springer Nature
The registered company is Springer Science+Business Media LLC New York

Preface

The nuclear envelope (NE) is a double membrane system enclosing the nucleus and is the central distinguishing feature of all eukaryotes. In addition to protecting the genetic material, the NE regulates the trafficking of proteins, RNAs, and ribosomes between the nucleus and cytoplasm. More recently, the identification of hundreds of NE transmembrane proteins (NETs) has revealed that the NE is also involved in diverse structural and signaling networks, both within the nucleus and through connections between the nucleus and the cytoskeleton. The importance of these networks is highlighted by the involvement of NETs and NE intermediate filaments, the nuclear lamins, in a wide range of inherited diseases. Indications that NE composition is highly tissue-specific further implicate the NE in enabling the level of complexity in gene regulation required to support tissue evolution in higher organisms.

Despite its considerable importance, the NE is among the least understood cellular organelles. This largely reflects the inherent difficulties in studying the NE and its component proteins. For example, lamins, as intermediate filaments, are highly insoluble. NETs embedded in the outer membrane tend to bind cytoskeletal proteins and these properties together make them difficult to work with. NETs in the inner membrane have these same issues and in addition often bind chromatin. Thus, even fragments lacking membrane-spanning regions tend to be insoluble. On top of this, the complex organization of the NE and its dynamic nature, undergoing disassembly and reassembly with each cell cycle, makes standard methodologies such as FRAP, coIP, ChIP, and quantification by Western blot subject to additional constraints that require modifications of procedures. The extraordinary complexity of the nuclear pore complex (NPC)—the largest complex in biology—also leads to specific refinement of standard protocols.

This volume provides a wide range of protocols used in studying the NE, with special attention to the experimental adjustments that may be required to successfully investigate this complex organelle in cells from various organisms. Many of these modifications have been only passed on within the laboratories working for many years in the field. We feel this volume is particularly timely now that many new laboratories have joined this extremely dynamic and rapidly growing field.

Leicester, UK — *Sue Shackleton*
Oslo, Norway — *Philippe Collas*
Edinburgh, UK — *Eric C. Schirmer*

Contents

Contributors

STEPHEN A. ADAM • *Department of Cell and Molecular Biology, Northwestern University Feinberg School of Medicine, Chicago, IL, USA*

MAI HASSAN AHMED • *Division of Biosciences, College of Life and Health Sciences, Brunel University London, Uxbridge, UK*

MANFRED ALSHEIMER • *Department of Cell and Developmental Biology, Biocenter, University of Würzburg, Würzburg, Germany*

SUSUMU ANTOKU • *Department of Pathology and Cell Biology, College of Physicians and Surgeons, Columbia University, New York, NY, USA*

TOBIAS ANTON • *Department of Biology II, Biozentrum, Ludwig-Maximilians-Universität München, Planegg-Martinsried, Germany*

PETER ASKJAER • *Andalusian Center for Developmental Biology (CABD), CSIC/JA/ Universidad Pablo de Olavide, Seville, Spain*

FLAVIA AUTORE • *Randall Division of Cell and Molecular Biophysics, King's College London, London, UK; British Heart Foundation Centre of Research Excellence, Cardiovascular Division, King's College London, London, UK*

MAFALDA AZEVEDO • *Graduate Program in Basic and Applied Biology (GABBA), Institute of Biomedical Sciences Abel Salazar, University of Porto, Porto, Portugal; Program in Developmental Biology, Sloan Kettering Institute, New York, NY, USA*

MRIDULA BALAKRISHNAN • *Program in Developmental Biology, Sloan Kettering Institute, New York, NY, USA; Weill Graduate School at Cornell Medical College, New York, NY, USA*

ALICE BARATEAU • *Unit of Functional and Adaptive Biology (BFA) CNRS UMR 8251, Université Paris Diderot, Sorbonne Paris Cité, Paris, France*

MARY BAYLIES • *Program in Developmental Biology, Sloan Kettering Institute, New York, NY, USA; Weill Graduate School at Cornell Medical College, New York, NY, USA*

RICARDO BENAVENTE • *Department of Cell and Developmental Biology, Biocenter, University of Würzburg, Würzburg, Germany*

URAL BIKKUL • *Division of Biosciences, College of Life and Health Sciences, Brunel University London, Uxbridge, UK*

MATTHEW G. BRAY • *Space Exploration Sector, Johns Hopkins Applied Physics Laboratory, Laurel, MD, USA*

JOANNA M. BRIDGER • *Division of Biosciences, College of Life and Health Sciences, Brunel University London, Uxbridge, UK*

BRIGITTE BUENDIA • *Unit of Functional and Adaptive Biology (BFA) CNRS UMR 8251, Université Paris Diderot, Sorbonne Paris Cité, Paris, France*

RICHARD D. BYRNE • *Signaling Programme, The Babraham Institute, Cambridge, UK*

JAMES CARTHEW • *School of Biological and Biomedical Sciences, University of Durham, Durham, UK*

BRIAN T. CHAIT • *Laboratory of Mass Spectrometry and Gaseous Ion Chemistry, The Rockefeller University, New York, NY, USA*

WAKAM CHANG • *Department of Pathology and Cell Biology, College of Physicians and Surgeons, Columbia University, New York, NY, USA*

CRAIG S. CLEMENTS • *Division of Biosciences, College of Life and Health Sciences, Brunel University London, Uxbridge, UK*

PHILIPPE COLLAS • *Department of Molecular Medicine, Institute of Basic Medical Sciences, University of Oslo, Oslo, Norway*

SUSAN COX • *Randall Division of Cell and Molecular Biophysics, New Hunt's House, King's College London, London, UK*

MAREK DROZDZ • *Sir William Dunn School of Pathology, Oxford, UK*

ISABELLE DUBAND-GOULET • *Pathophysiology of Striated Muscles Laboratory, Unit of Functional and Adaptive Biology, University of Paris Diderot-UMR CNRS 8251, Paris, France*

GREGORY FEDORCHAK • *School of Biomedical Engineering, Weill Institute for Cell and Molecular Biology, Cornell University, Ithaca, NY, USA*

MARK C. FIELD • *Division of Biological Chemistry and Drug Discovery, University of Dundee, Dundee, UK*

LAURENCE FLORENS • *The Stowers Institute for Medical Research, Kansas City, MO, USA*

ERIC FOLKER • *Program in Developmental Biology, Sloan Kettering Institute, New York, NY, USA; Biology Department, Boston College, Chestnut Hill, MA, USA*

HELEN A. FOSTER • *Division of Biosciences, College of Life and Health Sciences, Brunel University London, Uxbridge, UK*

LARRY GERACE • *Department of Cell and Molecular Biology, The Scripps Research Institute, La Jolla, CA, USA*

LAUREN S. GODWIN • *Division of Biosciences, College of Life and Health Sciences, Brunel University London, Uxbridge, UK*

MARTIN W. GOLDBERG • *School of Biological and Biomedical Sciences, University of Durham, Durham, UK*

ROBERT D. GOLDMAN • *Department of Cell and Molecular Biology, Feinberg School of Medicine, Northwestern University, Chicago, IL, USA*

GEORGINA GÓMEZ-SALDIVAR • *Andalusian Center for Developmental Biology (CABD), CSIC/JA/Universidad Pablo de Olavide, Seville, Spain*

SUSANA GONZALO • *Edward A. Doisy Department of Biochemistry and Molecular Biology, St Louis University School of Medicine, St. Louis, MO, USA*

KATJA GRAUMANN • *Department of Biological and Medical Sciences, Oxford Brookes University, Oxford, UK*

GREGG G. GUNDERSEN • *Department of Pathology and Cell Biology, College of Physicians and Surgeons, Columbia University, New York, NY, USA*

ERIKA L.F. HOLZBAUR • *Department of Physiology, The Pennsylvania Muscle Institute, Perelman School of Medicine at the University of Pennsylvania, Philadelphia, PA, USA*

HENNING F. HORN • *College of Science and Engineering, Qatar Foundation, Hamad bin Khalifa University, Doha, Qatar*

SAMUEL C. JENSEN • *Sanford Research, Sanford Children's Health Research Center, Sioux Falls, SD, USA*

IAKOWOS KARAKESISOGLOU • *School of Biological and Biomedical Sciences, University of Durham, Durham, UK*

RALPH H. KEHLENBACH • *Faculty of Medicine, Institute of Molecular Biology, University of Göttingen, Göttingen, Germany*

JOSEPH M. KELICH • *Department of Biology, Temple University, Philadelphia, PA, USA*

DAE IN KIM • *Sanford Research, Sanford Children's Health Research Center, Sioux Falls, SD, USA*
NADIA KORFALI • *The Wellcome Trust Centre for Cell Biology and Institute of Cell Biology, University of Edinburgh, Edinburgh, UK*
RAY KREIENKAMP • *Edward A. Doisy Department of Biochemistry and Molecular Biology, St Louis University School of Medicine, St. Louis, MO, USA*
ULRIKE KUTAY • *Department of Biology, Institute of Biochemistry, ETH Zurich, Zurich, Switzerland*
JAN LAMMERDING • *School of Biomedical Engineering, Weill Institute for Cell and Molecular Biology, Cornell University, Ithaca, NY, USA*
BANAFSHÉ LARIJANI • *Cell Biophysics Laboratory, Ikerbasque Basque Foundation for Science, Unidad de Biofísica (CSIC UPV/EHU), Leioa, Bizkaia, Spain; Research Center for Experimental Marine Biology and Biotechnology (PiE), University of the Basque Country (UPV), Leioa, Bizkaia, Spain*
HEINRICH LEONHARDT • *Department of Biology II, Biozentrum, Ludwig-Maximilians-Universität München, Planegg-Martinsried, Germany*
JANA LINK • *Department of Cell and Developmental Biology, Biocenter, University of Würzburg, Würzburg, Germany; Department of Chromosome Biology, Max F. Perutz Laboratories (MFPL), Vienna, Austria*
JIONG MA • *Department of Biology, Temple University, Philadelphia, PA, USA*
YOLANDA MARKAKI • *Department of Biology II, Biozentrum, Ludwig-Maximilians-Universität München, Planegg-Martinsried, Germany*
IRIS MEIER • *Department of Molecular Genetics, The Ohio State University, Columbus, OH, USA*
PETER MEISTER • *Cell Fate and Nuclear Organization, Institute of Cell Biology, University of Bern, Bern, Switzerland*
ROSE-MARIE MINAISAH • *British Heart Foundation Centre of Research Excellence, Cardiovascular Division, James Black Centre, King's College London, London, UK*
KRISHNA C. MUDUMBI • *Department of Biology, Temple University, Philadelphia, PA, USA*
SAMSON O. OBADO • *Laboratory of Cellular and Structural Biology, The Rockefeller University, New York, NY, USA*
ANJA R. OLDENBURG • *Department of Molecular Medicine, Institute of Basic Medical Sciences, University of Oslo, Oslo, Norway*
CHAN-GI PACK • *ASAN Institute for Life Sciences, ASAN Medical Center, University of Ulsan College of Medicine, Seoul, South Korea*
SUMIT PAWAR • *Department of Biology, Institute of Biochemistry, ETH Zurich, Zurich, Switzerland; Molecular Life Sciences Ph.D. Program, Zurich, Switzerland*
DOMINIC L. POCCIA • *Department of Biology, Amherst College, Amherst, MA, USA*
SARAH A. PORT • *Faculty of Medicine, Institute of Molecular Biology, University of Göttingen, Göttingen, Germany*
MICHAEL I. ROBSON • *The Wellcome Trust Centre for Cell Biology, University of Edinburgh, Edinburgh, UK*
MICHAEL P. ROUT • *Laboratory of Cellular and Structural Biology, The Rockefeller University, New York, NY, USA*
KYLE J. ROUX • *Sanford Children's Health Research Center, Sanford Research, Sioux Falls, SD, USA; Department of Pediatrics, Sanford School of Medicine, University of South Dakota, Sioux Falls, SD, USA*

ERIC C. SCHIRMER • *Wellcome Trust Centre for Cell Biology, University of Edinburgh, Edinburgh, UK*
VICTORIA K. SCHULMAN • *Program in Developmental Biology, Sloan Kettering Institute, New York, NY, USA; Department of Genetics, Yale University School of Medicine, New Haven, CT, USA*
SUE SHACKLETON • *Department of Molecular and Cell Biology, University of Leicester, Leicester, UK*
CATHERINE M. SHANAHAN • *British Heart Foundation Centre of Research Excellence, Cardiovascular Division, King's College London, London, UK*
SHIMI TAKESHI • *Department of Cell and Molecular Biology, Feinberg School of Medicine, Northwestern University, Chicago, IL, USA; Department of Human Genetics, University of Chicago, Chicago, IL, USA*
KENTARO TAMURA • *Department of Botany, Kyoto University, Kyoto, Japan*
OLGA TAPIA • *Department of Cell and Molecular Biology, The Scripps Research Institute, La Jolla, CA, USA*
ROSEMARIE UNGRICHT • *Department of Biology, Institute of Biochemistry, ETH Zurich, Zurich, Switzerland; Molecular Life Sciences Ph.D. Program, Zurich, Switzerland*
DAVID J. VAUX • *Sir William Dunn School of Pathology, University of Oxford, Oxford, UK*
DEREK T. WARREN • *British Heart Foundation Centre of Research Excellence, Cardiovascular Division, James Black Centre, King's College London, London, UK*
MEREDITH H. WILSON • *Department of Physiology, The Pennsylvania Muscle Institute, Perelman School of Medicine at the University of Pennsylvania, Philadelphia, PA, USA*
GRAHAM D. WRIGHT • *Institute of Medical Biology, A*STAR, Singapore, Singapore*
WEI XIE • *Institute of Medical Biology, A*STAR, Singapore, Singapore*
WEIDONG YANG • *Department of Biology, Temple University, Philadelphia, PA, USA*
QIUPING ZHANG • *British Heart Foundation Centre of Research Excellence, Cardiovascular Division, King's College London, London, UK*
XIAO ZHOU • *Department of Molecular Genetics, The Ohio State University, Columbus, OH, USA*

Part I

Nuclear Envelope Isolation

Chapter 1

Isolation, Proteomic Analysis, and Microscopy Confirmation of the Liver Nuclear Envelope Proteome

Nadia Korfali, Laurence Florens, and Eric C. Schirmer

Abstract

Nuclei can be relatively easily extracted from homogenized liver due to the softness of the tissue and crudely separated from other cellular organelles by low-speed centrifugation due to the comparatively large size of nuclei. However, further purification is complicated by nuclear envelope continuity with the endoplasmic reticulum, invaginations containing mitochondria, and connections to the cytoskeleton. Subsequent purification to nuclear envelopes is additionally confounded by connections of inner nuclear membrane proteins to chromatin. For these reasons, it is necessary to confirm proteomic identification of nuclear envelope proteins by testing targeting of individual proteins. The proteomic identification of nuclear envelope fractions is affected by the tendencies of transmembrane proteins to have extreme isoelectric points, strongly hydrophobic peptides, posttranslational modifications, and a propensity to aggregate, thus making proteolysis inefficient. To circumvent these problems, we have developed a MudPIT approach that uses multiple extractions and sequential proteolysis to increase identifications. Here we describe methods for isolating nuclear envelopes, determining their proteome by MudPIT, and confirming their targeting to the nuclear periphery by microscopy.

Key words Multidimensional protein identification technology (MudPIT), Nuclear envelope, Transmembrane protein, Inner nuclear membrane, Nuclear lamina, Digitonin permeabilization, Fluorescence microscopy

1 Introduction

The nuclear envelope (NE) is comprised of a double membrane system with many associated proteins and is perforated by nuclear pore complexes (NPCs). The outer nuclear membrane (ONM), continuous with the endoplasmic reticulum (ER), shares many ER proteins including ribosomes [1]. Other ONM proteins connect the cytoplasmic filaments to the NE [2]. The inner nuclear membrane (INM) has many of its own unique NE proteins, many of which are integral and connect both to chromatin and the nuclear lamin polymer that underlies the inner membrane and is also considered a core component of the NE. These many interconnections render biochemical isolation of "pure" NEs effectively impossible; however,

Sue Shackleton et al. (eds.), *The Nuclear Envelope: Methods and Protocols*, Methods in Molecular Biology, vol. 1411,
DOI 10.1007/978-1-4939-3530-7_1, © Springer Science+Business Media New York 2016

they also make it impossible to come to a clear consensus as to what constitutes the boundaries of theoretically "pure" NEs. Thus here we will present a method for isolation of NEs that first separates nuclei from most cellular contaminants taking advantage of their mass and density and removing most contaminating membranes by floating on sucrose cushions that are penetrated by the dense nuclei [3]. Next, nucleoplasmic contents are removed by swelling nuclei, digesting chromatin, and washing away released chromatin fragments by floating it on sucrose [4, 5]. These steps reduce the amount of weakly bound "contaminants," but further extractions for enrichment can be made based on the individual goals.

One approach to minimize "contaminants" is to isolate separate fractions of the connected cellular domains, analyze these independently, and subtract the proteins that occur in both fractions. This was done previously when the focus was on identifying transmembrane proteins because the principal membrane contaminants of NEs should be from the ER [6, 7]. Because nuclei and mitochondria remain intact and can be readily separated from other membranes with low-speed centrifugation [3], a microsomal fraction rich in ER and lacking NE contamination can be readily prepared [8] as the subtractive fraction. A second approach is to subsequently extract the NEs prior to mass spectrometry analysis with chemical treatments to remove particular types of potential contaminants. For example, after treatment with chaotropes or NaOH, principally those proteins embedded in the membrane should remain. On the other hand treatment with high salt and detergents should principally extract those proteins that are not connected in some way to the intermediate filament lamin polymer that was originally defined based on its resistance to such treatment [9, 10]. When we previously applied both these treatments to crude NEs, 14 % of proteins we identified in the fraction remaining after salt/detergent extraction and 54 % of those in the fraction remaining after a NaOH extraction had predicted membrane spanning domains, yet there was only minimal overlap between the transmembrane proteins identified in both fractions. Thus, the two different types of extraction are needed to recover most of the transmembrane proteins.

To comprehensively identify all proteins in the liver NE fraction, particularly with respect to transmembrane proteins, we use multidimensional protein identification technology (MudPIT) [11, 12]. Many transmembrane proteins have extreme isoelectric points and a strong tendency to aggregate due to hydrophobic stretches, and this makes their separation and isolation prior to mass spectrometry analysis problematic. However, if a complex protein mixture is first digested into peptides, the even more complex mixture generated should contain some from these transmembrane proteins. Such peptides have more homogeneous physicochemical properties than proteins, thus enabling their separation by reverse phase chromatography before application to a tandem mass spectrometer. Amino acid sequences are deduced from peptide fragmentation patterns using software such as SEQUEST [13], and the resulting peptide level information is reassembled into protein lists using software such as DTASelect [14]. Multiple analyses are then compared using software such as CONTRAST [14]. Label-free quantitative values can be calculated for each protein in each run using software such as NSAF v. 7 [15]. Various datasets from different extractions can then be combined or subtracted depending on the approach being used to generate a combined direct and in silico-purified list of NE proteins. These are then confirmed directly by testing for NE targeting as tagged fusions. This approach has thus far confirmed the localization of 31 liver NE transmembrane proteins (NETs) [6, 16–19].

2 Materials

2.1 Preparation of Liver Tissue

2.1.1 Animals

1. Ten 6–8-week-old rats (e.g., Sprague-Dawley or equivalent) or mice (e.g., CB6F1/J or equivalent) (*see* **Note 1**).
2. Volumes given are based on grams of liver or OD of nuclei. To estimate how many animals to use: ~5 g of liver can be obtained from one rat and ~1.25 g of liver can be obtained from one mouse (*see* **Note 2**). We generally produce 1000–2000 OD of nuclei (1 OD = 3,000,000 nuclei) from the livers of 10 rats (*see* **Note 3**).

2.1.2 Hardware

1. Dissection equipment: scissors, scalpel, and forceps/tweezers.
2. Two beakers on ice, one with 200 mL ddH_2O and another with 200 mL 0.25 M SHKM buffer.
3. Appropriate materials for covering surfaces during procedure and for cleaning and waste disposal.

2.1.3 Solutions

1. 100 mM phenylmethylsulfonyl fluoride (PMSF) in ethanol (*see* **Note 4**).
2. 1 M dithiothreitol (DTT) in water.
3. 0.25 M SHKM: 250 mM sucrose, 50 mM HEPES pH 7.4, 25 mM KCl, 5 mM $MgCl_2$, and freshly added 2 mM DTT (dithiothreitol) and 1 mM PMSF (*see* **Note 5**).

2.2 Preparation of NEs

2.2.1 Hardware

1. Scissors and beaker with diameter similar to the spread of the scissors blades for crudely chopping livers.
2. Potter-Elvehjem homogenizer with a motor-driven Teflon pestle providing 0.1 to 0.15 mm clearance and the drive motor capable of 1500 rotations per minute (e.g., Potter S Homogenizer from Sartorius: catalog numbers 853 3032

(motor), 854 2600 (60 mL homogenizer cylinder), and 854 3003 (Plunger made of PTFE)) (*see* **Note 6**).

3. Loose fitting (e.g., Wheaton type B pestle) glass dounce homogenizer with clearance between ~0.1 and 0.15 mm.
4. Assorted beakers, 2 funnels, and several spatulas.
5. Sterile cheesecloth (*see* **Note 7**).
6. Large bore Luer lock stainless steel needles (e.g., 14 gauge) of greater length than centrifuge tubes and glass Luer lock syringes (*see* **Note 8**).
7. Standard centrifuge with swinging-bucket rotor for low-speed ($1000 \times g$) pelleting of relatively large volumes (total ~200 mL, though this can be distributed to multiple tubes) of nuclei (e.g., Beckman Avanti J-25 floor model centrifuge with JS-13.1 rotor or Heraeus Tabletop Multifuge 3S-R centrifuge with swinging-bucket rotor) and corresponding tubes (*see* **Note 9**).
8. Pipet-Aid and 25 mL pipettes or other gentle vacuum-based systems for aspiration.
9. Ultracentrifuge with a swinging-bucket rotor that has a tube capacity of at least 200 mL if processing 10 rats (e.g., Beckman Coulter SW28 rotor with Beckman Coulter 344058 Ultra-Clear 25×89 mm centrifuge tubes).
10. Local standard light microscope, glass slides, and coverslips.

2.2.2 Solutions

DTT and protease inhibitors should always be added fresh. Only two parameters vary among the solutions: sucrose concentration and salt concentration. Solution names start with a molarity for the sucrose concentration and include the initials for the primary components: S for sucrose, H for HEPES, K for KCl, and M for $MgCl_2$ (*see* **Note 10**).

1. 1 M dithiothreitol (DTT) in H_2O (*see* Subheading 2.1.3).
2. 100 mM phenylmethylsulfonyl fluoride (PMSF) in methanol (*see* Subheading 2.1.3).
3. Other protease inhibitors (*see* **Note 11**): in addition to 1 mM PMSF, all subsequent solutions require freshly added 1 μg/mL aprotinin (from a 1 mg/mL stock in H_2O), 1 μM pepstatin A [from a 1 mM stock in DMSO (dimethyl sulfoxide)], and 10 μM leupeptin hemisulfate (from a 10 mM stock in H_2O) (*see* **Note 12**).
4. 0.25 M SHKM: 250 mM sucrose, 50 mM HEPES pH 7.4, 25 mM KCl, 5 mM $MgCl_2$, and freshly added 2 mM DTT and protease inhibitors. This is the same solution used to wash the freshly isolated livers, except that additional protease inhibitors are added to the fresh buffer in which homogenization is engaged (*see* **Note 13**).

5. 2.3 M SHKM: 2.3 M sucrose, 50 mM HEPES pH 7.4, 25 mM KCl, 5 mM $MgCl_2$, and freshly added 2 mM DTT (*see* **Note 14**).
6. 0.9 M SHKM: 0.9 M sucrose, 50 mM HEPES pH 7.4, 25 mM KCl, 5 mM $MgCl_2$, and freshly added 2 mM DTT and protease inhibitors (*see* **Note 15**).
7. 0.9 M SHM buffer: 0.9 M sucrose, 10 mM HEPES pH 7.4, 5 mM $MgCl_2$, and freshly added 2 mM DTT and protease inhibitors.
8. 0.3 M SHM buffer: 0.3 M sucrose, 10 mM HEPES pH 7.4, 5 mM $MgCl_2$, and freshly added 2 mM DTT and protease inhibitors.
9. DNase (e.g., we use Sigma DNase I D4527) resuspended at 10 U/μL in H_2O. If other DNases are used, the initial digestions should be more carefully monitored as the activities can vary considerably.
10. RNase resuspended in H_2O at 10 mg/mL. We typically also boil it for 20 min to inactivate other nuclease activities; however, in theory this should not be required since the goal is to destroy all nucleic acids in the nucleus.

2.3 Extraction of Fractions

2.3.1 Hardware

1. Refrigerated table-top microfuge.
2. TLA100.3 rotor for table-top ultracentrifuge or equivalent and corresponding tubes (e.g., Beckman Coulter 343778 polycarbonate 11 × 34 mm tubes).

2.3.2 Solutions

1. Salt/detergent extraction (for Subheading 3.3.2): octyl ß-d-glucopyranoside (also called n-octyl glucoside) resuspended at 1 % in a solution containing 25 mM HEPES pH 7.5, 400 mM KCl (*see* **Note 16**).
2. Alkaline extraction (for Subheading 3.3.3): 0.1 N NaOH, 1 mM DTT.
3. Chaotrope extraction (for Subheading 3.3.4): 8 M urea, 200 mM Na_2CO_3 (2×; *see* **Note 17**).

2.4 Preparation and Digestion of Proteins for MudPIT

2.4.1 Hardware

1. Thermomixer and block for 1.5 mL tubes (e.g., Eppendorf, 5355 000.011).
2. Mini microcentrifuge (e.g., VWR, 37000-700).
3. Refrigerated microcentrifuge with rotor.

2.4.2 Solutions

Other vendors can be used than those indicated, provided the purity of chemical/solution is similar. High-grade reagents are required for mass spectrometry analysis.

1. Hydrochloric acid, HCl (purity equivalent to J.T. Baker, JT9535-2).

2. Tris base (purity equivalent to EMD Chemicals, EM-9210), as a 1 M solution in HPLC grade water, pH adjusted to 8.5 with HCl, stored at 4 °C.
3. Benzonase (purity equivalent to Sigma-Aldrich, E8263), as a 0.1 U/μL stock in HPLC grade H_2O, stored at −20 °C.
4. Urea, solid (purity equivalent to Sigma-Aldrich, U1250).
5. HPLC grade water (purity equivalent to EMD Chemicals, EM-WX0008-1).
6. Tris(2-carboxylethyl)-phosphine hydrochloride, TCEP (purity equivalent to Pierce, 20490), as a 1 M stock in HPLC grade H_2O, stored at −20 °C.
7. 2-Chloroacetamide, CAM (purity equivalent to Sigma-Aldrich, C0267-100G), made fresh weekly as a 500 mM stock in HPLC grade H_2O and stored at −20 °C.
8. Endoproteinase Lys-C, sequencing grade (purity equivalent to Roche Applied Science, 11047825001), as a 1 μg/μL stock in double-distilled H_2O, stored at −20 °C.
9. Calcium chloride (purity equivalent to EMD Chemicals, EM-3000), as a 500 mM stock in HPLC grade H_2O, stored at room temperature.
10. Trypsin, modified sequencing grade (purity equivalent to Promega, V5111) as a 0.1 μg/μL stock in the provided stability-optimized resuspension buffer, stored at −20 °C.
11. Proteinase K (purity equivalent to Roche Applied Science, 1413783) diluted from 15.6 mg/mL to 0.25 μg/μL in H_2O (1 μL in 60 μL), stored at 4 °C.
12. Sodium carbonate (purity equivalent to Sigma-Aldrich, 451614).
13. 90 % formic acid (purity equivalent to J.T. Baker, JT0129-1).

2.5 Packing and Loading of Microcapillary Columns

The following equipment/reagents work together for our particular setup, and, as most are not common, details for each part are given. Others could potentially be used; however, it is important to us to pour a fresh column for each sample to be analyzed by mass spectrometry as failure to do so can result in contamination from a previous sample.

2.5.1 Hardware

1. Micropipette laser-based puller (Sutter Instrument Co, P-2000; http://www.sutter.com/).
2. Polyimide-coated fused silica capillary, 100 μm inner diameter (i.d.) × 365 μm outer diameter (o.d.) (Polymicro Technologies, TSP100375; http://www.polymicro.com/).
3. Polyimide-coated fused silica, 250 μm i.d. × 365 μm o.d. (Polymicro Technologies, TSP250350).

4. M-520 Inline MicroFilter Assembly (UpChurch Scientific, M-520; www.hplcsupply.com).
5. 0.5 μm PEEK filter end fitting (e.g., UpChurch Scientific, M-120X).
6. Microtight 365 μm sleeves (e.g., UpChurch Scientific, F-185X).
7. Column scribe (Chromatography Research Supplies, 205312; http://www.chromres.com/crs/).
8. Helium pressure cell (custom-made, MTA for blueprints available by request from John Yates, Scripps Research Institute, La Jolla, CA; or Brechbuehler, Inc., 1 100 110; http://www.brechbuehler.com/).

2.5.2 Solutions

1. HPLC grade methanol (purity equivalent to OMNISOLV*, 99.9 % min. by GC, EMD Chemicals, EM-MX0488-6).
2. HPLC grade acetonitrile (purity equivalent to LC-MS CHROMASOLV, >=99.9 %, Sigma-Aldrich, 34967).
3. 90 % formic acid (purity equivalent to J.T. Baker, JT0129-1).
4. Buffer A: acetonitrile/formic acid/double-distilled H_2O, 5/0.1/95, v/v.
5. Aqua 5 μm C18, 125 Ang. pore, bulk material (e.g., Phenomenex, 04A-4299; http://www.phenomenex.com/Phen/Home.htm).
6. 5 μm Partisphere strong cation exchange column (e.g., Whatman, WC4621-1507) (*see* **Note 18**).

2.6 Liquid Chromatography Coupled to Tandem Mass Spectrometry

2.6.1 Hardware

1. Degasser: e.g., Agilent 1100 series G1379A degasser, G1311A quaternary pump, and G1323B controller (Agilent Technologies).
2. Mass spectrometer.
3. Nano electrospray stage (custom-made, MTA for blueprints available by request from John Yates, Scripps Research Institute, La Jolla, CA; or Thermo Electron Nanospray II ion source; or Brechbuehler, Inc., 1 2000 1000).
4. Microtight tee (e.g., UpChurch Scientific, P-775).
5. MicroFerrule for 360 μm o.d. tubing (e.g., UpChurch Scientific, F-152).
6. Polyimide-coated fused silica, 50 μm i.d. × 365 μm o.d. (Polymicro Technologies, TSP 050375; http://www.polymicro.com/).
7. Gold wire 0.025″ diameter (Scientific Instrument Services, W352; http://www.sisweb.com/).

2.6.2 Solutions

1. Buffer A: acetonitrile/formic acid/double-distilled H_2O, 5/0.1/95, v/v.
2. Buffer B: acetonitrile/formic acid/double-distilled H_2O, 80/0.1/20, v/v.
3. Ammonium acetate 99.999 % metal basis (purity equivalent to Sigma-Aldrich, 372331).
4. Buffer C: 500 mM ammonium acetate in buffer A, filtered.

2.7 Data Analysis

2.7.1 Hardware

1. Linux computer cluster (over 100 nodes) dedicated to SEQUEST analysis.

2.7.2 Software

1. For searching MS/MS dataset: SEQUEST™ [13] (Thermo Electron and John Yates, Scripps Research Institute, La Jolla, CA).
2. For assembling and comparing protein lists: DTASelect/CONTRAST [14] (available by request from John Yates, Scripps Research Institute, La Jolla, CA).
3. For label-free spectral count quantitation: NSAF v. 7 [15].
4. For appending transmembrane domain predictions: TMHMM [20] (http://www.cbs.dtu.dk/services/TMHMM-2.0/).
5. For signal peptide predictions: SignalP [21] (http://www.cbs.dtu.dk/services/SignalP/).

2.8 Confirmation of NE Residence and INM Localization

2.8.1 Hardware

1. A standard tissue-culture setup including tissue-culture fume hoods, CO_2 incubators, and general tissue-culture consumables such as pipettes and cell culture media.
2. Tissue-culture 24-well plates (*see* **Note 19**).
3. 13 mm glass 1.5 oz coverslips and microscopy slides (*see* **Note 19**).
4. Tweezers extended tip (e.g., Dumont #5, Epoxy Coated) (*see* **Note 20**).
5. Fluorescent microscope setup. For regular (Subheading 3.8.1), Triton X-100 (Subheading 3.8.2), and digitonin (Subheading 3.8.3) procedures, any microscope setup will work from a standard epiflourescence microscope to a confocal; however, for the determination of outer from inner nuclear membrane (Subheading 3.8.4), a high degree of resolution is required. We have typically used structured illumination (OMX) microscopy for this [18, 22, 23] that requires an elaborate setup that is not widely available [24]. We have also successfully used more standard DeltaVision (Applied Precision Instruments) microscopes that are widely available to distinguish inner from outer nuclear membrane [18]. Though inferior to these others, we have also obtained adequate resolution using a more standard setup (Nikon TE-2000 microscope

equipped with a Xenon light source, Sedat quad filter set, 1.45 N.A. 100× Nikon objective, PIFOC Z-axis focus drive (Physik Instruments), and CoolSnapHQ High Speed Monochrome CCD (charge coupled device) camera (Photometrics)). Most critical when using a standard microscope setup is having a high numerical aperture objective and being able to take z-series and deconvolve the images (we use Autoquant X for this).

6. Aspirator setup with weak vacuum (*see* **Note 21**).

2.8.2 Solutions

1. Tissue-culture system with transfectable cells and culture medium. A brief discussion of advantages and disadvantages of different cell types is given at the beginning of Subheading 3.8.
2. Transfection reagents (*see* **Note 22**).
3. Phosphate-buffered saline (PBS) (*see* **Note 23**).
4. 4 % paraformaldehyde in PBS, either freshly made or diluted from vials stored under nitrogen.
5. Prefixation extraction buffer: 1 % Triton X-100 solution in PBS (for Subheading 3.8.2).
6. 100 mg/mL high purity digitonin (at least as high level purity as catalog number 300410 from EMD4Biosciences) in dimethyl sulfoxide (DMSO) (*see* **Note 24**) (for Subheading 3.8.3).
7. Trypan blue cell vital dye.
8. 4 % BSA solution in PBS (*see* **Note 25**).
9. DNA dyes such as DAPI (4,6-diamidino-2 phenylindole, dihydrochloride) or Hoechst 33342.
10. Control antibodies for a NE protein such as lamins or other NETs (for Subheadings 3.8.1–3.8.3. targeting experiments). A good commercially available lamin antibody is mouse monoclonal 119D5-F1 (e.g., catalog number MAB3213) (Chemicon Europe). Examples of commercially available NET antibodies are those against SUN2 (catalog number 06-1038) (Millipore) and LAP2β (catalog number 06-1002) (Millipore) (*see* **Note 26**).
11. Antibodies or expression constructs to mark the ER and nucleus (for Subheading 3.8.3). For a good commercially available ER marker, we recommend antibodies to calreticulin (e.g., catalog number 2891) (Cell Signaling Technology) and for a good nuclear marker any antibody or stain for general chromatin that labels throughout the nucleus as well as antibodies to proteins such as Ran.
12. Antibodies for distinguishing ONM from INM targeting (for Subheading 3.8.4). It is necessary to have an antibody to a protein from the cytoplasmic filaments of the nuclear pore complex

(NPC) (e.g., Nup358 or Nup214) and also one to the nuclear basket of the NPC (e.g., Nup153 or TPR). An example of an NPC cytoplasmic filament antibody is catalog number SC-28577 (Santa Cruz Biotechnology) against Nup358 (also called RanBP2). An example of a nuclear basket antibody is catalog number MMS-102P (Covance) against Nup153.

13. Secondary antibodies. If two different primary antibodies will be used, be certain to obtain secondary antibodies with minimal cross-reactivity to other species of primary antibody used. We recommend either Jackson ImmunoResearch Laboratories, Inc donkey or Molecular Probes-Invitrogen goat highly cross-adsorbed secondary antibodies.
14. Mounting medium: Fluoromount-G (catalog number 0100-01) (Southern Biotech) is recommended with our specific microscope setup. However, it is critical for colocalization studies to ensure that the mounting medium does not cause chromatic aberrations with the particular objective being used. Thus it is important to use a medium with a refractive index similar to the numerical aperture of the objective being used and to test it with microbeads containing multiple fluorophores to ensure that (near) perfect colocalization occurs.
15. Nail varnish that does not negatively impact on the glue holding lenses in microscope objectives.

3 Methods

The first step in NE enrichment is the isolation of nuclei. Critical to this step and all subsequent steps is the fact that nuclei from different tissues have distinct densities; thus the concentration of sucrose in buffers may need to be altered or centrifugation steps lengthened if NEs are to be isolated from tissues other than liver. Chapters detailing modifications of the NE protocol for human blood lymphocytes and rat muscle have been published in other volumes in the *Methods in Molecular Biology* series [25, 26].

3.1 Preparation of Tissue (Rodent Livers)

Most of the NE preparation procedure can be efficiently performed by one individual; however, euthanizing and dissecting the animals should be done quickly, and it is very helpful to have assistance at this point.

1. Overnight-fast the animals the night before the procedure (*see* **Note 27**).
2. Pre-weigh beaker containing 0.25 M SHKM, place on ice, and add PMSF.
3. Euthanize rats or mice according to local animal protocols.

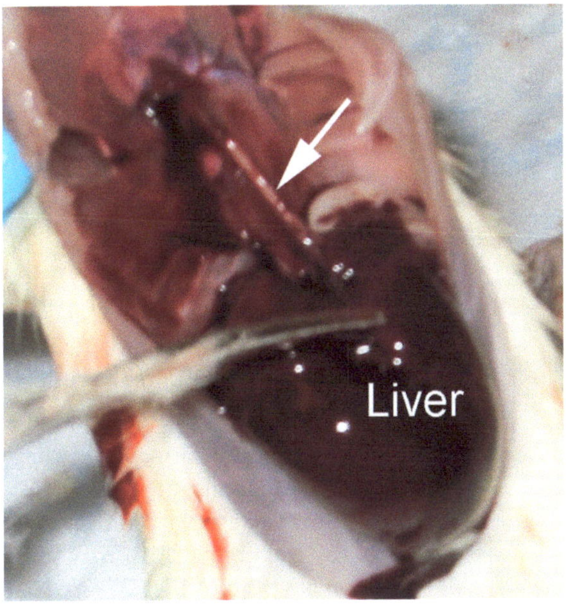

Fig. 1 Excision of livers from rats. It is critical during preparation of tissue to avoid cutting the yellowish tube (*arrow*) just beneath the liver as this is extremely rich in proteases (the liver has been pulled downward in the image shown)

4. Immediately pull up on the ventral abdominal skin. Make an incision anterior to posterior and two perpendicular incisions above the thoracic vertebrae and below the abdomen. Peel the skin back to access the liver.
5. Remove liver with a scalpel, being extremely careful to avoid the yellowish tube directly behind (Fig. 1). If this tube is cut, many proteases are released that can reduce the quality of NE preparations.
6. While clasped in the forceps, rinse fresh livers quickly in beaker containing H_2O, and then immediately place in the pre-weighed beaker on ice containing SHKM.
7. Return to the laboratory with livers as soon as local animal protocols have been satisfied. If possible, have two people working at this point so that one can begin processing the material while the other deals with disposal and cleanup.
8. Weigh the beaker containing the livers and subtract the previously measured weight from before adding livers to determine the total weight of the livers.

3.2 Preparation of NEs

1. Pour off buffer, and resuspend in fresh, ice-cold 0.25 M SHKM with freshly added protease inhibitors at 2 mL buffer for every gram of liver, e.g., 50 g of livers should be resuspended in 100 mL buffer (*see* **Note 28**).

2. Use scissors to chop livers into small pieces in beaker with buffer on ice.
3. Pour into 55 mL Potter-Elvehjem homogenizer, and homogenize at 1500 rpm in the cold, bringing the pestle to the bottom three times (*see* **Note 29**). Make sure that the Potter-Elvehjem homogenizer is immersed in ice-water container to prevent protein degradation.
4. Rinse the homogenizer with buffer and add to the homogenate (*see* **Note 30** if also preparing microsomes or mitochondria).
5. In a cold room, fold cheesecloth over four times and lay in funnel. Pour ~40 mL crude homogenate through cheesecloth (*see* **Note 31**).
6. As the flow slows, fold the cheesecloth over and roll a sterile pipette along the outside from top to bottom to squeeze fluid out. A wash with buffer poured into the central cavity formed by the cheesecloth may increase yield slightly, but use a minimal volume (Fig. 2) (*see* **Note 30**).
7. Remove to 50 mL round-bottom centrifuge tubes (*see* **Note 32**), and underlay with a 3 mL cushion of 0.9 M SHKM using a 14-gauge needle and syringe. Pellet nuclei at 1000 × *g* in a swinging-bucket rotor (e.g., 2000 rpm in a Beckman Coulter J6MI floor model centrifuge) for 10 min at 4 °C.
8. Remove the supernatant carefully with a 25 mL pipette as the pellets are very soft. Keep this supernatant if microsomal membranes are going to be prepared at the same time from the same tissue as they will be in this fraction (*see* **Note 30**).

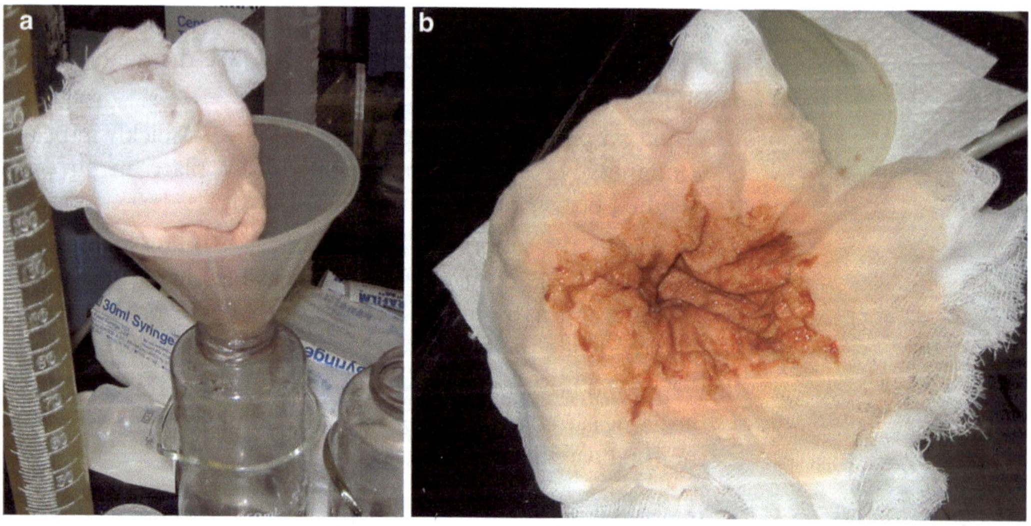

Fig. 2 Sieving nuclei through cheesecloth. (**a**) The cheesecloth is folded so that nuclei pass through four layers and the speed of sieving can be increased by folding the cheesecloth like a sac and applying gentle pressure to the side with a sterile pipette. (**b**) After sieving, much collagen remains on the cheesecloth

9. Resuspend pellets in 2.3 M SHKM and homogenize in Potter-Elvehjem homogenizer with three more strokes of the pestle at 1500 rpm. Rinse with the same buffer, accrue, and dilute with 0.25 M SHKM to a concentration of 1.9 M sucrose.
10. Aliquot 25 mL into each SW28 ultracentrifuge tube and underlay with 5 mL of 2.3 M SHKM using a 14-gauge needle in a Luer lock syringe (*see* **Note 8**).
11. Balance tubes from top (*see* **Note 33**).
12. Centrifuge in SW28 rotor for 60 min at 82,000 × *g* (25,000 rpm).
13. Carefully remove the tubes from SW28 rotor in a cold (4 °C) room. Scrape red layer at the top off with a spatula; then pour off the rest of the supernatant by rapid inversion. Keep the tubes upside down in the cold for 10 min to drain them. Then gently wipe out the inside walls of tubes with a folded Kimwipe (or equivalent towel), being very careful not to touch the pellet. Expected pellet size is shown in Fig. 3a.
14. Insert a clean dry spatula without touching the walls of the tube and scrape out the nuclear pellet (Fig. 3c). It is important that the spatula is dry so that the pellet will cling to it. Remove, again avoiding touching the walls of the tube, and resuspend pellet in 0.25 M SHKM with freshly added 2 mM DTT and protease inhibitors.

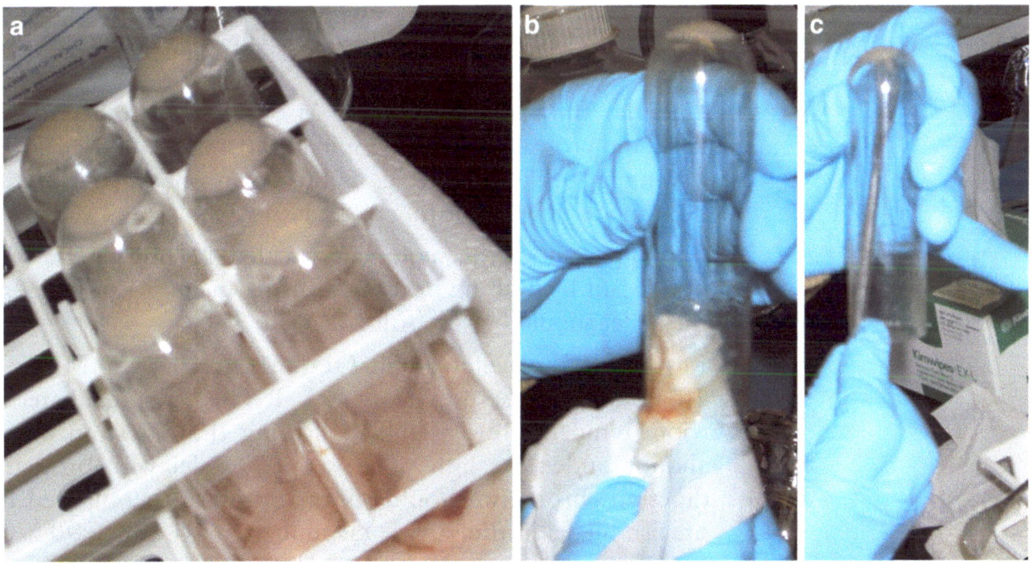

Fig. 3 Nuclear pellets. (**a**) The nuclear pellets obtained after **step 13** in Subheading 3.2 are very large. Shown are six tubes obtained from 10 rats. (**b**, **c**) After decanting the upper phases, it is critical to keep the tubes upside down; first wipe the sides of the tubes to remove proteases (**b**), and then carefully scrape the nuclei from the bottom of the tube without touching the sides of the tube (**c**)

15. Keep on ice and resuspend the pellet using a loose (Wheaton B-type pestle, ~0.1–0.15 mm clearance) dounce homogenizer until all aggregates are disassembled. Wash the homogenizer with the same buffer, and remove to round-bottom centrifuge tubes. Underlay with 5 mL of 0.9 M SHKM with freshly added 2 mM DTT and protease inhibitors as in **step 6**. Pellet nuclei by centrifugation at 1000 × *g* for 10 min. Decant by inversion as the pellet should be compact (*see* **Note 34**).
16. Resuspend in 0.25 M SHKM, dounce again on ice, and take a small aliquot for counting nuclei in a hemacytometer and also remove a small amount to a microfuge tube for observing under the microscope in **steps 18**, **19**, and **22**. Then repeat pelleting as in **step 15**.
17. During centrifugation count nuclei. The number of nuclei in the squares on the field should be multiplied by 10,000 (unless a different correction factor applies to your hemocytometer) and by the total volume (mL) in which cells were resuspended for centrifugation in **step 15**. This number should be divided by 3×10^6, which is the number of nuclei in an OD.
18. Resuspend in 10 % SHM with freshly added 2 mM DTT and protease inhibitors at 20 OD/mL. Take an aliquot to compare nuclei to the aliquot saved in **step 16**. The nuclei should be observed to swell in the hypotonic SHM buffer.
19. Add 4 U/mL DNase and 1 μg/mL RNase and incubate at room temperature for 20 min. Observe digestion on the microscope during the incubation. The grayish tint of the nuclei should diminish slightly (*see* **Note 35**).
20. Underlay the solution with 0.9 M SHM with freshly added DTT and protease inhibitors. Spin for 30 min at 6000 × *g* using a swinging-bucket rotor (e.g., 5000 rpm in a Beckman Coulter floor model J6MI centrifuge) (*see* **Note 36**).
21. Carefully remove the supernatant with a 25 mL pipette (do not decant by pouring or using a strong vacuum aspirator), as the pellet should be very soft (*see* **Note 37**).
22. Resuspend the pellet at 50 OD/mL in 10 % SHM. Add 20 U/mL DNase and 5 μg/mL RNase, and incubate at room temperature, carefully following the digestion in an aliquot under the light microscope. Nuclei with properly digested chromatin should appear translucent compared to grayish nuclei in the sample taken in **step 16**. When 90 % of nuclei are no longer phase gray, aliquot 4 mL of NE solution (200 OD) each to centrifuge tubes (chosen for desired storage method), and spin at 6000 × *g* (5000 rpm) for 30 min (no cushion).

23. If at least 90 % of nuclei are not phase clear within 20 min, instead of pelleting without a cushion, repeat **steps 20–22**, and increase the amount of nucleases.
24. Carefully aspirate the supernatants and immediately flash-freeze in liquid nitrogen and store at −80 °C.

3.3 Extraction of Fractions

The NEs generated will still have contaminants from chromatin and the cytoskeleton and are not advised to use for direct mass spectrometry analysis without further extraction/purification. However, any individual further extraction method is likely to remove true NE proteins along with contaminants. Thus it is important to carefully consider the choice for further extraction methods according to the NE characteristics of primary interest. The use of multiple extraction methods can provide relatively comprehensive coverage of NE proteins and can also be applied to increase confidence in those protein identifications that occur in fractions generated by different extraction methods. We always use at least two extraction methods. For example, with our primary focus on transmembrane proteins, we extracted with NaOH that solubilizes most proteins that are not in the membrane (including most of the intermediate filament nuclear lamin polymer) while not solubilizing membranes and thus enriches for proteins embedded in the membrane. We also extracted with detergent in high salt buffer (1 % ß-octylglucoside with 400 mM KCl) that solubilizes membranes and with them most membrane proteins that are not strongly bound to the intermediate filament lamin polymer, leaving for analysis the lamin polymer and the membrane proteins most tightly associated with it. Proteins identified in both fractions are the strongest candidate NETs; however, because of the wide range of transmembrane proteins in the NE, this method identified many now confirmed NETs [6, 22, 23, 27] that were only in one or the other fraction. Here we detail our methods for salt/detergent extraction (1 % octyl ß-d-glucopyranoside, 40 mM KCl), alkali extraction (NaOH), and chaotrope extraction (8 M urea, 200 mM Na_2CO_3).

Prior to starting the extractions, it is important to make sure that the ultracentrifuge chamber and rotor have been prechilled to 4 °C. It is crucial to control the amount of time that NEs are in NaOH such that the total time measured includes not just the centrifugation time but also the time it takes to resuspend, carry to the centrifuge, etc. If dealing with a large number of samples, it is advisable to either minimize the number of samples in NaOH at a given time or obtain additional assistance in processing samples.

3.3.1 Optimization of Extraction Conditions

If using a different buffer for extraction from those mentioned below (e.g., different detergent or salt, *see* **Note 16**), it is important to perform a pilot experiment using a small amount of NEs to confirm that known proteins are not lost under the new extraction conditions.

1. Separate out 4 OD of isolated NEs and pellet in a microfuge.
2. Extract using the novel method.
3. Load half of what remains from the 4 OD on an SDS-PAGE mini-gel and process through to a standard Western blot using antibodies to a known protein that is expected to fractionate with the NE components of interest. Antibodies to some possible proteins are noted in Subheading 2.8.2.

3.3.2 Salt/Detergent Extraction

1. Resuspend NE pellet in salt/detergent buffer (*see* **Note 16**). Use no more than 50 OD per mL. Pipette up and down until no visible particles can be observed.
2. Spin briefly (15 s at 10,000 ×*g*) in a chilled microcentrifuge to pellet large insoluble material. If the pellet represents a reasonable fraction of the starting material, fresh buffer can be added to this pellet and the process repeated.
3. Incubate on ice for 15 min.
4. Remove supernatant(s) to TLA100.3 ultracentrifuge tubes and pellet the lamina-NE protein complex at 66,000 ×*g* (35,000 rpm) for 35 min.
5. Rinse packed pellet with double-distilled H_2O, and either freeze at −80 °C or directly process to digest for mass spectrometry.

3.3.3 Alkali Extraction

1. Resuspend no more than 100 OD of NE pellet per 1 mL of NaOH/DTT solution on ice by rapid up and down pipetting, and immediately move to TLA100.3 ultracentrifuge tubes, and pellet insoluble material at 104,000 ×*g* (44,000 rpm) for 35 min (*see* **Note 38**).
2. Rinse packed pellet rapidly with double-distilled H_2O, and either freeze at −80 °C or directly process to digest for mass spectrometry.

3.3.4 Chaotrope Extraction

1. To a solution of 100 OD of NEs in 1 mL of SHKM, add an equal volume of 2× chaotrope solution to bring to a final concentration of 4 M urea/100 mM Na_2CO_3 solution (*see* **Note 39**). Mix by gentle pipetting.
2. Incubate on a shaker in a cold (4 °C) room for 15 min.
3. Pellet insoluble material at 135,000 ×*g* (50,000 rpm) in Beckman Tabletop Ultracentrifuge for 30 min.
4. As the urea and carbonate are compatible with subsequent buffers for protein digestion to peptides, the pellet can be directly frozen at −80 °C or directly processed for mass spectrometry.

3.4 Preparation and Digestion of Proteins for MudPIT

Dried membrane pellets from NaOH-extracted NEs (NaOH-NE), salt-and-detergent-extracted NEs (SD-NE), or other fractions from rat livers are first digested with endoproteinase Lys-C and trypsin; then any membranous material leftover after trypsin digestion is further digested by proteinase K at high pH [28].

3.4.1 Endoproteinase Lys-C + Trypsin

1. Resuspend NE pellets from rat livers in 120 μL of 0.1 M Tris–HCl, pH 8.5; disturb pellet by pipetting, and transfer from centrifuge tube to microfuge tube (*see* **Note 40**).
2. Add 0.1 U of benzonase at 0.1 U/μL; incubate at 37 °C for 30 min (*see* **Note 41**).
3. Add solid urea to 8 M; vortex.
4. To reduce disulfide bonds, add 0.1 M TCEP to 5 mM final concentration, and incubate at room temperature for 30 min.
5. To carboxyamidomethylate free cysteines, add 0.5 M CAM to 10 mM final concentration, and incubate at room temperature for 30 min in the dark.
6. Add endoproteinase Lys-C (at 0.5 μg/μL) to an approximate final substrate to enzyme ratio of 100:1 (w/w), and incubate at 37 °C overnight while shaking.
7. Dilute sample to 2 M urea with 100 mM Tris–HCl, pH 8.5.
8. Add 0.5 M $CaCl_2$ to a final concentration of 2 mM.
9. Add modified trypsin at 0.5 μg/μL (approximate substrate to enzyme ratio of 100:1), and incubate overnight at 37 °C while shaking.
10. Spin samples at 17,500 × *g* for 30 min.
11. Pull off supernatants and transfer to a new tube (keep pelleted membranous material for Subheading 3.4.2 below).
12. Add 90 % formic acid to 5 % to supernatants.
13. Combine all trypsin-digested supernatants from one sample type (e.g., NaOH-NE or SD-NE), mix well, and then split into four technical replicates to be analyzed independently by MudPIT.

3.4.2 Proteinase K at High pH

1. Resuspend membranous pellets obtained after spinning down the trypsin digests, in 60 μL 0.1 M sodium carbonate, pH 11.5, 8 M urea; mix well by pipetting and vortexing.
2. Bring solution to 5 mM TCEP with 1 M stock, at room temperature for 30 min.
3. Bring solution to 10 mM CAM with 0.5 M stock, at room temperature for 30 min in dark.
4. Add proteinase K; incubate for 4 h, at 37 °C while shaking.

5. Add 90 % formic acid to 5 % to supernatants.
6. Combine all proteinase K-digested pellets from one sample type (e.g., NaOH-NE or SD-NE), mix well, and then split into two technical replicates to be analyzed independently by MudPIT.

3.5 Packing and Loading of Microcapillary Column

Samples digested from membrane preparations may still contain particulate material at this step, which could clog microcapillary columns. For such samples, we use a "split" column approach (Fig. 4), in which the sample is loaded onto larger-diameter open-ended microcapillary columns packed with reverse phase and strong cation exchange resins, before being connected to a

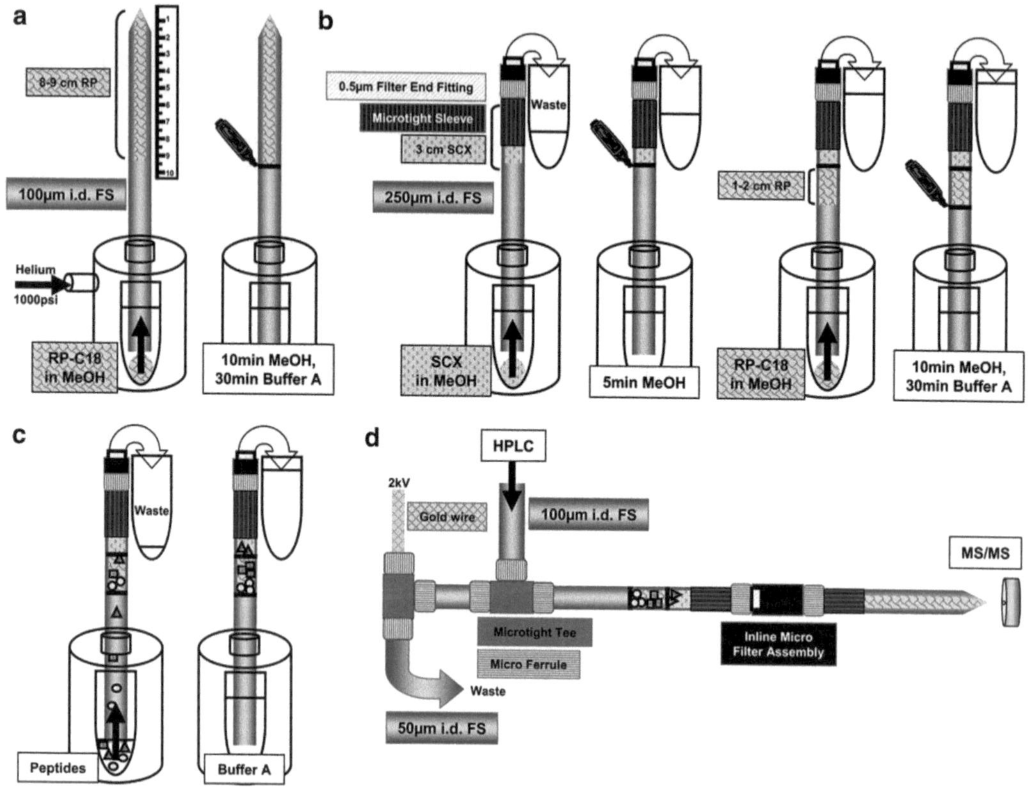

Fig. 4 Column packing, loading, and setup. (**a**) A 100 μm inner diameter (i.d.) fused silica capillary with a pulled 5 μm tip is inserted into a high-pressure device, packed using helium pressure with a slurry of Aqua C-18 reverse phase (RP) in methanol, and washed in methanol for 5 min, and then equilibrated in buffer A for 30 min. (**b**) A 250 μm i.d. fused silica capillary is fitted on one end with a microtight sleeve and a 0.5 μm filter end fitting, packed with strong cation exchange (SCX) material in a slurry, and then washed with methanol for 5 min. After marking the lower limit of the SCX resin with a marker, the column is further packed with Aqua C-18, then washed in methanol, and equilibrated in buffer A for 30 min. (**c**) The complex peptide mixture is pressure loaded onto the 250 μm i.d. column, which is subsequently washed in buffer A. (**d**) The loaded and washed 250 μm i.d. capillary is connected with the 100 μm i.d. column using a microfilter assembly. The split column is installed in line with a quaternary HPLC pump and a tandem mass spectrometer via two microtight tees

resolutive 100 μm column packed with reverse phase (*see* ref. 29) for a very detailed description of the following steps. This approach allows for extensive yet fast column washing after peptide loading.

3.5.1 Single-Phase Fused Silica 100 μm Microcapillary Column

1. Place about 40 cm of 100 μm i.d. × 365 μm o.d. fused silica into P-2000 laser puller, and use heating/pulling cycle settings (*see* **Note 42**) such as to pull the capillary to about a 5 μm opening.
2. Make slurries of 5 μm Aqua C18 Reverse Phase and of 5 μm strong cation exchange SCX material, both at about 15 mg/mL in 500 μL of methanol (*see* **Note 43**).
3. Pack a 100 μm fused silica column with 8–9 cm of 5 μm Aqua C18 RP using the high-pressure loading device (Fig. 4a).
4. Wash with methanol for at least 10 min; mark resin level in column with a marker (*see* **Note 44**).
5. Equilibrate in buffer A for at least 30 min.

3.5.2 Double-Phase Fused Silica 250 μm Microcapillary Assembly

1. Place an Inline MicroFilter Assembly with a microtight sleeve on one end of a 250 μm i.d. × 360 μm o.d. fused silica capillary cut to about 20 cm in length.
2. Pack the 250 μm capillary assembly with 3–4 cm (i.e., about 1 cm past the end of the green microtight sleeve) of 5 μm strong cation exchange material using the high-pressure loading device (Fig. 4b).
3. Wash with methanol for at least 5 min; mark resin level in column with a marker.
4. Pack fused silica column with 1–2 cm of 5 μm Aqua C18 RP (Fig. 4b); mark resin level in column with a marker.
5. Wash with methanol for at least 10 min.
6. Equilibrate in buffer A for at least 30 min.

3.5.3 Off-Line Loading and Desalting

1. Spin sample down at 17,500 × *g* for 30 min, and transfer to a new microfuge tube to remove any particulate.
2. Load the peptide sample onto the microcapillary 250 μm column assembly by placing the sample-containing microfuge tube in the high-pressure device (Fig. 4c).
3. Wash with buffer A for at least 30 min (*see* **Note 45**).

3.5.4 Connecting 100 μm Resolutive Column with Peptide-Loaded 250 μm Capillary

1. Connect the packed and washed 100 μm i.d. column to the loaded and washed 250 μm i.d. capillary assembly using the 2 μm filtered union and microtight sleeves (Fig. 4d).

3.6 Liquid Chromatography In Line With Tandem Mass Spectrometry

3.6.1 Multidimensional Liquid Chromatography

1. Install the loaded and washed split three-phase microcapillary column on the nanoelectrospray stage.
2. Connect the microcapillary column, quaternary HPLC pump, and gold wire through which a 2.4 kV voltage is applied to the liquid phase, and overflow tubing using two microtight tees (Fig. 4d) (*see* **Note 46**).
3. Keep the HPLC flow rate constant at 0.1 mL/min throughout the chromatography. However, to achieve a slower flow rate at the tip of the column of about 200–300 nL/min, split the flow using a waste line consisting of 50 μm i.d. fused silica capillary cut to about 40 cm (i.e., back pressure of ~40 bar).
4. Run a 15-step chromatography run (30 h) on samples with the gradient parameters described in Fig. 5. The chromatography is set up through and controlled by the Xcalibur™ instrument software.

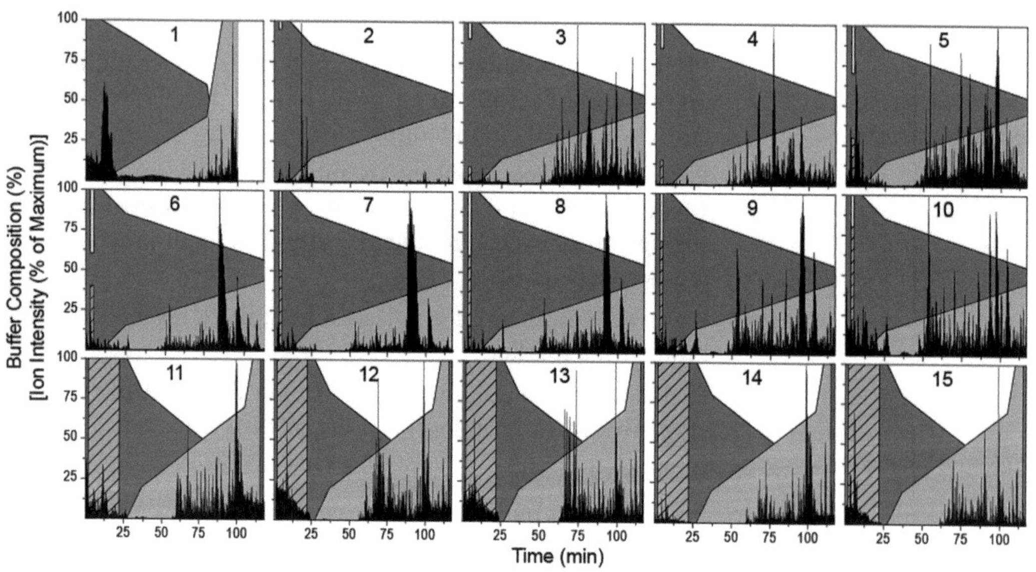

Fig. 5 Gradient profiles for a 15-step MudPIT run. Varying concentrations in buffers A, B, and C are represented by the dark gray areas, the light gray areas, and the striped gray bars, respectively. A representative baseline ion chromatograph is shown in the forefront (*black peaks*) for each step (from the analysis of the complex peptide mixture obtained from rat NaOH-extracted NEs). Each chromatographic step lasts 117 min, with the exception of the first one that lasts 100 min. Step 1 consists in a slow increase in acetonitrile concentration to 40 % buffer B over 80 min and then a sharp increase to 100 % B over 10 min. No ammonium acetate is used in step 1, while the salt concentration is equal to 5, 10, 15, 30, 40, 50, 60, 70, and 80 % C in steps 2 through 10, respectively, and 100 % C for steps 11–15. In steps 2 through 10, the salt bump starts after 3 min and lasts for 2 min, while it lasts for 20 min in the last five chromatographic steps. In steps 2 through 10, a rapid increase from 0 to 15 % buffer B occurs between 5 and 25 min, followed by a slow ramp to 45 % B over 92 min. For steps 11 through 15, buffer B concentration increases rapidly to 20 % between 22 and 37 min, followed by a slow ramp to 70 % B over 68 min, followed by a sharp increase to 100 % B maintained for 5 min, and followed by a return to 100 % for 2 min

3.6.2 Tandem Mass Spectrometry

1. Using Xcalibur™, set up the collision energy at 35 %.
2. Implement an acquisition scheme such as where a cycle of one full MS scan (from 400 to 1600 *m/z*) followed by five MS/MS events on the top five most intense ions is repeated continuously throughout the chromatographic elution time.
3. To allow less intense ions to be analyzed, enable dynamic exclusion for 120 s [30].

3.7 Data Analysis

3.7.1 Searching the MS/MS Dataset

1. Extract RAW files into ms2 file format [31] using RAW_Xtract [32].
2. Search MS/MS spectra without specifying differential modifications against a protein database, preferably matched to the species sourced for the liver (e.g., for rat we used a database consisting of 28,400 rat proteins [non-redundant NCBI sequences on July 10, 2006], plus 197 human and mouse homologs of previously identified NETs [6] and 172 sequences from usual contaminants [e.g., human keratins, IgGs, and proteolytic enzymes]).
3. To estimate false discovery rates, each non-redundant protein entry is randomized and added to the database (this brought our database total search space to 57,538 sequences).
4. Set the sequest.params file such as (1) the peptide mass tolerance is three; (2) no enzyme specificity is required; (3) parent ions are calculated with average masses, while fragment ions are modeled with monoisotopic masses; and (4) cysteine residues are considered fully carboxyamidomethylated and searched as a static modification of +57 Da.

3.7.2 Assembling and Comparing Protein Lists

1. To assemble and parse the peptide information contained in the SEQUEST output files, run DTASelect on sqt files [14].
2. To compare the proteins detected in salt-and-detergent-extracted NE, NaOH-extracted NE, and other extracted samples, create a CONTRAST table.
3. Select spectrum/peptide matches such as peptides are at least 7 amino acids long and their ends comply with the specificity of the proteolytic enzymes used, when appropriate. For trypsin-digested samples, peptides have to be fully tryptic, with DeltCn of at least 0.08; minimum XCorr of 1.8 for singly, 2.0 for doubly, and 3.0 for triply charged spectra; and maximum Sp rank of 10. For the proteinase K-digested samples, no specific peptide ends are imposed, but the DeltCn cutoff is increased to 0.15 [33], while XCorr minima are increased to 2.5 for doubly and 3.5 for triply charged spectra.
4. Merge peptide hits from all compared analyses to establish a master list of proteins identified by at least 2 peptides or 1 pep-

tide with 2 independent spectra. Based on the merged detected peptides, proteins may fall into three categories following the parsimony principle:

(a) Proteins detected by the exact same peptides are grouped together because they cannot be distinguished based on the available peptide data; only one arbitrarily selected representative protein entry is reported for such group of proteins, and its length is used to calculate NSAF values (see below).

(b) Proteins with at least one peptide uniquely mapping to them are considered unique entries.

(c) Subset proteins for which no unique peptides are detected are removed from the final list of identified proteins since the detection of their peptides can be explained more simply by other proteins with additional unique peptides.

5. Use NSAF7 (Tim Wen) to create the final report on all detected proteins across the different runs, calculate their respective normalized spectral abundance factor (NSAF) values, and estimate false discovery rates (FDR).

Spectral FDR is calculated as:

$$\mathrm{FDR} = \frac{2 \times \mathrm{SHUFFLED\_SpectralCounts}}{\mathrm{Total\_SpectralCounts}} \times 100.$$

Protein level FDR is calculated as:

$$\mathrm{ProteinFDR} = \frac{\mathrm{SHUFFLED\_Proteins}}{\mathrm{Total\_Proteins}} \times 100.$$

To estimate relative protein levels, NSAFs are calculated for each non-redundant protein, as described in Refs. 34–36:

$$(\mathrm{NSAF})_i = \frac{(\mathrm{SpectralCount} / \mathrm{Length})_i}{\sum_{k=1}^{N} (\mathrm{SpectralCount} / \mathrm{Length})_k}.$$

6. To deal with peptides shared between multiple proteins, implement the dNSAF approach to refine spectral counts [15]. For each run, dNSAFs are calculated based on distributed spectral counts, in which shared spectral counts are distributed based on spectral counts unique to each isoform [15].

3.7.3 Appending Signal Peptide and Transmembrane Domain Predictions

1. Predict the number of transmembrane segments using TMHMM on a fasta file containing the amino acid sequences for all detected proteins.

2. Predict the presence of a cleaved signal peptide or a membrane anchor using SignalP on a fasta file containing the 70 first amino acids for all detected proteins.
3. Append signal peptides and transmembrane domain predictions to the table containing detected protein quantitative information.

3.7.4 Generating a List of Putative NE Transmembrane Proteins

1. Average quantitative information (dNSAF values) across technical replicates.
2. Calculate ratio between averaged dNSAF values measured for proteins in NE fractions and in contaminating microsomal membrane ("MM") fractions ("NE to MM" ratio).
3. Rank detected proteins by decreasing dNSAF values and their frequency of detection in replicate analyses.
4. Sort proteins in different groups such as proteins previously annotated as NETs such as the 13 original NETs [37], as well as the NETs we have previously characterized from liver [6], blood [22], and muscle [23] are listed first.
5. Define membrane proteins enriched in NEs as proteins with at least one predicted transmembrane domain or membrane anchor, not detected in mitochondria proteomics studies [38] or MudPIT analyses of contaminating MMs fractions, or detected at lower levels in such MMs fractions (i.e., with NEs to MMs ratio greater than five).
6. Putative novel liver NETs should also have a plausible annotation when known, for example, some major histocompatibility complex proteins or solute carriers passed the NE enrichment selection criteria defined above, yet are not plausible as NETs based on their functional annotation.

3.8 Confirmation of NE Residence by Fluorescence Microscopy

Roughly 80 % of the proteins identified by NE proteomics that we have tested target to the NE as exogenously expressed tagged fusions. The 20 % that fail to target could indicate (1) errors in the protein identification, (2) contaminants that co-purified with the NEs, (3) mislocalization of a protein that is normally at the NE due to the tag interfering with its localization, (4) a protein that has a minor population at the NE but other populations in other cellular regions that make detection of the NE signal difficult (an estimated 40 % of the proteome has multiple cellular localizations; *see* ref. 39), and (5) a protein that only targets to the NE in a particular cell type (many of the NETs we identified are tissue specific and only target in certain cell types [18, 22, 23]). The methods described below use fluorescence microscopy of tagged exogenously expressed proteins and controls with well-defined cellular positions to confirm NE residence for individual proteins identified in the proteomic analysis. The same methods can be applied using antibodies by simply leaving out the tagged NE protein transfection; however, it is suggested to first test the tagged exogenously expressed proteins because (1) antibodies are comparatively time consuming to generate; (2) due to the dense packing of many proteins on the lamina, many antibodies to NE proteins are masked [40]; (3) many NE proteins we identified by proteomics are of low abundance, and so antibodies often do not give a strong signal above background; (4) as noted above many NE proteins are only expressed in a subset of tissues and cell lines, and though we have generally had success using the HepG2 liver hepatoma cell line for liver NETs, we have not always found correspondence between microarray data for expression in liver and in HepG2 cells using the BioGPS transcriptome database [41, 42]; and (5) due to the last three points, we have had only a ~30 % success rate generating peptide antibodies to NETs that work for immunofluorescence compared to a 90 % success rate for other peptide antibodies we have generated.

To easily distinguish nuclear rim fluorescence from the ER, cell lines used for targeting studies should be relatively flat with a large ER volume in two dimensions surrounding the nucleus (Fig. 6a). Particularly good and readily transfectable cell lines with these characteristics include the COS-7 monkey kidney cell line, the U2OS human osteosarcoma cell line, and the HT1080 human fibrosarcoma cell line. Though the nucleus-ER ratio is not as optimal, the original HeLa human cervical tumor cell line can also be used; however, be careful to avoid the HeLa-S3 subline that is comparatively rounded. Several laboratories have stably transfected various cell lines to express lamin A-GFP [43–45], and, if available, these can facilitate targeting studies by removing the need to stain for a NE marker. In cells that are comparatively small and round, the ER can appear as a rim around the nucleus, and so it is impossible to distinguish NE targeting without applying the more complex procedures described in Subheadings 3.8.2–3.8.4. When choosing a cell line, it can also be important to select one that matches the tissue and organism that was used for the original proteomic analysis as some NETs only target to the NE in certain tissues and cell lines, presumably due to the presence in these lines of partners that contribute to their targeting.

Selection of the tag to be used and its placement are also important. Monomeric red fluorescent protein (mRFP) is recommended because it eliminates background from antibody staining for the tag. Green fluorescent protein (GFP) can also be used, but we have found that GFP-fused NETs tend to express higher and are more likely to accumulate in aggregates in the cytoplasm, yielding a high background. In some cases targeting is much better if the tag is on one end of the protein rather than the other (Fig. 6b). Out of ~70 tested, we have found only a couple NETs that failed to target to the NE with an mRFP or GFP tag, yet targeted with an HA epitope tag. In these cases the comparatively large size of

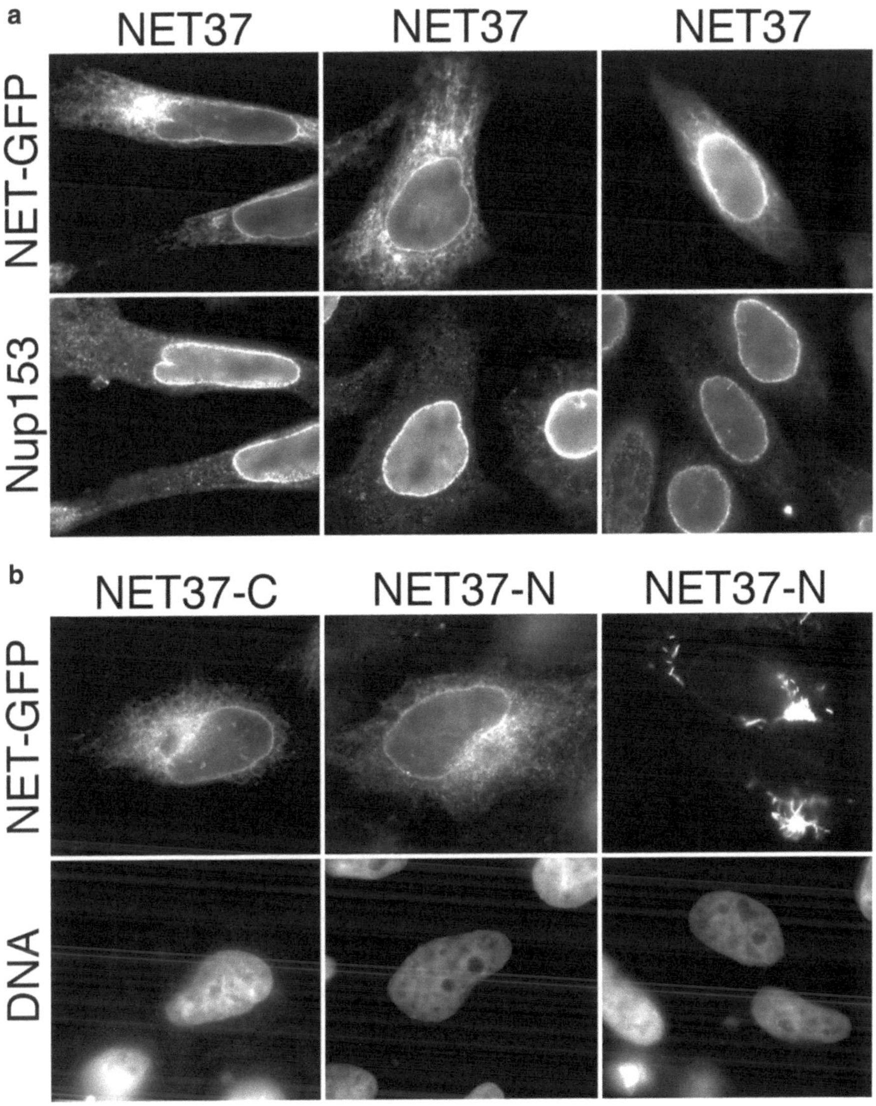

Fig. 6 NET targeting. (**a**) It is not always easy to distinguish the NE from ER packed against the nucleus in very round or flat thin cells (*left panels*). Thus, targeting studies are aided by selection of wide flat cell lines with a significant two-dimensional ER volume (*middle panel*) that better enable distinguishing a crisp rim around the nucleus from overexpressed material accumulated in the ER. Moreover, the same NET often exhibits a wide range of expression distributions with much in the ER in some cells (*left* and *middle panels*) and little in the ER in other cells (*right panel*). Confirmation of targeting is aided by co-staining for NE-specific proteins such as NPC proteins (shown in *lower panels*, i.e. Nup153) or lamins. The two images can be further overlayed to confirm colocalization. (**b**) NET37 targets nicely to the NE in nearly all cells examined if it is tagged with GFP at its carboxyl-terminus (NET37-C). However, when it is tagged at its amino-terminus (NET37-N), very few cells exhibit nice NE staining. Rather most cells have predominantly a cytoplasmic accumulation in rodlike structures (*right panels*). This is likely because the carboxyl-terminus of NET37 is located in the NE lumen and so unlikely to be blocked by the tag to interfere with its correct targeting, whereas the amino-terminus is located in the nucleoplasm and is likely involved in its tethering to partners so that the amino-terminal tag interferes with its localization. In the absence of other NE markers, DAPI staining can be used to distinguish the periphery of the nucleus (*lower panels*)

GFP (238 amino acid residues) compared to the HA epitope (11 amino acid residues) likely interfered with targeting. Though we originally used HA for a short tag, we now preferentially use FLAG or myc tags because we have found the antibodies to be more reliable. Another critical parameter is the placement of the tag at the amino- or carboxyl-terminus. While a majority of NETs target to the NE with tags on either end, failure to target or a reduction in the quality of NE targeting has been observed in ~30 % of those we have tested. Finally, as NETs tend to saturate binding sites at the NE and accumulate in the ER when highly overexpressed, it is ideal if possible to use vectors with lower expressing promoters compared to the standard CMV promoter: the PGK promoter sometimes makes a difference in visualizing clear rim staining against the cytoplasmic background.

3.8.1 Testing Tagged Fusions for NE Targeting

1. Clone NE coding sequences into tagged vectors (*see* **Note 47**).
2. Plate cells in a 24-well plate on 13 mm sterile coverslips to be at ~20 % confluency for the next day at transfection (*see* **Note 48**).
3. On the next day, check cell density using a light microscope, and if cells are at the expected confluency, proceed to transfect cells with fused NET constructs (*see* **Note 22**) and place back in incubator.
4. Between 24 and 48 h post-transfection, check if cells are expressing the fluorescent-tagged proteins using a tissue-culture fluorescent microscope if available (*see* **Note 49**).
5. If the cells are expressing fluorescent tags for both the novel NE protein and the control NE protein, the membrane permeable DNA dye Hoechst 33342 can be added to the tissue-culture media and incubated at 37 °C for 20 min to 1 h, and thus ignore **steps 7–12**.
6. Wash cells once with PBS (*see* **Note 50**) and fix with 4 % paraformaldehyde at room temperature for 10 min.
7. Wash again with PBS to remove excess formaldehyde. At this point the cells can either be maintained at 4 °C in PBS (*see* **Note 51**) or continue to be processed for immunofluorescence with antibody staining.
8. Incubate the cells for 4 min in PBS containing 0.2 % Triton X-100 to remove cell membranes followed by another wash with PBS.
9. Replace PBS with 4 % BSA solution and incubate at room temperature for 20 min.
10. Add primary antibodies in 4 % BSA solution (*see* **Notes 23, 25** and **52**). If using epitope tags or other commercial antibodies than those suggested here, apply at recommended concentrations from product literature. If using control antibodies to

lamin B1 or NETs suggested here, use at 1:100 dilution. Incubate for 45 min at room temperature unless otherwise specified in product literature.

11. Wash 3× with PBS containing 0.1 % NP-40 or Tween-20.
12. Add secondary antibodies in 4 % BSA solution that includes the DNA dye (either DAPI or Hoechst 33342 at 5 μg/mL. It is important to use secondary antibodies with minimal cross-reactivity to ensure that any colocalization observed is not due to bleedthrough. If using Molecular Probes Alexa secondary antibodies, we recommend 1:1000 dilution. If using the Jackson ImmunoResearch Laboratory donkey secondary antibodies, we recommend 5 μg/mL (*see* **Note 53**).
13. Wash 3–4× with PBS containing 0.1 % NP-40 or Tween-20, at least one time with a 10 min incubation.
14. Invert cover slips on glass microscope slides over a drop of Fluoromount-G or appropriate mounting medium for your microscope setup. Press down gently to remove excess mounting medium, and gently sponge with a paper towel.
15. Affix coverslips with nail varnish. Allow to air-dry and then clean bottom of coverslips (side that cells were *not* grown on that is now on the outside) with a wet paper towel and dry.
16. Acquire images using a fluorescent microscope.
17. Overlay the images from the separate channels for the putative NE protein and the control lamin or NET using either the microscope image acquisition software or an image processing software such as Photoshop. A novel protein that targets to the NE should give a rim staining that colocalizes with lamins or other markers.

3.8.2 Triton Extractions Prior to Fixation

Many inner NE proteins strongly interact with the intermediate filament lamin polymer, while several outer NE proteins interact with the cytoskeleton where it connects to the nucleus. Therefore, these tend to resist a prefixation detergent extraction that removes most cellular membranes and associated proteins.

1. Follow **steps 1–5** in Subheading 3.8.1, transfecting some coverslips with plasmid for just the novel putative NE protein, others with this construct plus controls for the NE that should resist extraction and ER controls that should not, and others with just the control constructs. If using antibodies for the controls, then be certain to have untransfected controls in case the overexpressed putative NE protein affects the control proteins.
2. Gently replace medium with room temperature PBS and place tissue-culture dishes on ice.

3. Gently add prechilled on ice prefixation extraction buffer (*see* **Note 54**). Incubate on ice for 1 min, gently tilting slide to mix (do not shake harshly or most cells will come off depending on the cell type used).
4. Gently rinse with cold PBS.
5. Gently add cooled formaldehyde, and replace dishes on the bench to slowly warm to room temperature during a 15 min incubation.
6. Continue processing and microscopy as in Subheading 3.8.1, except use control antibodies for an ER protein that should not resist extraction and a NE protein that should if using antibodies rather than control expression plasmids.
7. If a novel putative NE protein targets to the NE and resists the detergent prefixation extraction, then there should be a signal colocalizing with the NE control, and the ER control should have been removed (Fig. 7). If a signal is observed for the NE control, but no signal is observed for the novel putative NE protein, then it has likely been extracted; however, it is also possible that transfected cells were less adherent due to the novel protein's expression and preferentially lifted off during the extraction. In this case testing a tagged exogenously expressed fusion can be a benefit because often some cells have aggregated material in the cytoplasm that is not extracted and so can confirm that transfected cells remained on the coverslip.

3.8.3 Digitonin Permeabilization to Determine Membrane Topology

Digitonin removes cholesterol from membranes, leaving holes behind. Thus due to the relatively high content of cholesterol in the plasma membrane, digitonin preferentially permeabilizes the plasma membrane compared to ER and nuclear membranes. After formaldehyde fixation the cell membranes must be removed with a detergent extraction step (*see* **step 8** in Subheading 3.8.1) in order for antibodies to have access throughout the cell. If digitonin is used properly, the antibodies can access the cytoplasm due to permeabilization of the plasma membrane, but they cannot access the nucleus. One then compares the staining pattern for digitonin-permeabilized cells to that of cells permeabilized with Triton X-100. Thus there are four possible outcomes: (1) if the epitope recognized by the antibody is in the nucleoplasm (e.g., the amino-terminus of a type II transmembrane protein) and only in the INM, no staining should occur; (2) if the same epitope is in the nucleoplasm and in both the INM and in the ONM/ER, the inner nuclear membrane staining should be lost so that the NE signal would be roughly half of that of a Triton X-100-permeabilized cell (Fig. 8); (3) if the same epitope is only found in the ONM/ER, no difference should be observed between Triton- and digitonin-permeabilized cells (Fig. 8); and (4)

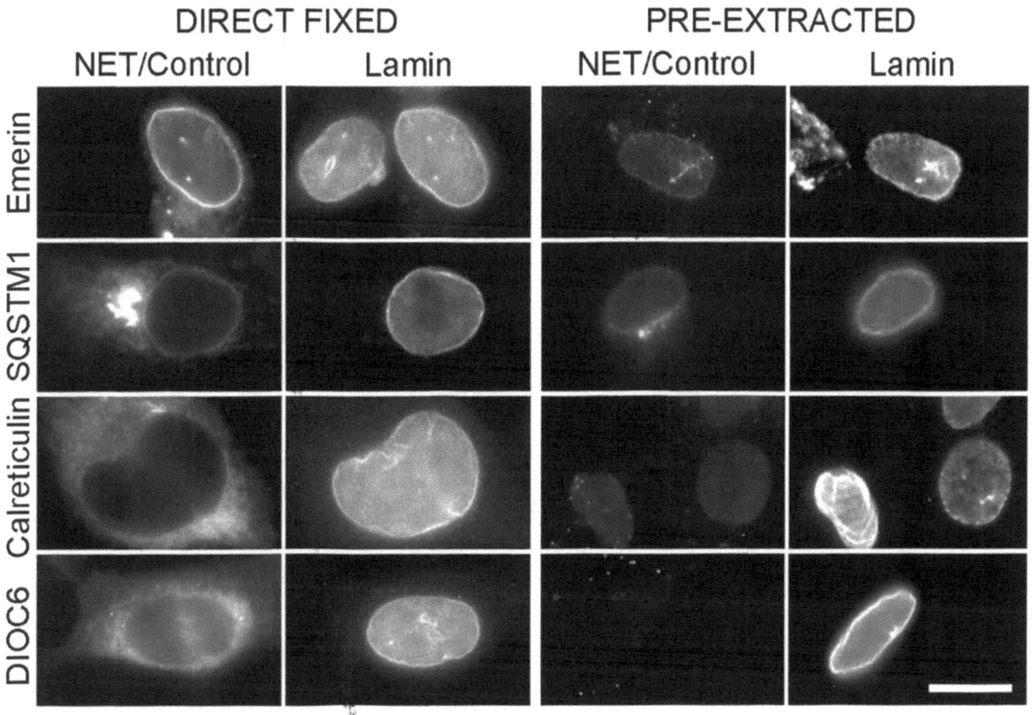

Fig. 7 Proteins that retain a nuclear rim staining after a prefixation detergent extraction with Triton X-100 are more strongly confirmed at the NE. This often indicates interactions with the lamin polymer or cytoplasmic filaments connected to the nucleus. *Left panels*. Two NETs and two negative controls are shown as they appear if not pre-extracted. Both the well-characterized NET emerin and the novel NET have a lot of material in the cytoplasm, but both have nuclear rim populations that colocalize with lamins. The appearance of a weak rim against the strong ER staining is often observed with the ER controls, which probably comes from the continuity of the ER with the ONM. *Right panels*. If the cells are pre-extracted with detergent, the NE protein population in the ER is largely removed except for some minor aggregates, while the nuclear rim signal remains. In contrast, both nuclear rim and ER populations of the ER control proteins are removed by the pre-extraction, while in the same cells the lamins are retained

if instead the epitope recognized by the antibody is in the lumen (e.g., at the carboxyl-terminus of a type II transmembrane protein), then whether it is in the ER or the NE, the signal should be lost with digitonin permeabilization because there is little cholesterol in both the NE and ER. Examples of the successful use of this approach can be found in [46–48].

A mapped antibody available to the endogenous protein has advantages in these experiments because overexpressed protein can inappropriately accumulate also in the ER. However, there is also a powerful advantage to transfecting cells with GFP-tagged proteins as cells can be tested with the tag on either end to learn about membrane topology, and the comparison of the GFP signal intensity distribution with the anti-GFP antibody signal intensity

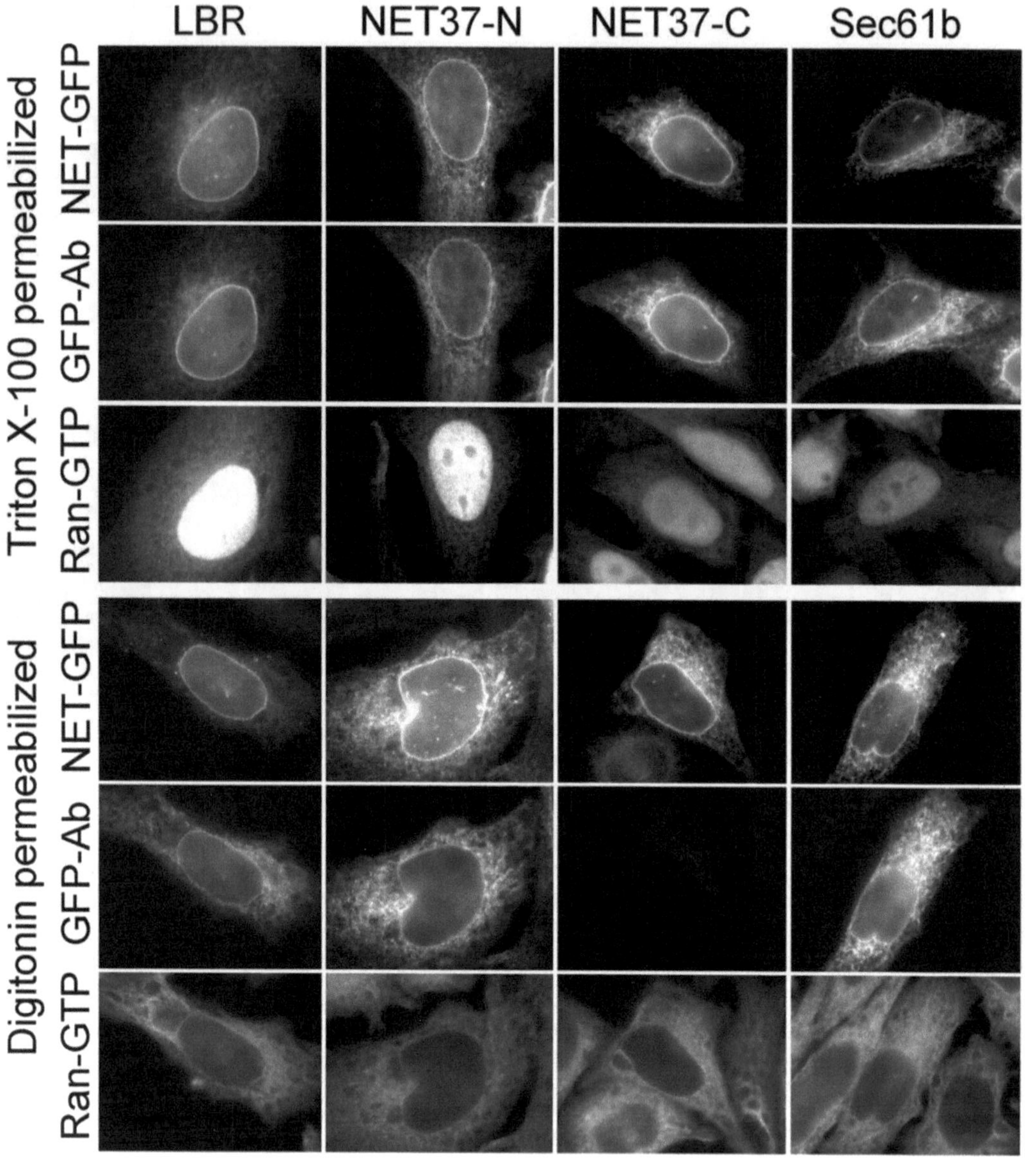

Fig. 8 Digitonin permeabilization can distinguish inner nuclear membrane populations and membrane topology. *Upper panels* show triton permeabilized cells expressing GFP fusions of LBR, NET37 with tags on either end, or Sec61b. Ran is used as a marker for nuclear membrane permeabilization as the antibody would not be able to access the predominantly nuclear population of Ran if the nuclear membrane is intact. The signal intensity and distribution are the same for the GFP itself and also for the signal from the Cy5 channel for an antibody to GFP. In the *lower panels*, the same transfections and stainings are shown for cells permeabilized with digitonin that only permeabilizes the plasma membrane. For LBR and NET37-N, the intensity of the nuclear rim signal compared to the ER signal is diminished with the GFP antibody staining because the population of GFP protein in the INM is not accessible to the antibodies. In both cases, the GFP tag is in the nucleoplasm for the INM population and in the cytoplasm for the ONM and ER populations. In contrast, the NET37-C has the GFP tag in the lumen of the NE for the INM population and in the lumen of the ER for the ONM and ER populations. Thus there is only background signal for the antibody staining in the digitonin-permeabilized cells. Sec61b is only in the ER, and the tag is not in the lumen, so that the signal intensity is similar even though material in the ONM gives the appearance of a weak rim staining. Failure of the antibodies to access the nucleus is confirmed by lack of nuclear Ran staining

distribution clearly reveals whether protein is at least partly in the INM. However, as a caution, it should be noted that sometimes results are not clear, which in some cases may be due to mixed membrane topologies as a potential overexpression artifact.

1. Follow **steps 1–7** in Subheading 3.8.1, except generate several additional untransfected coverslips for testing the optimal digitonin concentration (nine coverslips recommended). Transfected coverslips will also be needed for unpermeabilized, digitonin-permeabilized, and Triton X-100-permeabilized controls all in duplicate so that some can be co-stained for a nuclear marker and others for an ER marker (e.g., a minimum of six transfected coverslips per putative NE protein being tested).
2. Prepare three dilutions of the digitonin in PBS (*see* **Note 55**). For HeLa cells, a working concentration of 40 μg/mL for 10 min works well for us, though others have reported using 80 μg/mL for 6 min [49]. Thus a recommended starting point is 20, 40, and 80 μg/mL. Cool on ice.
3. Place tissue-culture dish containing coverslips on ice, and once chilled, use three coverslips for each digitonin dilution. Remove digitonin with 3 PBS washes from one of the three coverslips after 2.5 min, from another after 5 min, and the last one after 10 min.
4. Stain each of the coverslips with trypan blue dye, and check under a standard light tissue-culture microscope. If trypan blue dye has accumulated inside the nucleus, the treatment was too strong/too long. If trypan blue has not entered the cytoplasm, then the treatment was too weak/too short. Choose the best condition for the experimental samples (if none are ideal, then try additional dilutions or incubation times).
5. Incubate two coverslips with digitonin using the best condition, two others using the standard 4 min 0.2 % Triton X-100 extraction, and perform no extractions on the remaining two.
6. Quickly block cells with PBS containing BSA.
7. Process all coverslips for microscopy as in Subheading 3.8.1, staining one set with the Ran GTPase nuclear control antibody and the other with the ER control antibody. If the putative NE protein was transfected as a GFP fusion construct, then co-stain with GFP antibodies. If no transfections were performed, then co-stain with antibodies to the putative NE protein.
8. For secondary antibody step, note that when using GFP fusions, we use Cy5 secondary antibodies for the GFP antibody at 1:500 dilution in order to ensure that no similarities can be attributed to filter bleedthrough.

3.8.4 Determination of Inner vs. Outer Nuclear Membrane Localization

The NPC contains cytoplasmic filaments that extend roughly 80–100 nm above the nuclear membrane and a nuclear basket that extends a similar distance below the nuclear membrane. As the lumen of the NE is 50 nm, the total distance from the tip of the cytoplasmic filaments to the base of the nuclear basket is roughly at the resolution limits of light microscopy. Thus, comparing the degree of colocalization between a NE protein and an NPC cytoplasmic filament protein or an NPC nuclear basket protein can frequently distinguish an INM from an ONM protein (Fig. 9; for examples of this, *see* refs. 18, 22–24). However, when using this method, it is important to bear in mind that if the epitope for a NE protein is in the lumen, the degree of overlap will be too similar between cytoplasmic filament and nuclear basket markers to distinguish.

1. Prepare transfected cells and coverslips as in Subheading 3.8.1 with two coverslips for each NE protein.
2. Process for immunocytochemistry as in Subheading 3.8.1 using a cytoplasmic filament antibody (e.g., Nup358/RanBP2 or Nup214/CAN) for one coverslip and a nuclear basket antibody (e.g., Nup153 or TPR) for the other.
3. Analyze cells by microscopy using a high-resolution setup that can take z-series images.
4. Deconvolve images. If the nuclear basket signal appears to be concentrated in an inner plane compared to the putative NE protein signal but the cytoplasmic filament signal is in the same plane

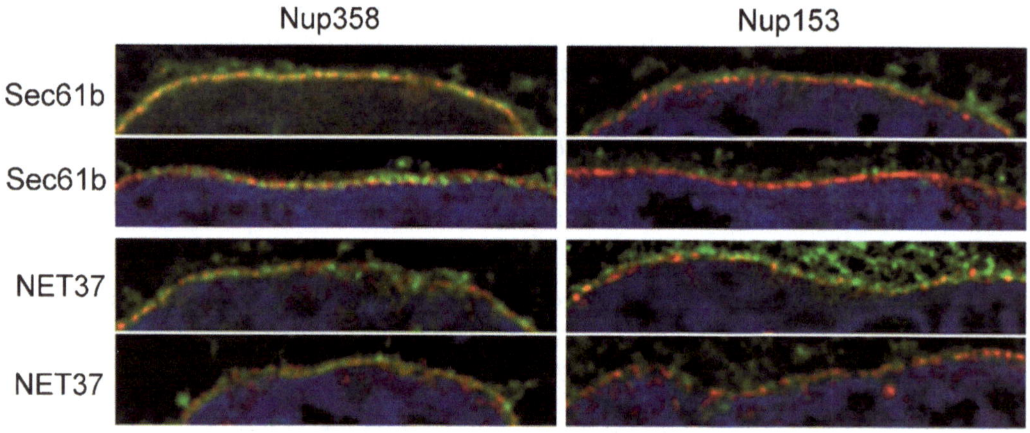

Fig. 9 Distinguishing INM from ONM targeting by high-resolution microscopy. If a putative NE protein is co-stained with antibodies to cytoplasmic and nucleoplasmic NPC components, it is sometimes possible to distinguish whether it is in the INM or ONM. When it concentrates in the same plane as the cytoplasmic nucleoporin Nup358 while the nucleoplasmic nucleoporin Nup153 is internal to it, then it is in the ONM (Sec61b). When it is in the same plane as Nup153 and is internal to Nup358, then it is in the INM (NET37). Because the distance between Nup153 and Nup358 is close to the resolution limits of light microscopy, this method does not work if the tag on the NE protein being investigated is in the lumen of the NE as the signal then similarly overlaps with both nucleoporins

with it, then the protein is in the ONM and not the INM. If the nuclear basket signal is in the same plane as the putative NE protein signal, then the protein is in the INM. If the cytoplasmic filament signal also overlaps, then it is in both INM and ONM, but if the cytoplasmic filament signal is in a separate outer plane, then the putative NE protein is just in the INM.

4 Notes

1. This protocol has been optimized for mouse and rat, but it should be applicable to most mammals.
2. The CB6F1/J mouse strain was chosen to obtain more tissue per animal because it has much greater than average body weight (~35 g at 9 weeks) while free of obesity-/diabetes-type defects. When choosing a heavier strain, it is important to check that the mice do not have diabetes-related problems as this could bias results, some NE diseases having associated diabetes defects.
3. As with most protocols, there is an optimal middle ground with too little or too much starting material resulting in lower yields. In our hands, 10–12 rats produce optimal yields without saturating the sucrose gradients in a Beckman Coulter SW28 rotor.
4. We have found that some preparations of PMSF precipitate and form large crystals that co-purify with NEs in the sucrose buffers. Therefore it is important to ensure that the PMSF is adequately in solution and that the batch of PMSF does not precipitate when diluting it. This should be directly tested before beginning the procedure as we have found that even different batches of PMSF from the same company have different propensities to precipitate. Resuspending the PMSF in ethanol and then clearing it by centrifugation appears to better avert this problem than resuspending in DMSO.
5. The addition of PMSF at this step is crucial as during dissection and crude lysis many serine proteases are released. Other protease inhibitors that are critical at later stages of the procedure could also be added, but proteins appear to be largely intact with just adding PMSF at this step.
6. It is essential to use the dounce for gentle lysis as blade tissue homogenizers tear the nuclei, resulting in extremely poor yields and an inability to separate clean microsomes or mitochondria if one is using them as a subtractive fraction. Our conditions have been optimized with the particular unit described; however, if this is not available, a large drill can be wall mounted in the cold room and used with a Teflon dounce homogenizer.

7. Muslin will also do if not chemically treated: make certain to ask supplier.
8. Due to the high viscosity of the 2.3 M sucrose solution, it takes several min for each tube if an 18-gauge needle is used. In contrast, with the wide bore size of a 14-gauge needle, this same procedure can be performed in 30 s. It is important to use a Luer lock syringe because the viscosity of the solution can produce high pressure on the connections. It is ideal to use syringes that have flat, non-sharp ends as it is both safer and easier. We had a difficult time sourcing these and so had our local workshop cut the tips from regular 14-gauge syringes.
9. While a wide variety of centrifuges can be used, it is essential that a swinging-bucket rotor is used for the sucrose cushions and gradients to be effective and for pellets to be loosely packed. If pellets are not loosely packed, yields will be greatly reduced due to aggregation of nuclei and NEs.
10. $MgCl_2$ concentration in the original procedure was 5 mM throughout; however, if NEs are being prepared for viewing by electron microscopy, dropping the concentration through most of the procedure to 0.1 mM will yield better structure. However, during DNase and RNase treatment, it is important to increase the $MgCl_2$ concentration back to 5 mM or the digestion and removal of nucleoplasmic contents will be poor.
11. The optimal protease inhibitors will vary according to the tissue being investigated. The choice for liver focuses on inhibiting serine, trypsin, cysteine, and aspartic proteases present in this tissue.
12. If general protease cocktails are used, it is important to make certain that they do not contain EDTA as the amounts of EDTA present typically chelate the Mg and inhibit nuclease digestion for removal of nucleoplasmic contents.
13. When preparing solutions 4–8, it is advisable to prepare them initially without protease inhibitors or DTT and add these fresh before use. As sucrose concentrations are critical for many of these steps, if the stock solutions of the protease inhibitors being used are less concentrated than 500×, it may be necessary to calculate the volume of the protease inhibitors into the solution volumes.
14. The solution can be prepared by adding 230 mL of an 85 % sucrose stock to 12.5 mL 1 M HEPES, 6.25 mL 1 M KCl, and 1 mL 1 M $MgCl_2$ and freshly added 2 mM DTT and protease inhibitors. Other concentrations of sucrose can then be obtained by mixing the 2.3 M SHKM with the 0.25 M SHKM.
15. The stock solution can be prepared by mixing 67 mL 0.25 M SHKM to 33 mL 2.3 M SHKM.

16. We have used octyl ß-d-glucopyranoside and KCl because we have found that a wide variety of the original NETs directly tested are retained with the nuclear lamin polymer in the pelleted fraction after extraction with this detergent and intermediate salt strength and concentration. However, others have used 1 % Triton X-100 [27], and NPC proteins have been collected by their preferential solubilization with Empigen BB [50]. Thus a wide variety of detergents can be used either individually or sequentially both to wash away contaminants and to extract protein sets with increasing stringencies of association with the lamin polymer. Similarly harsher salts such as NaCl or increasing salt concentrations can be used alone or in combination with detergents to the same effect. Different subtypes of the NET LAP1 were extracted with increasing salt concentrations in the presence of 1 % Triton X-100 [51].
17. When making urea/carbonate solution if having difficulty getting urea to 8 M, the concentration can be lowered; however, in doing so one must correspondingly change the dilution factor in Subheading 3.3.2 so that the final concentration is 4 M urea, 100 mM Na_2CO_3.
18. Bulk material is not available for the Partisphere strong cation exchange resin. The material is extracted by sawing the HPLC column in half with a hacksaw, then washed with methanol, dried, and stored as a powder. This SCX resin may now be replaced with Luna 5 μm SCX, 100 Ang. pore, bulk material (Phenomenex, Part # 04A-4398).
19. We recommend 24-well dishes with 13 mm coverslips because this setup lends itself to intermediate-scale analysis. However, one can use any type of tissue-culture dish and coverslip or microscope slide that is readily available. A note of caution with growing cells on coverslips is that until one gets used to recovering the coverslips from the wells, a reasonable probability exists that some will be broken. Choosing a coverslip size that is at least three millimeters smaller (and preferably five) than the tissue-culture well/dish greatly increases the success of coverslip recovery without breaking.
20. Bending the tips of the tweezers slightly can increase the success of coverslip recovery for some people. In our laboratory, one researcher had the best success without bending, while three researchers benefited from bending just one of the tips and four researchers benefited from bending both tips.
21. Too strong of a vacuum during aspiration will result in many of the cells being removed when performing a detergent prefixation extraction. Simply putting a 10 μL pipette tip on the end of a vacuum aspirator can largely eliminate this problem, but it is still necessary to be gentle and aspirate from the edge with the plate tilted.

22. Any transfection reagent can be used; however, it is advised to use those with low toxicity of which FuGENE is among the best we have tried. As the output is visual at the single cell level, the transfection efficiency is not critical, and higher toxicity transfection reagents can interfere with interpretation due to an increase in nuclear lobulation and apoptotic bodies. When relatively toxic reagents are used, it can also be difficult to find cells that do not have extremely bright debris in the same field. Also be careful to avoid transfection reagents that deposit autofluorescing material that sticks to the coverslip. In particular, calcium phosphate precipitation tends to generate too much background autofluorescence for effective image analysis.
23. Many laboratories have preferential methodologies for immunohistochemistry that involve other buffers such as Tris-buffered saline instead of PBS. There is no detriment to the procedure if one of these other buffers is used in place of PBS provided it does not result in precipitation of detergents added.
24. Different batches of digitonin can vary in efficacy. Thus it is advised to prepare a large amount in DMSO and store it at −80 °C until use. In this way batch variation should not add further to the already substantial experimental variation.
25. As for buffers the specific blocking reagent used is not critical. Although our laboratory commonly uses 4 % BSA in PBS for blocking in immunohistochemistry, the procedure is unlikely to be negatively affected if one prefers fish gelatin or milk as a blocking agent. The important point is to use the best blocking agent for a particular antibody. BSA and fish gelatin have been tested and work fine for all the antibodies specifically mentioned here.
26. When choosing a control antibody, it is critical to check for recent papers where commercially available antibodies were utilized as many commercially available antibodies we have tested have not worked. This in some cases may be because hybridoma cells sometimes lose their efficacy in liquid nitrogen and many companies apparently do not check for antibody quality when a new batch is made from the hybridoma. Regardless, many of the best antibodies in the NE area are not commercially available. The best lab resource for antibodies to NE proteins was generated by Professor Glenn E. Morris at the Wolfson Centre for Inherited Neuromuscular Disease, and these are made generally available through http://www.glennmorris.org.uk/mabs.htm.
27. Removing food the night before the procedure has been experimentally shown to increase yields by 30–50 % using this procedure with rats and reduces the RNA/DNA ratio [3]. However, an important caution before using another model system is that the length of time for starving may increase for animals with lower metabolism.

28. One can avoid the weighing step and estimate 5 g per liver.
29. This requires a reasonable amount of physical strength, and one must take care to keep the homogenizer straight with the direction of the pestle or the homogenizer can break. Also never stop the pestle rotation while it is inserted inside the homogenizer with liquid or this also can become stuck or break due to the vacuum produced during homogenization. To remove the pestle when homogenization is complete, lower the speed of rotation to the minimum, and pull up gently until the pestle is removed from the homogenizer. Turn the rotor off as soon as the pestle is removed from the homogenizer; it would be covered with tissue parts that would "fly off" if still rotating. It is advisable to wear protective eye goggles at this step.
30. If one is simply preparing NEs, as this step is followed by a low-speed centrifugation to pellet nuclei, the volume added here is irrelevant. However, if microsomes or mitochondria are being prepared in the same experiment, the supernatant will have to be subjected to high-speed centrifugation where large volumes could become counterproductive. Thus it is recommended to minimize the volumes added for washing if preparing other organelles from the same experiment.
31. There is an enormous amount of collagen in rat liver, and this will clog the cheesecloth and reduce the yield if the cheesecloth is overloaded. Therefore the homogenate from no more than 3 rats should be filtered through one four times-folded cheesecloth. Mice have much less collagen than rats, and therefore, roughly twice the amount of material can be used for the same amount of cheesecloth.
32. We prefer these because the pellet tends to distribute widely, but conical tubes could also be used.
33. Be careful when balancing the tubes and placing them in the centrifuge to prevent the cushion and layer from mixing.
34. This is done to wash away some of the high amount of contaminating collagen prior to extraction of nucleoplasmic contents. These washes are necessary to prevent saturation of sucrose cushions when chromatin is released, but are only necessary for liver preparations and other tissues that are particularly high in collagen. If not restricting food the night before the procedure, the amount of collagen will be greater, and additional washes will be necessary.
35. Do not be concerned if the nuclei do not become phase lucent at this point as this first treatment might be viewed as generally loosening rather than fully digesting the chromatin, and, moreover, the digestion will continue during centrifugation.

36. It is very important to use a swinging-bucket rotor when spinning through the sucrose cushion at this point in order to float any chromatin that is released away from the NEs.
37. The supernatant will appear cloudy, but this is mostly chromatin that has been ejected and should give a dark, worm-like appearance under the microscope quite distinct from NEs.
38. It is critical to get this spinning immediately as loss of membrane proteins was observed by Western and membrane vesicles looked very fragmented by electron microscopy even after just 10 min sitting on ice prior to centrifugation (Tinglu Guan, personal communication).
39. This procedure is adapted from Ref. 27.
40. Not all membranes will be resuspended at this stage.
41. This step is to digest any nucleic acids that might be contaminating the NE pellets.
42. A typical four-step parameter setup for pulling approx 3 to 5 μm tips from a 100 μm i.d. × 365 μm o.d. fused silica capillary is [heat = 290, velocity = 40, and delay = 200], [heat = 280, velocity = 30, and delay = 200], [heat = 270, velocity = 25, and delay = 200], and [heat = 260, velocity = 20, and delay = 200], with all other values set to zero.
43. This concentration roughly corresponds to an amount of resin powder covering the tip of a small spatula, about 2–3 mm^3.
44. Using a black background behind the capillary being packed helps seeing the reverse phase and SCX resin levels inside the column.
45. It is not recommended to wash columns extensively after loading if one's goal is to detect phosphorylated peptides.
46. Applying the voltage via the microtight tee most distal to the microcapillary column helps stabilize the electrospray.
47. Most cDNAs are available as clones from the IMAGE collection through vendors such as ATCC, Geneservice, and RZPD. Using these as templates simplifies cloning.
48. 20 % confluency ensures that cells will not overgow 48 h post-transfection for the HT1080 and HeLa cell lines. If using other cell types, then doubling times need to be determined and the % confluency accordingly reduced or increased.
49. If transfection efficiencies are so low that one cannot clearly identify transfected cells at this stage when using fluorescent protein tags, it may be better to start over with a better transfection procedure rather than processing the cells further.
50. The pre-wash with PBS step can be removed if using a cell type that is not strongly adherent or trying to visualize NET distribution in mitotic cells as washing the coverslips prior to fixation

typically removes most mitotic cells for standard cell lines. In these cases it is also beneficial to add a 2× formaldehyde solution directly to the culture media. The only difference is that there may be some mild autofluorescence from the phenol red of the media that does not get washed away if not engaging in further washes for antibody staining and if doing antibody staining, the serum proteins cross-linked outside the cells can, depending on the antibody used, result in increased background.

51. There is usually no noticeable loss of structure when maintained at 4 °C for 2 weeks; however, it should be remembered that formaldehyde is a reversible cross-linker, and so if maintained for more than a week, we typically perform another formaldehyde cross-linking step before proceeding.

52. We prefer going through all the steps while leaving the coverslips in the 24-well dish; however, if antibodies are precious, the amount of diluted antibody solution used can be decreased from ~250 μL per well to ~30 μL by performing these steps with the coverslips inverted over the antibody mix on a piece of parafilm. When doing so, it is important to use a humidified chamber so that coverslips do not dry out because their drying will ruin the experiment.

53. This protocol assumes that the user has general knowledge of microscopy and histochemical techniques; thus if one does not know how to use control and tag antibodies from distinct species and secondary antibodies with different fluorophores, please consult other chapters dealing with basic microscopy. If using an mRFP- or GFP-labeled novel putative NE protein and just one antibody, the issues of minimal cross-reactivity become irrelevant.

54. The 1 % Triton X-100 method is given because it is the most accepted for publication. However, the amount and type of detergent and salt can be changed. Aberrant nuclear morphologies are common when extracting cells prior to fixation, and performing two subsequent 0.1 % Triton X-100 extractions for 2 min each sometimes reduces this background. However, the additional processing steps can result in loss of cells from the coverslip even more readily than the high detergent concentration. The most stringent conditions tested were for the lamin B receptor (LBR), which resisted 2 % Triton X-100 and 1 M NaCl [52]; thus one can also increase the stringency to determine how tightly the protein of interest is bound.

55. Batches of digitonin vary, and each cell type has differences in cholesterol content; thus, each cell type needs to be directly tested with a particular batch of digitonin. The same concentration that in our hands took 10 min to be optimal for HeLa cells took only 3 min for 3T3-L1 cells.

Acknowledgments

This work was supported by a Senior Research Fellowship to E.C.S from the Wellcome Trust (095209) and core funding for the Centre for Cell Biology (092076) and by the Stowers Institute for Medical Research for L.F. We thank Professor Tom Rapoport for the control Sec61b expression construct.

References

1. D'Angelo MA, Hetzer MW (2006) The role of the nuclear envelope in cellular organization. Cell Mol Life Sci 63:316–332
2. Starr DA, Fischer JA (2005) KASH 'n Karry: the KASH domain family of cargo-specific cytoskeletal adaptor proteins. Bioessays 27:1136–1146
3. Blobel G, Potter VR (1966) Nuclei from rat liver: isolation method that combines purity with high yield. Science 154:1662–1665
4. Dwyer N, Blobel G (1976) A modified procedure for the isolation of a pore complex-lamina fraction from rat liver nuclei. J Cell Biol 70:581–591
5. Gerace L, Ottaviano Y, Kondor-Koch C (1982) Identification of a major polypeptide of the nuclear pore complex. J Cell Biol 95:826–837
6. Schirmer EC, Florens L, Guan T, Yates JR, Gerace L (2003) Nuclear membrane proteins with potential disease links found by subtractive proteomics. Science 301:1380–1382
7. Florens L, Korfali N, Schirmer EC (2008) Subcellular fractionation and proteomics of nuclear envelopes. Methods Mol Biol 432:117–137
8. Walter P, Blobel G (1983) Preparation of microsomal membranes for cotranslational protein translocation. Methods Enzymol 96:84–93
9. Gerace L, Burke B (1988) Functional organization of the nuclear envelope. Annu Rev Cell Biol 4:335–374
10. Stuurman N, Heins S, Aebi U (1998) Nuclear lamins: their structure, assembly, and interactions. J Struct Biol 122:42–66
11. Washburn MP, Wolters D, Yates JRR (2001) Large-scale analysis of the yeast proteome by multidimensional protein identification technology. Nat Biotechnol 19:242–247
12. Wolters DA, Washburn MP, Yates JRR (2001) An automated multidimensional protein identification technology for shotgun proteomics. Anal Chem 73:5683–5690
13. Eng J, McCormack A, Yates J r (1994) An approach to correlate tandem mass spectral data of peptides with amino acid sequences in a protein database. J Am Soc Mass Spectrom 5:976–989
14. Tabb DL, McDonald WH, Yates JRR (2002) DTASelect and contrast: tools for assembling and comparing protein identifications from shotgun proteomics. J Proteome Res 1:21–26
15. Zhang Y, Wen Z, Washburn MP, Florens L (2010) Refinements to label free proteome quantitation: how to deal with peptides shared by multiple proteins. Anal Chem 82:2272–2281
16. Brachner A, Reipert S, Foisner R, Gotzmann J (2005) LEM2 is a novel MAN1-related inner nuclear membrane protein associated with A-type lamins. J Cell Sci 118:5797–5810
17. Chen IH, Huber M, Guan T, Bubeck A, Gerace L (2006) Nuclear envelope transmembrane proteins (NETs) that are up-regulated during myogenesis. BMC Cell Biol 7:38
18. Malik P, Korfali N, Srsen V, Lazou V, Batrakou DG, Zuleger N, Kavanagh DM, Wilkie GS, Goldberg MW, Schirmer EC (2010) Cell-specific and lamin-dependent targeting of novel transmembrane proteins in the nuclear envelope. Cell Mol Life Sci 67:1353–1369
19. Wilhelmsen K, Litjens SH, Kuikman I, Tshimbalanga N, Janssen H, van den Bout I, Raymond K, Sonnenberg A (2005) Nesprin-3, a novel outer nuclear membrane protein, associates with the cytoskeletal linker protein plectin. J Cell Biol 171:799–810
20. Krogh A, Larsson B, von Heijne G, Sonnhammer EL (2001) Predicting transmembrane protein topology with a hidden Markov model: application to complete genomes. J Mol Biol 305:567–580
21. Bendtsen JD, Nielsen H, von Heijne G, Brunak S (2004) Improved prediction of signal peptides: SignalP 3.0. J Mol Biol 340:783–795

22. Korfali N, Wilkie GS, Swanson SK, Srsen V, Batrakou DG, Fairley EA, Malik P, Zuleger N, Goncharevich A, de Las Heras J, Kelly DA, Kerr AR, Florens L, Schirmer EC (2010) The leukocyte nuclear envelope proteome varies with cell activation and contains novel transmembrane proteins that affect genome architecture. Mol Cell Proteomics 9:2571–2585
23. Wilkie GS, Korfali N, Swanson SK, Malik P, Srsen V, Batrakou DG, Zuleger N, de Las Heras J, Kerr AR, Florens L, Schirmer EC (2011) Several novel nuclear envelope transmembrane proteins identified in muscle have cytoskeletal associations. Mol Cell Proteomics 10:M110.003129
24. Schermelleh L, Carlton PM, Haase S, Shao L, Winoto L, Kner P, Burke B, Cardoso MC, Agard DA, Gustafsson MG, Leonhardt H, Sedat JW (2008) Subdiffraction multicolor imaging of the nuclear periphery with 3D structured illumination microscopy. Science 320:1332–1336
25. Korfali N, Fairley EAL, Swanson SK, Florens L, Schirmer EC (2009) Use of sequential chemical extractions to purify nuclear membrane proteins for proteomics identification. Methods Mol Biol 528:201–225
26. Wilkie GS, Schirmer EC (2008) Purification of nuclei and preparation of nuclear envelopes from skeletal muscle. Methods Mol Biol 463:23–41
27. Dreger M, Bengtsson L, Schoneberg T, Otto H, Hucho F (2001) Nuclear envelope proteomics: novel integral membrane proteins of the inner nuclear membrane. Proc Natl Acad Sci U S A 98:11943–11948
28. Wu CC, MacCoss MJ, Howell KE, Yates JR 3rd (2003) A method for the comprehensive proteomic analysis of membrane proteins. Nat Biotechnol 21:532–538
29. Florens L, Washburn MP (2006) Proteomic analysis by multidimensional protein identification technology. Methods Mol Biol 328:159–175
30. Zhang Y, Wen Z, Washburn MP, Florens L (2009) Effect of dynamic exclusion duration on spectral count based quantitative proteomics. Anal Chem 81:6317–6326
31. McDonald WH, Tabb DL, Sadygov RG, MacCoss MJ, Venable J, Graumann J, Johnson JR, Cociorva D, Yates JR 3rd (2004) MS1, MS2, and SQT-three unified, compact, and easily parsed file formats for the storage of shotgun proteomic spectra and identifications. Rapid Commun Mass Spectrom 18:2162–2168
32. Venable JD, Dong MQ, Wohlschlegel J, Dillin A, Yates JR (2004) Automated approach for quantitative analysis of complex peptide mixtures from tandem mass spectra. Nat Methods 1:39–45
33. Zybailov BL, Florens L, Washburn MP (2007) Quantitative shotgun proteomics using a protease with broad specificity and normalized spectral abundance factors. Mol Biosyst 3:354–360
34. Florens L, Carozza MJ, Swanson SK, Fournier M, Coleman MK, Workman JL, Washburn MP (2006) Analyzing chromatin remodeling complexes using shotgun proteomics and normalized spectral abundance factors. Methods 40:303–311
35. Paoletti AC, Parmely TJ, Tomomori-Sato C, Sato S, Zhu D, Conaway RC, Conaway JW, Florens L, Washburn MP (2006) Quantitative proteomic analysis of distinct mammalian Mediator complexes using normalized spectral abundance factors. Proc Natl Acad Sci U S A 103:18928–18933
36. Zybailov B, Mosley AL, Sardiu ME, Coleman MK, Florens L, Washburn MP (2006) Statistical analysis of membrane proteome expression changes in Saccharomyces cerevisiae. J Proteome Res 5:2339–2347
37. Schirmer E, Gerace L (2002) Organellar proteomics: the prizes and pitfalls of opening the nuclear envelope. Genome Biol 3:REVIEWS1008
38. Mootha VK, Bunkenborg J, Olsen JV, Hjerrild M, Wisniewski JR, Stahl E, Bolouri MS, Ray HN, Sihag S, Kamal M, Patterson N, Lander ES, Mann M (2003) Integrated analysis of protein composition, tissue diversity, and gene regulation in mouse mitochondria. Cell 115:629–640
39. Foster LJ, de Hoog CL, Zhang Y, Zhang Y, Xie X, Mootha VK, Mann M (2006) A mammalian organelle map by protein correlation profiling. Cell 125:187–199
40. Tunnah D, Sewry CA, Vaux D, Schirmer EC, Morris GE (2005) The apparent absence of lamin B1 and emerin in many tissue nuclei is due to epitope masking. J Mol Histol 36:337–344
41. Su AI, Cooke MP, Ching KA, Hakak Y, Walker JR, Wiltshire T, Orth AP, Vega RG, Sapinoso LM, Moqrich A, Patapoutian A, Hampton GM, Schultz PG, Hogenesch JB (2002) Large-scale analysis of the human and mouse transcriptomes. Proc Natl Acad Sci U S A 99:4465–4470
42. Wu C, Orozco C, Boyer J, Leglise M, Goodale J, Batalov S, Hodge CL, Haase J, Janes J, Huss JW 3rd, Su AI (2009) BioGPS: an extensible and customizable portal for querying and organizing gene annotation resources. Genome Biol 10:R130

43. Broers JL, Machiels BM, van Eys GJ, Kuijpers HJ, Manders EM, van Driel R, Ramaekers FC (1999) Dynamics of the nuclear lamina as monitored by GFP-tagged A-type lamins. J Cell Sci 112:3463–3475
44. Chaturvedi P, Parnaik VK (2010) Lamin A rod domain mutants target heterochromatin protein 1alpha and beta for proteasomal degradation by activation of F-box protein, FBXW10. PLoS One 5, e10620
45. Favreau C, Higuet D, Courvalin JC, Buendia B (2004) Expression of a mutant lamin A that causes Emery-Dreifuss muscular dystrophy inhibits in vitro differentiation of C2C12 myoblasts. Mol Cell Biol 24:1481–1492
46. Furukawa K, Fritze CE, Gerace L (1998) The major nuclear envelope targeting domain of LAP2 coincides with its lamin binding region but is distinct from its chromatin interaction domain. J Biol Chem 273:4213–4219
47. Soullam B, Worman HJ (1993) The amino-terminal domain of the lamin B receptor is a nuclear envelope targeting signal. J Cell Biol 120:1093–1100
48. Soullam B, Worman HJ (1995) Signals and structural features involved in integral membrane protein targeting to the inner nuclear membrane. J Cell Biol 130:15–27
49. Melchior F (1998) Nuclear protein import in a permeabilized cell assay. Methods Mol Biol 88:265–273
50. Cronshaw J, Krutchinsky A, Zhang W, Chait B, Matunis M (2002) Proteomic analysis of the mammalian nuclear pore complex. J Cell Biol 158:915–927
51. Foisner R, Gerace L (1993) Integral membrane proteins of the nuclear envelope interact with lamins and chromosomes, and binding is modulated by mitotic phosphorylation. Cell 73:1267–1279
52. Worman HJ, Yuan J, Blobel G, Georgatos SD (1988) A lamin B receptor in the nuclear envelope. Proc Natl Acad Sci U S A 85: 8531–8534

Chapter 2

Exploring the Protein Composition of the Plant Nuclear Envelope

Xiao Zhou, Kentaro Tamura, Katja Graumann, and Iris Meier

Abstract

Due to rather limited sequence similarity, targeted identification of plant nuclear envelope and nuclear pore complex proteins has mainly followed two routes: (1) advanced computational identification followed by experimental verification and (2) immunoaffinity purification of complexes followed by mass spectrometry. Following candidate identification, fluorescence recovery after photobleaching (FRAP) and fluorescence resonance energy transfer (FRET) provide powerful tools to verify protein–protein interactions in situ at the NE. Here, we describe these methods for the example of *Arabidopsis thaliana* nuclear pore and nuclear envelope protein identification.

Key words *Arabidopsis thaliana*, KASH protein, Nuclear pore complex, Bioinformatics, Immunoaffinity purification, Fluorescence resonance energy transfer (FRET), Fluorescence recovery after photobleaching (FRAP)

1 Introduction

The plant nuclear envelope (NE) proteome has remained fairly elusive until rather recently [1]. Due to limited sequence similarity of NE and nuclear pore complex (NPC) proteins, identification by sequence similarity alone has only revealed a handful of proteins [2–11]. Thus, computational approaches searching for patterns instead of sequences and reiterative complex purification coupled with mass spectrometry—starting with the few known proteins—have to date been the most successful strategies to identify plant NPC and NE proteins [12]. Once identified, candidates must be verified as members of NE and NPC complexes. Fluorescence recovery after photobleaching (FRAP) and fluorescence resonance energy transfer (FRET) complement biochemical methods to verify protein–protein interactions in situ at the NE.

KASH proteins are outer nuclear membrane (ONM) proteins that contain a single transmembrane domain (TMD) spanning the outer nuclear membrane followed by a short C-terminal KASH

Sue Shackleton et al. (eds.), *The Nuclear Envelope: Methods and Protocols*, Methods in Molecular Biology, vol. 1411, DOI 10.1007/978-1-4939-3530-7_2,

© Springer Science+Business Media New York 2016

domain that interacts in the perinuclear space with the SUN domain of SUN proteins [1, 13–16]. Although the N-terminal cytoplasmic domain of KASH proteins varies, the C-terminal KASH domain is relatively conserved, especially the C-terminal four amino acids [1, 13, 14, 17]. For example, the C-terminal four amino acids of animal KASH proteins can be summarized to a "[PATHQL]PP[QTVFILM]" motif (square brackets enclose alternative amino acid residues at the respective position), suggesting that a similar pattern could also be found in plants [18].

The program DORY was developed to search for a putative KASH (pKASH) domain that should (1) be immediately C-terminal of a TMD, (2) be short (less than 40 amino acids based on known KASH proteins), and (3) terminate in a given four-amino-acid pattern [18]. DORY contains two functional units—the KASHFilter and the HomologyFilter. The KASHFilter collects protein sequences that contain a pKASH domain and the HomologyFilter divides these protein sequences into homologous groups (Fig. 1). Proteins in each group potentially belong to one protein family.

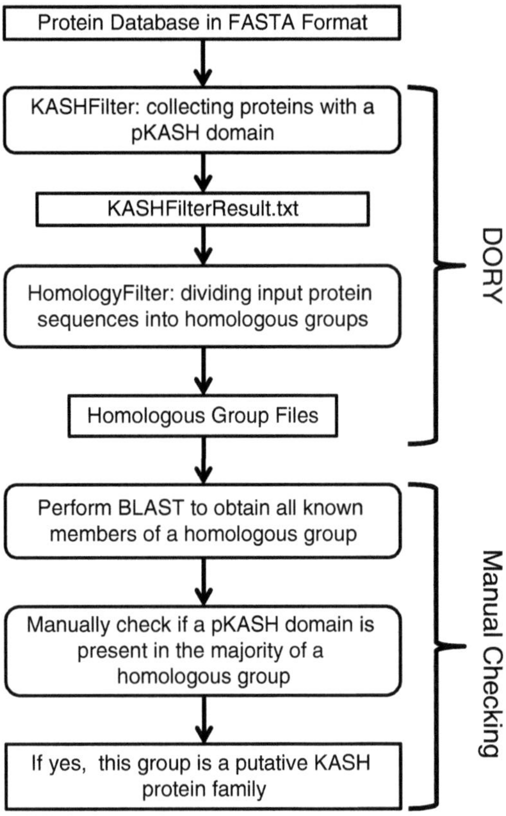

Fig. 1 Workflow of identifying KASH proteins using DORY. The steps can be divided into two sections—DORY search and manual checking. Input files, output files, and results are indicated by *rectangle frames*, and the searching steps are indicated in *round rectangular frames*

Considering that a protein sequence that is not a KASH protein may contain a pKASH domain by chance, we argue that if a pKASH domain is conserved in a protein family, then this protein family is more likely a functional KASH protein family. Therefore, the presence of a pKASH domain needs to be analyzed at the protein family level. Since DORY filters out proteins without a pKASH domain, researchers need to choose a protein sequence from each homologous group to collect all of its known homologs by BLAST and check if the pKASH domain is conserved in its homologs. The whole process is depicted in Fig. 1.

Although genome projects have provided large lists of genes, many gene products remain functionally uncharacterized. Determining the composition of protein complexes and the interaction networks in an organelle of interest establishes a framework, which generates strategies and hypotheses relating to the function, mechanism, and regulation of the organelle dynamics [19, 20]. We describe methods to isolate the *Arabidopsis* nuclear pore complex (NPC), which is one of the largest macromolecular protein complexes in the cell. By using immunoaffinity purification, NPC components are effectively purified from lysates of transgenic plants expressing GFP-tagged nucleoporins [12]. Subsequent mass spectrometry comprehensively identifies protein components in the affinity-purified complexes.

In the *Arabidopsis* genome, very few homologs of yeast and animal nuclear envelope proteins have been found. In such a case, the biochemical identification of protein complexes, which is independent of homology-based approaches, is useful. This approach is also a convenient and powerful method for revealing proteome-wide interactome maps that provide significant insights into functions of unknown proteins.

Fluorescence recovery after photobleaching (FRAP) is a live-cell imaging technique that enables the study of the mobile behavior of a protein and from this to draw conclusions about the functional properties of the protein. The protein under investigation needs to be fused to a fluorescent protein for confocal imaging. Low-power lasers are used to visualize the fusion proteins while causing as little damage as possible to the cells and tissue. The principle behind FRAP is that the laser not only excites the fluorophore, thus causing it to fluoresce and become visible, but at a higher output the laser causes irreversible structural damage to the fluorophore, which blocks the molecule from becoming excited. During FRAP, a selected area of fluorescence inside a cell is bleached followed by fluorescence recording. If fluorescence signal recovers in the selected area, excitable molecules have entered the area post-bleach, indicating protein movement (Fig. 2a).

Parameters measured and calculated include maximum fluorescence recovery (*MFR*), half time ($T_{1/2}$), mobile fraction, immobile fraction (Fig. 2b), and diffusion coefficient (D) [21].

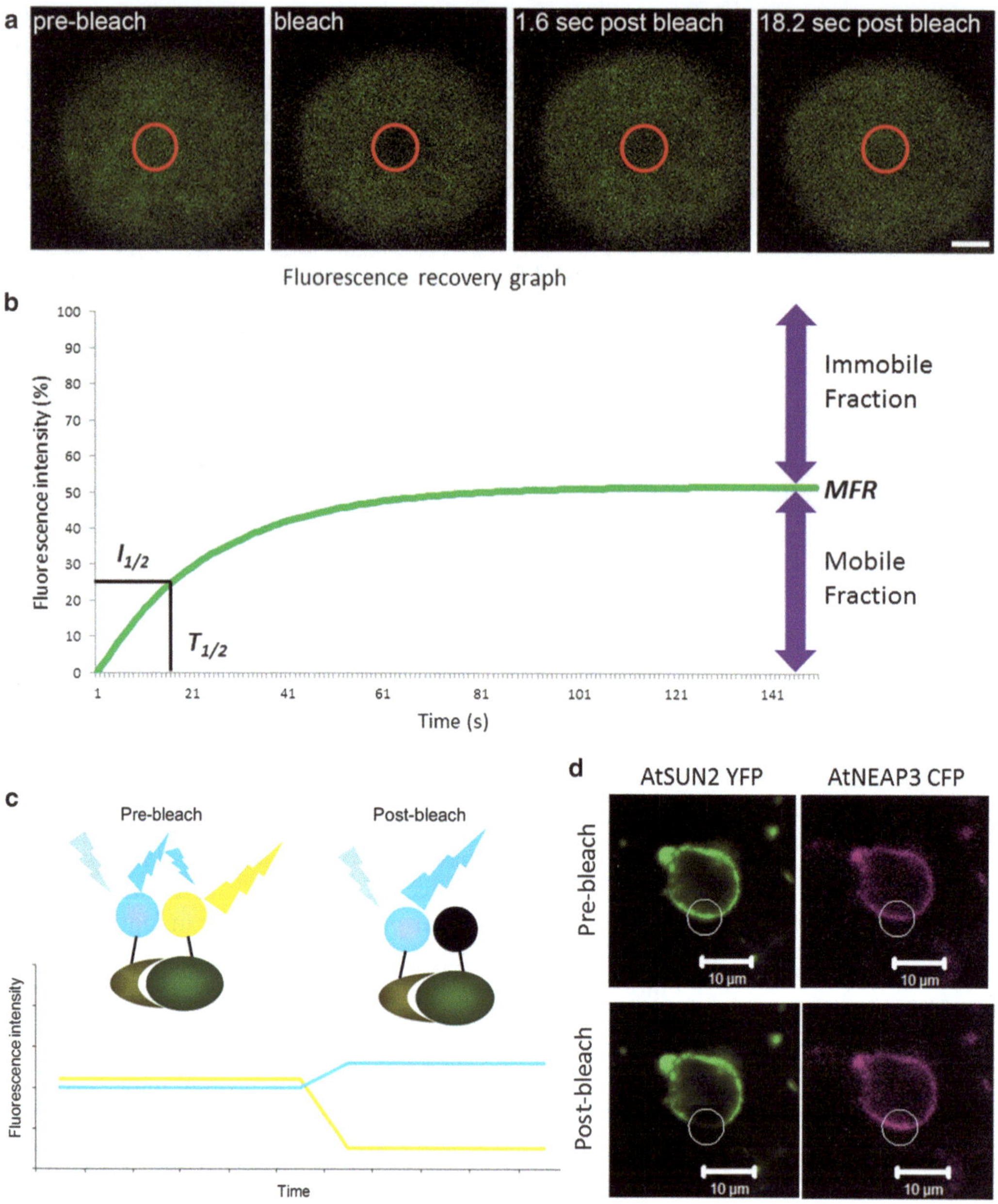

Fig. 2 Overview of FRAP and apFRET working concepts. (**a**) FRAP of LBR-GFP-labeled NE sheet; *red circle* is bleach ROI; fluorescence recovers quickly as LBR-GFP is highly mobile in NE membranes [27]. (**b**) Fitted fluorescence recovery curve (in this case of AtSUN1-YFP) displaying FRAP parameters typically analyzed. (**c**) Concept of apFRET; *green* structures are interacting proteins CFP and YFP are fused to; size of lightning indicates fluorescence intensity; the graph below shows fluorescence intensity of CFP (*blue line*) and YFP (*yellow line*) during apFRET; after the YFP bleach, YFP fluorescence diminishes and CFP fluorescence increases as YFP can no longer be excited with energy emitted by CFP. (**d**) Example of apFRET occurring between AtSUN2-YFP and AtNEAP3-CFP indicating interactions between the two NE proteins AtSUN2 and AtNEAP3 (Pawar, Evans, and Graumann, unpublished observations). YFP fluorescence in the white ROI diminishes after the bleach while CFP fluorescence in the ROI increases

The *MFR* is the highest post-bleach fluorescence intensity value and signifies the mobile fraction of a protein population. Subtraction of the mobile fraction from the pre-bleach fluorescence intensity value (100 %) results in the immobile fraction, the protein population that was bleached but not replaced by unbleached molecules. The assumption is that immobilized proteins are kept in place by binding interactions that anchor the proteins. Thus, *MFR*, mobile fraction, and immobile fraction are used to quantify mobile proteins. $T_{1/2}$ and D, in turn, are used to describe the quality of movement. $T_{1/2}$ is the time point at which half of the fluorescence has recovered. It is an indirect measure of protein velocity. D can be easily calculated if FRAP has been carried out in a two-dimensional structure such as the nuclear envelope (NE) sheet or an endoplasmic reticulum (ER) cisternae as it describes the protein movement more directly in $\mu m^2/s$ in the selected area. The velocity of protein movement can be affected by weaker binding interactions and by molecular obstacles hindering free protein movement. For membrane proteins, these hindrances can be caused by protein crowding of the membrane, lipid composition and density of the membrane, as well as structural networks interacting with the membrane and its embedded proteins [21]. These structural networks include the cell wall at the cell membrane [22] and the lamina at the NE [23]. FRAP has been used successfully to characterize NE proteins in various eukaryotic organisms and cell types [23–28].

Likewise, fluorescence resonance energy transfer or Förster resonance energy transfer (FRET) can be used in combination with live-cell confocal imaging and fluorescent protein fusions to study NE proteins. Specifically, FRET visualizes direct protein interactions in situ. It complements other protein interaction assays such as co-immunoprecipitation and yeast two-hybrid assays to confirm protein interactions in situ at the observed cellular location (in this case the plant NE).

In FRET, the emission energy of the donor fluorophore is used to excite the acceptor fluorophore (Fig. 2c). This transfer of energy can only occur when (1) the emission spectrum of the donor fluorophore and the excitation spectrum of the acceptor fluorophore overlap and (2) donor and acceptor fluorophores are in very close proximity—1 to 10 nm [29–31]. The two compatible fluorophore pairings for FRET commonly used are GFP–mRFP and CFP–YFP, where GFP and CFP are the donors and mRFP and YFP are the respective acceptors (Fig. 2c, d). The distance between the two fluorophores is affected by protein interactions but also protein folding and localization. Donor and acceptor are fused to two proteins, whose interaction can bring donor and acceptor into sufficiently close proximity for FRET to occur. However, this depends on the conformation and compartmentalization of the protein pair. For instance, SUN and KASH domains interact with each other [1, 14]. However, if the SUN protein contains the fluorophore at its nucleoplasmic N-terminus and the KASH protein contains the fluorophore on its cytoplasmic N-terminus, then the interaction is not observed by FRET because the distance between the two fluorophores is too great for energy transfer to occur.

FRET can be carried out in several ways—this chapter will focus on acceptor photobleaching FRET (apFRET), in which the acceptor fluorophore is bleached as described for FRAP. By bleaching the acceptor fluorophore, it can no longer use the donors' emission energy for excitation, which results in the donor emission fluorescence increasing [29, 31]. Hence, the donor fluorescence is monitored before and after the acceptor is bleached. If the donor fluorescence increases after the acceptor bleach, the two proteins fused to the donor and acceptor are considered to interact with each other (Fig. 2c, d). The difference in donor fluorescence before and after acceptor bleach is termed the FRET efficiency (E_F) and is expressed as a percentage. The E_F does not give indications on the strength of the interaction, i.e., a high E_F does not mean a strong binding interaction as the E_F is primarily dependent on the distance between the donor and acceptor fluorophore and their spectral overlap [31]. The apFRET technique has been successfully used to demonstrate protein interactions for both soluble and membrane-bound proteins in various experimental systems. Karpova et al. [29] provide a comprehensive description of apFRET to investigate binding of soluble nuclear proteins. In plant NE biology, apFRET has been used to study SUN proteins [25, 26, 32]. FRAP and apFRET techniques will be considered separately below. For apFRET, the conditions described will be for the donor–acceptor pairing YFP–CFP.

2 Materials

2.1 Computational Identification of Plant KASH Proteins

1. DORY: This program can be downloaded from the supplemental files of Zhou et al. [18] or from the following link: http://sourceforge.net/projects/doryforkash/.
2. Computer specifications: DORY is written in Java, and therefore a personal computer with Java installed is needed. A 64-bit computer system and 64-bit Java are recommended.
3. Jalview: a multiple sequence alignment analysis tool that can be downloaded at http://www.jalview.org/Web_Installers/install.htm.

   Internet connection is required because the following online resources will be needed:
4. BLAST (http://blast.ncbi.nlm.nih.gov/Blast.cgi).
5. Phobius (http://phobius.sbc.su.se/).

6. MAFFT (http://mafft.cbrc.jp/alignment/server/).
7. Protein database file: This must be in FASTA format. DORY can only read FASTA format files and does not check the input file format. To generate this file, the NCBI nonredundant protein sequences used in BLAST can be downloaded at ftp://ftp.ncbi.nlm.nih.gov/blast/db/FASTA/nr.gz.

2.2 Proteomic Identification of New Plant Nuclear Envelope and Nuclear Pore Proteins

2.2.1 Immuno-precipitation

1. Lysis buffer: 50 mM Hepes–KOH pH 7.5, 150 mM NaCl, 0.5 % (v/v) Triton X-100, 0.1 % (v/v) Tween-20. Store at 4 °C.
2. Wash buffer: 50 mM Hepes–KOH pH 7.5. Store at 4 °C.
3. Elution buffer: 100 mM Tris–HCl pH 6.8, 2 % (w/v) SDS, 20 % (w/v) glycerol, 5 % (v/v) 2-mercaptoethanol, 0.005 % (w/v) bromophenol blue.
4. Matrix for pulldowns conjugated to anti-tag antibody. This protocol has been optimized for magnetic beads conjugated to an anti-GFP antibody, specifically μMACS Anti-GFP MicroBeads (Miltenyi Biotec, Germany) (*see* **Note 1**).
5. Support materials for pulldowns, e.g., the μMACS system requires μColumns and μMACS Separator (both from Miltenyi).
6. Mortar and pestle.
7. Liquid nitrogen.

2.2.2 SDS-PAGE and Flamingo Staining

1. Running gel buffer: 1.5 M Tris–HCl pH 8.8.
2. Stacking gel buffer: 1 M Tris–HCl pH 6.8.
3. Ammonium persulfate: 10 % (w/v) solution in water. Make aliquots and store at −20 °C (*see* **Note 2**).
4. *N*,*N*,*N*,*N*′-tetramethylethylenediamine (TEMED). Store at 4 °C.
5. SDS-PAGE running buffer: Dissolve 3.03 g Tris, 14.41 g glycine, and 100 g SDS in 1 L water.
6. Fixing solution: 40 % (v/v) ethanol, 10 % (v/v) acetic acid (*see* **Note 3**).
7. Gel staining solution: We use the Flamingo fluorescent gel stain (Bio-Rad, USA) for which 1 volume of the stock should be diluted with 9 volumes of water (*see* **Note 4**).

2.2.3 Plant Materials

1. Seven- to ten-day-old transgenic *Arabidopsis* seedlings (fresh weight 0.3 g–3 g) expressing GFP-tagged nucleoporin (*see* **Note 5**).
2. Seven- to ten-day-old transgenic *Arabidopsis* seedlings (fresh weight 0.3 g–3 g) expressing free GFP as a negative control (*see* **Note 5**).

2.2.4 Trypsin Digestion for MS Analysis

1. Acetonitrile (mass spectrometry quality reagent).
2. Ammonium bicarbonate, 100 mM stock solution.
3. Reducing buffer: 10 mM DTT in 50 mM ammonium bicarbonate (*see* **Note 6**).
4. Alkylating buffer: 55 mM iodide acetamide in 50 mM ammonium bicarbonate (*see* **Note 7**).
5. Trypsin solution: 0.01 mg/mL trypsin in 50 mM ammonium bicarbonate. A high grade is required for MS analysis; we use "sequence grade" (Promega, USA). Make aliquots and store at −20 °C (*see* **Note 8**).
6. Peptide extraction buffer: 5 % (v/v) formic acid in 50 % (v/v) acetonitrile.

2.3 Imaging Techniques to Identify Protein–Protein Interactions at the Plant Nuclear Envelope

2.3.1 FRAP

1. Plant material expressing a NE protein fused to GFP or its variants CFP and YFP (*see* **Note 9**). Plant material can be either stably or transiently expressing and can be from any tissue that is normally easy to image (e.g., leaf, root, anthers) or cell culture (e.g., BY-2 cells).
2. Confocal microscope with laser, filter, beam splitter, and channel settings to image GFP, CFP, or YFP (Table 1).
3. Mounting materials including razor blade, pipette, microscope slide, cover slip, water, and lens oil (if oil-dipping lens used; *see* **Note 10**).
4. Microsoft Excel and GraphPad Prism for data analysis (*see* **Note 11**).

2.3.2 apFRET

1. Plant material co-expressing two NE proteins: one fused to CFP and the other to YFP. Plant material can be either stably or transiently expressing and can be from any tissue that is normally easy to image (e.g., leaf, root, and anthers) or cell culture (e.g., BY-2 cells).
2. Plant material expressing only the CFP-fused protein as control. Carry out the same apFRET experiment with CFP only

Table 1
Beam path settings for imaging GFP, CFP, and YFP with a Zeiss LSM confocal microscope

Fluorophore	Excitation laser wavelength (nm)	Emission wavelength captured (nm)	Beam splitters and filters for Zeiss LSM
GFP	488	505–530	HFT488
CFP[a]	458	470–500	HFT458/514; NFT515
YFP[a]	514	530–600	HFT458/514; NFT515

[a]Settings for simultaneous imaging of CFP and YFP

expressing tissue to ensure that an increase in CFP fluorescence in the co-expressing sample is due to the YFP bleach and hence protein interactions. Changes in CFP fluorescence when only CFP is expressed can be due to laser fluctuations, sample drifting, or membrane/organelle movement.

3. Confocal microscope with laser, filter, beam splitter, and channel settings that enable the simultaneous imaging of CFP and YFP (Table 1).
4. Mounting materials including razor blade, pipette, microscope slide, cover slip, water, and lens oil (if oil-dipping lens used; *see* **Note 12**).
5. Microsoft Excel for data analysis (*see* **Note 11**).

3 Methods

3.1 Computational Identification of Plant KASH Proteins

DORY has a user-friendly interface and can be easily set up. In the following text, the parameters adjustable in DORY will be in italic.

3.1.1 Setting Up "KASHFilter" Parameters

1. Set up "*TMD Frame Length*" and "*TMD Hydrophobic Threshold*" to identify proteins with a single TMD. DORY identifies potential TMDs by reading an amino acid sequence using a frame with a certain amino acid number (can be set in "*TMD Frame Length*"), sums up the hydrophobic value of each amino acid in this frame, and compares the sum with the "*TMD Hydrophobic Threshold.*" If the sum is not less than the "*TMD Hydrophobic Threshold,*" then the sequence in the frame is considered a TMD. The default values of "*TMD Frame Length*" and "*TMD Hydrophobic Threshold*" (20 and 32, respectively) have been tested to work best for identifying TMDs. Proteins with a single TMD will be kept, and the sequence C-terminal of the TMDs (the tail) will be subjected to further analysis (see next step).
2. Set up "*Maximum KASH Tail Length,*" "*Minimum KASH Tail Length,*" and "*Regex for KASH Tail*" to identify pKASH domains. If the length of the tail from the previous step is not more than "*Maximum KASH Tail Length*" and not less than "*Minimum KASH Tail Length,*" then it will be determined whether this tail terminates in a given C-terminal four-amino-acid pattern. This is done by comparing the tail to a regular expression set in "*Regex for KASH Tail.*" The C-terminal four-amino-acid pattern can be summarized from known KASH proteins, and some presets can be chosen from the dropdown menu of the "*Regex for KASH Tail.*" To customize "*Regex for KASH Tail,*" knowledge of regular expression is needed, and

the details can be found at http://www.regular-expressions.info. Some basics of regular expression are explained below.

Symbol	Match
\S	Any non-space characters
\S+	One or multiple non-space characters
[...]	Any character inside the square brackets
\Z	The end of a sequence

If the tail passes the regular expression test, it is considered as a pKASH domain and the protein sequence will enter an output file called "KASHFilterResult.txt."

If "*Output potential KASH tail in a file during the KASHFilter search*" is checked, then the pKASH domain will be output to a file named "KASHTail.txt." If "*In the output file, left pad KASH tail to the Maximum KASH Tail Length*" is also checked, the output pKASH domain sequences will be right aligned.

3. To confine a search to proteins within an amino-acid-length range, use the "*Protein Length Cutoff from to*" parameters.
4. If the protein names in the database contain species names, there are two ways to confine a search to proteins that belong to certain species.
   (a) Check the "*During KASHFilter search, keep the proteins whose protein names contain:*" checkbox. Click "*Choose Species Name File (one line one name)*" and choose a text file containing species names. In this text file, each line should contain only one species name. DORY will load the species names in this text file and simply check whether a protein name contains any of these species names. If yes, then this protein will be kept for further analysis; otherwise, it will be ignored.
   (b) Use "*Query NCBI Taxonomy Browser to filter non-eukaryotic proteins out.*" This is specifically designed for the NCBI nonredundant protein sequences file. DORY will read out the species name from the name of a protein and send a request to the NCBI Taxonomy server. If the response text contains the text set in "*Being positive, server return text should contain*" textbox, then this protein will be kept for further analysis; otherwise, it will be ignored.

3.1.2 Setting Up "HomologyFilter" Parameters

The KASHFilter will generate a file containing proteins with pKASH domains (KASHFilterResult.txt). Then the HomologyFilter can read this file, group the proteins into homologous groups, and save each group in separated text files with numbered names.

HomologyFilter uses two parameters, the "*E-value Cutoff*" and the "*Homolog Cutoff*."

1. "*E-value Cutoff*." Two proteins are considered homologs if the E-value of aligning these two proteins is less than the "*E-value Cutoff*" (*see* Subheading 3.1.5, **step 2**). DORY calculates the E-value using $Kmne^{-\lambda S}$. S is the score of an alignment using the Smith–Waterman algorithm. K and λ are Karlin–Altschul parameters whose values are obtained from the BLAST source code. Parameters m and n are the effective lengths of the query sequence and database, respectively. They are calculated by a modified "BLAST_ComputeLengthAdjustment" function from the BLAST source code.
2. "*Homolog Cutoff*." If an output group contains homologs less than the "*Homolog Cutoff*" value, then output files will be labeled with the prefix "belowHomoCutOff."

3.1.3 Setting Up Running Parameters

Either a full search can be performed, or only the KASHFilter can be run. A KASHFilterResult.txt file from a previous run of the KASHFilter can be directly run with the HomologyFilter, but the "*Database Total Protein Length*" and "*Database Total Protein Number*" have to be provided, which can be found in the log file of the previous run of the KASHFilter.

3.1.4 Running DORY

1. Click "*Open Database File*" button and choose a database file.
2. Click the "*Run*" button.
3. DORY will create a folder in the directory of the database file.
4. Inside this folder DORY will output the results and a log file that documents all parameters and steps of this run (*see* **Note 14**, if DORY takes too long to finish).

3.1.5 Manual Confirmation of the Candidates

After DORY outputs the homologous groups, the user must manually check whether a pKASH domain is present in the majority of a homologous group (*see* **Note 15**).

1. Choose one or more proteins from each group and perform protein BLAST against the nonredundant protein sequences. Since KASH proteins are eukaryotic proteins, the "Organism" parameter can be set to "Eukaryota." "Expect threshold" should be to set at 1e–4 or lower. However, if the threshold is too low, some true homologs will be filtered out. "Max target sequences" can be set at 500 at first and increased to a higher number if the maximum target sequence number is reached at the first round of BLAST. Click the "BLAST" button to run.
2. In the result webpage of BLAST, click "All" in the "Select" section. Manually uncheck undesired sequences if necessary (*see* **Note 13**). In the "Download" dropdown menu, choose

"FASTA (compete sequence)" to obtain the protein sequences of the selected homologs. The homologs identified by DORY and BLAST depend on the E-value threshold. DORY and BLAST use local alignment methods, which means that proteins which contain domains homologous to parts of the query protein may pass the E-value threshold. For example, Nesprin-1 contains actin-binding domains and spectrin repeats. Proteins containing any of these two domains may be classified as "homologs." However, they may not belong to the same protein family. Therefore, a large homologous group output by DORY may need to be further analyzed, especially when it contains large proteins having multiple domains.

The following steps are to determine whether a homologous group output by DORY belongs to a single protein family. Before manually performing the following analysis steps, it is worth trying the HomologyFilter again with a more stringent "E-value Cutoff."

(a) Use MAFFT to align the protein sequences of a homologous group. Set "Output order" to "Aligned" before starting the alignment.
(b) Download the alignment in "Clustal format," and open the alignment in Jalview.
(c) In Jalview menu, choose "ClustalX" in the "Colour" menu, uncheck "Wrap" in the "Format" menu, and adjust font in the "Format" menu to obtain an overview of the alignment.
(d) Scroll to manually check whether the sequences can be divided into sub-homologous groups based on the alignment.
(e) If yes, choose one sequence in a group by clicking it. Choose "Remove All Gaps" in the "Edit" menu. Then right-click, and choose "Selection"->"Edit"->"Copy." Paste the sequence in the BLAST webpage and set up a BLAST search. After this, in Jalview, choose "Undo Remove Gaps" to restore the alignment. Do this for all the sub-homologous groups identified. Compare the BLAST results of these subgroups to assess whether they belong to the same homologous group. If multiple homologous groups are identified, perform the next step for each one, respectively.

3. The presence of a C-terminal TMD can be tested using Phobius [33]. Phobius is also a prediction program and its prediction may not be accurate for a particular homolog, but it will provide an overview of whether a C-terminal TMD is predicted in the majority of a homologous group. If a C-terminal TMD is predicted, then check whether the C-terminal four amino acids

of the majority of this homologous group follow the pattern set in "*Regex for KASH Tail.*" If yes, then this homologous group is probably a KASH protein family within the larger superfamily of KASH domain-containing proteins. It is noteworthy that not all of the proteins in a homologous group will terminate in four amino acids exactly following the pattern. For example, when using the preset C-terminal four-amino-acid pattern "[PATHQL]PP[QTVFILM]" to identify animal KASH proteins, in the homologous groups obtained by BLAST, proteins terminating in PLPV and PSPT can also be found. "PLPV" and "PSPT" are quite similar to the known KASH domain C-terminal four-amino-acid patterns "PPPV" and "PPPT," respectively. Therefore, these outliers are probably also KASH proteins, and their C-terminal four amino acids can be used to improve the pattern used in "*Regex for KASH Tail*" (*see* below).

3.1.6 Improve the "Regex for KASH Tail"

After the manual confirmation, a new pattern of the C-terminal four amino acids may be summarized from the proteins believed to be KASH proteins. This new pattern can be used as an improved "*Regex for KASH Tail*" to perform a new round of searching.

3.1.7 Use DORY for Other Purposes

DORY can also be used to identify proteins that contain one TMD followed by a short conserved C-terminal sequence, as long as the conserved C-terminal sequence can be summarized by a generalized consensus. The steps to do this are the same as identifying KASH proteins, except with replacing the consensus sequence in the search parameters. In addition, with the source code available, DORY can be modified to perform searches according to individual needs.

3.2 Proteomic Identification of New Plant Nuclear Envelope and Nuclear Pore Proteins

Carry out all procedures at 4 °C to protect from protein degradation unless otherwise specified.

3.2.1 Immunoprecipitation with Anti-GFP Antibody Beads

1. Grind *Arabidopsis* seedlings to a powder in liquid nitrogen with a mortar and pestle.
2. Add ice-cold lysis buffer (3× volumes/gram fresh weight) to the powder and mix well. Transfer the lysates to a 1.5 mL tube and centrifuge the lysates at 20,400 × *g* for 5 min to remove cellular debris.
3. Transfer the supernatant to a new tube and centrifuge at 20,400 × *g* for 5 min to remove cellular debris completely.
4. Transfer the supernatant to a new tube and add 50 μL of magnetic beads conjugated to an anti-GFP antibody (e.g., μMACS Anti-GFP MicroBeads). Incubate for 30 min on ice.
5. Place μColumn in the magnetic field of the μMACS Separator. Equilibrate the column by applying 200 μL of lysis buffer.

6. Apply the lysates onto the column and flow through by gravity.
7. Rinse the column with 4× 200 μL lysis buffer.
8. Rinse the column with 1× 100 μL wash buffer (*see* **Note 16**).
9. Apply 20 μL of preheated 95 °C hot elution buffer to the column and incubate for 5 min at room temperature. Apply 50 μL of preheated 95 °C hot elution buffer to the column and collect eluate as the immunoprecipitate.

3.2.2 SDS-PAGE and Flamingo Staining

1. Prepare a 12.5 % gel for SDS-PAGE: Mix 5 mL of running gel buffer, 8.3 mL of 30 % acrylamide/bis mixed solution (29:1), 0.2 mL of 10 % SDS, 3.8 mL of water, and 0.2 mL of ammonium persulfate (*see* **Note 17**). Add 8 μL of TEMED to begin polymerization of acrylamide gel and cast gel within a 7.5 cm × 15 cm × 1 mm gel cassette.
2. Mix 0.375 mL of stacking gel buffer, 1 mL of 30 % acrylamide/bis mixed solution, 60 μL of 10 % SDS, 4.475 mL of water, and 60 μL of ammonium persulfate. Add 6 μL of TEMED and gently pour into gel caster. Insert a gel comb immediately without introducing air bubbles. Incubate at room temperature for 30 min to allow polymerization.
3. Apply the samples and electrophorese at 30 mA until bromophenol blue dye front from the elution buffer has reached the bottom of the gel.
4. Place the gel in a clean tray with 200 mL of fixing solution and incubate with gentle agitation for at least 2 h (*see* **Note 18**).
5. Pour off fixing solution and add 200 mL of Flamingo staining solution. Cover the gel tray with aluminum foil to limit light exposure and incubate with gentle agitation for at least 3 h (*see* **Note 19**).
6. Image the stained gels with fluorescent laser light (470–530 nm) and a longpass emission filter. Excise the protein bands of interest and transfer to a 1.5 mL tube (*see* **Note 20**).

3.2.3 In-Gel Digestion and Peptide Extraction for Mass Spectrometry

1. Wash the gel bands twice with 200 μL of 25 mM ammonium bicarbonate in 30 % (v/v) acetonitrile for 10 min followed by 100 % (v/v) acetonitrile for 15 min. Dry in a vacuum concentrator.
2. Add 200 μL of reducing buffer and incubate with shaking at 56 °C for 45 min.
3. Remove reducing buffer from the tube and add 200 μL of alkylating buffer. Incubate in dark at room temperature for 30 min.
4. Wash the gel bands with 200 μL of 50 mM ammonium bicarbonate followed by three times with 200 μL of 50 % (v/v)

acetonitrile in 50 mM ammonium bicarbonate. Dry in a vacuum concentrator.

5. Add 20 μL of trypsin solution and incubate at 37 °C overnight.
6. Recover the digested peptide twice with 20 μL of 5 % (v/v) formic acid in 50 % (v/v) acetonitrile. Combine the extracted peptide solutions and evaporate to 10 μL in a vacuum concentrator (*see* **Note 20**).
7. Subject the digested peptide to mass spectrometry.

3.3 Imaging Techniques to Identify Protein–Protein Interactions at the Plant Nuclear Envelope

3.3.1 FRAP

1. Set up the microscope for a FRAP experiment. Use a 100× or 63× oil or water dipping lens and minimal digital zoom factor. Keep laser transmission low, typically at 1–10 % to avoid photobleaching while measuring fluorescence recovery (transmission settings are dependent on microscope and fluorophore). Set up appropriate imaging setting for fluorophore (Table 1) and select a region of interest (ROI), in which fluorescence will be bleached and recovery measured. Select a control ROI, which is not bleached to monitor stability of fluorescence during the FRAP experiment. Keep the size of the ROI between different samples constant.
2. Set up bleaching options for bleach ROI: at least 5 pre-bleach scans to establish average pre-bleach fluorescence followed by bleach and post-bleach scans. The bleach should be carried out with the laser transmission set at 100 %, and the number of bleaching iterations is dependent on the strength of the signal; start at 2–3. The number of scans depends on the time period fluorescence recovery will be observed for. Diffusion is a fast process, which occurs in a matter of seconds, whereas protein turnover occurs over minutes and hours. If a long time course is selected, keep larger time intervals between post-bleach scans so as not to bleach recovering fluorescence. Keep bleach parameters between different samples constant. Use the same pinhole for bleach and recovery.
3. Mount plant tissue by excising the tissue with a razor blade and mounting it in water on a microscope slide. Alternatively, mount cultured cells in their culture medium on a microscope slide. Place the cover slip, and if necessary seal it on the slide with either double-sided tape or Valap (*see* **Notes 10**, **12**, and **21**).
4. Image the tissue to find appropriate nuclei. Before carrying out the FRAP experiment, choose whether to bleach the NE sheet (necessary if D needs to be calculated) or the NE rim in the nuclear midsection.
5. Place the bleach ROI over the area to be bleached and a control ROI over a fluorescence area not to be bleached. Carry out the bleach with settings defined in **step 2**.

6. Observe fluorescence intensity values in both ROIs over time. Typically, we take images every 1 s over a total period of 90 s for plant NE proteins. However, if proteins move more rapidly, shorter time periods and images per second can be used. Vice versa, slower movement (such as import) requires observations over longer time periods, and so, to avoid photobleaching, the time between each scan may be longer (*see* **step 2**).
7. If the fluorescence intensity stays constant in the control ROI, no uncontrolled bleaching has occurred and fluorescence values of the bleach ROI can be saved as a text file for subsequent analysis.
8. For each sample, carry out at least 30–50 bleach experiments and do not "reuse" the same nucleus.
9. Export raw fluorescence intensity values into an Excel file. In order to allow analysis and comparison of all samples, fluorescence intensity values need to be normalized to a percentage scale using the following formula:

$$I_N = \left[\left(I_T - I_{MIN} \right) / \left(I_{MAX} - I_{MIN} \right) \right] \times 100$$

   where I_N is the normalized fluorescence, I_T is the fluorescence intensity at a given time point, I_{MIN} is the fluorescence intensity immediately after the bleach, and I_{MAX} is the average pre-bleach fluorescence intensity [27].
10. Export the normalized fluorescence intensity values for each sample into GraphPad Prism or a similar curve fitting software. Fit the data with a nonlinear regression. Useful equations in the GraphPad Prism library include one-phase association, one-site binding, and one-phase exponential association, but user-defined equations can also be used. Use the equation with the best fit (highest R^2 value). Use the fitted values to plot the recovery curve graph. The fitted value for the last time point is the *MFR*.
11. Calculate $T_{1/2}$ by dividing the *MFR* by 2 ($I_{1/2}$). In GraphPad Prism, use $I_{1/2}$ to interpolate $T_{1/2}$. Alternatively, the curve equation can be used to calculate $T_{1/2}$ manually by using $I_{1/2}$.
12. Calculate *D* if fluorescence recovery was measured in a two-dimensional structure (e.g., the NE sheet). Use the following equation for this:

$$D = 0.88 \times \left(\omega^2 / 4 \times T_{1/2} \right)$$

   where ω is the radius of the bleach ROI [34].
13. Determine the average *MFR*, $T_{1/2}$, and, if calculated, *D* from the approximately 30–50 bleach experiments of each data set and carry out statistical analysis of these parameters. These can typically include F-tests for variance and Student's *t*-test for differences between data sets.

3.3.2 apFRET

1. Set up the microscope for an apFRET experiment. Use a 100× or 63× oil or water dipping lens and a minimal digital zoom factor. Keep laser transmission low, typically at 1–10 % to avoid uncontrolled photobleaching/activation of the acceptor and donor fluorophores. It is also wise to avoid white light on the samples as this can activate background FRET. Set up appropriate imaging settings for CFP and YFP (Table 1) and select a ROI in which acceptor fluorescence will be bleached and donor fluorescence will be measured. Keep the size of the ROI between different samples constant.
2. Set up bleaching options for YFP bleach: at least 5 pre-bleach scans to establish average pre-bleach fluorescence followed by bleach and post-bleach scans. Only one post-bleach scan is necessary. The bleach should be carried out with the YFP excitation laser transmission set at 100 %, and the number of bleaching iterations is dependent on the strength of the signal; start at 2–3. Keep bleach parameters between different samples constant.
3. Mount plant tissue by excising the tissue with the razor blade and mounting it in water on the microscope slide. Alternatively, mount cultured cells in their culture medium on the microscope slide. Place the cover slip and if necessary seal it on the slide with either double-sided tape or Valap (*see* **Notes 10**, **12**, and **21**).
4. Image the tissue to find appropriate nuclei.
5. Place the bleach ROI over the area to be bleached. Carry out the bleach with settings defined in **step 2**.
6. Observe the *first* fluorescence intensity values of both YFP and CFP *immediately after the bleach*. The YFP value should have significantly decreased (significantly compared to non-bleach YFP changes due to laser fluctuations), and the CFP fluorescence intensity may have changed after the bleach depending on interactions (to determine this, see steps below).
7. Fluorescence values of the bleach ROI can be saved as text file for analysis.
8. For each sample, carry out at least 30–50 bleach experiments and do not "reuse" the same nucleus. For analysis, only use the CFP fluorescence intensity values.
9. Export the raw CFP fluorescence intensity values into Excel. In order to allow analysis and comparison of all samples, fluorescence intensity values need to be normalized to a percentage scale using the following formula:

$$I_{\mathrm{N}} = \left[\left(I_{\mathrm{T}} - I_{\mathrm{MIN}}\right) / \left(I_{\mathrm{MAX}} - I_{\mathrm{MIN}}\right)\right] \times 100$$

where I_N is the normalized fluorescence, I_T is the fluorescence intensity at the given time point, I_{MIN} is the fluorescence intensity immediately after the bleach, and I_{MAX} is the average pre-bleach fluorescence intensity [27].

10. Calculate the average pre-bleach fluorescence intensity value for each sample.
11. For each sample, subtract the average pre-bleach fluorescence intensity value from the CFP fluorescence intensity value immediately after the bleach. The resulting value equals E_F.
12. Calculate the difference between two pre-bleach values to determine a no-bleach control E_F. In addition, carry out apFRET for the CFP only expressing sample and calculate the E_F value as a CFP only control E_F.
13. Calculate the average E_F, no-bleach control E_F and CFP only control E_F from the 30–50 experiments per data set and carry out statistical analysis. These can typically include F-tests for variance and Student's *t*-test for differences between controls and non-controls. If the two control (no YFP bleach and CFP only) E_F values are significantly lower than the no-control E_F value, the CFP fluorescence intensity has increased because FRET has occurred and the two fusion proteins interact with each other.

4 Notes

1. Miltenyi provides several different types of magnetic beads for other epitope tags.
2. Do not freeze/thaw more than twice.
3. Prepare just before using.
4. Prepare just before using.
5. Weight of plant material depends on the expression level of the GFP-fusion protein. In case of using the constitutive 35S promoter for expressing GFP fusions, 0.3 g of *Arabidopsis* seedlings is sufficient for this experiment.
6. Prepare just before using.
7. Prepare just before using.
8. Do not freeze/thaw more than twice.
9. mRFP can be used but bleaching efficiency of the mRFP laser is low due to the longer wavelength. GFP and its variants are advisable to use.
10. To avoid drifting of cells under the microscope, once the sample has been mounted, the slide can be sealed with either Valap or double-sided tape. Valap is made from equal parts Vaseline, lanolin, and paraffin wax.

11. OpenOffice or other curve fitting software might be used instead.
12. The nucleus can be quite mobile and might drift out of focus during the FRAP and apFRET experiments. To avoid nuclear movement, tissue sections can be treated with latrunculin B before mounting. For this, excise tissue and submerge in 25 μM latrunculin B. Incubate for approximately 20–30 min (for leaf tissue) and then mount tissue as described. Note that this should not be used when interactions/protein mobility in connection with actin are examined.
13. The same issue needs to be taken into consideration when obtaining homologs of a protein using BLAST. The "Distribution of Blast Hits on the Query Sequence" section in the BLAST result webpage needs to be consulted. Only the protein sequences that have good whole-sequence alignment should be chosen to download.
14. DORY takes too long to finish. The workload of the HomologyFilter increases exponentially with the number of input proteins. If DORY takes too long to finish, it is very likely that the result of the KASHFilter contains too many protein sequences. In this case, stop DORY, refine the parameters of the KASHFilter, and try again.
15. Reasons that not all members of a putative KASH protein family contain a pKASH domain: First, protein sequences predicted from DNA sequences may have miss-predicted C-termini. Second, partial proteins may lack their C-termini. Third, splice variants of some KASH genes do not encode a KASH domain, for example, *Caenorhabditis elegans* zyg-12 [35]. Therefore, if the majority of a homologous group contains a pKASH domain, then this group can be considered as a putative KASH protein family.
16. To eliminate the possibility of interference of SDS-PAGE, it is an important step to remove high concentrations of residual salts and detergents from the immune complex before the elution step.
17. The final concentration of acrylamide here is 12.5 %; however, this can be varied according to the size of the proteins of interest.
18. Gels can be left in fixing solution for up to 24 h. Shortened fixation time may reduce sensitivity.
19. Gels can be stored in Flamingo staining solution for up to 6 months and imaged without significant loss of sensitivity. For long-term storage, the gels should be placed in the dark at 2–8 °C.

20. We highly recommend using the low-binding tubes and tips that ensure best recovery rates of the peptides.
21. If a long time course is used, plant material may deteriorate on the microscope slide. Use culture dishes or ½ MS agar on slides.

Acknowledgments

I.M. thanks the National Science Foundation for financial support of research in this area. K.T. thanks the Grants-in-Aid for Scientific Research (nos. 15K14545 and 26711017) for supporting this work. K.G. thanks the Leverhulme Trust for an Early Career Fellowship supporting her research.

References

1. Zhou X, Graumann K, Meier I (2015) The plant nuclear envelope as a multifunctional platform LINCed by SUN and KASH. J Exp Bot 66:1649–1659
2. Jacob Y, Mongkolsiriwatana C, Veley KM et al (2007) The nuclear pore protein AtTPR is required for RNA homeostasis, flowering time, and auxin signaling. Plant Physiol 144: 1383–1390
3. Xu XM, Rose A, Muthuswamy S et al (2007) NUCLEAR PORE ANCHOR, the Arabidopsis homolog of Tpr/Mlp1/Mlp2/megator, is involved in mRNA export and SUMO homeostasis and affects diverse aspects of plant development. Plant Cell 19:1537–1548
4. Reeves PH, Murtas G, Dash S et al (2002) Early in short days 4, a mutation in Arabidopsis that causes early flowering and reduces the mRNA abundance of the floral repressor FLC. Development 129:5349–5361
5. Murtas G, Reeves PH, Fu Y-F et al (2003) A nuclear protease required for flowering-time regulation in Arabidopsis reduces the abundance of SMALL UBIQUITIN-RELATED MODIFIER conjugates. Plant Cell 15:2308–2319
6. Parry G, Ward S, Cernac A et al (2006) The Arabidopsis SUPPRESSOR OF AUXIN RESISTANCE proteins are nucleoporins with an important role in hormone signaling and development. Plant Cell 18:1590–1603
7. Muthuswamy S, Meier I (2011) Genetic and environmental changes in SUMO homeostasis lead to nuclear mRNA retention in plants. Planta 233:201–208
8. Zhang Y, Li X (2005) A putative nucleoporin 96 is required for both basal defense and constitutive resistance responses mediated by suppressor of npr1-1, constitutive 1. Plant Cell 17:1306–1316
9. Cheng YT, Germain H, Wiermer M et al (2009) Nuclear pore complex component MOS7/Nup88 is required for innate immunity and nuclear accumulation of defense regulators in Arabidopsis. Plant Cell 21:2503–2516
10. Kanamori N, Madsen LH, Radutoiu S et al (2006) A nucleoporin is required for induction of Ca2+ spiking in legume nodule development and essential for rhizobial and fungal symbiosis. Proc Natl Acad Sci U S A 103: 359–364
11. Saito K, Yoshikawa M, Yano K et al (2007) NUCLEOPORIN85 is required for calcium spiking, fungal and bacterial symbioses, and seed production in Lotus japonicus. Plant Cell 19:610–624
12. Tamura K, Fukao Y, Iwamoto M et al (2010) Identification and characterization of nuclear pore complex components in *Arabidopsis thaliana*. Plant Cell 22:4084–4097
13. Razafsky D, Hodzic D (2009) Bringing KASH under the SUN: the many faces of nucleo-cytoskeletal connections. J Cell Biol 186: 461–472
14. Starr DA, Fridolfsson HN (2010) Interactions between nuclei and the cytoskeleton are mediated by SUN-KASH nuclear-envelope bridges. Annu Rev Cell Dev Biol 26:421–444
15. Gundersen GG, Worman HJ (2013) Nuclear positioning. Cell 152:1376–1389
16. Tatout C, Evans D, Vanrobays E et al (2014) The plant LINC complex at the nuclear envelope. Chromosome Res 22:241–252

17. Kim DI, Birendra KC, Roux KJ (2015) Making the LINC: SUN and KASH protein interactions. Biol Chem 396:295–310
18. Zhou X, Graumann K, Wirthmueller L et al (2014) Identification of unique SUN-interacting nuclear envelope proteins with diverse functions in plants. J Cell Biol 205:677–692
19. Braun P, Aubourg S, Van Leene J et al (2013) Plant protein interactomes. Annu Rev Plant Biol 64:161–187
20. Jones AM, Xuan Y, Xu M et al (2014) Border control—a membrane-linked interactome of Arabidopsis. Science 344:711–716
21. Sprague BL, McNally JG (2005) FRAP analysis of binding: proper and fitting. Trends Cell Biol 15:84–91
22. Martinière A, Runions J (2013) Protein diffusion in plant cell plasma membranes: the cell-wall corral. Front Plant Sci 4:515
23. Ellenberg J, Siggia ED, Moreira JE et al (1997) Nuclear membrane dynamics and reassembly in living cells: targeting of an inner nuclear membrane protein in interphase and mitosis. J Cell Biol 138:1193–1206
24. Östlund C, Sullivan T, Stewart CL et al (2006) Dependence of diffusional mobility of integral inner nuclear membrane proteins on A-type lamins. Biochemistry (Mosc) 45:1374–1382
25. Graumann K, Runions J, Evans DE (2010) Characterization of SUN-domain proteins at the higher plant nuclear envelope. Plant J 61: 134–144
26. Graumann K (2014) Evidence for LINC1-SUN associations at the plant nuclear periphery. PLoS One 9, e93406
27. Graumann K, Irons SL, Runions J et al (2007) Retention and mobility of the mammalian lamin B receptor in the plant nuclear envelope. Biol Cell 99:553–562
28. Zuleger N, Kelly DA, Richardson AC et al (2011) System analysis shows distinct mechanisms and common principles of nuclear envelope protein dynamics. J Cell Biol 193: 109–123
29. Karpova T, Baumann C, He L et al (2003) Fluorescence resonance energy transfer from cyan to yellow fluorescent protein detected by acceptor photobleaching using confocal microscopy and a single laser. J Microsc 209:56–70
30. Chen Y, Mauldin JP, Day RN et al (2007) Characterization of spectral FRET imaging microscopy for monitoring nuclear protein interactions. J Microsc 228:139–152
31. Sparkes IA, Graumann K, Martinière A et al (2011) Bleach it, switch it, bounce it, pull it: using lasers to reveal plant cell dynamics. J Exp Bot 62:1–7
32. Graumann K, Vanrobays E, Tutois S et al (2014) Characterization of two distinct subfamilies of SUN-domain proteins in Arabidopsis and their interactions with the novel KASH-domain protein AtTIK. J Exp Bot 65: 6499–6512
33. Kall L, Krogh A, Sonnhammer EL (2007) Advantages of combined transmembrane topology and signal peptide prediction—the Phobius web server. Nucleic Acids Res 35: W429–W432
34. Axelrod D, Koppel D, Schlessinger J et al (1976) Mobility measurement by analysis of fluorescence photobleaching recovery kinetics. Biophys J 16:1055
35. Wilhelmsen K, Ketema M, Truong H et al (2006) KASH-domain proteins in nuclear migration, anchorage and other processes. J Cell Sci 119:5021–5029

Chapter 3

High-Efficiency Isolation of Nuclear Envelope Protein Complexes from Trypanosomes

Samson O. Obado, Mark C. Field, Brian T. Chait, and Michael P. Rout

Abstract

Functional understanding of the nuclear envelope requires the identification of its component proteins and their interactions. Trypanosomes cause human and livestock diseases worldwide but are so divergent from animals and fungi that in silico searches for homologs of proteins are frequently of low value. Here we describe a strategy for the straightforward identification of nuclear envelope proteins from trypanosomes that classifies proteins and their interaction networks in the nuclear pore complex. Milling frozen whole cells into a powder and rapid screening of buffer conditions for optimization of complex isolation is described. The method is inexpensive and potentially applicable to many organisms, providing fast access to functional information.

Key words Nuclear envelope, Nuclear pore complex, Affinity isolation, Cryomilling, *Trypanosoma*, Proteomics, Molecular evolution

1 Introduction

Trypanosomes are pathogenic Protozoa and may be one of the earliest branching taxa following radiation of the eukaryotes from the last eukaryotic common ancestor [1, 2]. Extreme evolutionary divergence represents a challenge to functional understanding of proteins, as many are novel and lack obvious homologs in model organisms [3]. The trypanosome nucleus harbors several novel systems, including polycistronic transcription of most protein-coding genes, and a lamina and kinetochores comprised of completely distinct proteins to the functionally analogous complexes of higher eukaryotes [4–6]. Extreme sequence divergence in the case of the nuclear pore complex (NPC) proteins and the lamina demands direct approaches for identification of proteins and protein–protein interactions.

We describe procedures that together allow the rapid isolation of protein complexes from African trypanosomes (Fig. 1). Similar approaches have been successfully applied to a broad range of

Sue Shackleton et al. (eds.), *The Nuclear Envelope: Methods and Protocols*, Methods in Molecular Biology, vol. 1411, DOI 10.1007/978-1-4939-3530-7_3, © Springer Science+Business Media New York 2016

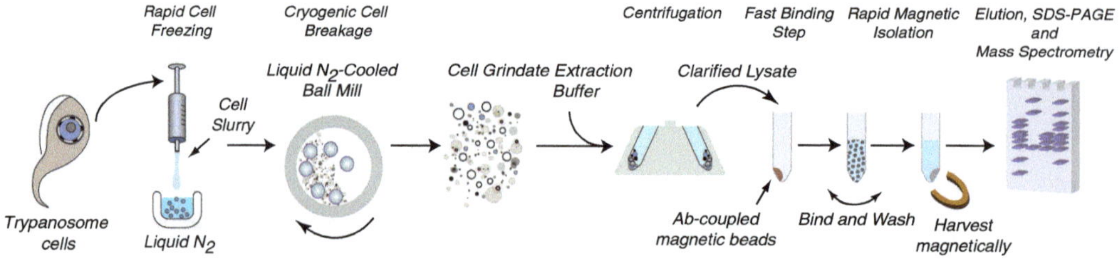

Fig. 1 Overall schematic of biophysical cell lysis and proteomics. Cells are harvested, flash frozen in liquid nitrogen, and added to precooled steel jars containing steel balls. As the jar is centrifuged in a planetary ball mill, it rotates, causing the balls to cascade with high force and mill the frozen cells into a powder. Various extraction solutions can then be tested to find the most suitable conditions for the protein complex of interest. The cell powder is resuspended in an extract solution and clarified by centrifugation. Affinity capture is performed in the presence of anti-GFP antibody coupled to magnetic beads and then eluted and fractionated by SDS–PAGE. Protein bands are excised, digested with trypsin, and identified by mass spectrometry. This figure is adapted and modified from Oeffinger et al. [7]

organisms, and we believe that these methods have major utility in deciphering the nuclear and cellular biology of trypanosomes. This protocol exploits the ability to grow parasites in semi-defined media, routine genetic modification, and specifically facile introduction of an epitope tag into one allelic copy of the diploid genome. The epitope tag is typically GFP, although a wide range of common epitopes and/or ectopic expression can also be used successfully. Parasites are frozen in liquid nitrogen and milled to a powder using a Retsch Planetary Ball Mill, providing highly efficient breakage of cells without a need for solubilization or denaturing reagents. This powder can be conveniently stored at −80 °C for several months. Individual aliquots of this stock powder can then be tested in affinity isolation procedures using magnetic beads with covalently coupled antibody allowing rapid screening of buffers, detergents, and additives, to arrive at an optimal, or acceptable, set of conditions suitable for electrospray ionization mass spectrometry. The cryomilling approach can provide rapid and inexpensive access to a large constellation of interaction data.

2 Materials

It is crucial to consult and follow the Material Safety Data Sheets and your institute's health and safety procedures for the appropriate handling of equipment and potentially hazardous materials used in these protocols. If in doubt, we advise contacting the authors for advice, which will be gladly given.

2.1 Preparation of Frozen Trypanosome Pellets

1. Refrigerated centrifuge and rotor for spinning large cell volumes (e.g., Beckman JLA 8.100 rotor).

2. Phosphate-buffered saline (PBS) if working with procyclic form trypanosomes or trypanosome dilution buffer (TDB) if working with bloodstream form trypanosomes. TDB—pH 7.4: 5 mM KCl, 80 mM NaCl, 1 mM $MgSO_4$, 20 mM Na_2HPO_4, 2 mM NaH_2PO_4, and 20 mM glucose.
3. Resuspension buffer: Ice cold 1× PBS or TDB containing 100 mM dithiothreitol (DTT) and protease inhibitors without EDTA.
4. 50-mL polypropylene Falcon-type tubes.
5. 20-mL syringe and syringe caps.
6. Liquid nitrogen.
7. Styrofoam container.
8. Pipettor.
9. 10- and 20-mL pipettes.

2.2 Conjugation of Magnetic Beads

1. Rotator for microfuge tubes at 30 °C.
2. Vacuum aspirator or equivalent.
3. Anti-GFP antibody (*see* **Note 1**).
4. 0.1 M sodium phosphate buffer ($NaPO_4$)—pH 7.4: 2.62 g $NaH_2PO_4 \times H_2O$ (MW 137.99) (2 mM) and 14.42 g $Na_2HPO_4 \times 2H_2O$ (MW 177.99) (20 mM). Adjust pH if necessary.
5. 3 M ammonium sulfate (stock solution): 39.6 g $(NH_4)_2SO_4$ (MW 132.1). Dissolve in 0.1 M sodium phosphate buffer (pH 7.4) and adjust to 100 mL.
6. Phosphate-buffered saline (PBS).
7. PBS + 0.5 % Triton X-100: Include 0.5 % (w/v) Triton X-100 in 100 mL PBS solution.
8. 100 mM glycine HCl pH 2.5—make fresh.
9. 10 mM Tris–HCl pH 8.8.
10. 100 mM triethylamine—make fresh.
11. 1× PBS/50 % glycerol.

2.3 Cryomilling

1. Liquid nitrogen.
2. Retsch Planetary Ball Mill PM100.
3. Retsch steel jar, lid, and 20-mm steel balls (*see* **Note 2**).
4. Teflon puck (custom made) and Teflon insulating jar (available from Retsch, a division of Verder Scientific that manufactures equipment for homogenizing laboratory samples) (Fig. 2a).
5. Cryoprotective gloves.
6. Safety goggles.

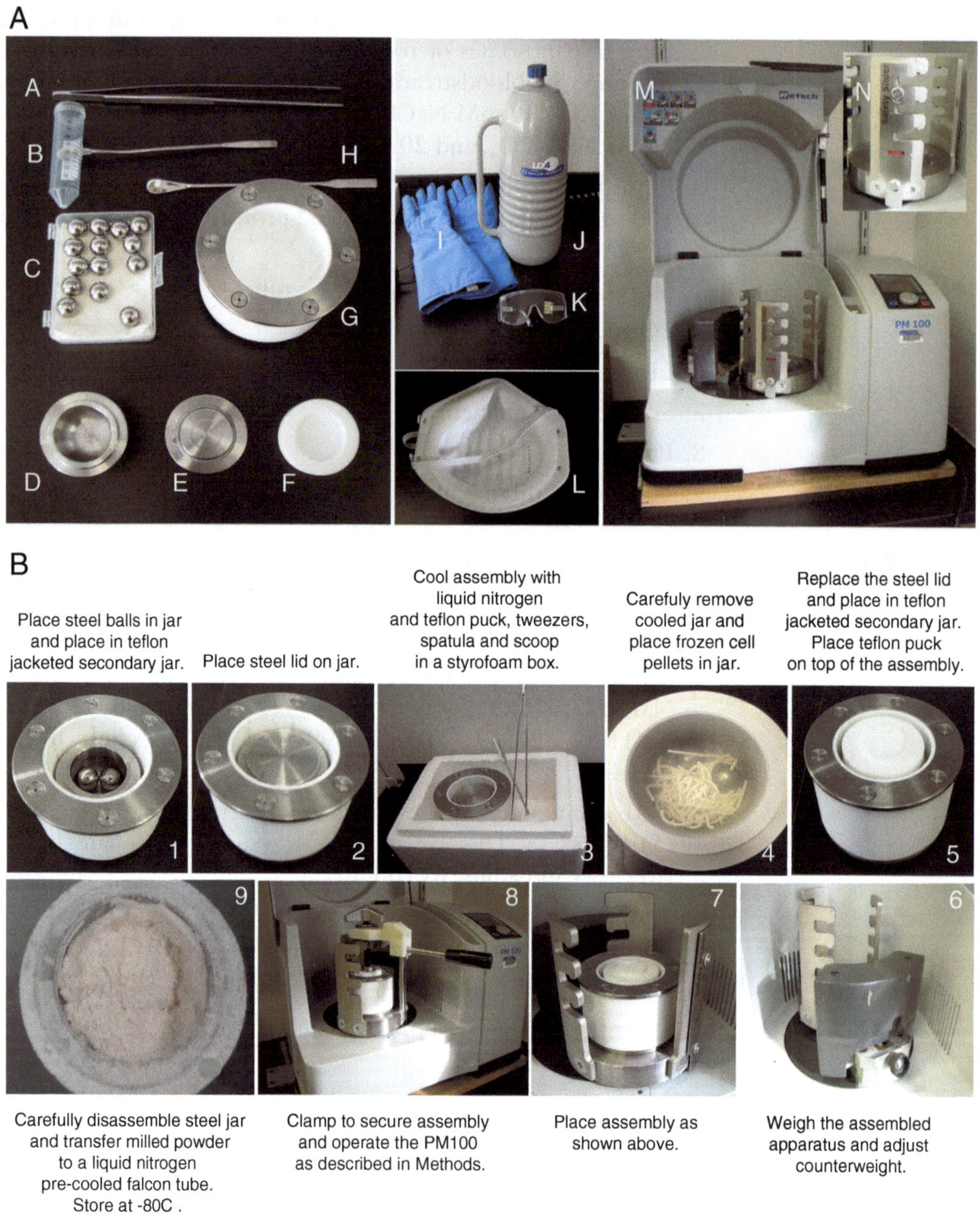

Fig. 2 Requirements and setup of the cryomilling procedure. (**a**) *Panels* showing required items: *A*, tweezers; *B*, liquid nitrogen scoop for pouring 2–15 mL (made by making an incision on a polypropylene tube using a razor blade and then inserting the spatula to create a handle); *C*, 20-mm steel balls; *D*, 50-mL steel jar; *E*, steel jar lid; *F*, Teflon puck; *G*, Teflon jar; *H*, spatula; *I*, cryo-gloves; *J*, liquid nitrogen dewar; *K*, goggles; *L*, face mask; *M*, Retsch PM100 Planetary Ball Mill; and *N*, *inset* showing a recommended safety adaptation to prevent accidental use of the PM100 without clamping the assembled milling apparatus. (**b**) A step-by-step guide to assembly of components

7. Large tweezers.
8. Spatula.
9. Liquid nitrogen scoop. This can be generated by fusing the handle to a 50-mL polypropylene tube (Fig. 2a).

2.4 Affinity Capture

1. Magnetic beads conjugated with anti-GFP (or other tag used) antibody.
2. Magnetic rack for microcentrifuge tubes.
3. Extraction buffers for affinity capture that contain a buffering agent such as Tris or HEPES, salt for ionic strength such as NaCl/KCl, detergents such as Triton, and protease inhibitors. Divalent cations such as Magnesium may be added as required.
4. Microcentrifuge.
5. Cryomilled powder.
6. Elution buffer: protein gel loading dye buffer without reducing agents added or 40 mM Tris–HCl pH 8.0 and 2 % SDS.
7. Reducing agent: 50 mM DTT.

3 Methods

3.1 Preparation of Frozen Trypanosome Pellets

1. For optimal results, we use 2×10^{10} cells as the minimum number of cells for efficient cryomilling. Harvest cells by pelleting at 3000 × *g* in a centrifuge at 4 °C (*see* **Note 3**).
2. Discard the supernatant and resuspend the cell pellet in 40 mL of ice-cold 1× PBS for procyclic form trypanosomes or TDB for bloodstream form trypanosomes.
3. Pellet the resuspended cells at 1500 × *g* for 5 min at 4 °C.
4. Discard supernatant and add ice-cold resuspension buffer equivalent to the volume of the cell pellet.
5. Remove the piston/plunger from a 20-mL syringe and add a cap to the nozzle to prevent flow through of the cell suspension (Fig. 3).
6. Transfer resuspended cells into the 20-mL syringe and place in a 50-mL polypropylene Falcon-type tube as illustrated (Fig. 3).
7. Pellet the resuspended cells at 5000 × *g* for 5 min at 4 °C to compact the cells and minimize buffer carryover.
8. During **step 7**, fill a Styrofoam box with liquid nitrogen and cool a pre-labeled 50-mL polypropylene Falcon-type tube to liquid nitrogen temperature.
9. Discard the supernatant from the pelleted cells and remove the syringe cap.
10. Fill the cooled 50-mL tube with liquid nitrogen.

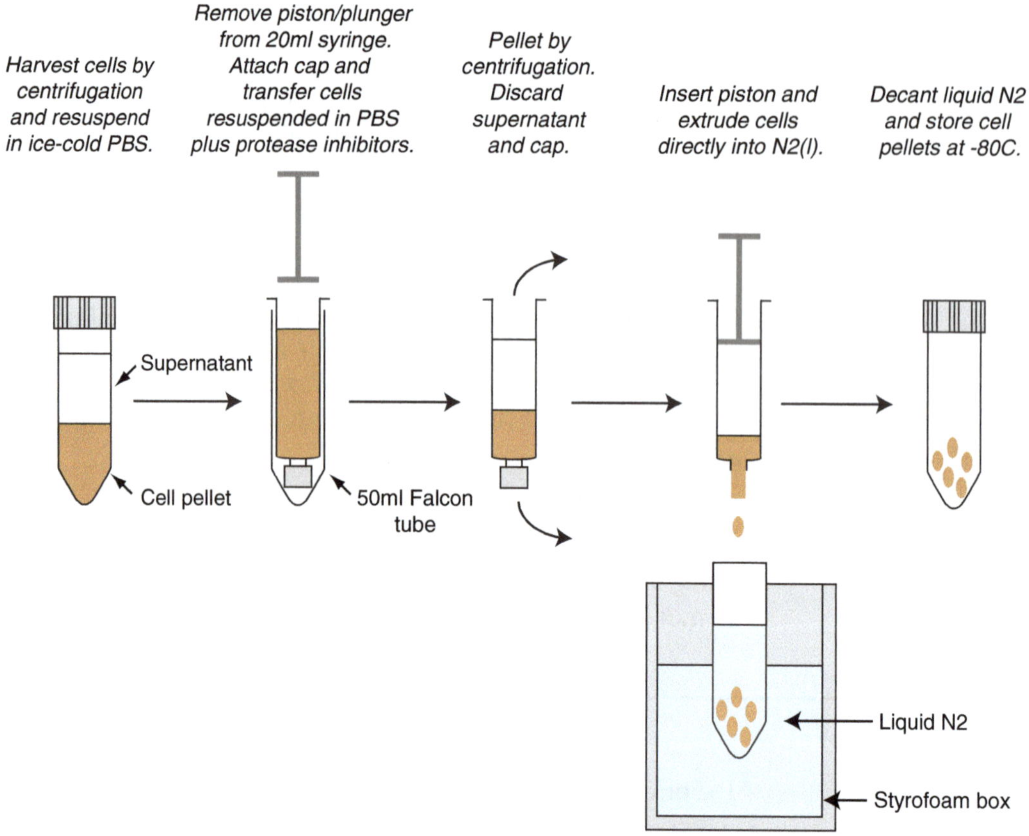

Fig. 3 Illustration of the making of frozen cell pellets. *Highlighted* are the key steps to facilitate the making of frozen cell pellets. Additional information can be found on www.ncdir.org/public-resources/protocols

11. Gently place the piston/plunger back onto the syringe that contains the cell pellets and apply very gentle pressure extruding the trypanosome cell slurry into the 50-mL tube containing the liquid nitrogen (Fig. 3).
12. Decant the remaining liquid nitrogen from the 50-mL tube that contains the frozen cell pellets/noodles and store at −80 °C until required for cryomilling (*see* **Note 4**).

3.2 Cryomilling Frozen Trypanosome Pellets

This protocol is heavily modified from an earlier version [7]. A modification of the earlier protocol has been shown to greatly facilitate mammalian cell affinity isolation procedures [8]. Here we describe a modified form for trypanosomes.

Extensive notes and description of any modifications and instructions that facilitate the cryomilling technique are available on the website of the National Center for Dynamic Interactome Research (www.ncdir.org/public-resources/protocols). It is important to visit the website for the latest updates and additional information, particularly with respect to safety procedures, prior to attempting the cryomilling procedure. *Protective clothing, goggles, and cryoprotective gloves must be used at all times during cryomilling.*

1. Precool the steel jar, lid, Teflon puck, and Teflon insulating jar and steel balls in a Styrofoam box containing liquid nitrogen.
2. Once cooled to liquid nitrogen temperature (very little bubbling), carefully remove all cooled units from the liquid nitrogen using tweezers.
3. Assemble as shown in Fig. 2b.
4. Put frozen cells into the assembled and cooled steel jar (Fig. 2).
5. Place the cooled lid and then the cooled Teflon puck on top of the steel jar (Fig. 2).
6. Set the milling setup into the PM100 (Fig. 2).
7. Clamp down the milling jar as per manufacturers' instructions (*be sure to clamp down the jar or serious injury may occur*).
8. Adjust the setting on the PM100 to mill your frozen material at 400 rpm for 3 min, with a 1-min interval (the PM100 will pause briefly after each minute and switch direction to allow counterrotation of the milling jar). Cool the jar after each 3-min cycle by adding 10–15 mL of liquid nitrogen in the space between the Teflon jar and the 50-mL cryomilling steel jar. Allow the liquid nitrogen to evaporate completely (10 s) and repeat. Alternatively, the whole setup can be unclamped and cooled in a Styrofoam box with liquid nitrogen (do not submerge the jars completely as this will result in the loss of cell material/powder) for about 30 s before re-clamping and resuming the milling procedure. *It is important for the operator to follow good standard practice and take extreme care when handling liquid nitrogen.*
9. Repeat **step 8** three times (*see* **Note 5**).
10. Unclamp the steel jar and add 2–5 mL of liquid nitrogen into the grinding jar using the liquid nitrogen scoop described in Fig. 2 (*see* **Note 6**). The liquid nitrogen helps dislodge any compacted powder and facilitates liquid cryomilling that results in a finer millate. *It is important for the operator to follow good standard practice and take extreme care when handling liquid nitrogen.*
11. Re-clamp and cool the jar by adding 10–15 mL of liquid nitrogen in the space between the Teflon jar and the 50-mL cryomilling steel jar.
12. Mill the cell powder for 1–2 min at 400 rpm (*see* **Note 7**). Cool the jar immediately as in **step 11** while it is still clamped on the PM100 by pouring about 10–15 mL of liquid nitrogen using the liquid nitrogen scoop into the space between the Teflon

jacketed insulating jar and the 50-mL cryomilling steel jar. The liquid nitrogen will vaporize while cooling the cryomilling jars but simultaneously enables any residual evaporated nitrogen gas to cool down and lower the internal pressure in the steel milling jar. *Thus, this step is important to lower the risk of injury or sample loss when opening the jar. Nonetheless, unclamp the lid very slowly to enable a slow pressure release to prevent a sudden venting of gas that can lead to powder loss or injury.* Add 2–5 mL of liquid nitrogen into the grinding jar and re-clamp. Re-clamp and cool the jars once again as in **step 11**.

13. Repeat **step 12** ten times. While the repetitive nature of these steps are somewhat arduous, these conditions have been optimized: attempting to engage fewer but longer grinding cycles results in pressure buildup and sample warming that are unacceptable.
14. Cool the milling jars as in **step 12** prior to unclamping and carefully transfer the milled powder into a precooled and labeled 50-mL polypropylene tube that is being maintained cooled in a Styrofoam box containing liquid nitrogen.
15. Store milled powder until needed at −80 °C (*see* **Note 4**).

3.3 Conjugating Proteins to Magnetic Beads

This protocol is designed for the use with Life Technologies Dynabeads with epoxy chemistry and is modified from an earlier protocol [9]. Other approaches can be used, but we have found that the combination of speed and avoiding centrifugation steps provided by magnetic beads is optimal of those approaches we have tried. Each vial contains 300 mg of Dynabeads which correspond to the amount sufficient for the analysis of ~120 g of cell powder. Do not prepare the whole batch unless you plan to use it within 6 months (*see* **Note 8**).

1. Weigh out the appropriate amount of Dynabeads required (*see above*) in a 15-mL Falcon-type tube.
2. Add 5 mL of 0.1 M sodium phosphate buffer, pH 7.4, and vortex the tube for 30 s.
3. Insert the tubes with beads into the magnet holder. Wait until all are attached to the magnet. The solution will appear clear. Aspirate off the buffer, taking care not to touch the magnetic beads.
4. Resuspend the beads in fresh 5 mL of 0.1 M sodium phosphate buffer, pH 7.4.
5. Agitate slowly for 10 min on a shaking platform or rotator.
6. Harvest the beads magnetically as in **step 3** and discard supernatant.
7. Setup the conjugation reaction once the incubation in **step 5** has been started. Add the reagents sequentially in the order as follows.

8. For 100 mg of beads (1/3 of a Dynabeads vial) the ideal reaction volume is 2 mL. Add the following reagents to a 5-mL round bottomed tube with a snap cap or an equivalent reaction vessel in the order indicated:
   (a) 10 μg of affinity purified polyclonal anti-GFP antibody per mg of magnetic beads is required for optimal antibody conjugation. For other epitope tags, the amount required for conjugating beads needs to be tested on a small scale (i.e., use 10 mg of beads for each test reaction).
   (b) Depending on the volume of the antibody used, add 0.1 M sodium phosphate pH 7.4 to make the volume up to 1 mL (1/2 of the total reaction volume of 2.0 mL).
   (c) Add 1 mL (based on a 2 mL reaction volume) of 3 M ammonium sulfate to a final concentration of 1.5 M (*see* **Note 9**). Add slowly while shaking the tube on a slow-speed vortex. If testing conditions for other epitopes, it is advisable to test a range of Ammonium Sulfate concentrations to optimize this step.
   (d) Transfer the antibody/Sodium Phosphate/1.5M Ammonium Sulfate solution onto the magnetic beads.
9. Wrap the 5-mL tube snap cap with Parafilm to prevent leaking and evaporation.

   Allow the conjugation/reaction to proceed on a rotating platform at 30 °C for 24 h (*see* **Note 10**).

Following day—wash the Dynabeads. Perform all washes as described above after transferring the Dynabead suspension to a 15 mL Falcon-type tube and by inserting the tubes into the magnet holder. You can aspirate the supernatant using a vacuum aspirator.

Wash once with 5 mL of 100 mM glycine–HCl pH 2.5—add and then remove as fast as possible. Prolonged exposure to low pH may denature your antibody.

Wash once with 5 mL of 10 mM Tris–HCl pH 8.8.

Wash once with 5 mL of fresh 100 mM triethylamine—add and then remove as fast as possible. Prolonged exposure to high pH may denature your antibody.

Wash the coated beads—a total of 4 washes with 1× PBS in 5-mL tubes with a 5-min incubation for each wash.

Wash once with PBS + 0.5 % Triton X-100 for 5 min.

Wash again with PBS + 0.5 % Triton X-100 but incubate 15 min on rocker.

Wash one last time with 1× PBS.

Resuspend all beads in a total of 0.7 mL of 1× PBS + 50 % glycerol for 100 mg of beads.

Store the coated beads at −20 °C.

3.4 Affinity Capture of Tagged Proteins

This method can be used with any affinity tag as long as good affinity isolation antibodies are available. We describe GFP-tagged proteins in this instance. This protocol is modified from an earlier protocol [7]. For any individual protein–protein complex being investigated, a trial with smaller volumes should be conducted comparing different affinity isolation buffers. The range of buffers we typically test is given in Table 1. It is recommended that an aliquot of the input, flow through, eluate, and cell pellet be saved for analysis by Western blotting to make sure that your affinity handle is efficiently captured from your sample. If not, adjust the amount of beads and/or length of affinity capture accordingly.

1. Remove the cryomilled powder from the –80 °C freezer and place it in a Styrofoam box containing liquid nitrogen with a rack to keep the Falcon-type storage tube cold and upright.
2. Weigh out 50 mg to 1 g of cell powder, depending on the abundance of your target protein, into a precooled tube. The type of tube depends on the amount weighed—1.5- or 2-mL microcentrifuge tubes for powder of 50–200-mg and larger tubes for greater powder amounts (up to a 50-mL Falcon-type tube for 1 g, rather than a 15-mL tube for ease of powder resuspension) (*see* **Note 11**).

Table 1
Buffer compositions for affinity isolation experiments

1.	20 mM HEPES, pH 7.4, 100 mM NaCl
2.	20 mM HEPES, pH 7.4, 100 mM NaCl, 0.1 % CHAPS
3.	20 mM HEPES, pH 7.4, 100 mM NaCl, 0.5 % Triton
4.	20 mM HEPES, pH 7.4, 250 mM NaCl
5.	20 mM HEPES, pH 7.4, 250 mM NaCl, 0.1 % CHAPS
6.	20 mM HEPES, pH 7.4, 250 mM NaCl, 0.5 % Triton
7.	20 mM HEPES, pH 7.4, 100 mM NaCitrate, 0.1 % CHAPS
8.	20 mM HEPES, pH 7.4, 100 mM NaCitrate, 0.5 % Triton
9.	20 mM HEPES, pH 7.4, 250 mM NaCitrate, 0.1 % CHAPS
10.	20 mM HEPES, pH 7.4, 250 mM NaCitrate, 0.5 % Triton
11.	20 mM HEPES, pH 7.4, 250 mM NaCl, 0.5 % Triton, 0.5 % deoxy Big CHAP
12.	20 mM HEPES, pH 7.4, 250 mM NaCitrate, 0.5 % Triton, 0.5 % deoxy Big CHAP
13.	40 mM Tris, pH 8.0, 250 mM NaCl, 0.5 % Triton, 0.5 % deoxy Big CHAP
14.	40 mM Tris, pH 8.0, 250 mM NaCitrate, 0.5 % Triton, 0.5 % deoxy Big CHAP

3. Transfer weighed powder to an ice bucket and sit on ice for 30 s, until it gets close to thawing. *Do not let it thaw as rapid protein degradation may occur.*
4. Resuspend powder in room temperature extraction buffer of choice, at a ratio of 1:9 (powder-extraction buffer), by vortexing and pipetting up and down. Place on ice.
5. Once all aliquots are resuspended, sonicate on ice with a microtip sonicator to break apart aggregates that may be invisible to the eye. For example, we use a Misonix Ultrasonic Processor XL at Setting 4 (~20-W output) for 2× 1 s.
6. Clarify the cell lysate by centrifugation in a microcentrifuge at 20,000 × g for 10 min at 4 °C.
7. During the centrifugation in **step 6**, wash the antibody-conjugated magnetic beads once with the affinity isolation buffer of choice. Harvest the beads magnetically and discard buffer.
8. Add the clarified cell lysate to the magnetic beads and rotate at 4 °C for 1 h (or longer if necessary).
9. Wash the magnetic beads 3× in the affinity isolation buffer used during affinity capture (*see* **Note 12**).
10. After the third wash, transfer the magnetic beads to a fresh tube. This ensures that cellular protein that may have been adsorbed onto the original tube prior to elution is discarded.
11. Harvest the beads magnetically and discard the buffer.
12. Spin down the magnetic beads, and aspirate residual buffer trapped between the beads during the magnetizing process.
13. Elute your affinity captured protein complex with nonreducing SDS–PAGE loading buffer or equivalent (*see* **Note 13**).
14. Harvest the beads magnetically and transfer the eluate to a fresh tube.
15. If appropriate add reducing agent and heat at 70 °C for 10 min.

3.5 Downstream Analysis

Depending on the experiment, the isolated protein complexes may be taken directly to mass spectrometry using standard methods. If so desired, the complexes may also be analyzed by SDS–PAGE. We routinely analyze complexes using precast gel systems that provide highly reproducible results. This provides a quality control prior to mass spectrometric analysis or can be used to identify/isolate individual protein bands. All of these methods are standard.

An example of the affinity isolation of most of the trypanosome nuclear pore complex—a huge structure that has stable, as well as dynamic, components. Isolating protein complexes from different cell compartments requires buffer optimization [10]. As an example of the flexibility of the cryomilling and affinity

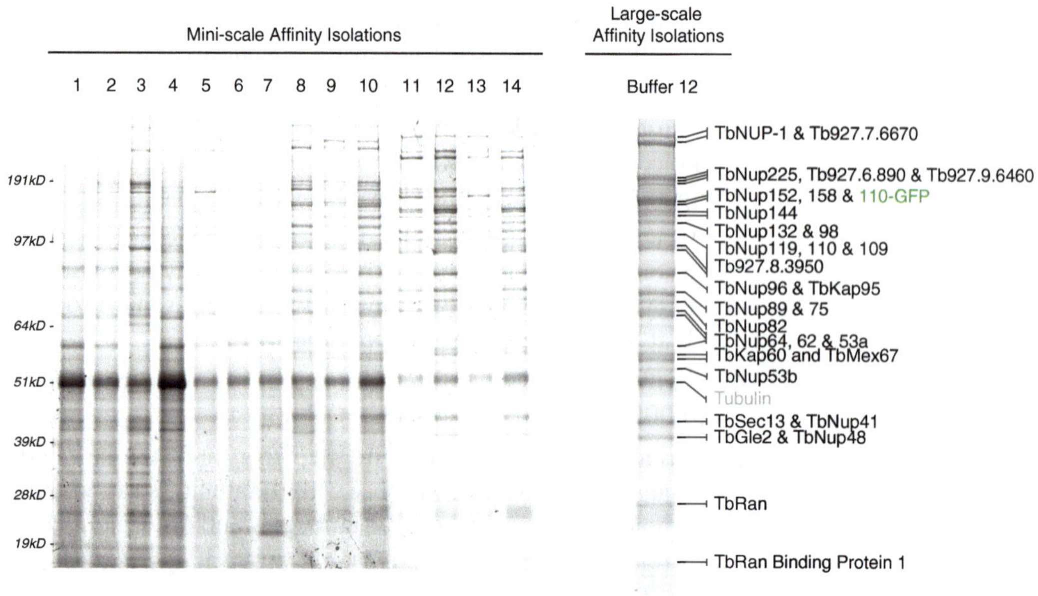

Fig. 4 Testing of various affinity isolation solvents on the nuclear basket protein TbNup110. 50 mg of TbNup110–GFP cell lysate powder was weighed out into 2-mL tubes and affinity purified in 14 different buffers containing various concentrations of salt and detergents (Table 1). Nup110 maintains the highest number of its interaction network in buffer 12. This condition was chosen for further exploration in a larger-scale pullout. Protein bands were excised out of the SDS–PAGE gel and identified by mass spectrometry. 18 of the original 22 Nups identified by DeGrasse et al. [11] in a classical biochemical fractionation and proteomic screen [11], as well as new Nups were identified with this method, demonstrating the power of this approach. Additionally, we observe interactions with the lamin-like NUP-1 [4] as well as transport factors

techniques, we highlight the ability to use low-stringency to high-stringency solutions to affinity isolate the trypanosome NPC (Fig. 4; Table 1). These methods are also applicable to dynamic protein complexes.

4 Notes

1. Antibody must not be stored in glycerol or Tris buffer or any solutions that contain amino, hydroxyl, or sulfhydryl groups, as these react with the epoxy groups on the magnetic beads. Dialyze to exchange buffer if this is the case.
2. The lid of the grinding jar we use has been modified with O-rings that are custom made by Marco Rubber (http://www.marcorubber.com) and consist of a stainless steel coil ensheathed in Teflon (PFA–PTFE-encapsulated spring-energized seal). These seals better withstand the extreme temperatures associated with liquid nitrogen use. To make these alterations, contact Marco Rubber and send them the original rubber O-ring that comes as standard with the steel lid

so that they may have the correct dimensions to manufacture the PFA–PTFE-encapsulated spring-energized seal. Pressure valves may be added to the lid of the cryomilling jar to relieve excess pressure buildup. We recommend pressure valves from the Lee Company (http://www.leeimh.com/). Specifically we use forward-venting valves that relieve pressures that exceed 5 bar/500 KPA. These are installed as to manufacturers' instructions (http://www.leeimh.com/metal/check-valves-axial/check-valve-558F.htm). Typically we install one valve in the center of a 50-mL jar lid or two valves in a 125-mL jar lid (Fig. 2).

3. A minimum volume of cell pellet is required to enable efficient cryomilling and recovery as there are losses to the walls of the jar and the steel balls. We find that 3 L of cells at $2–2.5 \times 10^7$ cells/mL generates a good-sized pellet.
4. Milled powder should be stored with a loosened cap at −80 °C in the first 24 h to allow any trapped nitrogen to escape or else tubes may explode.
5. It is necessary to do this initial "dry" milling to allow the cell pellets to be broken down to a coarse powder. If there is liquid in the jar, this initially coarse grinding is less or ineffective.
6. Do not add liquid nitrogen above the level of the steel balls, as this will cause excessive pressure during the cryomilling.
7. During liquid cryomilling, the balls should not make a rattling sound but rather should be quiet. If the balls make a loud rattling sound, then it means that the liquid in the grinding jar has evaporated into nitrogen gas.
8. It is not recommended to conjugate more magnetic beads than will be used within 6 months. Beads can be stored at −20 °C in 50 % glycerol to extend shelf life. However, there is a slow bleeding of the antibody from the beads which increases over time due to reversibility of the disulfide bridges in the antibody heavy and light chains.
9. Ammonium sulfate enhances the conjugation reaction by binding and excluding water, thus reducing reaction volume and promoting hydrophobic interactions between the bead surface and the antibody.
10. The coupling of the antibodies to the beads can be allowed to proceed up to 48 h without significant loss of antibody-binding activity. Indeed, if the antibodies being coupled are labile, it is recommended that the coupling reaction be performed at 4 °C for 48 h.
11. The amount of powder weighed out depends on the abundance of the tagged protein. Typically for mini-IP tests, 50 mg is sufficient initially. Due to difficulty weighing out several

small amounts of frozen powder, the use of metallic spice spoons is recommended. By testing out different sized spoons, on milled powder, one can have a standard spoon for weighing out fixed amounts of powder for each IP.

12. When affinity-isolating membrane proteins, it is recommended that the washes should be in the absence of detergent. The presence of detergent significantly reduces the yield of membrane proteins.

13. Reducing agents will cause antibody leakage into your eluate and must be avoided at this step.

References

1. Simpson AG, Roger AJ (2004) The real "kingdoms" of eukaryotes. Curr Biol 14:R693–R696
2. Dacks JB, Walker G, Field MC (2008) Implications of the new eukaryotic systematics for parasitologists. Parasitol Int 57:97–104
3. Berriman M, Ghedin E, Hertz-Fowler C et al (2005) The genome of the African trypanosome *Trypanosoma brucei*. Science 309: 416–422
4. DuBois KN, Alsford S, Holden JM et al (2012) NUP-1 Is a large coiled-coil nucleoskeletal protein in trypanosomes with lamin-like functions. PLoS Biol 10:e1001287
5. Akiyoshi B, Gull K (2014) Discovery of unconventional kinetochores in kinetoplastids. Cell 156:1247–1258
6. Holden JM, Koreny L, Obado SO et al (2014) Nuclear pore complex evolution: a trypanosome MLp analogue functions in chromosomal segregation but lacks transcriptional barrier activity. Mol Biol Cell 25:1421–1436
7. Oeffinger M, Wei KE, Rogers R et al (2007) Comprehensive analysis of diverse ribonucleoprotein complexes. Nat Methods 4:951–956
8. Domanski M, Molloy K, Jiang H et al (2012) Improved methodology for the affinity isolation of human protein complexes expressed at near endogenous levels. Biotechniques 0:1–6
9. Cristea IM, Williams R, Chait BT et al (2005) Fluorescent proteins as proteomic probes. Mol Cell Proteomics 4:1933–1941
10. Hakhverdyan Z, Domanski M, Hough LE et al (2015) Rapid, optimized interactomic screening. Nat Methods 12:553–560
11. DeGrasse JA, DuBois KN, Devos D et al (2009) Evidence for a shared nuclear pore complex architecture that is conserved from the last common eukaryotic ancestor. Mol Cell Proteomics 8:2119–2130

Part II

Nuclear Envelope Protein Interactions, Localization, and Dynamics

Chapter 4

Superresolution Microscopy of the Nuclear Envelope and Associated Proteins

Wei Xie, Henning F. Horn, and Graham D. Wright

Abstract

Superresolution microscopy is undoubtedly one of the most exciting technologies since the invention of the optical microscope. Capable of nanometer-scale resolution to surpass the diffraction limit and coupled with the versatile labeling techniques available, it is revolutionizing the study of cell biology. Our understanding of the nucleus, the genetic and architectural center of the cell, has gained great advancements through the application of various superresolution microscopy techniques. This chapter describes detailed procedures of multichannel superresolution imaging of the mammalian nucleus, using structured illumination microscopy and single-molecule localization microscopy.

Key words Superresolution microscopy (SRM), Structured illumination microscopy (SIM), Single-molecule localization microscopy (SMLM), Stochastic optical reconstruction microscopy (STORM), Photoactivated localization microscopy (PALM), Nuclear envelope (NE), Nuclear lamins, Nuclear pore complex (NPC), Synaptonemal complex, Linker of the nucleoskeleton and cytoskeleton (LINC) complex

1 Introduction

Superresolution microscopy (SRM) describes a variety of techniques enabling the acquisition of microscopic, or nanoscopic, images with a resolution surpassing the diffraction limit, first described by Ernst Abbe in 1873 [1]. This limit defines the resolution that can be achieved by an optical microscope and is based upon the emission wavelength and the numerical aperture (NA) of the objective lens (Eq. 1) [2, 3]. Essentially all conventional optical microscopes are limited to a resolution of approximately 220 nm; therefore, objects falling within this distance cannot be resolved. SRM techniques allow this limit to be overcome, to varying degrees, through modification of the illumination of the sample, exploitation of the fluorophore's behavior in different chemical and photophysical environments, and application of computational algorithms [2, 4]:

Sue Shackleton et al. (eds.), *The Nuclear Envelope: Methods and Protocols*, Methods in Molecular Biology, vol. 1411, DOI 10.1007/978-1-4939-3530-7_4, © Springer Science+Business Media New York 2016

$$d = \frac{\lambda}{2n \sin \Theta} \text{ where } n \sin \Theta \text{ is equal to the NA} \quad (1)$$

The 2014 Nobel Prize for Chemistry was awarded to three scientists, Eric Betzig, Stefan Hell, and William E Moerner, for their work on the "development of super-resolved fluorescence microscopy". A notable absence is of Mats Gustafsson, who sadly passed away in 2011, but whose work, along with others, was key in the development of structured illumination microscopy (SIM) and later 3D-SIM [5–7].

1.1 3D Structured Illumination Microscopy (3D-SIM)

3D-SIM gives a twofold increase in resolution over conventional microscopy in all three dimensions (lateral ~110 nm, axial ~300 nm). While this increase is not as significant as for other SRM techniques, it is perhaps the easiest to execute and most versatile, as the sample preparation is the same as for conventional fluorescence microscopy. The principle relies on illuminating the sample with a structured pattern, which is moved in phase and rotated during a *z*-stack acquisition. The interaction of the pattern with the architecture inherent within the sample produces moiré fringes. Subsequent computational reconstruction allows the higher frequency information contained within these moiré fringes to be extracted into a SIM image. It works effectively for objects or features that are discrete and/or structured (e.g., filaments and puncta) rather than diffusive signals. Commercial implementations are currently available from GE Healthcare (OMX), Zeiss (Elyra), and Nikon (N-SIM).

1.2 Single-Molecule Localization Microscopy (SMLM)

Photoactivated localization microscopy (PALM) [8] and stochastic optical reconstruction microscopy (STORM) [9, 10] are two types of SRM techniques commonly referred to as single-molecule localization microscopy (SMLM). SMLM can achieve an approximate lateral resolution of 20 nm. Although PALM and STORM both rely on fluorescent probes that are able to blink on and off, they differ in their approach and types of fluorophores used. PALM employs genetically modified fluorescent proteins that are both photostable and photoconvertible (e.g., mEos2) [11, 12]. STORM uses standard organic fluorescent dyes that can be induced to blink, either through dye pairing (STORM) [9] or using reducing reagents (dSTORM) [10]. Dissection of the principles for the two techniques can be found in the excellent reviews by Deschout et al. [13] and Small et al. [14]. In this chapter, we focus in detail on the methodology of dual-channel SMLM of the mammalian nucleus, by combining PALM (using mEos2-tagged lamin) with dSTORM (Alexa Fluor 647 labeled nuclear pore complexes). 3D SMLM solutions are increasingly available through the use of astigmatism and cylindrical lenses and PSF shaping and biplane illumination. Commercial implementations for SMLM are currently available from Zeiss (PALM), Nikon (n-STORM), Leica (GSD), GE Healthcare (DLM), and Bruker (Vutara).

1.3 Stimulated Emission Depletion (STED) Microscopy

Stimulated emission depletion (STED) microscopy can, in theory, offer an infinite increase in resolution [15]. However in practice the commercial implementations (e.g., Leica, Abberior, PicoQuant) are able to reliably produce approximately 30–70 nm lateral resolution. The principle relies on exciting the sample with a diffraction limited spot, which is then immediately followed by a second, donut-shaped laser of a longer wavelength. The donut-shaped beam depletes the fluorescence from around the periphery of the original focal spot, reducing the size (area and/or volume) from which emitted light is collected. The power of the depletion beam ultimately determines the resultant size of the focal spot and resolution achieved. These systems can be added on to existing confocal platforms, producing their images through raster scanning over a field of view (FoV).

1.4 Recent Applications of SRM on Nuclear Envelope Components

Studies employing conventional optical microscopy on the components of the nuclear envelope (NE) have been hindered by the sub-diffraction dimension of the structures. The inner and outer nuclear membranes (INM, ONM) are separated by the ~50 nm perinuclear space (PNS) [16]. Inside the PNS, the KASH proteins in the ONM interact with the SUN proteins in the INM, to form the linker of the nucleoskeleton and cytoskeleton (LINC) complex. The LINC complex physically tethers the nucleus to all three cytoskeletal networks. Beneath the INM is the 20–50 nm thick lamina, comprised of a class of proteins called nuclear lamins. The NE is also traversed by the nuclear pore complexes (NPCs), composed of eight proteinaceous asymmetric spokes assembled into a pseudo-eightfold symmetric pore of approximately 120 nm in diameter [17].

While electron microscopy can produce images with nanometer resolution, the sample preparation for nuclear components is complicated, lengthy, and artifact prone. SRM provides unique advantages in the study of NE components: (1) as an optical technique, it is less disruptive and less invasive to cellular structures; (2) many routinely used fluorophores are suitable for 3D-SIM and dSTORM [18], making simultaneous and unambiguous labeling of distinct nuclear components relatively easy. Recently, Löschberger et al. demonstrated that the integral NPC component membrane protein GP210 is distributed in an eightfold radial symmetry in the *Xenopus laevis* oocyte, by correlating dSTORM with scanning electron microscopy [19, 20]. The same octameric arrangement of NPCs was consistently demonstrated in both *Xenopus* and human cells by STED and GSD [21, 22]. By combining SRM and computer simulation, individual nucleosomes along chromatin fibers could be visualized and quantified to study their arrangement in vivo [23]. 3D-SIM, while lacking the lateral resolution when compared with other SRM techniques, is still able to resolve single NPCs. This, combined with the advantage of simultaneously labeling of chromatin, nuclear lamina, and the NPC, makes 3D-SIM a powerful tool for studying the architecture of the nucleus, revealing finer details beyond the resolution of conventional microscopy [7, 24]. Of particular relevance, Horn et al. conducted a functional analysis of a KASH protein forming a meiotic complex with SUN protein using 3D-SIM [25], and their protocols of sample preparation and image acquisition are described in detail in this chapter.

2 Materials

2.1 Sample Preparation

1. Spermatocyte spreads prepared from adult male mice [25].
2. Mouse fibroblast cells transfected with mEos2-lamin A [26].
3. Phenol red-free Dulbecco's modified eagle medium (DMEM) supplemented with 10 % fetal bovine serum (FBS), 4.0 mg/L L-glutamine.
4. Phosphate-buffered saline (PBS).
5. 0.01 % w/v poly-L-lysine.
6. 1, 3, and 4 % paraformaldehyde (PFA) in PBS.
7. 50 mM NH_4Cl in PBS.
8. 0.1, 0.15, 0.2, 0.4 % Triton X-100 in PBS.
9. 0.1 % Tween-20 in PBS.
10. Blocking buffers: 10 % normal donkey serum and 0.2 % Triton X-100 in PBS (for spermatocytes); 0.2 % bovine gelatin and 0.02 % NaN_3 as a disinfectant in PBS (for mouse fibroblasts).
11. Primary antibodies: mouse monoclonal anti-synaptonemal complex protein 3 (SCP3) (D-1; Santa Cruz Biotechnology, Inc), rabbit polyclonal anti-KASH5 (Yenzym Antibodies LLC) [25], mouse monoclonal QE5 antibody [27].
12. For 3D-SIM secondary antibodies against appropriate species conjugated with bright and photostable fluorescent dyes, which excite and emit at appropriate wavelengths for your microscope system (*see* **Note 1**).
13. For dSTORM secondary antibodies against appropriate species conjugated with Alexa Fluor 647 (Life Technologies) were exclusively used (*see* **Note 1**).
14. Clean high precision #1.5 (#1.5H) coverslips >18 mm in diameter and clean standard microscopy slides (*see* **Note 2**).
15. Non-setting glycerol-based anti-fade mounting media that do not contain any additional fluorophores as counterstains (*see* **Note 2**).
16. LabTek II Chamber Slide system (Thermo Fisher Scientific) (*see* **Note 3**).

17. Cysteamine/MEA in PBS (*see* **Note 4**).
18. 100 nm diameter multispectral beads, which excite and emit at appropriate wavelengths for your microscopes system, diluted 1:1000 in PBS and seeded in the imaging chamber at least 2 h prior to image acquisition, as fiducial markers (*see* **Note 5**).
19. Nail varnish or other methods for sealing coverslips.

2.2 3D-SIM

1. A 3D-SIM microscope (DeltaVision OMX v4 Blaze; GE Healthcare) equipped with 405, 488, 568, and 647 nm lasers for excitation (with intensity control by neutral density filters) and a BGR-FR filter drawer (emission wavelengths 436/31 for blue emitting dyes, 528/48 for green emitting dyes, 609/37 for red emitting dyes, and 683/40 for far-red emitting dyes). There is a separate light path for acquiring widefield fluorescence images that were subsequently deconvolved. The widefield light path uses a solid-state illuminator, but otherwise shares the same components.
2. A Plan Apochromat 100×/1.4 PSF oil immersion objective lens (Olympus) and three sensitive liquid-cooled EM-CCD cameras (Evolve; Photometrics), one for each of the fluorescence emission channels—blue, green, and red. The blue camera is equipped with a fast-switching emission filter wheel to switch between blue and far-red filters. This combination of lens and camera offers a fixed maximal field of view of 40 μm × 40 μm at 512 × 512 pixels. The SIM reconstruction process adds further pixels to result in a 1024 × 1024 image (0.039 μm per pixel).
3. An immersion oil of refractive index (RI) 1.514 was used for the experiments described in this chapter (*see* **Note 6**).
4. Two computers are operated by the user, one to control the acquisition parameters (Windows 7) and the second to reconstruct the images post-acquisition using the SoftWorx software (Linux, CentOS).

2.3 dSTORM and PALM

1. An inverted TIRF microscope (Olympus IX-81) with a sensitive, liquid-cooled EM-CCD camera (Evolve; Photometrics) used at 256 × 256 pixels and a frame rate of 60 frames per second (15 ms exposure). The setup of the microscope is described in detail by Ahmed et al. [28]. In this chapter PALM of mEos2 uses a 561 nm laser for imaging the converted form, and dSTORM of Alexa Fluor 647 uses a 647 nm laser, both of which also require a 405 nm laser for photoconversion/recovery. The power of the imaging laser reaching the sample is approximately 50 mW. Care is taken to reduce vibrations and drift, and the microscope is installed on an anti-vibration table.

2. A Plan Apochromat 100×/1.49 oil immersion TIRF objective lens (Olympus) is used together with a sensitive liquid-cooled EM-CCD camera (Evolve; Photometrics).
3. A high-performance computer to handle the large datasets and rapid data transfer rates (datasets are several GBs in size and the analysis is computationally intensive). Our configuration is a 3 GHz multicore processor, with 16 GB of fast RAM and a 1 TB hard drive (SSD recommended), running 64bit Windows 7. MetaMorph (Molecular Devices) is used for image acquisition, RapidStorm [29] is used for SMLM reconstruction, and Fiji [30] is used for subsequent image processing.

3 Methods

3.1 Sample Preparation

3.1.1 3D-SIM: Spermatocyte Spreads for Studying the Role of KASH5

1. Thoroughly clean #1.5H coverslips (*see* **Note 2**) and coat with a 0.01 % (w/v) poly-L-lysine solution. Incubate at room temperature (RT) for 10 min, and centrifuge for a further 10 min at 37 °C. Allow coverslips to air-dry completely. Dip into a 1 % PFA 0.15 % Triton X-100 PBS solution and allow the excess to drip off briefly before placing into a moist incubation chamber. It is important to keep a film of PFA solution on the coverslip and not allow it to dry.
2. Plate the spermatocyte spreads, prepared as in ref. 25, directly onto the coverslips rather than slides (*see* **Note 7**), by adding several drops of testes suspension onto each coverslip. Incubate for 2 h in a moist chamber.
3. After incubation, wash the coverslips by immersion for 3 min in 0.1 % Triton X-100 PBS and allow to air-dry (*see* **Note 8**).
4. Fix specimens for 10 min in 4 % PFA. Wash in PBS and permeabilize for 10 min in 0.2 % Triton X-100. Wash briefly in PBS, then invert samples onto a drop of blocking buffer on Parafilm in a moist chamber and incubate for 1 h at RT.
5. Remove excess blocking buffer and invert coverslips onto a drop of primary antibody solution (against SCP3 and KASH5 diluted at 1:100 in blocking buffer) on Parafilm in a moist chamber at 4 °C overnight.
6. Allow the moist chamber to come back up to RT and wash samples in 0.2 % Triton X-100 by dipping 5 times in each of three small beakers (30–50 mL volume) (15 times total).
7. Invert coverslips onto a drop of appropriate secondary antibody solution (diluted at 1:250 in blocking buffer) for 1 h at RT in the dark and then perform the same wash steps as in **step 6**.
8. After the last wash, fix samples in 4 % PFA for 10 min at RT. This post-staining (secondary) fixation step helps to reduce the

number of free-floating fluorescent particles in the sample (*see* **Note 9**). After fixation, wash samples in PBS 3 times.

9. Remove excess PBS, mount samples (*see* **Note 2**), and seal the coverslip. Before imaging, clean the surface of the glass coverslip thoroughly using chloroform and a cotton swab.

3.1.2 dSTORM and PALM: Mouse Fibroblasts

1. Seed approximately 5×10^4 mouse fibroblast cells in a LabTek Chamber (*see* **Note 3**), and incubate for 24 h at 37 °C in 5 % CO_2 using phenol red-free DMEM supplemented with 10 % FBS and 4.0 mg/L L-glutamine.
2. Perform all the following procedures at RT unless otherwise stated. Replace medium with 3 % PFA to fix cells for 10–15 min.
3. Replace PFA with 50 mM NH_4Cl to quench the residual PFA for 10 min.
4. Replace NH_4Cl with 0.2 % Triton X-100 to permeabilize the cells for 5 min, and then remove it.
5. Add 0.2 % bovine gelatin to block the samples for at least 20 min, and then remove it.
6. Incubate samples with primary QE5 antibody for 20–30 min.
7. Wash samples three times with 0.1 % Tween-20, each for 10 min.
8. Incubate samples with Alexa Fluor 647 secondary antibody (diluted at 1:500 in blocking buffer) for 20–30 min in the dark, and wash as in **step 7**.
9. After the last wash, keep samples in PBS for PALM, or incubate with MEA solution for dSTORM (*see* **Note 4**).

3.2 Image Acquisition

3.2.1 3D SIM

1. Mount the slides on the stage of the microscope with immersion oil (RI 1.514) and identify an appropriate sample. As the OMX does not have eyepieces, there are two options for locating the samples: (1) if they are relatively prevalent on the coverslip, then the *spiral mosaic* tool can be used to scan a large field of view (FoV, e.g., a 9×9 array) from within which samples can be selected; (2) for sparse samples, the slide can be mounted on a widefield DeltaVision microscope (e.g., CORE, PersonalDV, or Elite) and sample positions stored, which can be revisited on the OMX, after stage-stage mapping calibration.
2. EM-CCD camera gain should be fixed at 170 and laser powers set to either 1 or 10 %, with exposures typically varying between 10 and 50 ms to achieve a signal of between 3000 and 15,000 counts. Photobleaching must be reduced to a minimum for 3D-SIM acquisitions (*see* **Note 10**).
3. For each dataset, 15 images (three rotations by five-phase shifts of the structured illumination pattern) per *z*-position per channel

are acquired at a z-spacing of 0.125 μm. The OMX requires a minimum of a 1 μm stack (eight slices).

4. Complete SIM reconstruction and chromatic alignment (xyz; *see* **Note 11**) post-acquisition using the SoftWorX program. Acquiring both a SIM and widefield image of the same sample means the latter can be used as a reference to check for SIM artifacts, and establish how well the SIM reconstruction is performing (Fig. 1).

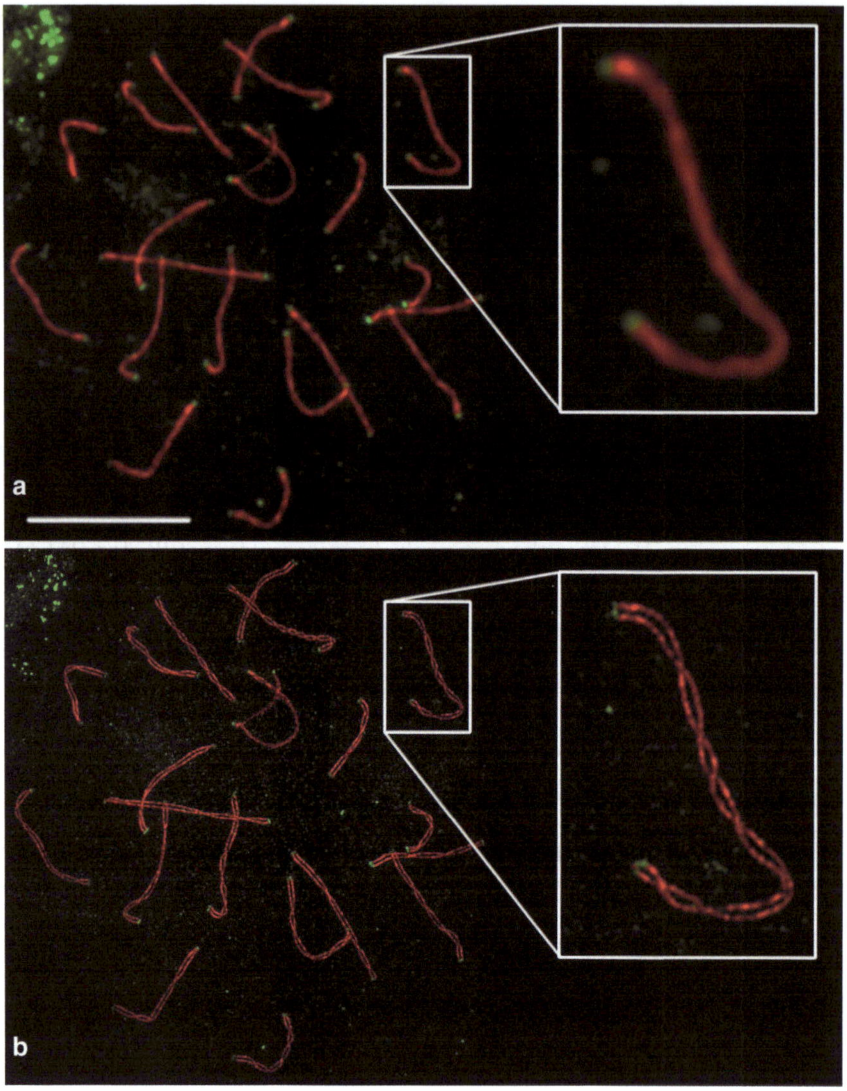

Fig. 1 Maximum intensity projections of a widefield-deconvolved (**a**) and 3D-SIM (**b**) image of the same spermatocyte spread. The samples were prepared as described and are labeled for SCP3 (*red*) and KASH5 (*green*). The *inset* images show a zoomed in example to illustrate the gain in resolution from 3D-SIM. Scale bar, 10 μm

3.2.2 dSTORM and PALM

1. Establish the chromatic alignment using 100 nm diameter multispectral beads seeded in a LabTek Chamber and mount on the microscope (*see* **Notes 5** and **6**) with immersion oil of RI 1.518.
2. Select a FoV with many uniformly distributed beads. Then, using the same frame size as used for imaging (256 × 256), focus on the beads and acquire 1500 frames (15 ms exposure, EM Gain set to 200) for each of the 647 and 568 nm lasers. Process the image stacks with RapidStorm to generate SRM reference images of each channel [29] (*see* **Note 12**).
3. To generate the elastic transformation matrix, open the two reference images in the "bUnwarpJ" plug-in of Fiji [30, 31]. Set the 568 nm image as the source and 647 nm image as the target, and set the parameters exactly as shown in Fig. 2. Uniformly spread landmarks for 20–30 corresponding beads must be identified and saved for future reference. Run the plug-in and save the elastic transformation matrix file at the end.
4. Replace the PBS in the sample chamber with the MEA solution pH 8.5 (*see* **Note 4**), and allow to settle on the lens (*see* **Note 6**).

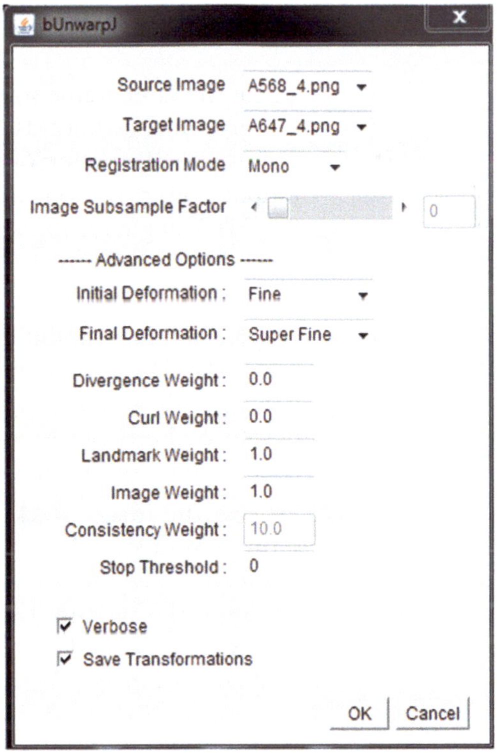

Fig. 2 Settings for bUnwarpJ plug-in in Fiji

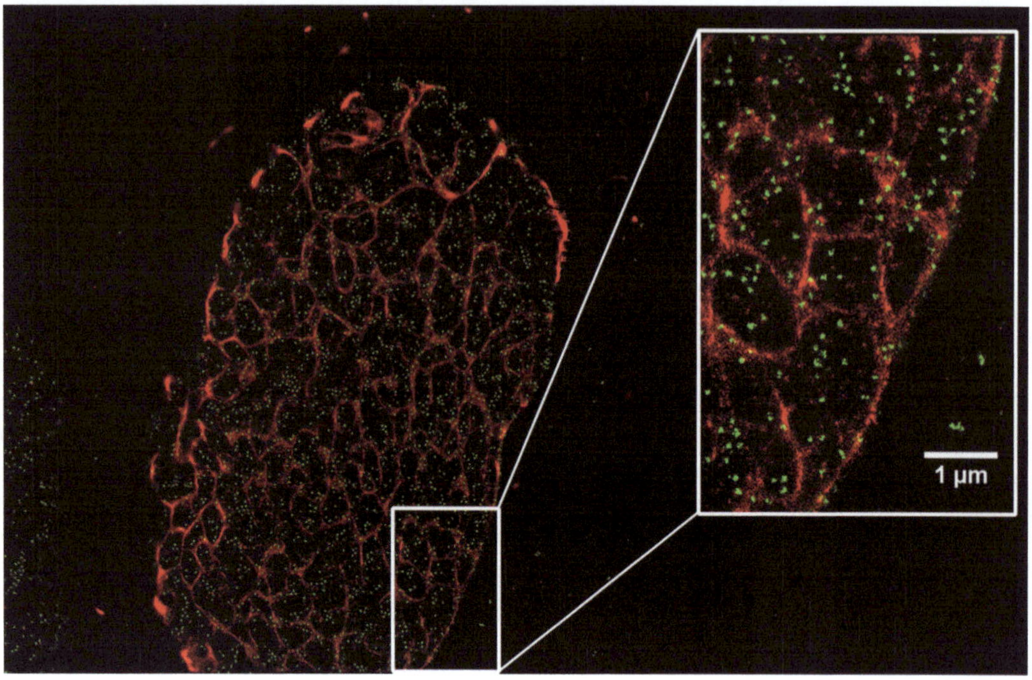

Fig. 3 Merged dual-channel SMLM image showing a mouse fibroblast cell nuclear envelope. The samples were prepared as described and are labeled for mEos2-lamin A by PALM (*red*) and NPCs (*green*) by dSTORM. The *inset* images show a zoomed in example to illustrate the individually localized NPCs. Scale bar, 1 μm

5. Using the same frame size in **step 2**, locate and focus on a cell nucleus, and acquire 10,000 frames (15 ms exposure, EM Gain set to 200) using 647 nm laser at a minimum of 100 mW output power, without the 405 nm laser (*see* **Note 13**). Tune the TIRF angle during acquisition to improve signal to noise ratio (SNR; *see* **Note 14**). Switch off the 647 nm laser when acquisition is complete.
6. Acquire another 10,000 frames of the mEos2-lamin A channel, using the 561 nm laser at a minimum of 100 mW output power, with 405 nm turned on and pulsing 2 mW output power to photoconvert mEos2 (*see* **Note 13**). Switch off both lasers when acquisition is complete.
7. Process the image stacks with RapidStorm, using the same settings as in **step 2**, to generate an SRM images of individual channels. Open the two processed images in Fiji, and apply the "bUnwarpJ" plug-in [31] using the elastic transformation matrix file generated in **step 3**. Save the chromatically corrected dual-channel image (Fig. 3).

4 Notes

1. In our experiments with 3D-SIM, we made use of the commercial fluorescent dyes Alexa Fluor 488, 568, and 647 (Life Technologies, now Thermo Fisher Scientific). Other commercially available dyes have also been found to be suitable. The selection of dyes should be based on their photophysical properties (bright, photostable with appropriate excitation and emission spectra), and their suitability should be tested on the microscope and optimized. For dSTORM Alexa 647 has been shown to have exceptionally good photophysical properties.
2. 3D-SIM requires spherical aberration to be kept to a minimum. As such, clean high-precision coverslips of exactly 170 μm (#1.5H) should be used, as the variations in standard coverslip thickness can be as high as 12 %. To clean the coverslips, we immersed them for three times for 3 min in xylene, then three times for 3 min in 100 % ethanol, and finally boiled in water in a pressure cooker. Non-setting glycerol-based anti-fade mounting media should be used (we used VECTASHIELD H-1000 from Vector Laboratories), and the RI of the immersion oil should be matched to that of the sample (also *see* **Note 6**). Care should be taken to ensure the sample is not tilted on the stage by carefully mounting the coverslip parallel to the slide, without any labeling stickers at either end as both could produce artifacts in the images. Microscopy slides with marks for the exact center are preferred for accurate mounting of the coverslip within the limited stage range of the OMX.
3. For SMLM of nuclear proteins, fibroblasts are preferred due to the relatively flat-shaped nucleus. During dSTORM/PALM imaging, cells need to be immersed in MEA solutions to induce photoswitching of the Alexa Fluor 647 dye; therefore, the growth, fixation, and staining of the cells should be performed in a LabTek Chamber to allow final submersion in MEA, which also limits the absorption of oxygen from the atmosphere.
4. For dSTORM, 100 mM MEA in PBS adjusted to pH 8.5 using HCl was found to be the preferred condition Alexa Fluor 647. However different fluorophores used in dSTORM require slightly varying MEA concentrations and pH in order to achieve optimal image acquisition [18]. MEA solution should be prepared fresh, with an oxygen depletion step recommended. Readers are encouraged to titrate the MEA buffer formula best suited for their own applications. Photoconversion of mEos2 by the 405 nm laser is not affected by MEA.

5. For SMLM, channel references should be calibrated regularly, especially on the day of image acquisition, to ensure accuracy for dual-channel alignment of dSTORM/PALM images. 100 nm multispectral beads are used for this purpose. As the beads are seeded directly on the bottom of the glass chamber, whereas the nucleus of fibroblasts are actually further up from the bottom, the outcome of the alignment is not considered perfect, but the accuracy is estimated to be within 20 nm.
6. For 3D-SIM on the OMX, the immersion oil must be selected from various oils of different RIs, 0.002 graduations, to match with the sample. Selection is empirical, based on analyzing the spherical aberration present in orthogonal (*xz* or *yz*) views. For both 3D-SIM and SMLM both excess oil and any bubbles within the oil should be avoided, and 5–10 min should be allowed for the slide/chamber to settle stably on the lens.
7. 3D-SIM relies on the preservation of the structured illumination pattern therefore the penetration depth of the technique is limited. Samples should be mounted on the coverslip, if possible kept to within 16 μm of the surface, rather than being mounted onto the slide.
8. For 3D-SIM, after plating the spermatocyte spreads, dried coverslips can either be processed immediately for staining or stored long term at −80 °C.
9. 3D-SIM images are acquired by sequentially taking images of the same sample as the illumination pattern moves and rotates; therefore, it is critical that objects within the sample remain perfectly still. If objects move, the reconstruction algorithm will create artifacts, manifested as flares or starbursts within the images. The secondary fixation step and subsequent washes help to minimize the number of free-floating particles and debris.
10. For 3D-SIM, photobleaching should be kept to a minimum by careful selection of photostable fluorophores and the use of an anti-fade mounting medium. The Alexa Fluor dyes are photostable dyes available in wavelengths to match the lasers and filters in our microscope (e.g., Alexa Fluor 568 is much preferred than Alexa Fluor 594). The acquisition settings for laser power and exposure time were tuned to produce sufficient SNR (a maximum intensity count of between 3000 and 15,000 over the background on our EM-CCDs) while minimizing photobleaching. A reduction in intensity of approximately 30 % through photobleaching is tolerable to produce artifact-free reconstructions.
11. For 3D-SIM reconstruction, the following major settings were used:
    - Weiner filter constant 0.0010
    - Background intensity offset 50

- Channel-specific optical transfer functions, derived from images of sub-diffraction beads acquired on this microscope
- Channel-specific K0 angles.

For chromatic alignment, a warp algorithm is first calculated using an alignment slide—a mirrored surface with an array of holes. A *z*-stack is acquired of this sample through the transmitted light path. The resultant warp algorithm is then verified against SIM images of sub-diffraction multispectral beads. The accuracy of this algorithm is checked before image acquisition and refined when necessary.

12. For SMLM reconstruction, the following major settings were used in RapidStorm:
    - Size of one input pixel 100
    - PSF FWHM 312
    - Minimum spot distance 3
    - Amplitude discarding threshold 1.0E4
    - Two-kernel distance threshold 500
    - Maximum two-kernel improvement 0.1.
13. During dSTORM image acquisition, the use of a pulsing 405 nm laser is able to enhance the blinking of Alexa Fluor dyes. However, it also irreversibly photoconverts mEos2 in PALM. Therefore in dual-channel applications involving both dSTORM and PALM, dSTORM image acquisition should be performed first without using the 405 nm laser and PALM image acquisition second using the 405 nm laser for mEos2 photoconversion.
14. TIRF microscopy greatly enhances the SNR during imaging of cellular components within 200 nm of the coverslip surface [32, 33]. In contrast, this effect is disadvantageous for imaging the nucleus as it is usually 1–2 μm away from the surface of the coverslip. However, a sub-optimal TIRF angle is able to generate a HILO illumination [34] allowing deeper penetration to the nucleus while maintaining SNR. Readers are encouraged to adjust the TIRF angle after focusing to achieve the best results as the ideal TIRF angle varies from cell to cell.

Acknowledgments

Srivats Hariharan (Olympus Singapore, formerly IMB, A*STAR) for building the SMLM microscope in the IMB Microscopy Unit together with the lab of Sohail Ahmed (IMB, A*STAR), John Lim Soon Yew (IMB, A*STAR) for assisting with the maintenance and operation of the SMLM microscope and for supporting the image processing and analysis, Declan Lunny (IMB, A*STAR) for the advice and help with the sample preparation, and Brian Burke (IMB, A*STAR) and Colin Stewart (IMB, A*STAR) for supervising the research projects within their labs.

References

1. Abbe E (1873) Beiträge zur Theorie des Mikroskops und der mikroskopischen Wahrnehmung. Archiv f mikrosk Anatomie 9: 413–418
2. Schermelleh L, Heintzmann R, Leonhardt H (2010) A guide to super-resolution fluorescence microscopy. J Cell Biol 190:165–175
3. Sousa AA, Kruhlak MJ (2013) Introduction: nanoimaging techniques in biology. Methods Mol Biol 950:1–10
4. Galbraith CG, Galbraith JA (2011) Super-resolution microscopy at a glance. J Cell Sci 124:1607–1611
5. Gustafsson MG (2000) Surpassing the lateral resolution limit by a factor of two using structured illumination microscopy. J Microsc 198:82–87
6. Gustafsson MG, Shao L, Carlton PM et al (2008) Three-dimensional resolution doubling in wide-field fluorescence microscopy by structured illumination. Biophys J 94:4957–4970
7. Schermelleh L, Carlton PM, Haase S et al (2008) Subdiffraction multicolor imaging of the nuclear periphery with 3D structured illumination microscopy. Science 320:1332–1336
8. Betzig E, Patterson GH, Sougrat R et al (2006) Imaging intracellular fluorescent proteins at nanometer resolution. Science 313: 1642–1645
9. Rust MJ, Bates M, Zhuang X (2006) Sub-diffraction-limit imaging by stochastic optical reconstruction microscopy (STORM). Nat Methods 3:793–795
10. Heilemann M, Van De Linde S, Schuttpelz M et al (2008) Subdiffraction-resolution fluorescence imaging with conventional fluorescent probes. Angew Chem Int Ed Engl 47: 6172–6176
11. Subach FV, Patterson GH, Manley S et al (2009) Photoactivatable mCherry for high-resolution two-color fluorescence microscopy. Nat Methods 6:153–159
12. Mckinney SA, Murphy CS, Hazelwood KL et al (2009) A bright and photostable photoconvertible fluorescent protein. Nat Methods 6:131–133
13. Deschout H, Cella Zanacchi F, Mlodzianoski M et al (2014) Precisely and accurately localizing single emitters in fluorescence microscopy. Nat Methods 11:253–266
14. Small A, Stahlheber S (2014) Fluorophore localization algorithms for super-resolution microscopy. Nat Methods 11:267–279
15. Klar TA, Jakobs S, Dyba M et al (2000) Fluorescence microscopy with diffraction resolution barrier broken by stimulated emission. Proc Natl Acad Sci U S A 97:8206–8210
16. Burke B, Stewart CL (2014) Functional architecture of the cell's nucleus in development, aging, and disease. Curr Top Dev Biol 109:1–52
17. Grossman E, Medalia O, Zwerger M (2012) Functional architecture of the nuclear pore complex. Annu Rev Biophys 41:557–584
18. Van De Linde S, Loschberger A, Klein T et al (2011) Direct stochastic optical reconstruction microscopy with standard fluorescent probes. Nat Protoc 6:991–1009
19. Loschberger A, Van De Linde S, Dabauvalle MC et al (2012) Super-resolution imaging visualizes the eightfold symmetry of gp210 proteins around the nuclear pore complex and resolves the central channel with nanometer resolution. J Cell Sci 125:570–575
20. Loschberger A, Franke C, Krohne G et al (2014) Correlative super-resolution fluorescence and electron microscopy of the nuclear pore complex with molecular resolution. J Cell Sci 127:4351–4355
21. Gottfert F, Wurm CA, Mueller V et al (2013) Coaligned dual-channel STED nanoscopy and molecular diffusion analysis at 20 nm resolution. Biophys J 105:L01–L03
22. Szymborska A, De Marco A, Daigle N et al (2013) Nuclear pore scaffold structure analyzed by super-resolution microscopy and particle averaging. Science 341:655–658
23. Ricci MA, Manzo C, Garcia-Parajo MF et al (2015) Chromatin fibers are formed by heterogeneous groups of nucleosomes in vivo. Cell 160:1145–1158
24. Markaki Y, Smeets D, Fiedler S et al (2012) The potential of 3D-FISH and super-resolution structured illumination microscopy for studies of 3D nuclear architecture: 3D structured illumination microscopy of defined chromosomal structures visualized by 3D (immuno)-FISH

opens new perspectives for studies of nuclear architecture. Bioessays 34:412–426

25. Horn HF, Kim DI, Wright GD et al (2013) A mammalian KASH domain protein coupling meiotic chromosomes to the cytoskeleton. J Cell Biol 202:1023–1039
26. Wang AS, Kozlov SV, Stewart CL et al (2015) Tissue specific loss of A-type lamins in the gastrointestinal epithelium can enhance polyp size. Differentiation 89:11–21
27. Pante N, Bastos R, Mcmorrow I et al (1994) Interactions and three-dimensional localization of a group of nuclear pore complex proteins. J Cell Biol 126:603–617
28. Ahmed S, Chou A, Sem KP et al. (2014) Using dSTORM to probe the molecular architecture of filopodia. Proc SPIE 8950
29. Wolter S, Loschberger A, Holm T et al (2012) rapidSTORM: accurate, fast open-source software for localization microscopy. Nat Methods 9:1040–1041
30. Schindelin J, Arganda-Carreras I, Frise E et al (2012) Fiji: an open-source platform for biological-image analysis. Nat Methods 9:676–682
31. Arganda-Carreras I, Sorzano CS, Marabini R et al (2006) Consistent and elastic registration of histological sections using vector-spline regularization. In: Beichel R, Sonka M (eds) Computer vision approaches to medical image analysis. Springer, Berlin, pp 85–95
32. Ambrose EJ (1956) A surface contact microscope for the study of cell movements. Nature 178:1194
33. Axelrod D (1981) Cell-substrate contacts illuminated by total internal reflection fluorescence. J Cell Biol 89:141–145
34. Tokunaga M, Imamoto N, Sakata-Sogawa K (2008) Highly inclined thin illumination enables clear single-molecule imaging in cells. Nat Methods 5:159–161

Chapter 5

Analyses of the Dynamic Properties of Nuclear Lamins by Fluorescence Recovery After Photobleaching (FRAP) and Fluorescence Correlation Spectroscopy (FCS)

Shimi Takeshi, Chan-Gi Pack, and Robert D. Goldman

Abstract

The major structural components of the nuclear lamina are the A- and B-type nuclear lamin proteins which are also present in the nucleoplasm. Studies of molecular movements of the lamins in both the lamina and nucleoplasm of living cell nuclei have provided insights into their roles in maintaining nuclear architecture. In this chapter, we present protocols for quantitatively measuring the mobilities of lamin proteins by fluorescence recovery after photobleaching (FRAP) and fluorescence correlation spectroscopy (FCS) in mammalian cell nuclei.

Key words Lamina, Lamin, FRAP, FCS

1 Introduction

In mammalian cells, the nuclear lamina (NL) lies between the inner membrane of the nuclear envelope (NE) and chromatin. There are four nuclear lamin isoforms which collectively are the major structural proteins in the NL [1]. The four isoforms include lamins A (LA), C (LC), B1 (LB1), and B2 (LB2). These proteins play important roles in regulating the size, shape, and stiffness of the nucleus, and they are thought to be involved in chromatin organization, transcription, DNA replication, and DNA repair [2]. The lamins contain an α-helical central rod domain with mainly non-α-helical N- and C-terminal domains [3]. These isoforms polymerize into filamentous structures as determined by electron microscopy [4] and more recently by three-dimensional structured illumination microscopy (3D-SIM) [5]. Confocal microscopic observations have shown that A- and B-type lamin fibrils appear to form separate but interacting meshworks [6]. A minor fraction of lamins is also present in the nucleoplasm [6, 7]. The mechanisms for regulating the assembly and disassembly of the NL involve

Sue Shackleton et al. (eds.), *The Nuclear Envelope: Methods and Protocols*, Methods in Molecular Biology, vol. 1411,
DOI 10.1007/978-1-4939-3530-7_5, © Springer Science+Business Media New York 2016

phosphorylation and dephosphorylation during both interphase and mitosis. In interphase nuclei, there is a fraction of LA/C in the nucleoplasm which is phosphorylated, apparently preventing the incorporation of these two isoforms from being incorporated into the filamentous structures comprising the NL [8].

Live cell imaging with fluorescently labeled lamins has provided important insights into their dynamic properties. Live cell imaging techniques that have been used for analyzing lamin dynamics include fluorescence recovery after photobleaching (FRAP) and fluorescence correlation spectroscopy (FCS). These two techniques have been used to quantitatively determine the mobilities of fluorescently labeled lamins within the nuclei of living cells. FRAP studies have revealed that the lamin subunit mobilities and exchange rates are much slower in the NL, while FCS has revealed that some lamin isoforms exhibit high mobilities within the nucleoplasm. Using GFP-tagged wild-type and mutant lamin isoforms, studies by FRAP and FCS have revealed the kinetics of their association and disassociation from the NL and their diffusion coefficients in the nucleoplasm. For example, we and others have shown by FRAP that <25 % of GFP-LA and LC in the NL are exchanged within 10 min [7–11], indicating that LA/C is immobile in the NL. On the other hand, LA/C in the nucleoplasm is highly mobile as their diffusion coefficients determined by FCS are between ~0.4 and ~5.6 $\mu m^2/s$ [6].

2 Materials

2.1 Microscope and Sample Setup for FRAP

1. Live cell imaging chamber: Glass bottom microwell dishes (35 mm petri dish, 20 mm microwell, No. 1.5 cover glass, No. P35G-1.5-20-C, Mat-Tek).
2. Culture medium for microscopy: Leibovitz's L15 medium (phenol-red free, Life Technologies) containing 10 % serum and antibiotics (i.e., penicillin and streptomycin).
3. HeLa cells transfected with GFP-LA cultured for 48–72 h after transfection [6].
4. Spinning disk confocal microscope. A setup such as a Nikon Ti Perfect Focus microscope (Nikon) with CF160 lenses and equipped with a Yokogawa CSU-X1 spinning disk unit (Yokogawa) and an iXon 3 14-bit EM CCD camera (Andor Technology) is ideal for FRAP studies.
5. Laser for photobleaching GFP-LA fluorescence: 488 nm argon laser line.
6. Temperature and CO_2 control must be maintained. For this purpose, we use a Tokai-HIT stage-top incubator set for 37 °C and 5 % CO_2.

2.2 Data Analysis for FRAP

Andor iQ2 (Andor Technology) and Prism 6 (GraphPad) software

2.3 Microscope and Sample Setup for FCS

1. Live cell imaging chamber: 8-well with No. 1.0 or No. 1.5 borosilicate chambered cover glass (Lab-Tek or Lab-Tek II, Nunc).
2. Culture medium for microscopy: Opti-MEM I reduced serum medium (phenol-red free, Life Technologies).
3. HeLa cells transfected with GFP-LA and cultured for 48–72 h after transfection.
4. Zeiss LSM 510 Microscope equipped with a water-immersion objective lens (C-Apochromat 40×,1.2 NA) and a Zeiss ConfoCor 2 spectrometer.
5. Laser for FCS: 488-nm argon laser line
6. Calibration for FCS: rhodamine 6G [Rh6G] fluorescent dye – 10^{-7} M in distilled water solution.

2.4 Data Analysis for FCS

LSM510 ConfoCor 2 (Carl Zeiss) software

3 Methods

3.1 Cell Culture and Transfection

1. Grow HeLa cells on a 10 cm cell culture plastic dish under a humidified 5 % CO_2 atmosphere at 37 °C in Dulbecco's modified Eagle's medium (DMEM) supplemented with 10 % fetal calf serum and antibiotics (penicillin and streptomycin).
2. Detach when cells are 80–90 % confluent using 0.05 % trypsin-EDTA solution and then centrifuge at 200 × *g* for 5 min followed by resuspension in 500 μL of DMEM.
3. Transfer the cell suspension into a 4 mm electroporation cuvette, and then add 10 μg of the 5 plasmid-encoding GFP-LA.
4. Electroporate at 260 V and 950 μF.
5. Add 5.0 mL of DMEM supplemented with 10 % FCS and antibiotics, pipette up and down, and place cells immediately into a 100 mm culture dish.
6. ~6 h after incubation in a tissue culture incubator, change the medium with fresh DMEM supplemented with 10 % FCS and antibiotics.
7. Incubate for ~24 h in a CO_2 incubator.
8. Trypsinize and replate the electroporated cells at ~15 % density (5×10^4 cells) onto a glass bottom microwell dish for FRAP or 8-well with No. 1.0 or No. 1.5 borosilicate chambered cover glass for FCS.
9. Incubate for ~24 h in a tissue culture incubator (*see* **Note 1**).

3.2 FRAP

1. Make sure that the objective and laser are perfectly aligned as per manufacturer's instructions.
2. Overlay cells with Leibovitz's L15 medium containing 10 % serum and antibiotics including penicillin and streptomycin.
3. Incubate cells 37 °C and under 5 % CO_2 for at least 30 min before microscopic observation.
4. Equilibrate a CO_2 incubator (see above) mounted on the microscope stage at 37 °C for 30 min.
5. Open imaging software (e.g., NIS-Elements) and setup FRAP conditions following manufacturer's protocol (laser power, dwelling time, timing of imaging pre- and post-bleaching and time-lapse setting).
6. Place a dish onto the stage.
7. Find the correct focus and switch on the stage control to maintain focus (e.g., Nikon Perfect Focus System). Maintaining focus is essential throughout FRAP measurements (Fig. 1).

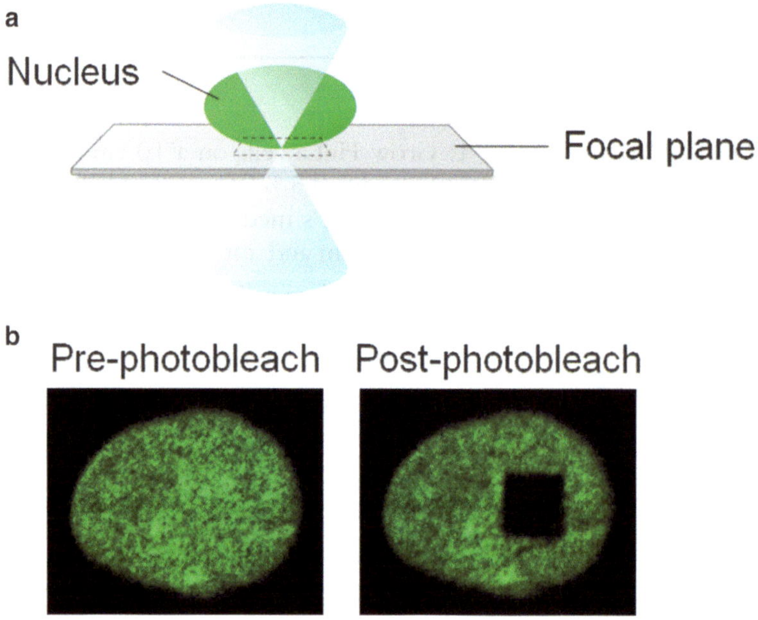

Fig. 1 Procedure of fluorescence recovery after photobleaching (FRAP) measurement on GFP-LA in the NL of living cell nuclei. (**a**) FRAP measurement is performed as shown in a schematic diagram. Fluorescence signals of GFP-LA are detected in the NL region at the bottom of the nucleus to visualize a maximally flat surface. A focus drift from the region is automatically corrected by a real-time correction system (see also text). (**b**) Representative fluorescence images of GFP-LA in living HeLa cells were taken at the lower nuclear surface by confocal microscopy immediately before and after a square region are photobleached

3.3 Performing a Test Bleach

The key to success for a FRAP experiment is the proper setting of photobleaching and acquisition conditions. It is essential that sufficient time is devoted to optimizing these conditions. First perform a test bleach to optimize bleach times, duration time, and region of interest (ROI) within the NL. Then optimize image acquisition conditions for monitoring the recovery of the fluorescent signal using time-lapse imaging with a well-equipped spinning disk confocal microscope (see above).

1. Find an area of interest using the eyepieces. Cells with moderate GFP-LA expression are recommended (*see* **Note 2**).
2. Make sure of the alignment of the optical axis and choose correct objective lens by following manufacturer's instruction.
3. Test photobleaching in several cells using a 488 nm laser at 100 % output to determine optimal duration time, size, and location of ROI for efficient photobleaching (*see* **Note 3**).
4. Roughly estimate the timescale required for the fluorescence recovery of GFP-LA in the mobile fraction.
5. Optimize image acquisition conditions including scanning speed, time intervals, and duration time for imaging and choose adequate laser wavelength and adjust the laser power at the low level to avoid any excess photobleaching by imaging. The acquisition speed should be fast enough to monitor the fluorescence recovery (*see* **Note 4**).

3.4 Performing a FRAP Experiment

1. Find an area of interest using the eyepieces. Cells with moderate GFP-LA expression are recommended, and if possible, other unbleached cells should be visible in the field of view (*see* **Note 5**).
2. Select the system to continuously acquire images of the NL (*see* **Note 6**). High-speed scan mode can be helpful for avoiding excess photobleaching caused by image acquisition.
3. Adjust the conditions for image acquisition (see above).
4. Choose an ROI area for photobleaching in a defined region of the NL on the acquired image. To have sufficient non-photobleached GFP-LA molecules for detecting the full fluorescence recovery in the mobile fraction, photobleached area should be <80 % of the entire lamina region.
5. Set the FRAP experiment parameters including image acquisition time, time-lapse interval, and total time period of the experiment. To measure only the fluorescence recovery of GFP-LA by the molecular exchanges but not the flow of newly synthesized GFP-LA, total time period should be at maximum 20 min.
6. Monitor cells including both non-bleached and bleached areas of the NL for the fluorescence recovery at 30 s intervals as indicated above: we use Andor iQ2 software for FRAP.

3.5 FRAP Analysis

We analyze our FRAP data using GraphPad Prism 6 software as previously described [8, 12]. The region of interest (ROI) is set on the photobleached region of the NL (ROI1), and ROI2 is selected from the background. Normalized fluorescence intensity (I) is obtained as follows:

$$100 \times (\text{GFP LA signals in ROI1 background signals in ROI2}) \Big/ \left(\begin{array}{l}\text{GFP LA signals in ROI1 immediately before photobleaching} \\ \quad \text{background signals in ROI2}\end{array}\right)$$

As shown in Fig. 2, the mobile fraction of GFP-LA is calculated by curve-fitting the fluorescence recovery data to the following formula to estimate one-phase disassociation-association kinetics:

$$I_{(t)} = I_0 + \left(I_{\text{plateau}} - I_0\right)\left(1 - \exp^{-kt}\right) \qquad (1)$$

$I_{(t)}$: Normalized fluorescence intensity in the bleached area at a given time t.

I_{original}: Normalized original fluorescence intensity in the photobleached area (i.e., 100 %, immediately before photobleaching).

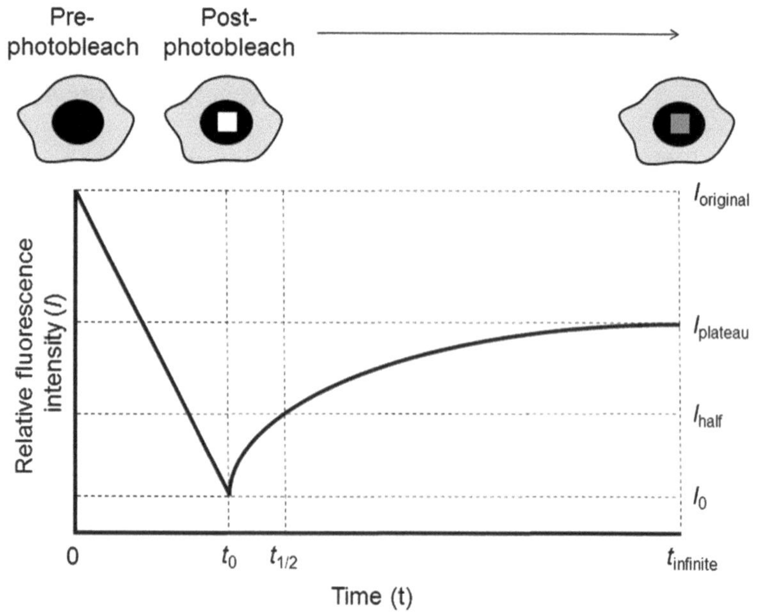

Fig. 2 An approach for curve-fitting the fluorescence recovery data is taken to estimate one-phase disassociation-association kinetics on the assumption that lamin molecules shuttle between the NL and nucleoplasm but do not laterally move along with the NL. $I_{(t)}$, I_{original}, I_0, I_{half}, and I_{plateau} are normalized fluorescence intensities in the bleached area at a given time t, 0 (before photobleaching), t_0 (immediately after photobleaching), $t_{1/2}$ as ln(2)/k, and t_{infinite} (if experimental time continues forever), respectively. These values are computationally calculated using the equation as described in Subheading 3.5

I_0: Normalized initial fluorescence intensity in the photobleached area when the t is t_0 (i.e., <20 %, immediately after photobleaching).

$I_{plateau}$: Normalized fluorescence intensity at plateau when the t is $t_{infinite}$ (if experimental time continues forever).

I_{half}: $(I_{plateau} - I_0)/2$. Normalized half fluorescence intensity when the t is $t_{1/2}$.

k: Constant. It is expressed as the reciprocal of time units.

t_{half}: $t_{1/2} - t_0$. The time unit of t as $\ln(2)/k$.

3.6 ConfoCor 2 Adjustment and FCS Measurement

Solution sample containing rhodamine 6G (Rh6G) as a standard dye both for 488 and 543 nm excitation lines must be daily measured for checking correct FCS optical setup and stable confocal detection volume. Select an optical pathway on ConfoCor 2 setup for Rh6G fluorescent dye or GFP (e.g., band-pass filter, BP505-550 nm). GFP-LA will be detected with the same condition after this calibration process.

1. Place 100 μL of Rh6G solution at a concentration of 10^{-7} M in a well of an 8-well chambered cover glass.
2. Adjust the center of the objective lens at 200 μm above the upper surface of the coverslip. Adjust the correction ring of the objective lens to maximal counts per molecule (CPM in kHz) setting.
3. Adjust the pinhole size and its position by auto-adjustment according the x and y-axes. The pinhole size typically selected is equivalent to 1 Airy unit (e.g. 70 μm for 488 nm excitation in the LSM 510-ConfoCor 2 system).
4. Adjust the excitation at 488 nm to 3 %. The emission signal is detected at BP505-550 nm (green channel) for Rh6G (or GFP-LA). Excitation power for Rh6G as control can be increased up to a twofold higher value for better CPM values (i.e., high signal to noise ratio) and autocorrelation function (*see* **Note 7**).
5. For FCS measurements, use ten sequential repetitions of single 20 s measurement. Hardware and software packages for the ConfoCor 2 provide ten fluorescence autocorrelation functions for the ten repetitions and additionally calculate and provide an averaged autocorrelation function.
6. Obtain the autocorrelation functions for Rh6G solution (*see* **Note 8**). The fluorescence autocorrelation functions (FAFs) from the BP505-550 nm channel, $G(\tau)$, are calculated by

$$G(\tau) = \frac{I(t)I(t+\tau)}{I(t)^2} \tag{1}$$

where τ denotes the time delay, I is the fluorescence intensity of the green channel, and $G(\tau)$ denotes the autocorrelation function of green channel.

7. The $G(\tau)$ functions are fitted with the ConfoCor 2 software installed in the LSM 510-ConfoCor 2 system. Set all parameters as free and fit the functions. After fitting the measured functions of Rh6G with a three-dimensional single-component model, determine the average structure parameter. The structure parameter, s, represents the ratio of the beam waist (wxy) and the axial radius (wz), $s = wz/wxy$. The averaged structure parameter must range from four to eight, which means that the confocal detection volume (e.g., optical setup for ConfoCor 2) is stable and best to use (*see* **Note 9**).

3.7 LSM and FCS Measurements for Cells

1. Observe the cells with a confocal microscope (Fig. 3a) to find a cell expressing GFP-LA at the low level, even though highly expressed cell also can be measured and analyzed (*see* **Note 10**).
2. Scan and image the whole cell using confocal scanning setup.
3. Select measuring positions in the nucleus of cellular image using the FCS positioning tool (Fig. 3a, cross hair).
4. For FCS measurements, set the laser intensity to 0.3 % for the 488 nm and the values of bleach time to 0 s, measuring time to 10 s, and repetition to 5. For fit condition in ConfoCor software, set the structure parameter into a value determined from Rh6G calibration measurement and set the value as fixed one.
5. Measure the selected positions and obtain autocorrelation functions for the green channel (Fig. 3b).
6. If necessary, calibrate for the real position of the FCS measurement to the position on the LSM images with a cover glass coated with dried Rh6G according to manufacturer's manual (*see* **Note 11**).

3.8 Curve Fitting and Data Analysis

FCS collects fluorescent signals from confocal detection volume with time for calculating fluorescence correlation function which contains parameters of molecular concentration and mobility (Fig. 4a–c). The acquired fluorescence autocorrelation functions (FAFs) are fitted using a one-component model (i, 1) for Rh6G in solution and a two-component model (i, 1 and 2) for GFP-LA in the nucleoplasm of cells, respectively, as follows:

$$G(\tau) = 1 + \frac{1 - T_{\text{triplet}} + T_{\text{triplet}} \exp^{\left(-\tau/\tau_{\text{triplet}}\right)}}{\left(1 - T_{\text{triplet}}\right)} \times \frac{1}{N} \sum_i y_i \left(1 + \frac{\tau}{\tau_i}\right)^{-1} \left(1 + \frac{\tau}{s^2 \tau_i}\right)^{-1/2} \quad (2)$$

Where T_{triplet} is the average fraction of triplet state molecules, τ_{triplet} is the triplet decay time (*see* **Note 12**), yi is the fraction of the components, and τi is the diffusion time of the components. N is the average number of fluorescent molecules in the confocal volume defined by beam waist (w_{xy}) and half of the long axis (w_z) of the volume, and s is the structure parameter defined by $s = w_z/w_{xy}$. The value of s in each confocal setup is determined experimentally from the

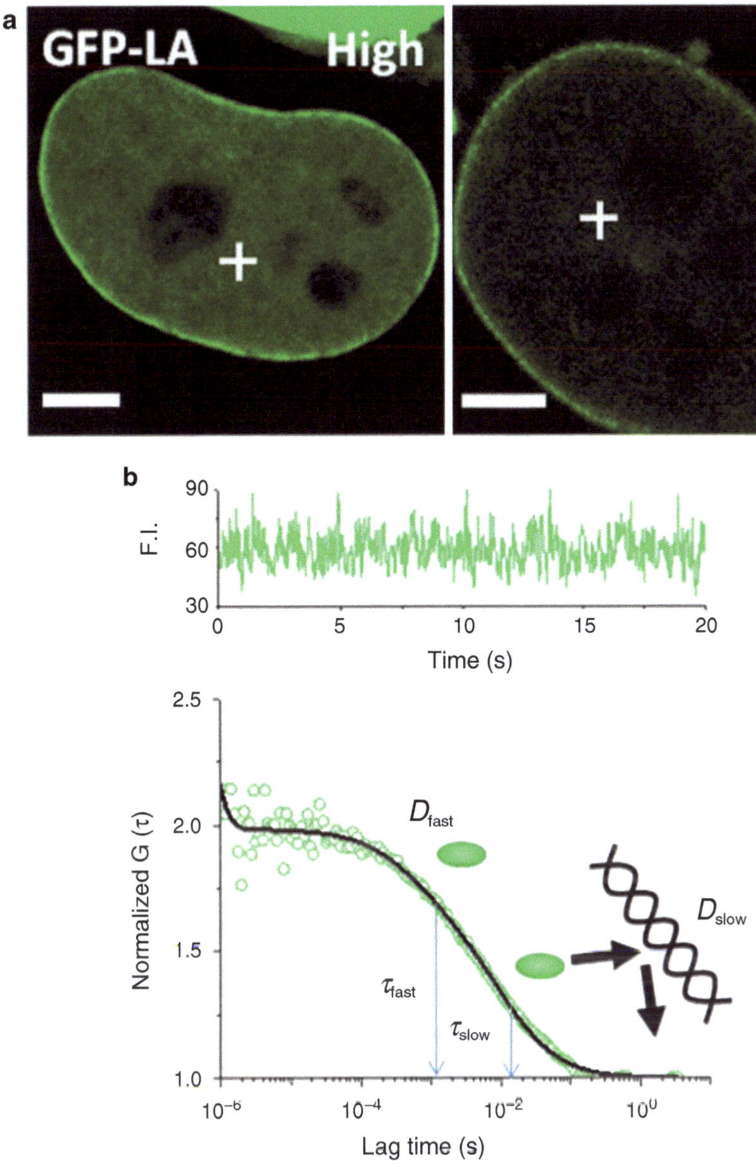

Fig. 3 Procedure of fluorescence correlation spectroscopy (FCS) measurement on GFP-LA in the nucleoplasm of living cell nuclei. (**a**) HeLa cells expressing GFP-LA at the high or low levels were selected and observed by confocal microscopy before FCS measurement. After imaging each cell, a position indicated with cross hair was measured by FCS for 10 s five times at 2-s intervals. Scale bar, 5 μm. (**b**) Averaged fluctuations of fluorescence intensity (F.I.) for GFP-LA during 20 s (*upper*, *green*), and the corresponding FAF (*lower*, *green*) were obtained from the cell expressing GFP-LA at the low level (**a**, *right*). Strongly expressed GFP-LA (**a**, *left*) was photobleached at the beginning of the measurement. A fitted function curve (*black line*) was obtained from the model equation with two-component model. It should be noted that two component model is adapted since one-component model is not fit the measured function (see also text)

measurement of the standard dye, Rh6G. The determined value of s is fixed for fitting the autocorrelation functions obtained from cell measurements. The diffusion time (τi) of component i is related to the translational diffusion coefficient (Di) of component i by

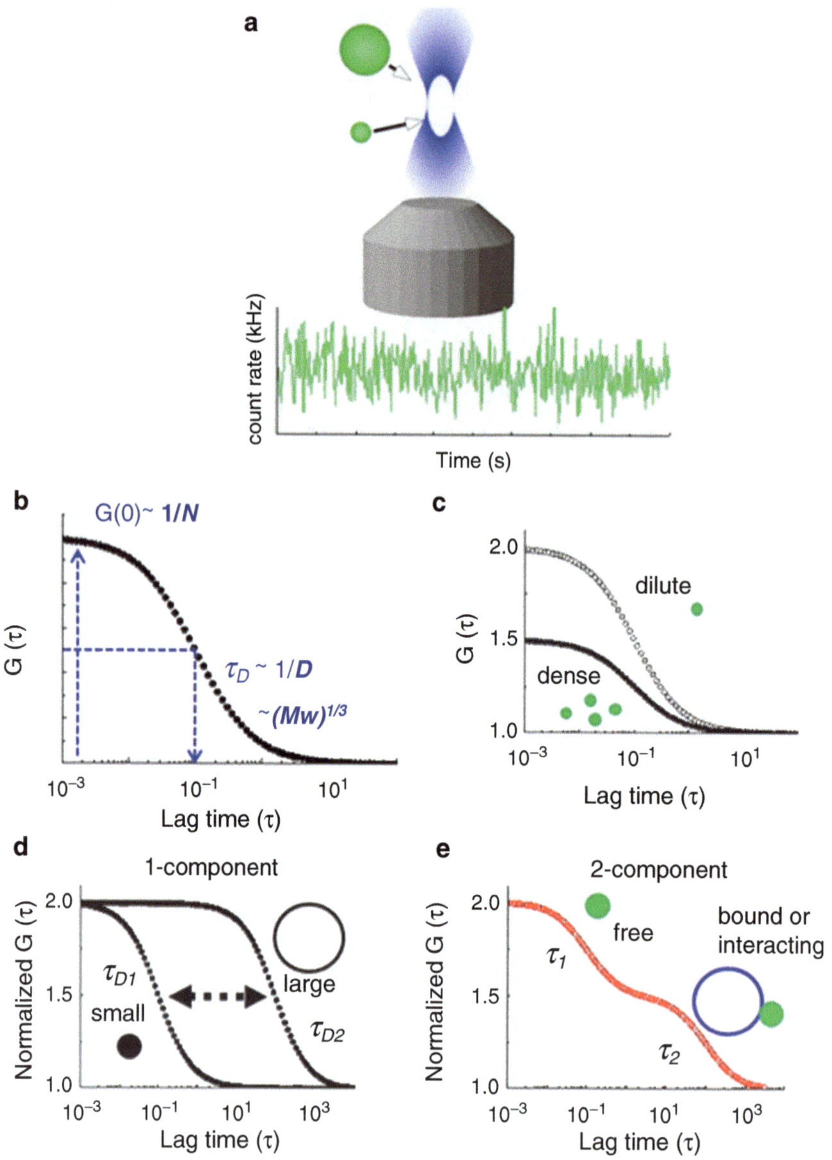

Fig. 4 Fluorescence autocorrelation functions and quantitative parameters. (**a**) FCS measurement based on fluctuation analysis of fluorescence intensity (Eq. 1). Small and large spheres tagged with fluorescent probes move in or out of the confocal volume (*white-out region in blue*) at different speeds (*top*), resulting in the characteristic fluctuation of fluorescence intensity (*bottom*). Fluorescence intensity is represented by count rate in kHz. (**b**) Measurement of single-color FCS provides a fluorescence autocorrelation function (FAF), $G(\tau)$. $G(\tau)$ allows analysis of two important physical parameters: the average number of fluorescent molecules (N) in the confocal volume and the translational diffusion time (τ_D) of the molecules through the open volume of detection (Eqs. 2 and 3). Diffusional motion of fluorescent molecules represented by τ_D or diffusion coefficient (D) reflects the molecular weight (MW) and the surrounding viscosity according to the Stokes-Einstein relation (Eq. 4). (**c**) Value of $G(0)$, correlation amplitude, is changed according to molecular number in the confocal volume as it is inversely proportional to N as shown in (**b**). For example, two FAFs with dilute (*open circle*) and dense (*closed circle*) molecule number of the same molecular size are plotted in the graph. (**d**) Left-shifted and right-shifted FAFs plotted in the graph are obtained from FCS measurements single species of small- and

$$\tau_i = \frac{w^2}{4D_i} \quad (3)$$

The diffusion of a free (i.e., non-interacting) and spheroidal molecule is related to various physical parameters by the Stokes-Einstein equation:

$$D_i = \frac{k_B T}{6\pi\eta r_i} \quad (4)$$

where T is the absolute temperature, ri is the hydrodynamic radius of the spheroidal molecule, η is the fluid-phase viscosity of the solvent, and k_B is the Boltzmann constant. The diffusion coefficients of GFP-LA ($D_{GFP\text{-}LA}$) in cells are calculated from the published diffusion coefficient of Rh6G, D_{Rh6G} (280 μm²/s), the measured diffusion times of Rh6G (τ_{Rh6G}), and GFP-LA ($\tau_{GFP\text{-}LA}$) as follows:

$$\frac{D_{GFP\ LA}}{D_{Rh6G}} = \frac{\tau_{Rh6G}}{\tau_{GFP\ LA}} \quad (5)$$

Single-measured diffusion times and calculated diffusion coefficients obtained from one-component model analysis are represented by τ_D and D, respectively (Fig. 4b, d). If one-component model cannot fit to a measured FAF, fit the FAF with two-component model (*see* **Note 13**). The two mobile components of τ_1 and τ_2 (Fig. 4e) can be represented by a fast mobile (τ_{fast} or D_{fast}) and a slow mobile (τ_{slow} or D_{slow}) component (*see* also Fig. 3b).

4 Notes

1. Because it takes at least 48 h after transfection for GFP-LA to completely localize to the NL, the transfected cells should be examined at 48–72 h post-transfection by FRAP and FCS experiments.
2. GFP-LA does not properly localize to the NL when it is expressed at the very high level.
3. Start from a short-duration time for photobleaching at the maximum laser power to avoid cellular damage.

Fig. 4 (continued) large-sized molecules. Value of $G(0)$ is normalized to 2 for comparison of two different diffusions (τ_{D1}, τ_{D2}). Measured FAFs from single species of fluorescent molecules can be fitted by one-component model (*see* Eq. 2, $i=1$). (**e**) A FAF representing a diffusion of two species of fluorescent molecules with small to large molecular number ratio, 1:1, is plotted in the graph. Measured FAFs obtained from a solution sample or a live cell measurement with two species of fluorescent molecules can be fitted by two-component model (*see* Eq. 2, $i=2$). The one component is a fast diffusion of a freely mobile species of fluorescent molecules (*green*), and the other is a very slow diffusion of fluorescent molecules bound to large mobile or immobile targets (*blue*)

4. Because GFP-LA is immobile in the NL, time intervals and duration time for imaging can be relatively long (e.g., every 30 s for 20 min).
5. Assign one cell for FRAP experiment and another cell for showing similar GFP expression as a control.
6. Keep finding a focus including region of interest until PFS system locks on to the focus.
7. High CPM value is also dependent on precise adjustments of pinhole position and correction ring of objective corresponding to the thickness of cover glass used.
8. Other standard dyes with high fluorescence yield such as rhodamine 123 and Alexa 488 can be also used. Averaged fluorescence intensity (count per second; cps in kHz) detected by ConfoCor 2 must be temporally stationary during measurement time. If it is not, diffusion time and structure parameter deviate from ideal values when one-component model analysis is adapted. Nonstationary fluctuations of fluorescence intensity can originate from fast or slow photobleach or absorption of dye molecules into cover glass.
9. If average *s* value is not unstable and far from 5, repeat processes 3, 4, and 5 until the *s* value becomes stable.
10. The quantity of DNA plasmids used for FCS must be minimized for preparing low-expressed cells comparing to LSM measurement which images highly fluorescent cells.
11. For LSM-ConfoCor 2 system, there is a difference between measured position for FCS and selected positions on an LSM image. By tuning the real position and measuring a cell on the center of LSM image, the difference can be minimized. Newly commercialized system such as LSM780 will not need such tuning process.
12. Excitation must be optimized to minimize effect of photochemical property of GFP such as triplet and blinking.
13. Check the deviation plot obtained from one-component model curve fit, which is provided by ConfoCor 2 software. If the deviation is systematically fluctuated, it is not likely to be best fitted. In that case, use two-component model fit and then tree-component model for better fit result.

Acknowledgments

This work was supported by grants from the National Institutes of Health (GM106023) and (CA03176) Award to R.D.G.

References

1. Adam SA, Goldman RD (2012) Insights into the differences between the A- and B-type nuclear lamins. Adv Biol Regul 52:108–113
2. Burke B, Stewart CL (2013) The nuclear lamins: flexibility in function. Nat Rev Mol Cell Biol 14:13–24
3. Parry DA, Conway JF, Steinert PM (1986) Structural studies on lamin. Similarities and differences between lamin and intermediate-filament proteins. Biochem J 238:305–308
4. Aebi U, Cohn J, Buhle L, Gerace L (1986) The nuclear lamina is a meshwork of intermediate-type filaments. Nature 323: 560–564
5. Schermelleh L, Carlton PM, Haase S, Shao L, Winoto L, Kner P, Burke B, Cardoso MC, Agard DA, Gustafsson MG, Leonhardt H, Sedat JW (2008) Subdiffraction multicolor imaging of the nuclear periphery with 3D structured illumination microscopy. Science 320:1332–1336
6. Shimi T, Pfleghaar K, Kojima S, Pack CG, Solovei I, Goldman AE, Adam SA, Shumaker DK, Kinjo M, Cremer T, Goldman RD (2008) The A- and B-type nuclear lamin networks: microdomains involved in chromatin organization and transcription. Genes Dev 22: 3409–3421
7. Moir RD, Yoon M, Khuon S, Goldman RD (2000) Nuclear lamins A and B1: different pathways of assembly during nuclear envelope formation in living cells. J Cell Biol 151:1155–1168
8. Kochin V, Shimi T, Torvaldson E, Adam SA, Goldman A, Pack CG, Melo-Cardenas J, Imanishi SY, Goldman RD, Eriksson JE (2014) Interphase phosphorylation of lamin A. J Cell Sci 12:2683–2696
9. Broers JL, Machiels BM, van Eys GJ, Kuijpers HJ, Manders EM, van Driel R, Ramaekers FC (1999) Dynamics of the nuclear lamina as monitored by GFP-tagged A-type lamins. J Cell Sci 112:3463–3475
10. Scaffidi P, Misteli T (2005) Reversal of the cellular phenotype in the premature aging disease Hutchinson-Gilford progeria syndrome. Nat Med 11:440–445
11. Dahl KN, Scaffidi P, Islam MF, Yodh AG, Wilson KL, Misteli T (2006) Distinct structural and mechanical properties of the nuclear lamina in Hutchinson-Gilford progeria syndrome. Proc Natl Acad Sci U S A 103:10271–10276
12. Shimi T, Koujin T, Segura-Totten M, Wilson KL, Haraguchi T, Hiraoka Y (2004) Dynamic interaction between BAF and emerin revealed by FRAP, FLIP, and FRET analyses in living HeLa cells. J Struct Biol 147:31–41

Chapter 6

Probing Protein Distribution Along the Nuclear Envelope In Vivo by Using Single-Point FRAP

Krishna C. Mudumbi and Weidong Yang

Abstract

Determining the locations of nuclear envelope transmembrane proteins and their concentrations across the outer and inner nuclear membranes has been a challenging and time-consuming process. Typically, this required the week-long process of fixing and immunogold staining of cells prior to analysis by electron microscopy. Here, we describe a method, single-point fluorescence recovery after photobleaching (spFRAP), which is able to quickly determine the localization and distribution of nuclear membrane proteins along the double nuclear envelope membranes with a precision of 10–15 nm in a matter of 10–20 min the day after transfection.

Key words Nuclear envelope transmembrane proteins, Fluorescence recovery after photobleaching, Single molecule, Super-resolution

1 Introduction

The nuclear envelope (NE) is comprised of three distinct membrane systems: the outer nuclear membrane (ONM) that is continuous with the endoplasmic reticulum, the inner nuclear membrane (INM) where many membrane proteins that organize and regulate the nucleus reside, and the pore membrane where ONM and INM are connected and nuclear pore complexes (NPCs) are inserted. Determining the distribution of NE transmembrane proteins (NETs) between these distinct membrane compartments has historically been challenging, requiring the counting of immunogold-labeled NETs by electron microscopy [1, 2], which yields no dynamic information and could be undercounted due to epitope masking or steric factors. NETs play many important roles in the organization and structure of the NE, nucleus, and cytoskeleton [3–10]. Mutations in these proteins have been linked with a variety of debilitating diseases [11–14]. Therefore, determining the localization and concentration of transmembrane proteins to the INM is extremely important. Recently, the groundbreaking

Sue Shackleton et al. (eds.), *The Nuclear Envelope: Methods and Protocols*, Methods in Molecular Biology, vol. 1411,
DOI 10.1007/978-1-4939-3530-7_6, © Springer Science+Business Media New York 2016

techniques of single-molecule tracking and super-resolution light microscopy are revolutionizing the field of molecular biology by providing unprecedented spatial resolutions. These techniques are able to break the diffraction-limit barrier of conventional light microscopy by localizing single molecules or reducing point spread function of fluorophores [15]. Fluorescence recovery after photobleaching (FRAP) is a widely used technique where a cellular membrane, which contains membrane proteins tagged with a fluorophore, is photobleached with a high-powered laser and the recovery of fluorescence, through the diffusion of the protein of interest, is recorded [16]. Here, we combine single-molecule tracking and FRAP techniques to determine the spatial distribution of NETs in live cells. In general, we first pre-photobleach fluorescently tagged proteins in a small detection area (~0.5 μm), then track fluorescent intact single protein molecules moving into this photobleached area, and finally reconstruct the locations of these protein molecules to form their super-resolution spatial distributions in the NE in vivo. In this way, the distribution of NETs along the NE was determined in vivo with a precision of 10–15 nm in just minutes on the day after cell transfection. Furthermore, by calculating the diffusion coefficient of the NETs in question and using FRAP to determine the immobilized fraction of NETs on the INM, the absolute concentration of NETs on both the ONM and INM can be quantified

2 Materials

All materials are prepared using deionized water and sterile techniques. We also provide specific details of our microscope setup and the detailed approach to achieve the resolution stated here.

2.1 Microscope Components

1. Microscope: Olympus IX81 equipped with a 1.4 NA 100× oil immersion objective (UPLSAPO 100XO, Olympus).
2. CCD camera: On-chip multiplication gain charge-coupled device camera (Cascade 128+, Roper Scientific).
3. Lasers: 35 mW 633-nm He-Ne laser (Melles Griot), 50 mW solid state coherent 488-nm He-Ne laser (Obis).
4. Lamp: Mercury lamp with GFP filter set up.
5. Filters: Dichroic filter (Di01-R405/488/561/635-25x36, Semrock) and an emission filter (NF01-405/488/561/635-25X5.0, Semrock), two neutral density filters (Newport).
6. Optical chopper (Newport) to generate an on-off mode of laser excitation (*see* **Note 1**).
7. Data acquisition and processing: Slidebook software package (Intelligent Imaging Innovations).

8. GLIMPSE software (Gelles lab).
9. ThunderSTORM software [17].

2.2 Tissue Culture

1. Cell line: we use wild-type (WT) HeLa cells.
2. Media: DMEM, high glucose, GlutaMAX™ Supplement (Life Technologies), 10 % fetal bovine serum, 1 % penicillin-streptomycin. DMEM with 1 % penicillin-streptomycin and no FBS was used for transfections (as per the specific manufacturer's protocol; adapt according to whichever transfection reagent is used).
3. 0.25 % trypsin EDTA.
4. 1× PBS: 137 mM NaCl, 2.7 mM KCl, 10 mM Na_2HPO_4, 2 mM KH_2PO_4, pH adjusted to 7.4 with HCl.
5. Flasks and dishes: 25 cm^2 flasks were used to grow WT HeLa cells. Glass bottom dishes (e.g., MatTek Corporation) were used for transfection and for recording single-molecule events after. If other dishes are used, they should have at least equivalent optical properties.
6. Plasmid preparation: Plasmids for transfection were encoding NETs fused to GFP, although other fluorescent protein fusions can be used with appropriate adjustments in laser wavelength and bleaching parameters. Plasmids were prepared using the plasmid miniprep (alkaline lysis) method [18].
7. Transfection: In our case, HeLa cells were transfected using the lipofection technique as specified by Mirus Bio TransIT-LT1 Transfection Reagent protocol (*see* **Notes 2** and **3**). If other transfection reagents are used, they should be certain to not leave autofluorescent material adhered to the glass as this background will interfere with imaging.
8. Transport buffer: 20 mM Hepes, 110 mM KOAc, 5 mM NaOAc, 2 mM MgOAC, 1 mM EGTA, pH adjusted to 7.3 with HCl.

3 Methods

3.1 Tissue Culture and Transfection

1. At least 1 week in advance, using sterile techniques, start a fresh culture of your cell line from a frozen stock (−80 °C) by thawing at 37 °C and putting in a 25 cm^2 culture flask with 5 mL DMEM media (at 37 °C). Place cells into an incubator and incubate at 37 °C with 5 % CO_2 for 24 h in the first instance and split the culture at least three times over the week for the cells to be at optimal health for transfection and optimal optical conditions for live cell microscopy.
2. Using sterile technique, 2 days before the single-molecule experiments, split the cells and plate them onto a glass

bottom dish. Return to the incubator for at least 12 h of growth (*see* **Note 4**).

3. Using the protocol provided by the manufacturer and sterile techniques, transfect cells (in our case, using LT1 transfection reagent) with plasmids coding for the protein of interest. In our case, serum-free DMEM media with the LT1 and plasmid complex was added dropwise to the glass bottom dishes and then gently mixed by moving the dishes side to side so that the solution can distribute homogeneously. Return glass bottom dishes to the incubator and grow for 18+ h at 37 °C with 5 % CO_2 so that the NET-fluorescent protein fusions are expressed at reasonable levels at the time of the experiment.
4. The following day, using sterile technique, remove the DMEM media from the glass bottom dishes. Wash twice using 1 mL of PBS warmed to 37 °C. Following the second wash, add 2 mL of pre-warmed (to 37 °C) transport buffer to the dishes. Note that it is important to use the transport buffer and to allow the cells to equilibrate in it (approximately 45 min) to reduce fluorescent backgrounds for single-molecule work: the phenol red in DMEM interferes with experiments and even some of the media specially designed to reduce background in FRAP experiments yield too high backgrounds for single-molecule work.

3.2 spFRAP Experiment

1. Prior to imaging the sample, both the 488-nm and the 633-nm lasers need to be aligned so that their light paths overlap exactly when they reach the microscope objective (*see* **Note 5**). Generate a diffraction-limit illumination volume (alternatively named illumination point spread function) of the 488-nm excitation laser (≈210 nm in the *x* and *y* directions and ≈500 in *z* direction) by focusing the laser through a high-NA microscope objective (1.4 NA in this particular setup) (Fig. 1).
2. Using a 100× oil immersion objective, visualize the cells using a mercury lamp with the filter setting adjusted for GFP visualization.
3. Only cells with good nuclear morphology (*see* **Note 6**) should be targeted. Adjust the focus so that the equator of the cell is targeted and such that the laser focus is on the left or right edge of the NE (tangent to the edge of NE that is being targeted).
4. Using the lamp, set the 500-ms exposure time to take a "before" photobleaching image of the cell while avoiding overexposure and saturation of the image.
5. Close the port to the mercury lamp and switch to the lasers. First, using the 633-nm laser, align the laser to the intended area of photobleaching. Close the port to the 633-nm laser and open the port to the 488-nm laser to begin photobleaching.

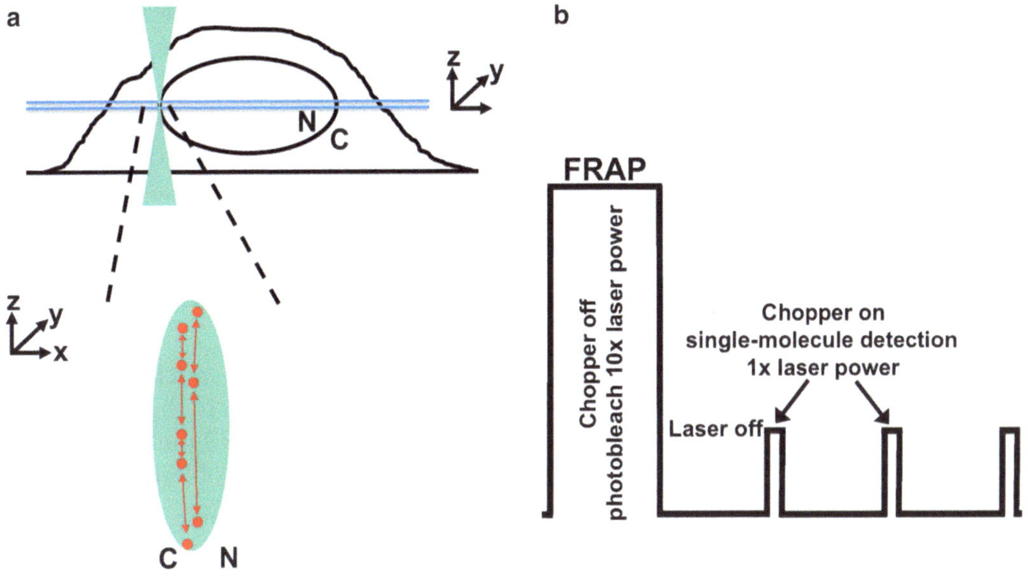

Fig. 1 Single-point illumination and FRAP used to detect transmembrane proteins on the NE. (**a**) Imaging of single-molecule events of NETs along the NE by single-point spFRAP microscopy. The focal plane is between the *two light blue lines. C* cytoplasm; *N* nucleus. (**b**) A laser power with tenfold difference was used to photobleach and detect in the experiments. An optical chopper was used to regulate the laser to have an on-off mode. The longer off time allowed GFP-NETs outside the photobleached area to have enough time to diffuse into the detection area

6. Using high laser power (*see* **Note 7**), photobleach a small region of the NE. The bleached region should be around 0.5 μm in diameter.
7. For subsequent detection, lower the laser power by at least tenfold using the neutral density filter between the laser and the sample (*see* **Note 8**).
8. Engage the optical chopper at 2-Hz rotation speed with an on time of 1/10 of the total frames recorded.
9. Record videos at a 0.4-ms frame rate for 30 s. Ten total videos were taken consecutively. In our setup, we use Slidebook software for recording images and videos.
10. Once recording is finished, switch back to the mercury lamp. Take an "after" picture while avoiding overexposure and oversaturation of the image and compare with the "before" picture to make sure the NE did not shift during our measurements (*see* **Note 9**).

3.3 Bulk FRAP Experiment

1. Prior to imaging the sample, both the 488-nm and the 633-nm lasers need to be aligned so that their light paths overlap exactly when they reach the microscope objective (*see* **Note 5**). Generate a 5-μm illumination area in the focal plane by inserting a defocus lens before the objective.

2. Using a 100× oil immersion objective, visualize the cells using a mercury lamp with the filter setting adjusted for GFP visualization.
3. Only cells with good nuclear morphology (*see* **Note 6**) should be targeted. Adjust the focus so that the equator of the cell is targeted and such that the illumination area is on the left or right edge of the NE.
4. Using the lamp, set the 500-ms exposure time to take a "before" photobleaching image of the cell while avoiding overexposure and saturation of the image.
5. Close the port to the mercury lamp and switch to the lasers. First, using the 633-nm laser, align the laser to the intended area of photobleaching. Close the port to the 633-nm laser and open the port to the 488-nm laser to begin photobleaching.
6. Using high laser power (*see* **Note 7**), photobleach a fraction of the NE. The bleached region should be around 5 μm in diameter.
7. Imaging parameters may be adjusted for a particular set of proteins being investigated based on their recovery times. In general, these should be based on obtaining images with a high-intensity signal output from the start of the time-lapse pictures to the end FRAP recovery. It is important to select an exposure time that will give a high-intensity signal output without causing too much photobleaching and a decrease in the overall signal by the end of the recovery phase (*see* **Note 10**).
8. After photobleaching, switch from the laser to the mercury lamp and take an "after" photobleaching picture. Compare the "before" and the "after" pictures to make sure there is no significant shift of NE during the detection.
9. Convert time-lapse data into individual .tiff files using Slidebook and then analyze them using the ImageJ plugin FRAP Profiler as described in Subheading 3.4.2.

3.4 Data Analysis

3.4.1 Analysis of spFRAP Data

1. Convert images from videos generated with Slidebook into multiple .tiff files.
2. Open the files using the ImageJ plugin ThunderSTORM and filter the raw data for a high signal-to-noise ratio (SNR) with parameters including single-molecule intensity (>2000) and Gaussian width (between 0.5 and 1.5) of single-molecule spots.
3. From the data prefiltered by ThunderSTORM, determine the average x and y pixel positions to find the appropriate ROI to be covered when further running the GLIMPSE software (*see* **Note 11**).
4. In GLIMPSE, use nine ROIs covering all the possible x and y pixel positions determined in the previous step and select just

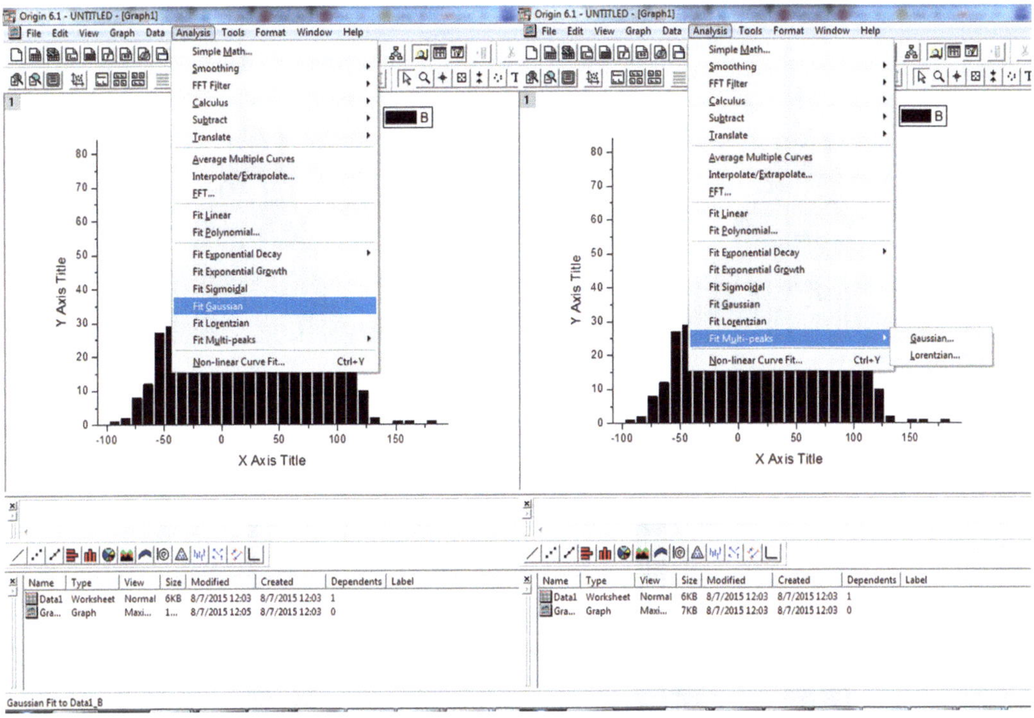

Fig. 2 Screen shots showing how to fit histogram data with single and two-peak Gaussian functions

the frames during which the optical chopper was "open" and the NE was exposed to the laser.

5. Run GLIMPSE software using nine ROIs to localize each point in the detection area.
6. Raw data from GLIMPSE is then filtered using a high SNR, determined by using the intensity of single-molecule spot and by selecting for points in the focal plane of the microscope by using the width of the Gaussian fittings from GLIMPSE (height and width respectively). Generally, data with a GLIMPSE-generated height value higher than 2000 and width between 0.5 and 1.5 is used (*see* **Note 12**).
7. Fit the resultant data with a single Gaussian function to remove the background noise (*see* **Note 13**) (Figs. 2 and 3).
8. Next, further fit the subtracted data with a two-peak Gaussian to determine the distribution of the NEt along the NE (Figs. 2 and 3).
9. By using the full width at half maximum (FWHM) as determined by data analysis software (a built-in function in Origin 6.1 is used here), the range between where points can lie along either side of the NE can be determined for further analysis such as diffusion coefficient and protein concentration along the NE.

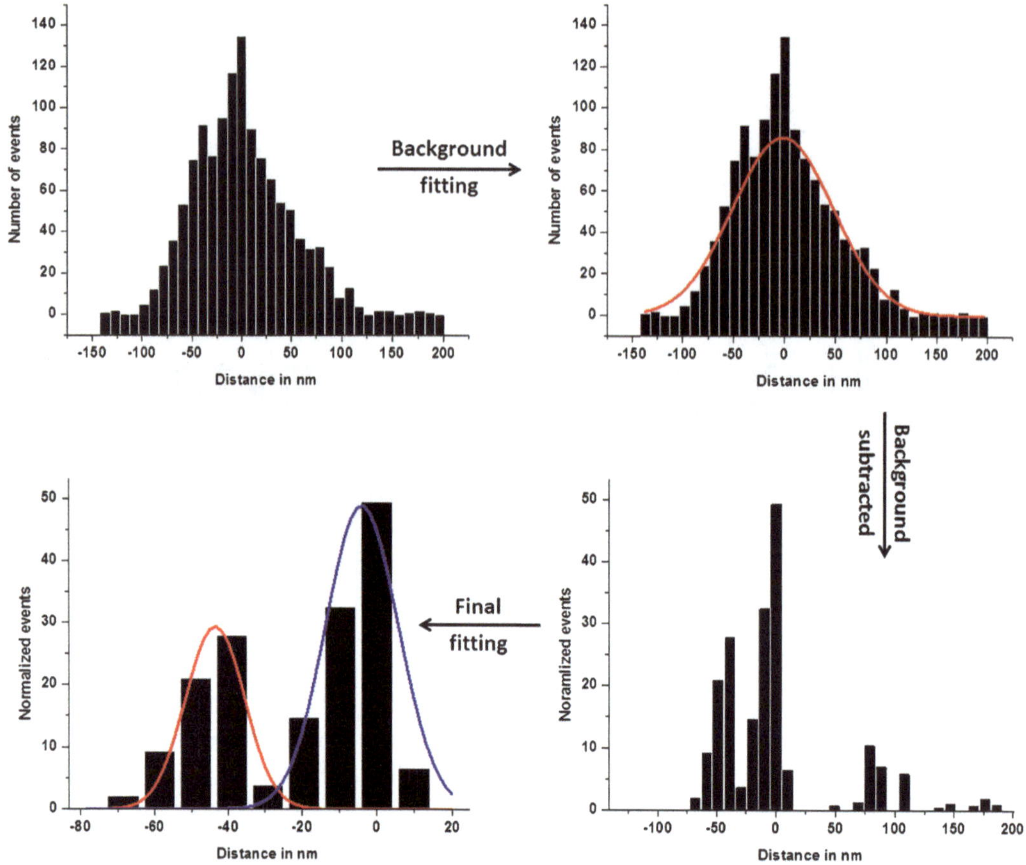

Fig. 3 Histograms were prepared from the raw data. The background noise (without including the data of two peaks) was fit with a single Gaussian function (shown in *red*). Next, the background was subtracted from the raw data to generate a histogram of the normalized data with two clear peaks. Finally, the resultant data was then fit with a two-peak Gaussian function to determine the localization of NETs on the NE (INM shown in *red* and ONM shown in *purple*)

3.4.2 Analysis of Regular FRAP Data

1. Open the FRAP recovery images in sequence starting with the "before" photobleach picture, followed by the "after" photobleach picture, and then the sequence of time-lapse images of recovery.
2. Open up the ROI manager (Analyze/Tools/ROI Manager) and select the two regions of interest. The first ROI is the region that was photobleached and the second ROI is the nuclear envelope, which must be carefully outlined with the freehand selection tool.
3. Run the FRAP Profiler plugin and use the normalized output to determine the immobilized fraction.

4 Notes

1. The speed of the chopper was set to 2 Hz rotation with an on time of 1/10 of the total frames recorded.
2. A 3:1 ratio of LT1 transfection reagent to plasmid resulted in the best transfection efficiencies for our HeLa cells.
3. Electroporation techniques were also used for transfection, resulting in similar final spFRAP results.
4. The cells plated here should be about 50 % confluent after the 12 h growth period in order to achieve optimal transfection results. If the cells from a 25 cm^2 dish are ~90–95 % confluent, trypsinize the cells and resuspend in 5 mL media before adding 300 μL of the suspended HeLa cells to the glass bottom dish.
5. Only the 488-nm laser that is being used to excite the fluorophore needs to be precisely focused, whereas the other laser, 633-nm, will be used only to align the 488-nm for single-molecule measurements. The 633-nm laser is used for alignment because it does not excite the GFP-tagged NET and maybe used as a reference point to know the alignment of the 488-nm laser on the NE in the region of interest. While a 633-nm laser is used in these experiments, any other laser that similarly doesn't excite GFP can be used. A glass slide coated with Alexa Fluor 647-labeled GFP is used to focus and align the lasers.
6. Only cells in interphase should be selected. Cells that seem to be in the middle of mitosis or cells with NEs that are not circular in shape should be avoided.
7. A laser power of 20 mW was used to photobleach the NE for 30 s.
8. You must select the best power for your system in order to get a high signal with reduced background noise. A range between 2 mW and 50 μW can be used in these experiments with our setup.
9. "Before" and "after" pictures can be overlapped quite easily and used to compare the movement of the NE during the imaging time. Ideally, there should be no movement of the NE during the video acquisition. Data from cells that exhibit obvious movements during data collection should be discarded. These pictures are taken with the mercury lamp to obtain wide-field epifluorescent images.
10. It is best to try and use a set exposure time for each group of NETs studied in an experiment. For instance, if three different NETs are being studied and compared, one should use the same exposure time for the time-lapse imaging of each NET. The authors have found that a 2 s exposure time with an 18 s delay between images works best.

11. The filtering in Subheading 3.4.1 **step 2** by using the ImageJ plugin ThunderSTORM was done so that the locations with good S/N ratios from which single-molecule data was collected by spFRAP can be determined. The GLIMPSE software uses ROIs in the region of data prefiltered by ThunderSTORM to further determine the detailed XY distribution with high spatial localization precision.
12. It has been determined that GLIMPSE-generated data with Gaussian width between −0.5 and 0.5 is noise and should be eliminated.
13. The Origin software package was used to perform data analysis, but any powerful data analysis software may be used in its place (refer to Fig. 3).

References

1. Ellenberg J et al (1997) Nuclear membrane dynamics and reassembly in living cells: targeting of an inner nuclear membrane protein in interphase and mitosis. J Cell Biol 138(6): 1193–1206
2. Wilhelmsen K et al (2005) Nesprin-3, a novel outer nuclear membrane protein, associates with the cytoskeletal linker protein plectin. J Cell Biol 171(5):799–810
3. Gruenbaum Y, Margalit A, Goldman RD, Shumaker DK, Wilson KL (2005) The nuclear lamina comes of age. Nat Rev Mol Cell Biol 6(1):21–31
4. Burns LT, Wente SR (2012) Trafficking to uncharted territory of the nuclear envelope. Curr Opin Cell Biol 24(3):341–349
5. Heessen S, Fornerod M (2007) The inner nuclear envelope as a transcription factor resting place. EMBO Rep 8(10):914–919
6. Hetzer MW, Wente SR (2009) Border control at the nucleus: biogenesis and organization of the nuclear membrane and pore complexes. Dev Cell 17(5):606–616
7. Wilson KL, Foisner R (2010) Lamin-binding proteins. Cold Spring Harb Perspect Biol 2(4):a000554
8. Zuleger N, Korfali N, Schirmer E (2008) Inner nuclear membrane protein transport is mediated by multiple mechanisms. Biochem Soc Trans 36(6):1373
9. Zuleger N, Kerr AR, Schirmer EC (2012) Many mechanisms, one entrance: membrane protein translocation into the nucleus. Cell Mol Life Sci 69(13):2205–2216
10. Arib G, Akhtar A (2011) Multiple facets of nuclear periphery in gene expression control. Curr Opin Cell Biol 23(3):346–353
11. Stewart CL, Roux KJ, Burke B (2007) Blurring the boundary: the nuclear envelope extends its reach. Science 318(5855):1408–1412
12. Schreiber KH, Kennedy BK (2013) When lamins go bad: nuclear structures and disease. Cell 152(6):1365–1375
13. Dauer WT, Worman HJ (2009) The nuclear envelope as a signaling node in development and disease. Dev Cell 17(5):626–638
14. Méndez-López I, Worman HJ (2012) Inner nuclear membrane proteins: impact on human disease. Chromosoma 121(2):153–167
15. Huang B, Bates M, Zhuang X (2009) Super resolution fluorescence microscopy. Annu Rev Biochem 78:993
16. Axelrod D, Koppel D, Schlessinger J, Elson E, Webb W (1976) Mobility measurement by analysis of fluorescence photobleaching recovery kinetics. Biophys J 16(9):1055
17. Ovesný M, Křížek P, Borkovec J, Švindrych Z, Hagen GM (2014) ThunderSTORM: a comprehensive ImageJ plug-in for PALM and STORM data analysis and super-resolution imaging. Bioinformatics 30(16):2389–2390
18. Klein RD, Selsing E, Wells RD (1980) A rapid microscale technique for isolation of recombinant plasmid DNA suitable for restriction enzyme analysis. Plasmid 3(1):88–91

Chapter 7

The Use of Two-Photon FRET–FLIM to Study Protein Interactions During Nuclear Envelope Fusion In Vivo and In Vitro

Richard D. Byrne, Banafshé Larijani, and Dominic L. Poccia

Abstract

FRET–FLIM techniques have wide application in the study of protein and protein–lipid interactions in cells. We have pioneered an imaging platform for accurate detection of functional states of proteins and their interactions in fixed cells. This platform, two-site-amplified Förster resonance energy transfer (a-FRET), allows greater signal generation while retaining minimal noise thus enabling application of fluorescence lifetime imaging microscopy (FLIM) to be routinely deployed in different types of cells and tissue. We have used the method described here, time-resolved FRET monitored by two-photon FLIM, to demonstrate the direct interaction of Phospholipase Cγ (PLCγ) by Src Family Kinase 1 (SFK1) during nuclear envelope formation and during male and female pronuclear membrane fusion in fertilized sea urchin eggs. We describe here a generic method that can be applied to monitor any proteins of interest.

Key words Nuclear envelope, Two-photon FLIM, FRET, Antibody, Membrane fusion

1 Introduction

FRET–FLIM techniques are used to study protein and protein–lipid interactions in cells [1, 2, 3]. We have applied these to study of nuclear envelope fusion events following fertilization [4]. Following fertilization of eggs of the sea urchin and many other organisms, several nuclear events occur requiring membrane fusion [5]. The sperm's nuclear envelope is disassembled prior to reassembly for the most part from fusion of membranes derived from the egg's endoplasmic reticulum and polar remnants of the sperm nuclear membranes resulting in the male pronuclear envelope [6, 7]. In many cells, the male pronucleus unites with the pronucleus of the egg to form the zygote nucleus, a process requiring the fusion of their nuclear envelopes. Following mitosis, either a new nuclear envelope is formed around all the chromosomes or, as is the case for the sea urchin and some other species, individual chromosomes

Sue Shackleton et al. (eds.), *The Nuclear Envelope: Methods and Protocols*, Methods in Molecular Biology, vol. 1411,
DOI 10.1007/978-1-4939-3530-7_7, © Springer Science+Business Media New York 2016

are enclosed by nuclear envelopes to form karyomeres which then fuse with each other to create the zygote nucleus [8, 9].

The control of these membrane fusion events remains incompletely understood. Study of the formation of the nuclear envelope around the male chromatin in fertilized eggs of the sea urchin, both in vivo and in vitro, suggest that it requires a fusogenic membrane vesicle population (MV1), rich in polyphosphoinositides, that serves as a source of diacylglycerol, a fusogenic lipid created by the hydrolysis of phosphatidylinositol(4,5)bisphosphate catalyzed by phospholipase C γ (PLCγ) [6, 10]. PLCγ is regulated in sea urchin eggs by SFK1 which activates it by adding a single phosphate group [11–14]. Activation requires that SFK1 and PLCγ are not only co-localized in vesicles or on the forming nuclear envelope but that they directly interact prior to completion of nuclear envelope formation.

A quantitative and objective method for assessing the activation status and the complex formation of these proteins was needed to complement current methods which assess the concentration of proteins rather than their activation status in different types of processes such as membrane fusion. This goal was achieved by use of FRET–FLIM, which can image molecular interactions. FRET can occur between a donor and an acceptor molecule only if they are within ~1–10 nm of one another and can therefore be used to monitor proximity and activation status such as phosphorylation state.

Two main types of determinations of FRET are possible: steady-state and time-resolved (detection of chromophore lifetime) measurements. Time-resolved measurements yield additional information about FRET. One of the major advantages of measuring fluorescence lifetime is that, unlike the steady-state method which is intensity based, it is independent of chromophore concentration [15, 16]. In contrast to steady-state measurements, where a change in intensity of the recorded emission is observed and thus becomes prone to artifacts that lead to overestimation of FRET efficiency (the parameter used to assess functional status and protein–protein interactions), time-resolved FRET detected by FLIM is a highly effective method for quantifying intracellular molecular interactions in cells and tissue [4, 12, 17].

Coincidence (two-site) FRET involves simultaneously labeling a single protein on two distinct sites with donor- and acceptor-conjugated probes and detecting the FRET between them. Although this method also relies on antigen–antibody immune complexes (IHC), when using two-site a-FRET (amplified FRET), the FRET is most efficient when the distance between the two antibodies is less than 10 nm within the same antigen. Two-site FRET is therefore necessarily more specific. The method has gained recent popularity due to this high specificity and its relative insensitivity to intensity artifacts. Studies of *endogenous* proteins using FLIM have been limited due to the lack of sensitivity. Two-site a-FRET solves this problem by amplifying the signal [17]. Moreover, in contrast to immunofluorescence, it has a dynamic range which can also be quantifiably classified [17].

We performed time-resolved FRET measurements during nuclear envelope formation and a-FRET during male and female pronucleus fusion in sea urchin fertilized eggs both in vitro and in vivo by using antibodies directed against PLCγ and SFK1. We showed that SFK1 and PLCγ directly interact in vivo in the cell cortex of unfertilized eggs and at 3 min postfertilization throughout egg cytoplasm and around the male pronucleus [14]. By 5 min postfertilization, the direct interaction was no longer detected, although the enzymes remained co-localized. In a cell-free system that assembles nuclear envelopes around exogenously added demembranated sperm nuclei, both enzymes were recruited to the nuclei on vesicles from cytoplasmic extracts. There, they transiently exhibited FRET until after the nuclear envelope was formed. They then remained associated with the nuclear membrane but not directly interacting.

Along with pharmacological data, these findings led to a model in which a membrane vesicle population (MV1) highly enriched in inactive PLCγ, SFK1, and PtdIns(4,5)P_2 binds to two regions of the nuclear surface and then initiates membrane fusion of other vesicles (mainly derived from the endoplasmic reticulum) following PLCγ phosphorylation and activation by SFK1. The elevated production of the fusogenic lipid DAG by PLCγ hydrolysis of PtdIns(4,5)P_2 facilitates this fusion of the vesicles surrounding the nucleus to complete NE formation after which PLCγ and SFK1 disengage though remaining in the membrane.

To analyze interactions during pronuclear fusion, tyramide signal amplification (a-FRET) was used [4]. It was shown than PLCγ and SFK1 directly interacted upon pronuclear contact and fusion but not during their approach. Additional information localized PLCγ at the site of contact. PLCγ was activated by phosphorylation at Tyr783 with an accompanying depletion of PtdIns(4,5)P_2 consistent with local production of diacylglycerol. Inhibitors of PLCγ and SFK1 blocked pronuclear fusion. Thus, similar regulation operates in fusions required for formation of the zygote nuclear envelope and for the de novo formation of the male pronuclear envelope.

Here, we report on the application of a-FRET with tyramide signal amplification (a-FRET) to assess the roles in the fusion of the male and female pronuclear envelopes that can be applied to monitor any proteins for which specific primary antibodies are available.

2 Materials

2.1 Sea Urchin Gametes

1. Artificial sea water: dissolve 1650 g Instant Ocean Reef Salt in 50 L distilled water and stir continuously for at least 16 h before adding to aquaria.
2. *Lytechinus pictus* sea urchins: maintain at 18 °C in a closed aquarium of artificial sea water.
3. 22 G 1½ in. needles.
4. 1 mL syringes.
5. 0.5 M KCl.
6. MPSW: filter artificial sea water through a 0.22-μm membrane filter.
7. 5 and 10 mL glass beakers.
8. 100 mL weighboat and tissue paper.

2.2 Fertilization and Fixation of Sea Urchin Eggs

1. Nytex mesh with 210 μm pore diameter.
2. 3-amino-1,2,4-triazole (ATA), 30 mM: prepare fresh as a 10× stock in distilled water no more than 30 min before use and store on ice.
3. 7.4 % paraformaldehyde (PFA): prepare in MPSW (*see* **Note 1**). Freeze in aliquots and store at −20 °C.
4. Rocking platform.
5. 0.05 % sodium azide in PBS.

2.3 Antibody–Fluorophore Conjugation

1. $F(ab')_2$ antibody.
2. Atto 488 dye.
3. 50 mM sodium borate, pH 8.5 (20× stock).
4. PD Minitrap™ G-25 column.
5. NanoDrop 2000 UV–Vis spectrophotometer or equivalent.

2.4 Antibody Labeling of Sea Urchin Eggs

1. 3 % fatty acid-free BSA (w/v)/0.5 % saponin (w/v) solution in PBS prepared fresh.
2. 1 mg/mL sodium borohydride solution in PBS, prepared fresh.
3. Primary antibodies.
4. Secondary $F(ab')_2$–Atto 488 conjugate (prepared in Subheading 3.3).
5. Secondary HRP conjugate antibody.
6. TSA–Cy5 reagent.
7. 2 mg/mL Hoechst 33342 stock in dH_2O.
8. ProLong® Gold mounting medium.

2.5 Two-Photon FLIM of Labeled Eggs

1. Nikon TE2000-E inverted microscope with 60×/1.49NA oil immersion objective or equivalent.
2. 6 W Mira laser or equivalent.
3. TRI2 custom software (*see* **Note 2**).

3 Methods

3.1 Obtaining Sea Urchin Gametes

1. Insert needle through the perioral membrane (*see* **Note 3**) into the coelomic cavity and inject sea urchin with 100 μL 0.5 M KCl. Repeat on opposite side of animal (*see* **Note 4**).
2. Invert females (shedding orange eggs) over a beaker filled with MPSW. Males should be inverted in a weigh boat. Keep animals moist with direct application of MPSW (females) or covering with MPSW soaked tissue paper (males).
3. Periodically aspirate sperm with a plastic Pasteur pipette and keep sperm on ice.

3.2 Fertilization and Fixation of Sea Urchin Eggs

All the following steps take place at 16 °C unless otherwise stated.

1. If more than one female is required, check eggs from individual animals to ensure correct fertilization. Dilute sperm approximately 500-fold in MPSW and add a few drops to eggs before viewing under a light microscope with a 10× objective. Pool fertilized eggs (90 % of eggs should raise a fertilization envelope within 30 s of sperm addition).
2. Pass eggs twice through a 210 μm Nytex mesh to remove the jelly coat (*see* **Note 5**).
3. Add the appropriate volume of 10x ATA to the egg suspension (*see* **Note 6**).
4. Dilute sperm 500-fold in MPSW and add to the egg suspension to give a 10–100:1 sperm/egg ratio.
5. Confirm fertilization as in Step 1.
6. Filter eggs through a 210 μm Nytex mesh to remove fertilization envelopes.
7. Place eggs in a beaker on an orbital shaker.
8. At the appropriate time point required for the experiment, remove an aliquot of eggs from the suspension and add to an equal volume of ice cold 7.4 % PFA in MPSW.
9. Shake for 1 h.
10. Centrifuge egg suspension at 100 ×*g*, 2 min, 4 °C (*see* **Note 7**).
11. Remove supernatant and wash eggs twice in PBS as above.
12. For long-term storage, resuspend the egg pellet in 0.05 % azide in PBS and store at 4 °C.

3.3 Atto 488 Conjugation to F(ab′)$_2$ Fragment

1. Add 100 μg of F(ab′)$_2$ antibody fragment to 25 μg of Atto 488 chromophore.
2. Adjust total volume to 100 μL with PBS; adjust to pH 8.5 with 50 mM sodium borate buffer. The conjugation reaction takes 1 h in the dark at 20 °C with shaking.
3. Equilibrate a PD Minitrap™ G-25 column with 15 mL PBS.
4. Dilute conjugation reaction in an equal volume of PBS and add to the column as soon as the last of the equilibration PBS is absorbed into the column.
5. Elute the dye–antibody conjugate with repeated additions of 300 μL PBS.
6. Collect the faster-running “colored” fraction (Fig. 1a). This is the dye-antibody conjugate (*see* **Note 8**).
7. Calculate the dye/protein ratio of the chromophore–antibody conjugate as per manufacturer’s instructions using a nanodrop device (*see* **Note 9**).
8. Store conjugates at 4 °C.

3.4 Antibody Labeling of Sea Urchin Eggs

All the chemical, antibody, and dye reagent incubation steps in this method should be separated by a wash with 0.5 mL PBS (Fig. 1b).

1. Transfer an aliquot of fixed eggs from the required time point to a 1.5 mL microcentrifuge tube and centrifuge at 150 × *g*, 1 min, 4 °C. Aspirate the supernatant.
2. Resuspend the eggs in 0.5 mL of 3 % (w/v) fatty acid-free BSA/0.5 % (w/v) saponin, and incubate for 15 min at RT on an overhead mixer (*see* **Note 10**).
3. Centrifuge eggs and aspirate as above.
4. Quench eggs with 1 mg/mL sodium borohydride for 2 min at RT (*see* **Note 11**).
5. Incubate eggs with the first primary antibody for 1 h at RT, e.g., anti-PLCγ (*see* **Note 12**).
6. Incubate eggs with the F(ab′)$_2$-Atto-488 conjugate for 1 h at RT (*see* **Note 12**).
7. Incubate eggs with the second primary antibody for 1 h at RT, e.g., anti-SFK1.
8. Incubate eggs with the secondary HRP conjugate for 1 h at RT (*see* **Note 13**).
9. Prepare the TSA-Cy5 reagent as per the manufacturer’s instructions and incubate with the eggs for 10 min in the dark (*see* **Note 13**).
10. Incubate eggs with 2 μg/mL Hoechst 33342 for 15 min at RT.
11. Place eggs onto 0.005 % poly-L-lysine-coated 13 mm coverslips, for 15 min, dip in dH$_2$O, and mount in 8 μL ProLong Gold on a glass slide as per manufacturer’s instructions (*see* **Note 14**).

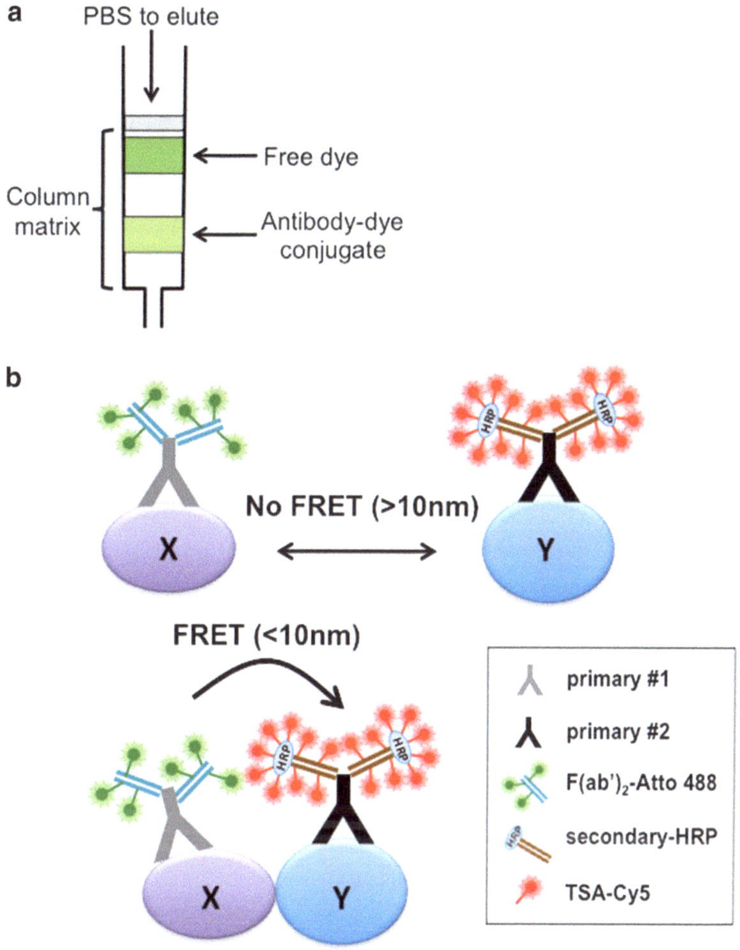

Fig. 1 Antibody dye conjugation and sample labeling. (**a**) Schematic of F(ab′)$_2$–Atto 488 purification. Upon completion of the conjugation step, column loading, and elution with PBS, free Atto 488 dye will separate from the F(ab′)$_2$–Atto 488 conjugate. The latter runs faster and is observed as a paler band when compared to free dye. (**b**) Principle of egg labeling and FRET. Eggs are labeled sequentially as described in the methods section, resulting in the first protein of interest being labeled with a primary antibody and the F(ab′)$_2$–Atto 488 conjugate. The second protein will be labeled with a specific primary antibody, HRP-conjugated secondary, and subsequently reacted with TSA–Cy5 reagent. When proteins X and Y are more than 10 nm apart, FRET cannot take place between the donor (Atto-488) and acceptor (Cy5) chromophores. However, when X and Y interact, they will be sufficiently close for FRET to occur between the Atto-488 and Cy5 chromophores. This results in a reduction in the donor chromophore lifetime

3.5 Two-Photon FLIM of Labeled Eggs

1. Place the glass slide on the stage of a Nikon TE2000-E inverted microscope (or equivalent).
2. Using the Hg lamp and appropriate filter sets for Hoechst/DAPI, select an egg in which the male and female pronuclei are in the same focal plane and acquire image (Fig. 2a).

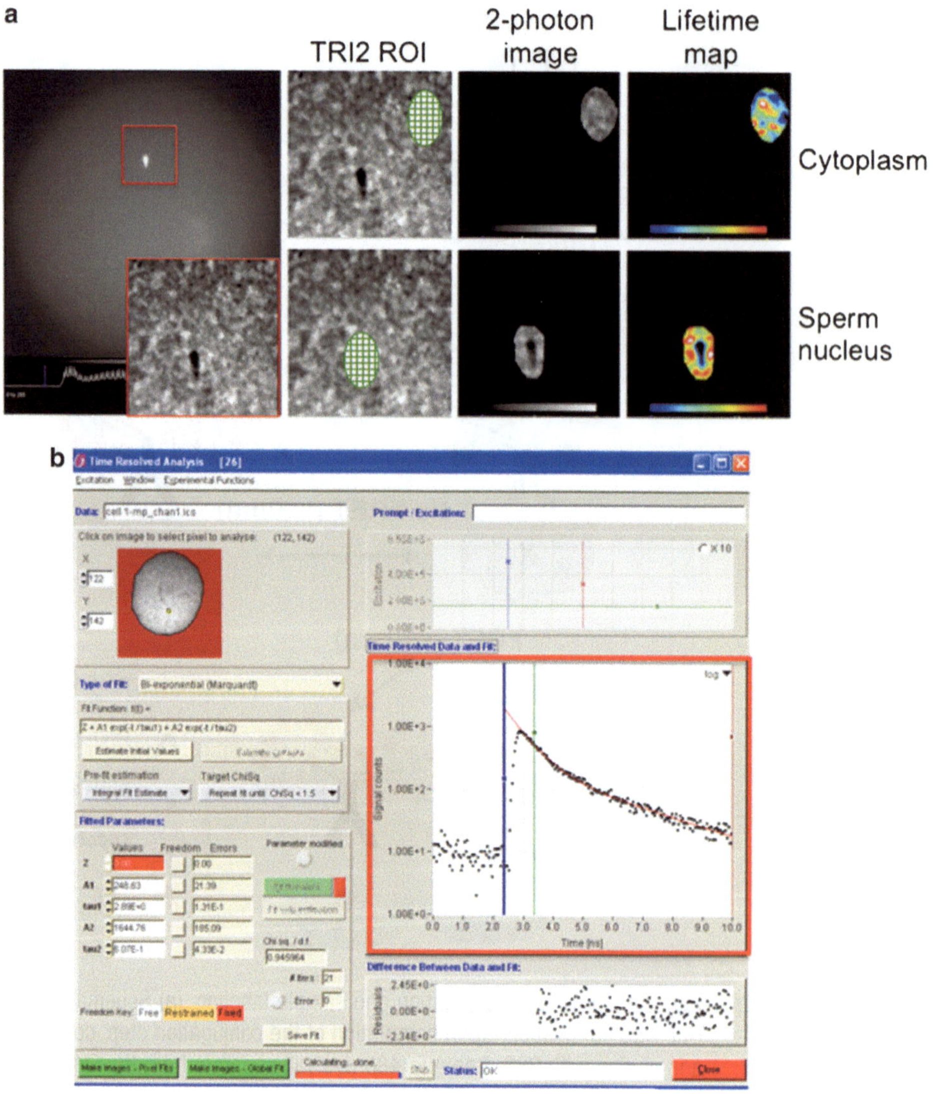

Fig. 2 Example of data acquired during two-photon FLIM to measure FRET. (**a**) A sea urchin egg fixed shortly after fertilization. The *left-hand panel* shows an image acquired with the Hg source using the DAPI filter set. The Hoechst-stained sperm nucleus is visible and having just penetrated the egg is conical in shape. The *inset* shows a magnified image of the sperm nucleus, acquired with a 920 nm laser line, used to excite the primary antibody–Atto 488 chromophore. Upon opening this two-photon image in TRI2, an ROI can be drawn as required (*green hashed oval*). The second column of images shows two ROIs, one around the sperm nucleus, the other in the cytoplasm. These specific ROIs (third column) are used to curve fit the chromophore decay and derive an average lifetime (τ) value for the ROI. The lifetime can be illustrated as a heatmap (*right-hand column*), showing lifetime distributions within the ROI. (**b**) An example of a bi-exponential Marquardt curve fit to a whole egg in TRI2. The initial image of an egg labeled with antibody–Atto 488 is shown in the *top left corner*, and the curve fit is highlighted in the *red box* (*right*). A curve fit value as close to 1.00 as possible is desirable

3. Acquire image of the donor (Atto 488) and acceptor (Cy5) with the Hg lamp (*see* **Note 15**).
4. Excite samples with the 920 nm laser line of a 6 W Mira laser at a power of 27 mW, and subject to a maximum of 800 scans (*see* **Note 16**).
5. Use TRI2 to calculate the donor chromophore (Atto488) lifetime (τ). Fit the photon intensity/time plot with a bi-exponential Marquardt curve (Fig. 2b).

4 Notes

1. PFA has a toxic vapor. Adding the powder to hot MPSW should only be performed in a fume hood.
2. This software is available from Paul Barber, Dept. of Oncology, University of Oxford.
3. The perioral cavity is the soft membrane that surrounds the sea urchin mouth.
4. Animals shed gametes from the gonopores on the aboral (top) surface.
5. Egg jelly can also be removed with a brief (1 min) wash in pH 5.0 MPSW, followed by two washes in MPSW.
6. This step should be performed approximately 5 min prior to fertilization to allow ATA to permeate the eggs.
7. This force of centrifugation will allow collection of the eggs, but the sperm present will be removed in the supernatant.
8. The upper slower moving band of color (usually brighter) is the unconjugated dye. This can also be eluted for quality control purposes.
9. Conjugates should have a dye/protein ratio of <3:1.
10. This solution is best made fresh on the day of the sample preparation.
11. When solubilized in PBS, sodium borohydride should effervesce. If not, effective quenching will not occur.
12. Primary and secondary antibody dilutions should be first optimized by confocal microscopy.
13. Given that the TSA step is an amplification one, it is very important to optimize the required dilution of the HRP secondary antibody and incubation period with TSA-Cy5 to avoid creating a nonspecific signal.
14. Poly-L-lysine-coated coverslips are made fresh on the day of sample preparation.

15. It is important that an acceptor image is acquired for every egg analyzed to ensure sample-to-sample consistency in acceptor concentration. The donor image is used as a before-and-after test to ensure no excessive photobleaching occurs during the acquisition.
16. 800 scans is enough to collect a sufficient number of photons (>800) for a bi-exponential Marquardt curve fit. The 920 nm excitation does not directly excite Cy5. The scanning number was controlled to ensure no photobleaching occurred (*see* **Note 15**).

Acknowledgments

We are grateful to Paul Davies (Macrae and Co.) for sea urchin procurement, Selvaraju Veeriah for optimization of the TSA protocol, and Christopher Applebee for sample acquisitions.

References

1. Stöckl MT, Herrmann A (2010) Detection of lipid domains in model and cell membranes by fluorescence lifetime imaging microscopy. Biochim Biophys Acta 1798:1444–1456
2. Fernandes F, Coutinho A, Prieto M et al (2015) Electrostatically driven lipid–protein interaction: answers from FRET. Biochim Biophys Acta 1848:1837–1848
3. Wallrabe H, Periasamy A (2005) Imaging protein molecules using FRET and FLIM microscopy. Curr Opin Biotechnol 16:19–27
4. Byrne RD, Veeriah S, Applebee CJ et al (2014) Conservation of proteo-lipid nuclear membrane fusion machinery during early embryogenesis. Nucleus 5:441–448
5. Poccia D, Collas P (1996) Transforming sperm nuclei into male pronuclei *in vivo* and *in vitro*. Curr Top Dev Biol 34:25–88
6. Collas P, Poccia D (1996) Distinct egg membrane vesicles differing in binding and fusion properties contribute to sea urchin male pronuclear envelopes formed *in vitro*. J Cell Sci 109:1275–1283
7. Dumas F, Byrne RD, Vincent B et al (2010) Spatial regulation of membrane fusion controlled by modification of phosphoinositides. PLoS One 5:e12208
8. Longo F (1972) An ultrastructural analysis of mitosis and cytokinesis in the zygote of the sea urchin, *Arbacia punctulata*. J Morphol 138: 207–238
9. Domart M-C, Hobday TMC, Peddie CJ et al (2012) Acute manipulation of diacylglycerol reveals roles in nuclear envelope assembly & endoplasmic reticulum morphology. PLoS One 7:e51150
10. Barona T, Byrne RD, Pettitt R et al (2005) Diacylglycerol induces fusion of nuclear envelope membrane precursor vesicles. J Biol Chem 280:41171–41177
11. Byrne RD, Garnier-lhomme M, Han K et al (2007) PLCγ is enriched on poly-phosphoinositide-rich vesicles to control nuclear envelope assembly. Cell Signal 19:913–922
12. Byrne RD, Zhendre V, Larijani B et al (2009) Nuclear envelope formation *in vitro*: a sea urchin egg cell-free system. Methods Mol Biol 464:207–223
13. Byrne RD, Larijani B, Poccia DL (2009) Tyrosine kinase regulation of nuclear envelope assembly. Adv Enzyme Regul 49:148–156
14. Byrne RD, Applebee C, Poccia DL et al (2012) Dynamics of PLCγ and Src family kinase 1 interactions during nuclear envelope formation revealed by FRET-FLIM. PLoS One 7:e40669
15. Calleja V, Alcor D, Laguerre M et al (2007) Intramolecular and intermolecular interactions of protein kinase B define its activation *in vivo*. PLoS Biol 5:e95
16. Calleja V, Laguerre M, Parker PJ et al (2009) Role of a novel PH-kinase domain interface in PKB/Akt regulation: structural mechanism for allosteric inhibition. PLoS Biol 7:e17
17. Veeriah S, Leboucher P, de Naurois J et al (2014) High throughput time-resolved-FRET reveals Akt/PKB activation as a poor prognostic marker in breast cancer. Cancer Res 74: 4983–4995

Chapter 8

Identifying Protein-Protein Associations at the Nuclear Envelope with BioID

Dae In Kim, Samuel C. Jensen, and Kyle J. Roux

Abstract

The nuclear envelope (NE) is a critical cellular structure whose constituents and roles in a myriad of cellular processes seem ever expanding. To determine the underlying mechanisms by which the NE constituents participate in various cellular events, it is necessary to understand the nature of their protein-protein associations. BioID (proximity-dependent *bio*tin *id*entification) is a recently established method to generate a history of protein-protein associations as they occur over time in living cells. BioID is based on fusion of a bait protein to a promiscuous biotin ligase. Expression of the BioID fusion protein in a relevant cellular environment enables biotinylation of vicinal and interacting proteins of the bait protein, permitting isolation and identification by conventional biotin-affinity capture and mass-spec analysis. In this way, BioID provides unique capabilities to identify protein-protein associations at the NE. In this chapter we provide a detailed protocol for the application of BioID to the study of NE proteins.

Key words BioID, Biotinylation, Protein-protein interactions, Nuclear envelope, Nuclear lamina, Nuclear pore complex

1 Introduction

The nuclear envelope (NE) is a highly organized organelle consisting of the inner nuclear membrane (INM) and outer nuclear membrane (ONM), separated by a lumen called the perinuclear space (PNS). The ONM and INM are connected to form annular junctions of high membrane curvature where large protein complexes, the nuclear pore complexes (NPCs), regulate nucleo-cytoplasmic trafficking. In Metazoa a protein meshwork called the nuclear lamina underlies the INM. Consisting of type V intermediate filament proteins called lamins, the nuclear lamina serves as a structural framework for many NE-associated proteins as well as elements of chromatin. Enriched in the INM is a class of transmembrane proteins called NETs (*n*uclear *e*nvelope *t*ransmembrane proteins), retained at the NE in part through associations with nuclear proteins. Spanning the NE is the LINC complex (*l*inker of *n*ucleoskeleton and *c*ytoskeleton) that physically couples the cytoskeleton to the

Sue Shackleton et al. (eds.), *The Nuclear Envelope: Methods and Protocols*, Methods in Molecular Biology, vol. 1411,
DOI 10.1007/978-1-4939-3530-7_8, © Springer Science+Business Media New York 2016

nucleus. This complex is formed by interactions in the PNS lumen between INM SUN-domain proteins bound to nucleoskeletal proteins and ONM resident KASH-domain proteins that interact with elements of the cytoskeleton. An overview of key NE structures is depicted in Fig. 1.

The NE plays important roles in various cellular mechanisms including nuclear positioning and migration (reviewed in ref. 1), chromatin organization (reviewed in ref. 2), transcriptional regulation (reviewed in ref. 3), cell signaling (reviewed in ref. 4), cytoskeleton organization (reviewed in ref. 5), and mechanotransduction (reviewed in ref. 6). Thus, perhaps not surprisingly, mutations in many constituents of the NE lead to a myriad of human diseases that impact a variety of cells and tissues (reviewed in ref. 7). An important goal is to define the interactions of the various protein networks at the NE, which will help determine the mechanisms of human diseases associated with the NE and shed light on the various normal cellular functions of the NE.

There are currently two primary approaches widely used to identify protein-protein interactions (PPIs): yeast two-hybrid (Y2H, or protein complementation) and affinity-protein complex purification. Fundamentally, Y2H utilizes bait (a protein of interest) and prey (binding partners) each conjugated with a fragment of a functional protein. Upon direct binding of the bait and prey, the fragments are joined together to create a functional protein that generates a selectable biological effect. Y2H is a powerful method for monitoring PPIs since it can be employed for genome-wide screening and has the potential to identify weak PPIs. However, Y2H frequently suffers from high false-positive and false-negative rates, perhaps in part because it often requires using fragments of proteins as bait and/or prey and screens them in an unnatural setting [8]. Protein complex purification relies on the solubilization of stable intact protein complexes to allow for their selective enrichment by affinity purification. However, this method needs to be optimized to balance the solubility and stability of the complexes. Transient and weak interactors can easily be lost by harsh lysis/wash conditions while false positives are increased with milder lysis/wash conditions. Given that NE proteins are frequently insoluble under more physiological conditions, often due to direct/indirect association with the nuclear lamina and/or membrane integration, it can be difficult to accomplish solubilization of meaningful protein complexes. To overcome the solubility issue of the NE proteins, chemical crosslinking has been employed to the protein complex purification approach [9]. However, the crosslinking approach intrinsically has a potential drawback to complicate interpretation of the identified candidates since they might just exist in a larger meshwork of protein associations. In addition, protein complex purification permits only a snapshot of PPIs at the time of lysis which can mask meaningful associations that are

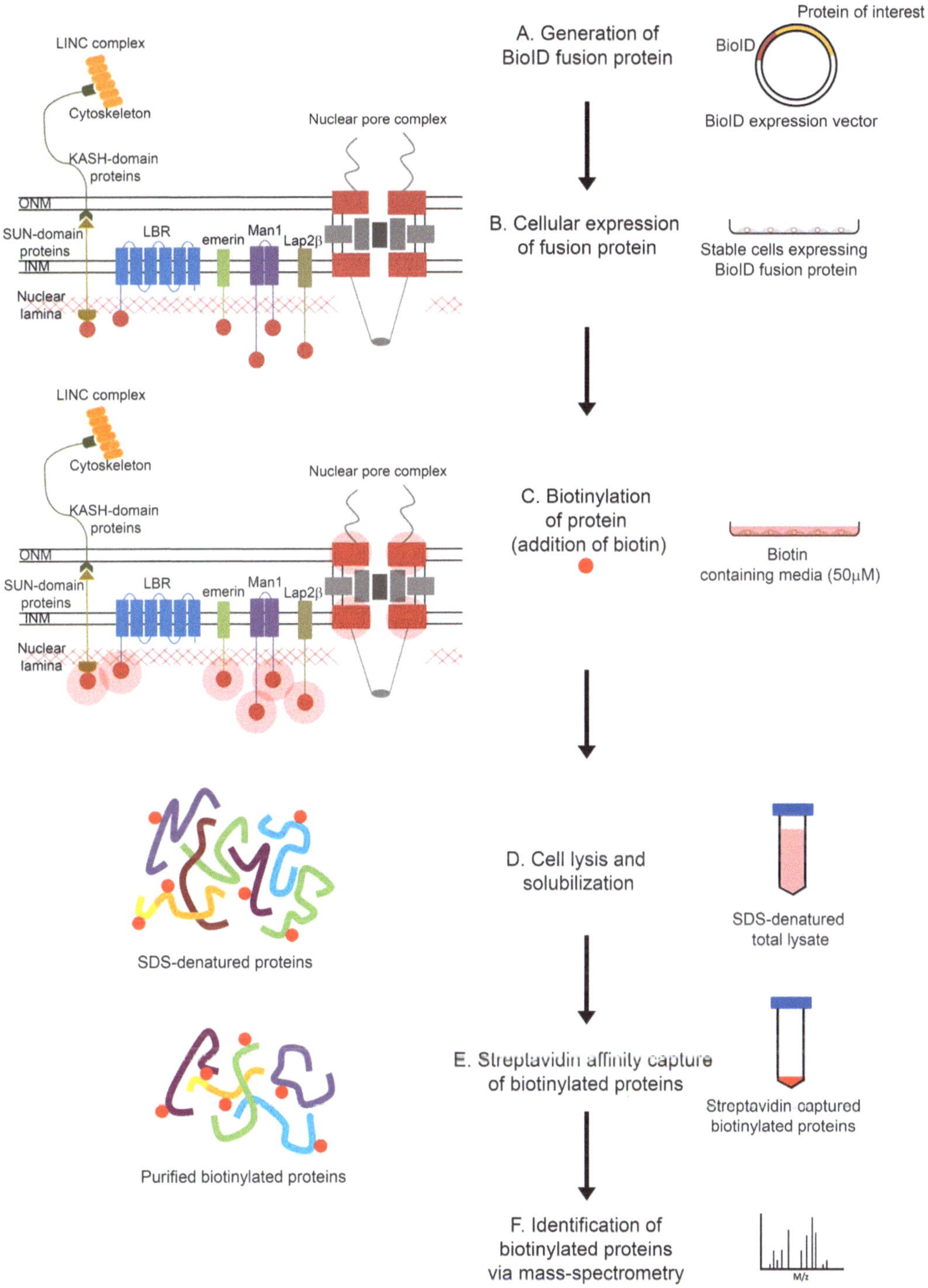

Fig. 1 Overall scheme of the BioID method at the NE. The status of BioID fusion proteins and biotinylated proteins is depicted on the *left*. Experimental procedures are depicted on the *right*. The NE proteins that have been analyzed by BioID are indicated with BioID fusion (*red*). Nup107-160 complex at the NPC is colored in *red*. (*A*) Generate expression BioID vector. (*B*) Stably express BioID fusion in cells. (*C*) Induce biotinylation (*pink circle*) by adding 1× biotin. (*D*) Lyse stable cells. At this point, cell lysate contains mixture of biotinylated proteins and non-biotinylated proteins. (*E*) Isolate biotinylated proteins using biotin-affinity capture. (*F*) Identify isolated proteins using mass spectrometry

transient or of low frequency but nevertheless biologically relevant. Collectively, the limitations with existing approaches to study PPIs stimulated development of a new approach.

BioID (for proximity-dependent *bio*tin *id*entification) was developed to circumvent some of the limitations of current methods for PPI detection and provide a complementary approach [10]. Based on cellular expression of a protein of interest fused with a promiscuous bacterial biotin ligase, BioID enables biotinylation of proximal proteins over time in living cells. Proteins biotinylated in this way can be selectively analyzed using conventional biotin-affinity capture methods followed by mass spectrometry (Fig. 1). In *E. coli*, biotin ligase (BirA) biotinylates acetyl-CoA carboxylase and regulates transcriptional activity of the biotin operon [11]. To this end, BirA generates and delivers reactive biotin (biotinoyl-5′-AMP) to a specific substrate. BioID overcomes this specificity to generate a promiscuous biotinylation by utilizing a mutant BirA (R118G) that lacks DNA-binding ability and has substantially decreased affinity to reactive biotin, leading to its release and ability to react with primary amines on proximate proteins [12, 13]. Using a stable subcomplex of the NPC as a molecular ruler, the practical biotinylation range of BioID appears to be ~10 nm [14].

The BioID method provides practical advantages over traditional methods in PPI detection. First, BioID can overcome the solubility issue since it allows for biotin labeling to occur in living cells prior to cell lysis. Secondly, since the BioID fusion proteins can theoretically be expressed in any cells, PPIs can be screened in more relevant contexts, thus reducing false positives/negatives. In addition, since the biotin labeling occurs over time to generate a history of protein associations, BioID has the potential to detect weak PPIs and transient PPIs. However, BioID is not without limitations. Like other methods, the candidates identified by BioID need to be validated in order to determine the nature of their association. Since the BioID system is known to biotinylate primary amines of lysines, these posttranslational modifications may impact the functions of both bait and candidates. Perhaps the most appreciated limitation of BioID is that this system relies on overexpression of BioID fusion proteins. Abnormally expressed fusion proteins may induce aberrant localization, something especially relevant for INM proteins that have nucleoplasmic-domain size constraints to permit proper targeting [15]. To improve the targeting ability of BioID, the Roux lab has developed a smaller variant of a promiscuous biotin ligase (manuscript in press).

Due to its unique capabilities, BioID is becoming a widely accepted method to identify PPIs. At the NE, BioID has been used for the nuclear lamina [10], NPC [14], isoforms of lamins (A, B1, B2, and C), and other INM proteins including LBR, emerin, TMPO, MAN1, SUN1, SUN2, and LAP1 (unpublished data, Kim and Roux; Fig. 1). At the time we prepared this manuscript, BioID has also been applied to the study of PPIs at chromatin [16], cell junctions [17–22], centrosomes [23, 24], Hippo signaling [25], HIV-host cell interactions [26], *Chlamydia psittaci*-host cell interactions [27], and in cell culture and mouse xenografted tumors to study c-Myc [28]. In addition to these mammalian applications, BioID has been used to study protein associations in *Trypanosoma brucei* [29] and *Toxoplasma gondii* [30]. Here we provide a detailed protocol for the application of BioID to the study of the NE.

2 Materials

All reagents should be prepared using autoclaved DNAse/RNase-free tubes while wearing gloves to minimize excess keratin contamination. It is also suggested that ultrapure water be used when preparing the various solutions. It is recommended to store reagents at room temperature (RT) unless otherwise specified.

2.1 Induction of Biotinylation

1. 20× biotin: 1 mM biotin. Bring final volume up with serum-free medium, sterilize with 0.22 μm filter. Store at 4 °C for up to 8 weeks (*see* **Note 1**).

2.2 BioID Pulldown

1. Lysis buffer: 50 mM Tris (pH 7.4), 500 mM NaCl, 0.2 % SDS (w/v), store up to 2 weeks. Freshly add 1 mM DTT and 1× combined protease inhibitor solution at the time of cell lysis (*see* **Note 2**).
2. Wash buffer 1: 2 % SDS (w/v). Store up to 2 weeks (*see* **Note 2**).
3. Wash buffer 2: 0.1 % (w/v) deoxycholic acid, 1 % (w/v) Triton X-100 (TX-100), 1 mM EDTA, 500 mM NaCl, 50 mM Hepes (pH 7.5). Store up to 2 weeks (*see* **Note 3**).
4. Wash buffer 3: 250 mM LiCl, 10 mM Tris (pH 7.4), 1 mM EDTA, 0.5 % (w/v) NP-40, 0.5 % (w/v) deoxycholic acid. Store up to 2 weeks (*see* **Note 3**).
5. Tris wash buffer: 50 mM Tris (pH 7.4).
6. Ammonium bicarbonate (NH_4HCO_3): 50 mM ammonium bicarbonate. Store at 4 °C for up to 2 weeks.
7. Dynabeads® MyOne™ Streptavidin C1 (Life Technologies).

2.3 Validation of Biotinylation

1. SDS-PAGE sample buffer: 50 mM Tris–Cl (pH 6.8), 12 % (w/v) sucrose, 2 % (w/v) SDS, 20 mM dithiothreitol (DTT), 0.004 % bromophenol blue. Make fresh.
2. BSA blocking buffer: 1 % (w/v) bovine serum albumin, fraction V, 0.2 % (w/v) TX-100. Bring final volume up with 1× PBS. Store at 4 °C for up to 2 weeks.
3. ABS blocking buffer: 10 % (w/v) adult bovine serum, 1 % (v/v) TX-100. Bring final volume up with 1× PBS. Store at 4 °C for up to 2 weeks.

4. Quenching solution: 3 % (v/v) hydrogen peroxide. Bring the final volume up with 1× PBS.
5. PFA fixing buffer: 3 % (w/v) paraformaldehyde (PFA), bring the final volume up with 1× PBS. Dissolve PFA while heating at 80 °C in a fume hood. Once the buffer is cooled, add $MgCl_2$ and $CaCl_2$ each to 0.1 M final concentration. Store at 4 °C.

3 Methods

All steps should be conducted at room temperature unless otherwise specified. To minimize keratin contamination, wear gloves and use DNase/RNase-free materials.

3.1 Generation of BioID Fusion Protein

1. It is critical not to disrupt any known targeting motifs and post-translational modifications by the fusion of the BioID ligase. If GFP has previously been fused to the bait (BioID is about the size of GFP) and shown not to perturb targeting and/or function, then this can inform BioID placement (*see* **Note 4**).
2. Cloning a BioID fusion protein typically requires PCR. Using recombinational cloning techniques, PCR products can be inserted to the BioID vectors without restriction enzyme digestions. This technique provides more freedom to choose restriction enzymes in the vector.
3. The biotinylation range can be modulated by flexible linkers inserted between the BioID ligase and bait proteins (e.g., BioID-flexible linker-bait). The length of the flexible linker may vary depending on the goals of the experiment. Addition of short flexible linkers may also relieve any steric hindrance between the bait and BioID ligase.
4. If ectopic expression of the fusion protein needs to be inducibly regulated due to unwanted biological effects, consider using an inducible expression system. The BioID system itself is functional without inducible expression since biotinylation is regulated by the addition of excess biotin.

3.2 Validation of BioID Fusion Protein

1. Assess the expression of the BioID fusion protein and biotinylation of endogenous proteins using transient transfection. Since abnormal expression may lead to aberrant targeting or aggregation of the fusion protein, care must be taken to monitor the localization and biotinylation capability of the fusion protein. Often proper targeting may not become clear until a lower level of expression is obtained by stable expression.
2. Plate the target cells, including non-transfected (or mock transfected) controls. Deliver the BioID construct(s) to the cells using an appropriate high-efficiency transfection reagent. Apply 20× biotin to a final concentration of 50 μM (1×) at the

time of transfection. If the transfection protocol requires media change, re-add 1× biotin at the time of media change.

3. After overnight incubation, analyze the cells by immunofluorescence (IF) and Western blot (WB) (Fig. 2) (*see* **Note 5**).
4. For IF, fix the cells for 10 min using 3 % PFA fixing buffer, followed by permeabilization with 0.4 % TX-100 for 15 min.

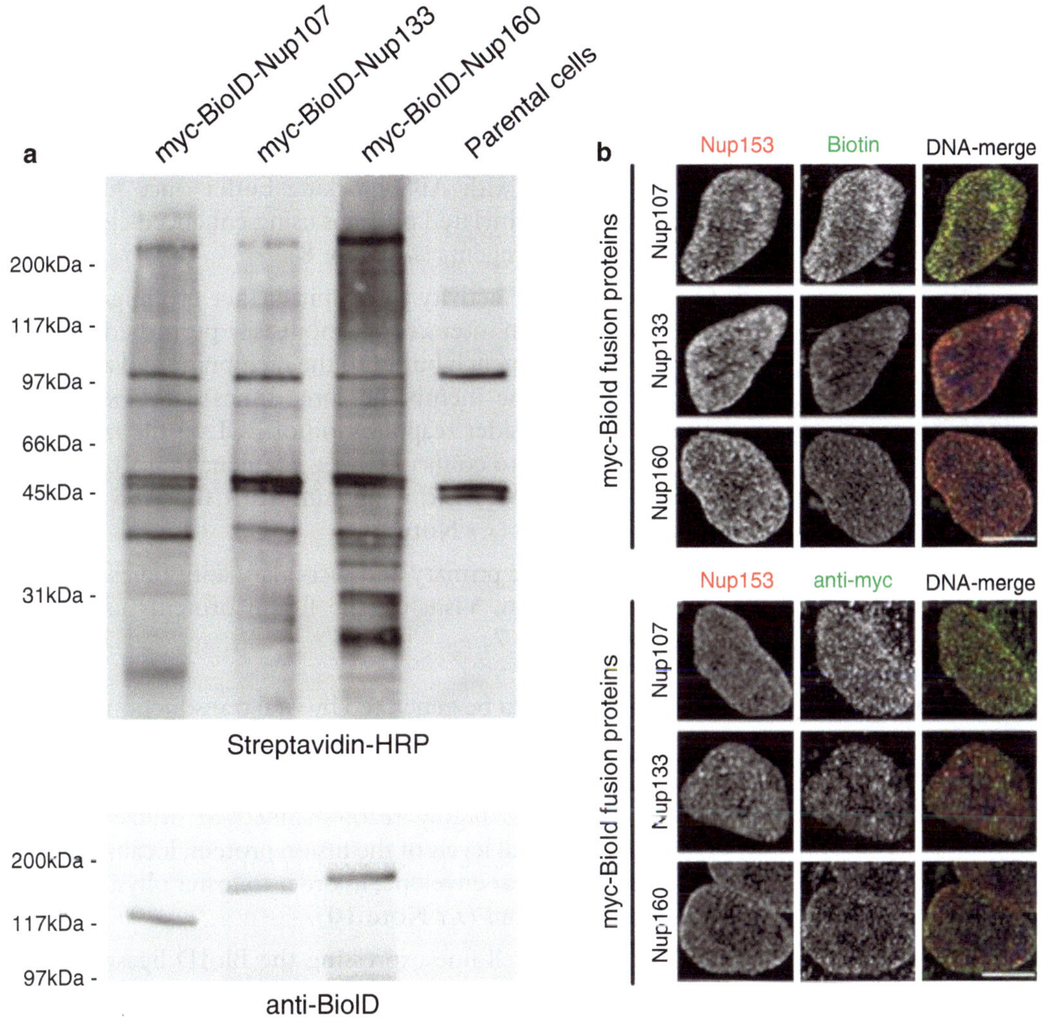

Fig. 2 Validation of BioID fusion proteins. HEK293 cells stably expressing myc-BioID fused with the NPC constituents Nup107, Nup133, and Nup160 were analyzed using WB and IF. (**a**) Following SDS-PAGE, biotinylated proteins were probed with Streptavidin-HRP (*top*). The same blot was quenched and subsequently probed with anti-BioID antibody (*bottom*). Compared to the few naturally biotinylated proteins in the cells that do not express BioID proteins (parental), extensive biotinylation was observed from the lysate of cells expressing BioID fusion protein. (**b**) Biotinylation and localization of the BioID fusion proteins were monitored by confocal microscopy. Biotinylated proteins were labeled with streptavidin-Alexa Fluor 488 (*green*, *top*). In parallel, the BioID fusion proteins were detected with anti-myc antibody (*green*, *bottom*). Biotinylated proteins and the fusion protein colocalizes with the NPC, labeled with anti-Nup153 antibody (*red*). DNA was detected with Hoechst dye (*blue*). Scale bar is 10 μm

Label the fusion protein with appropriate primary and secondary antibodies. Detect biotinylated proteins using fluorescently labeled streptavidin (Streptavidin–Alexa Fluor). To label the nucleus, add fluorescent DNA-binding reagents (*see* **Notes 6** and 7).

5. For WB, wash the cells briefly with PBS to remove BSA in the culture medium. Lyse the cells using SDS sample buffer. Boil the samples at 95 °C for 5 min and subsequently sonicate the samples to shear the genomic DNA. Perform SDS-PAGE and transfer to nitrocellulose membranes. After transfer, incubate the membrane with BSA blocking buffer for 20 min while agitating the membrane. Incubate the membrane with streptavidin-HRP for 40 min. Wash membrane with PBS twice for 5 min each and with ABS blocking buffer once for 5 min. Visualize the biotinylated proteins using enhanced chemiluminescence (ECL) regents (*see* **Note 8**).
6. Quench the HRP activity to permit further probing since the streptavidin-biotin interaction is not easily perturbed by stripping. Apply the quench buffer to the membrane and agitate for 20 min. Wash the membrane three times with 1× PBS for 5 min each. Consider reapplication of ECL to the membrane and visualization to confirm success of quenching. It is recommended to incubate the membrane with the ABS blocking buffer for 20 min (*see* **Note 9**).
7. Apply appropriate primary and secondary antibodies to detect the fusion protein. Visualize the fusion proteins using ECL regents (*see* **Note** 7).

3.3 Stable Cell Line Generation

1. Stable cell lines can be generated by viral transduction or clonal selection, dependent on cell type and expression vector.
2. Carefully monitor the expression level of the fusion proteins by IF and WB. It is highly recommended to utilize the cells expressing minimal levels of the fusion protein, localized exclusively to the nuclear envelope, in order to foster physiologically relevant associations (*see* **Note 10**).
3. Also generate a cell line expressing the BioID ligase only for use as a control.

3.4 Induction of Biotinylation

1. The following BioID pulldown protocol is optimized for 4×10^7 HEK293 cells (4 × 10 cm culture dishes) per pulldown. Cell numbers can be scaled depending on the purpose and the resolution of pulldown; however, the buffer volumes should also be scaled accordingly.
2. Set up the cells stably expressing BioID fusion protein. In parallel, set up control cells expressing the BioID ligase only.

3. Once these cells reach 80 % confluence, apply 20× biotin to a final concentration of 50 μM (1×).
4. Incubate the cells for 16–18 h in 1× biotin (*see* **Note 5**).

3.5 BioID Pulldown

1. Before starting lysis make sure all the reagents and equipment are readily set up.
2. Remove medium from the cells by aspiration and rinse the cells in each dish twice with 3 ml of fresh autoclaved 1× PBS. Rock the cells back and forth with 1× PBS prior to aspiration. Remove 1× PBS completely prior to lysis (*see* **Note 11**).
3. Apply 540 μL of lysis buffer to each plate of the cells.
4. Collect the cells by scraping the bottom of the dish and pool the lysates from the four culture dishes into a single 15 ml conical tube.
5. Add 240 μL of 20 % TX-100 and mix briefly by vortexing (*see* **Note 12**).
6. Immediately place the 15 ml conical tube on ice and keep on ice until **step 14**.
7. Sonicate the sample. If the sample is still cloudy, sonicate the sample until the sample is mostly clear (*see* **Note 13**).
8. Once the sample is clear, add 2.4 mL of 4 °C 50 mM Tris–HCl (pH 7.4) (*see* **Note 14**).
9. Sonicate once with the same conditions as **step 9** and evenly aliquot the lysate into three prechilled 2 mL microcentrifuge tubes (approximately 1.8 mL per tube).
10. Spin the samples at 16,500 × *g* for 10 min at 4 °C.
11. During the centrifugation, place new 15 mL conical tubes on ice to prechill.
12. Once the centrifugation is finished, transfer and pool together the supernatants from the three microcentrifuge tubes into the prechilled 15 mL conical tubes prepared at **step 13**. Do not disturb any insoluble pellet (if present) at the tube wall when transferring the supernatant (*see* **Note 15**).
13. In parallel, prepare Dynabeads® MyOne™ Streptavidin C1 for capture of biotinylated proteins. Place 1.5 mL microcentrifuge tubes in a magnetic stand and add 500 μL of lysis buffer and 500 μL of 50 mM Tris–HCl (pH 7.4) (*see* **Note 16**).
14. Resuspend the Dynabeads well. Once evenly distributed in the bottle, take 300 μL of the Dynabeads and transfer to the 1.5 mL microcentrifuge tubes set up at **step 15**. Wait for 3 min for the magnetic beads to be drawn to the wall of the tube (*see* **Note 17**).

15. Carefully remove the supernatant from the 1.5 mL microcentrifuge tubes prepared at **step 16**. Avoid disturbing the magnetic beads on the tube wall. Take 1 mL of lysate from the 15 mL conical tube from step 14 and resuspend the magnetic beads. Transfer the combined lysate and resuspended Dynabeads back to the original 15 ml conical tube. Do not dry out Dynabeads.
16. Recap the 15 mL conical tube and wrap the top with parafilm.
17. Incubate the magnetic beads on rotator overnight (~18 h) at 4 °C (*see* **Note 18**).
18. After incubation with the Dynabeads, place the 15 mL conical tubes onto the magnetic stand and allow the magnetic beads to be collected on the side of the tube for 3 min.
19. Remove the supernatant, paying close attention to not disturb the magnetic beads on the side of the 15 mL conical tube.
20. Resuspend the magnetic beads in 1 mL of wash buffer 1 and transfer from the 15 mL conical to a 1.5 mL microcentrifuge tube.
21. Wash the samples on a rotator at RT for 8 min.
22. Place tubes on the magnetic stand for 3 min and remove thc supernatant. Try not to disturb the magnetic beads at the tube wall (*see* **Note 19**).
23. Add wash buffer 1 and resuspend the magnetic beads with gentle pipetting.
24. Repeat **steps 24** and **25**.
25. Resuspend the magnetic beads with 1 mL of wash buffer 2.
26. Repeat **steps 24** and **25**.
27. Resuspend the magnetic beads with 1 mL of wash buffer 3.
28. Repeat **steps 24** and **25**.
29. Resuspend the beads with 1 mL of 50 mM Tris–HCl (pH 7.4).
30. During this step, take 100 μL (10 %) of the resuspended mixture and transfer it to a newly labeled 1.5 mL microcentrifuge tube. This will be reserved for WB analysis to monitor success of pulldown. The remaining 900 μL sample will be used for mass-spec analysis.
31. Spin all tubes, including the 10 % samples, at 2000 × *g* for 5 min to pellet the beads.
32. Remove supernatant from the samples for mass-spec analysis. Resuspend the magnetic beads with 150 μL of ammonium bicarbonate buffer. Quickly flash-freeze the samples in liquid nitrogen and store at −80 °C (*see* **Note 20**).
33. Analyze the BioID sample using mass-spec analysis following on-bead digestion (*see* **Note 21**).

3.6 Validation of BioID Pulldown

1. Perform WB (*see* Subheading 3.2) on the 10 % samples to ensure that the fusion proteins are present and biotinylation has occurred before sending the frozen samples off for mass spectrometry analysis.
2. Remove supernatant from the 10 % samples. Resuspend the magnetic beads with 100 μL of fresh sample buffer and boil at 95 °C for 5 min.
3. Spin the sample to separate the magnetic beads and SDS sample buffer. Do not disturb the magnetic beads when removing supernatant.
4. Perform WB analysis as described in Subheading 3.2, **steps 6–8**.

4 Notes

1. Use cell culture grade biotin. Dissolve biotin with pipetting or agitation.
2. SDS often precipitates below 25 °C. Warm the solutions at 37 °C to dissolve SDS again.
3. Deoxycholic acid precipitates under pH 7.1. Using a stock solution of deoxycholic acid (10 %) is recommended to make wash buffer 2 and 3. Deoxycholic acid must be protected from light.
4. The mammalian expression systems for N-terminal and C-terminal BioID are available at Addgene https://www.addgene.org/Kyle_Roux/.
5. Biotinylation generally reaches its maximum level approximately 16 h post incubation with 1× biotin solution. However, biotinylation can be observed after 2 h of incubation at a considerably reduced level. Biotin incubation time may vary depending on the purpose of experiment, abundance of the fusion protein, and nature of protein interactions.
6. Methanol fixation increases mitochondria background with fluorescently labeled streptavidin, likely by exposing the naturally biotinylated mitochondrial carboxylases.
7. The N-terminal BioID vector was designed with a myc-tag (myc-BioID) while the C-terminal one has a HA-tag (BioID-HA). Tagging peptides can be re-engineered if necessary. Fusion proteins can also be detected using a commercially available BirA antibody (BioFront Technologies).
8. A brief wash with high salt (300 mM in 1× PBS) after ABS wash can reduce background.
9. Since the quenching solution inactivates the HRP activity, it is important to remove the quenching solutions completely.

The residual quenching solution might impair the sensitivity of HRP-conjugated antibodies in further probings.

10. It is recommended to check the functionality of the fusion protein (e.g., subcellular targeting, biotinylation, activity if known, etc.).
11. This step ensures that any excess biotin is removed from the cells. Residual free biotin binds streptavidin magnetic beads, impairing the efficiency of the BioID pulldown. Remove residual PBS efficiently to prevent increase of the pulldown volume.
12. Adding a fivefold excess of Triton X-100 dilutes out the SDS and prevents its precipitation at 4 °C.
13. Sonication condition may vary depending on sonicators. Make sure that the sonicator's probe tip is just above the bottom of the 15 ml conical tube and centered. Prolonged sonication will generate heat; therefore make sure to keep the sample/tube in ice during sonication.
14. Dilution with 50 mM Tris (pH 7.4) will provide more favorable conditions for biotin capture.
15. The combined lysate can be quickly frozen with liquid nitrogen and stored at −80 °C if one wishes to capture biotinylated peptides at a later date.
16. Any streptavidin coupled with diverse matrix can be used for capturing biotinylated proteins, but further optimization might be needed.
17. This step is to remove azide from the Dynabeads storage solution and to equilibrate the magnetic beads with the conditions for affinity capture.
18. Incubation time may vary depending on the purpose of experiment. Shorter incubation times yield fewer candidates.
19. It is recommended to use a microcentrifuge tube opener to minimize keratin contamination.
20. Magnetic beads may be resuspended in different solutions depending of the analysis method. This protocol is optimized for on-bead digestion followed by 1D-LC MS/MS [10, 14]. At this point, the magnetic bead-containing samples can be stored at −80 °C.
21. Candidates can be identified by MS using proteins separated by SDS-PAGE. However, it is recommended to utilize on-bead digestion to bypass the difficulty of releasing biotinylated proteins from the biotin-capturing molecule coated materials. The on-bead digestion approach can also minimize keratin contamination introduced during SDS-PAGE. There are five recognized abundant naturally biotinylated proteins in eukaryotes: pyruvate carboxylase (PC), methylcrotonoyl-CoA carboxylase

subunit alpha (MCC1), acetyl-CoA carboxylase 1(ACACA), acetyl-CoA carboxylase 1(ACACB), and propionyl-CoA carboxylase alpha chain (PCCA). There are some commonly identified background proteins from BioID. The use of a BioID-only control will permit their identification. It is noted that number and abundance of candidates identified by BioID do not imply biological relevance. The candidates need to be validated by other methods.

Acknowledgments

These studies were supported by grants RO1GM102203, RO1GM102486, and RO1EB014869 (to K.J.R.) from the National Institutes of Health and by Sanford Research startup funds (K.J.R.)

References

1. Burke B, Roux KJ (2009) Nuclei take a position: managing nuclear location. Dev Cell 17:587–597
2. Wong X, Luperchio TR, Reddy KL (2014) NET gains and losses: the role of changing nuclear envelope proteomes in genome regulation. Curr Opin Cell Biol 28:105–120
3. Talamas JA, Capelson M (2015) Nuclear envelope and genome interactions in cell fate. Front Genet 6:95
4. Choi JC, Worman HJ (2014) Nuclear envelope regulation of signaling cascades. Adv Exp Med Biol 773:187–206
5. Kim DI, Kc B, Roux KJ (2015) Making the LINC: SUN and KASH protein interactions. Biol Chem 396:295–310
6. Osmanagic-Myers S, Dechat T, Foisner R (2015) Lamins at the crossroads of mechanosignaling. Genes Dev 29:225–237
7. Worman HJ, Foisner R (2010) The nuclear envelope from basic biology to therapy. Biochem Soc Trans 38:253–256
8. Bruckner A, Polge C, Lentze N, Auerbach D, Schlattner U (2009) Yeast two-hybrid, a powerful tool for systems biology. Int J Mol Sci 10:2763–2788
9. Kubben N, Voncken JW, Demmers J, Calis C, van Almen G, Pinto Y, Misteli T (2010) Identification of differential protein interactors of lamin A and progerin. Nucleus 1:513–525
10. Roux KJ, Kim DI, Raida M, Burke B (2012) A promiscuous biotin ligase fusion protein identifies proximal and interacting proteins in mammalian cells. J Cell Biol 196:801–810
11. Chapman-Smith A, Cronan JE Jr (1999) Molecular biology of biotin attachment to proteins. J Nutr 129:477S–484S
12. Choi-Rhee E, Schulman H, Cronan JE (2004) Promiscuous protein biotinylation by Escherichia coli biotin protein ligase. Protein Sci 13:3043–3050
13. Cronan JE (2005) Targeted and proximity-dependent promiscuous protein biotinylation by a mutant Escherichia coli biotin protein ligase. J Nutr Biochem 16:416–418
14. Kim DI, Birendra KC, Zhu W, Motamedchaboki K, Doye V, Roux KJ (2014) Probing nuclear pore complex architecture with proximity-dependent biotinylation. Proc Natl Acad Sci U S A 111:E2453–E2461
15. Antonin W, Ungricht R, Kutay U (2011) Traversing the NPC along the pore membrane: targeting of membrane proteins to the INM. Nucleus 2:87–91
16. Lambert JP, Tucholska M, Go C, Knight JD, Gingras AC (2015) Proximity biotinylation and affinity purification are complementary approaches for the interactome mapping of chromatin-associated protein complexes. J Proteomics 118:81–94
17. Fredriksson K, Van Itallie CM, Aponte A, Gucek M, Tietgens AJ, Anderson JM (2015) Proteomic analysis of proteins surrounding occludin and claudin-4 reveals their proximity to signaling and trafficking networks. PLoS One 10:e0117074
18. Guo Z, Neilson LJ, Zhong H, Murray PS, Zanivan S, Zaidel-Bar R (2014) E-cadherin interactome complexity and robustness

resolved by quantitative proteomics. Sci Signal 7:rs7

19. Steed E, Elbediwy A, Vacca B, Dupasquier S, Hemkemeyer SA, Suddason T, Costa AC, Beaudry JB, Zihni C, Gallagher E, Pierreux CE, Balda MS, Matter K (2014) MarvelD3 couples tight junctions to the MEKK1-JNK pathway to regulate cell behavior and survival. J Cell Biol 204:821–838
20. Ueda S, Blee AM, Macway KG, Renner DJ, Yamada S (2015) Force dependent biotinylation of myosin IIA by alpha-catenin tagged with a promiscuous biotin ligase. PLoS One 10:e0122886
21. Van Itallie CM, Aponte A, Tietgens AJ, Gucek M, Fredriksson K, Anderson JM (2013) The N and C termini of ZO-1 are surrounded by distinct proteins and functional protein networks. J Biol Chem 288:13775–13788
22. Van Itallie CM, Tietgens AJ, Aponte A, Fredriksson K, Fanning AS, Gucek M, Anderson JM (2014) Biotin ligase tagging identifies proteins proximal to E-cadherin, including lipoma preferred partner, a regulator of epithelial cell-cell and cell-substrate adhesion. J Cell Sci 127:885–895
23. Firat-Karalar EN, Rauniyar N, Yates JR 3rd, Stearns T (2014) Proximity interactions among centrosome components identify regulators of centriole duplication. Curr Biol 24:664–670
24. Comartin D, Gupta GD, Fussner E, Coyaud E, Hasegan M, Archinti M, Cheung SW, Pinchev D, Lawo S, Raught B, Bazett-Jones DP, Luders J, Pelletier L (2013) CEP120 and SPICE1 cooperate with CPAP in centriole elongation. Curr Biol 23:1360–1366
25. Couzens AL, Knight JD, Kean MJ, Teo G, Weiss A, Dunham WH, Lin ZY, Bagshaw RD, Sicheri F, Pawson T, Wrana JL, Choi H, Gingras AC (2013) Protein interaction network of the mammalian Hippo pathway reveals mechanisms of kinase-phosphatase interactions. Sci Signal 6:rs15
26. Ritchie C, Cylinder I, Platt EJ, Barklis E (2015) Analysis of HIV-1 Gag protein interactions via biotin ligase tagging. J Virol 89:3988–4001
27. Mojica SA, Hovis KM, Frieman MB, Tran B, Hsia RC, Ravel J, Jenkins-Houk C, Wilson KL, Bavoil PM (2015) SINC, a type III secreted protein of Chlamydia psittaci, targets the inner nuclear membrane of infected cells and uninfected neighbors. Mol Biol Cell 26(10): 1918–1934
28. Dingar D, Kalkat M, Chan PK, Srikumar T, Bailey SD, Tu WB, Coyaud E, Ponzielli R, Kolyar M, Jurisica I, Huang A, Lupien M, Penn LZ, Raught B (2015) BioID identifies novel c-MYC interacting partners in cultured cells and xenograft tumors. J Proteomics 118:95–111
29. Morriswood B, Havlicek K, Demmel L, Yavuz S, Sealey-Cardona M, Vidilaseris K, Anrather D, Kostan J, Djinovic-Carugo K, Roux KJ, Warren G (2013) Novel bilobe components in Trypanosoma brucei identified using proximity-dependent biotinylation. Eukaryot Cell 12:356–367
30. Chen AL, Kim EW, Toh JY, Vashisht AA, Rashoff AQ, Van C, Huang AS, Moon AS, Bell HN, Bentolila LA, Wohlschlegel JA, Bradley PJ (2015) Novel components of the Toxoplasma inner membrane complex revealed by BioID. MBio 6:e02357–02314

Chapter 9

In Situ Detection of Interactions Between Nuclear Envelope Proteins and Partners

Alice Barateau and Brigitte Buendia

Abstract

Proximity ligation assay (PLA) appears as a quick and easy technique to visualize within fixed cells the occurrence and in situ distribution of protein complexes. PLA has been validated to detect protein-protein interactions within the nuclear compartment. Here, we describe a protocol which allows the detection of interactions between A-type nuclear lamins and either LEM-domain proteins (such as emerin, integrated within the inner nuclear membrane, and LAP2α which accumulates within the nucleoplasm) or gene regulatory factors (e.g., the transcription factor SREBP1). The distinct amounts and patterns of PLA signals obtained for various complexes highlight the pertinence of using PLA to reveal in situ where and to which extent nuclear envelope proteins bind specific partners.

Key words Proximity ligation assay, Nuclear proteins, Lamins

1 Introduction

Pull-down and blot overlay assays using purified recombinant proteins produced in bacteria, rabbit reticulocyte, or wheat germ lysates routinely allow the detection of direct protein-protein interactions. In the case of A-type nuclear lamins, nuclear proteins that belong to the intermediate filament protein family, such methods have led to the identification of ~30 direct binding partners [1–4]. Still, in vitro interactions need to be validated in cells. Moreover, expression level, posttranslational modifications, and subcellular distribution of two partners of interest are expected to vary in different cell types or in particular extracellular environments, affecting the efficiency of their association to each other. Biochemical approaches such as co-immunoprecipitation (co-IP) are often used to detect interactions. Co-IP requires a significant amount of cellular material and a determination of cell lysis conditions enabling solubilization of the two partners without affecting their

Sue Shackleton et al. (eds.), *The Nuclear Envelope: Methods and Protocols*, Methods in Molecular Biology, vol. 1411,
DOI 10.1007/978-1-4939-3530-7_9, © Springer Science+Business Media New York 2016

interaction. The latter is particularly tricky in the case of nuclear lamins which, as components of the nuclear matrix, are largely resistant to biochemical extraction [5]. Lastly, although pull-down and co-IP assays are complementary, they are incomplete regarding knowledge of the subcellular site where the protein-protein interactions take place. To palliate to this limitation, fluorescence resonance energy transfer (FRET) and proximity ligation assays (PLA) can be used. In comparison to FRET which requires overexpression of two ectopic proteins with distinct fluorescent tags, e.g., [GFP-emerin–RFP-lamin A] or [GFP-emerin–RFP-LAP1] [6], [GFP-lamin A–DsRed-lamin B1] [7], or [CFP-lamin A–YFP-cfos] [8], PLA has the advantage to detect complexes composed of either ectopic proteins, such as [FLAG-lamin A–HA-SREBP1] [9] or endogenous proteins, such as [lamin A/C–SREBP1] [9], [lamin A/C–phosphorylated H2AX] [10], or [lamin B1–activated p38 MAPK] [11].

The PLA methodology was initially reported more than 10 years ago [12]. PLA consists in an adaptation of the proximity ligation technique for highly specific in situ detection coupled to a localized amplification reaction [13, 14]. A representation of the different steps leading to in situ detection of protein-protein complexes is shown in Fig. 1. PLA necessitates cell fixation, permeabilization, and incubation with a mix of two primary antibodies of different species, as for conventional immunofluorescence. In contrast to immunofluorescence however, fluorescently labeled secondary antibodies are replaced by two PLA probes. These consist of specific secondary antibodies, each with a unique short DNA strand (oligonucleotide) attached to it. When the two proteins of interest are in the same complex, proximity between the PLA probes (<40 nm) is sufficiently close to enable interaction between the oligonucleotide after addition and hybridization of two connector oligonucleotides. The latter are guided to form a circular structure covalently joined by enzymatic DNA ligation. Upon addition of the specific phi29 DNA polymerase, one of the oligonucleotides serves as a primer for rolling circle amplification (RCA). The RCA product is a randomly coiled, single-stranded chain composed of up to several hundreds of complements of the DNA circle, subsequently detected by hybridizing complementary fluorophore-labeled oligonucleotides.

Here, we describe a PLA protocol that enables in our hands successful detection of interactions between A-type lamins and either "lamin/emerin/MAN1" (LEM)-domain proteins such as emerin which is integrated within the inner nuclear membrane, LAP2α which accumulates in the nucleoplasm [15], or gene regulatory factors such as the transcription factor SREBP1 [9].

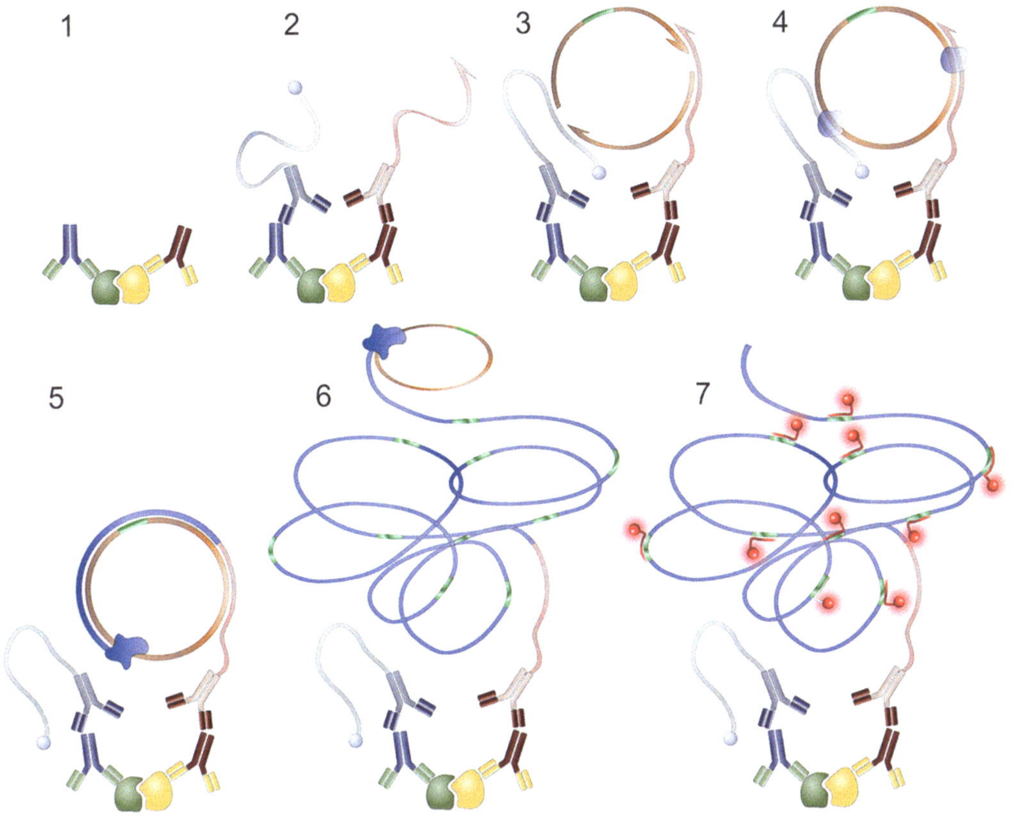

Fig. 1 Steps of the proximity ligation assay, leading to detection of proximities between protein partners in cells. (1) Two primary antibodies bind their specific targets. (2) Two PLA probes (secondary antibodies, each with a unique short oligonucleotide attached to it) bind to primary antibodies. (3) Two connector oligonucleotides allow interaction between the short DNA strands when they are in close proximity (<40 nm). (4) Action of a DNA ligase allows circularization of the connector oligonucleotides. (5) One of the oligonucleotides serves as primer for rolling circle amplification (RCA) by the phi29 DNA polymerase. (6) RCA produces a randomly coiled, single-stranded chain composed of up to several hundreds of complements of the DNA circle. (7) The RCA product is detected by hybridizing complementary fluorophore-labeled oligonucleotides

2 Materials

2.1 Laboratory Equipment

1. Sterile chamber slides for cell culture: Millicell EZ 8-well glass slides (0.7 cm^2/well; 0.2–0.7 mL/well), or Ibidi 15-well μ-slides (0.125 cm^2/well; 10–50 μL/well).
2. Heat transfer block, e.g., Thermomixer or equivalent.
3. Shaker.
4. Staining jars.
5. Vacuum pump.
6. Freeze block for enzymes.

7. Grease pen.
8. Glass coverslips (24 × 50 mm, 0.13–0.17 mm thick).
9. Nail varnish.

2.2 Reagents

1. High purity water.
2. Phosphate-buffered saline (PBS).
3. Triton X-100.
4. PBS containing 0.5 % Triton X-100.
5. PBS containing 0.1 % Triton X-100.
6. PBS containing 0.05 % Triton X-100.
7. Bovine serum albumin (BSA), IgG- and protease-free.
8. Paraformaldehyde solution at 4 % in PBS.
9. 3 % fixative solution in PBS. Prepare freshly and keep at room temperature.
10. Blocking solution for immunofluorescence: 2 % BSA in PBS containing 0.1 % Triton X-100. Dissolve 0.1 g BSA in 5 mL PBS containing 0.1 % Triton X-100. This can be stored for a few days at 4 °C.
11. Antibody dilution solution: mix 1 mL blocking solution with 1 mL PBS containing 0.1 % Triton X-100. This can be stored for a few days at 4 °C.
12. DuoLink II wash buffer A (no. 82047, Sigma; *see* **Note 1**). Keep at 4 °C for storage and bring to room temperature (RT) before use.
13. DuoLink II wash buffer B (no. 82048, Sigma; *see* **Note 2**). Keep at 4 °C for storage and bring to RT before use.
14. Mounting medium with DAPI. Keep at 4 °C.

2.3 Antibodies

1. Rabbit anti-lamin A/C [16].
2. Mouse anti-emerin (NCL, Leica).
3. Rabbit anti-LAP2α (IQ175, ImmuQuest).
4. Mouse anti-lamin A/C (clone 636, Novocastra).

2.4 PLA Probes and Detection Reagents

1. DuoLink II In Situ PLA Probe Anti-Rabbit PLUS (ref DUO92002, Sigma), containing a 5× stock of donkey anti rabbit IgG conjugated with an oligonucleotide PLUS. Keep at 4 °C.
2. DuoLink II In Situ PLA Probe Anti-Mouse MINUS (ref DUO92004, Sigma), containing a 5× stock of donkey anti mouse conjugated with an oligonucleotide MINUS. Keep at 4 °C.
3. DuoLink II Detection Reagents Orange (ref DUO92007, Sigma). This kit contains (1) the ligase and the 5× ligation reagent (including two connector oligonucleotides that

hybridize the PLA probes) required to prepare the ligation mix and (2) the DNA polymerase and the 5× rolling circle amplification reagent (including nucleotides and fluorescently labeled oligonucleotide probes that hybridize the RCA product) required to prepare the amplification mix. Keep at −20 °C.

2.5 Fluorescently Labeled Secondary Antibodies

Fluorescently labeled secondary antibodies are used to localize proteins by immunofluorescence (*see* **Note 3**).

1. Cy3-conjugated affinipure goat anti-mouse IgG (used at 1:400; 115-165-146, Jackson Immunoresearch).
2. Cy2-conjugated affinipure goat anti-rabbit IgG (used at 1:60; 111-225-144, Jackson Immunoresearch).

3 Methods

The methods described here are for the MilliCell EZ slide (8-well glass). Throughout the protocol, media and buffers are removed gently from the wells and from the reaction areas on the microscopic slides by aspiration using a pipette tip connected to a vacuum pump system. Special care is required at all steps to prevent slides from drying out. For incubations with blocking solution, antibodies, or PLA reagents, use open droplet reactions in a humid environment to prevent excessive evaporation. The entire protocol from cell fixation to mounting of slides with medium takes more than 6 h (Fig. 2).

3.1 Antibody Preparation

Antibodies are prepared just before use with the antibody dilution solution.

1. For detection of lamin A/C-emerin complexes: prepare a mix of rabbit anti-lamin A/C (use at 1:500) and mouse anti-emerin (use at 1:30).

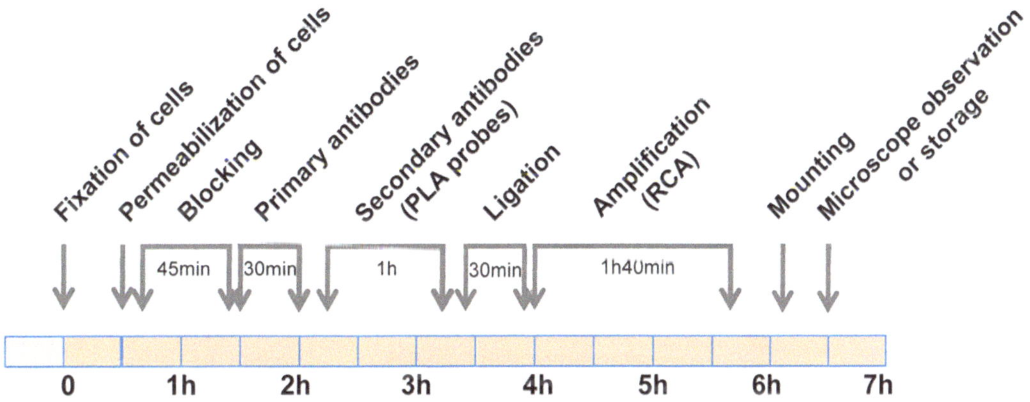

Fig. 2 Time course of the PLA protocol

2. Negative control for lamin A/C-emerin complexes detected by PLA: prepare rabbit anti-lamin A/C (use at 1:500).
3. In situ detection of lamin A/C-LAP2α complexes: prepare a mix of rabbit anti-LAP2α (use at 1:250) and mouse anti-lamin A/C (use at 1:150).
4. Negative control for lamin A/C-LAP2α complexes detected by PLA: prepare mouse anti-lamin A/C (use at 1:150).

3.2 Cell Preparation

1. Remove cell culture media from each well of the chamber slide.
2. Quickly wash cells in PBS (0.5 mL/well).
3. Add fixative solution (0.5 mL/well) and incubate for 12 min at RT (*see* **Note 4**).
4. Remove fixative solution and wash in PBS (0.5 mL/well) for 5 min at RT.
5. Permeabilize cells with PBS/0.5 % Triton X-100 (0.5 mL/well) for 5 min at RT.
6. Remove permeabilization solution and add PBS (0.5 mL/well).
7. Disassemble the MilliCell EZ slide unit to access the microscope slide (*see* **Note 5**; Fig. 3).
8. Remove PBS and delimitate eight reaction areas with the hydrophobic barrier pen.
9. Fill each reaction area with 100 μL PBS.

3.3 Blocking

1. Remove PBS and add blocking solution for immunofluorescence (100 μL/reaction area). Incubate 45 min at RT.

3.4 Primary Antibody Incubation

1. Prepare a fresh cocktail of primary antibodies (*see* Subheading 3.1), 50 μL/reaction area.
2. Remove blocking solution.
3. Add primary antibodies solution within the reaction areas. Incubate 30 min at RT (Fig. 1, step 1).

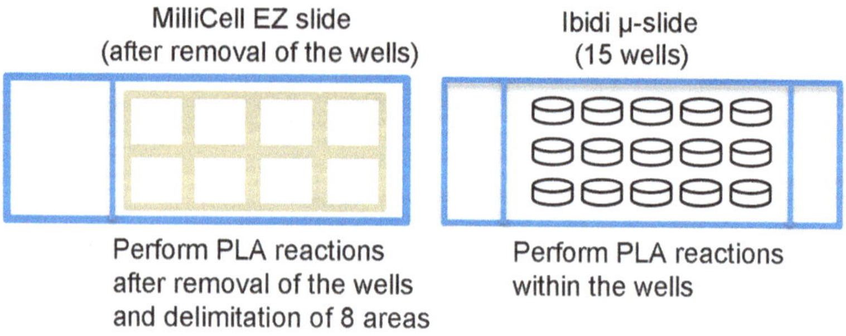

Fig. 3 The two models of slides used for PLA

4. In the last 5 min of incubation, prepare the mix of the two PLA probes: dilute the stocks 1:5 in antibody dilution solution (30 μL/reaction area: 6 μl PLA probe anti-Mouse MINUS + 6 μL probe anti-Rabbit PLUS + 18 μL of antibody dilution solution). Keep 20 min at RT before use.
5. Remove primary antibodies from slide.
6. Wash slide in a stain jar containing ~70 mL PBS/0.05 % Triton, 3 times 5 min at RT with gentle agitation.

3.5 Secondary Antibody Incubation

1. Drain off residual PBS from the slide.
2. Add the mix of PLA probes from Subheading 2.4 (30 μL/reaction area). Incubate 1 h at 37 °C (Fig. 1, step 2) (*see* **Note 6**; Fig. 4).
3. Thaw the 5× ligation reagent at RT.
4. Prepare the ligation buffer (29.25 μL/reaction area: 6 μL 5× ligation reagent + 23.25 μL high purity water).
5. Remove PLA probes from the slide.

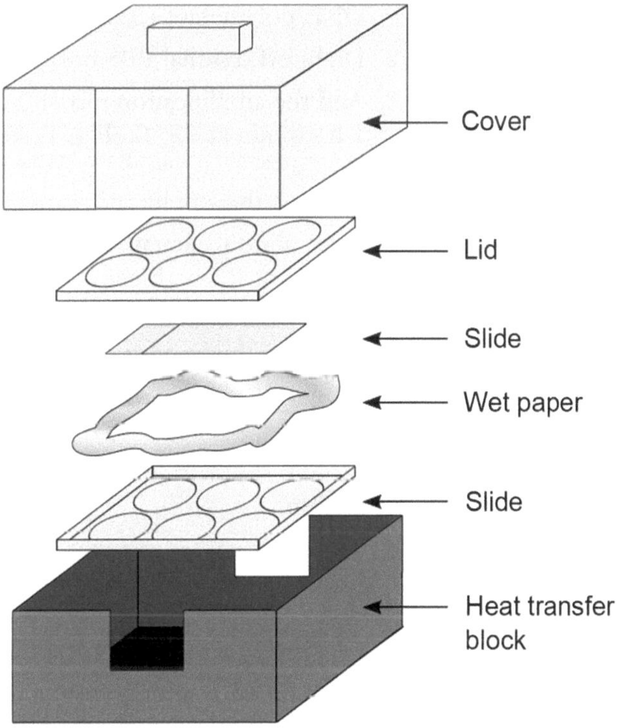

Fig. 4 Control of a humid environment within a heat transfer block. The microscope slide is transferred within a box (e.g., two lids of a 6–12 well plates positioned one on the other, enclosing wet paper) preheated at 37 °C by placing it in a heat transfer block

6. Wash the slide in a stain jar with ~70 mL of wash buffer A twice 5 min at RT with gentle agitation.

3.6 PLA: Ligation

1. Just before use, prepare the ligation mix from Subheading 2.4: remove the ligase from the freezer using a freezing block and dilute 1:40 in ligation buffer (30 μL/reaction area: 0.75 μL ligase + 29.25 μL ligation buffer). Vortex.
2. Drain off the residual PBS buffer from the slide.
3. Add the ligation mix (30 μL/reaction area). Incubate 30 min at 37 °C (Fig. 1, steps 3 and 4).
4. Thaw the 5× amplification reagent at RT.
5. Prepare the amplification buffer (29.62 μL/reaction area: 6 μL of 5× amplification reagent + 23.62 μL of high purity water).
6. Remove the ligation mix from the slide.
7. Wash slides in a stain jar with ~70 mL wash buffer A, twice 2 min at RT with gentle agitation.

3.7 PLA: Rolling Circle Amplification

1. Just before use, prepare the amplification mix: remove DNA polymerase from freezer using a freezing block and dilute 1:80 in amplification buffer (30 μL/reaction area: 0.38 μL DNA polymerase + 29.62 μL amplification buffer). Vortex.
2. Drain off residual PBS buffer from the slide.
3. Add the amplification mix (30 μL/reaction area). Incubate for 1 h 40 min at 37 °C (Fig. 1, steps 5–7) (*see* **Note 6**).

3.8 PLA: Final Washes and Mounting

1. Remove the amplification mix from slide.
2. Wash slide in a stain jar with ~70 mL wash buffer B, twice 10 min at RT with gentle agitation.
3. Wash slide in a stain jar with ~70 mL wash buffer B diluted 1:100 in high purity water, for 1 min at RT with gentle agitation.
4. Let slides dry at RT in the dark.
5. Mount slides in the dark with a coverslip using mounting medium (*see* **Note** 7; Fig. 5).
6. Wait at least 15 min before observation in the microscope. Keep slides at −20 °C in the dark for several months if needed.

3.9 Microscope Observation and Image Analysis

1. Observe slides in a confocal microscope (*see* **Note** 7).
2. Analyze PLA signals using suitable software packages (*see* **Note 8**).

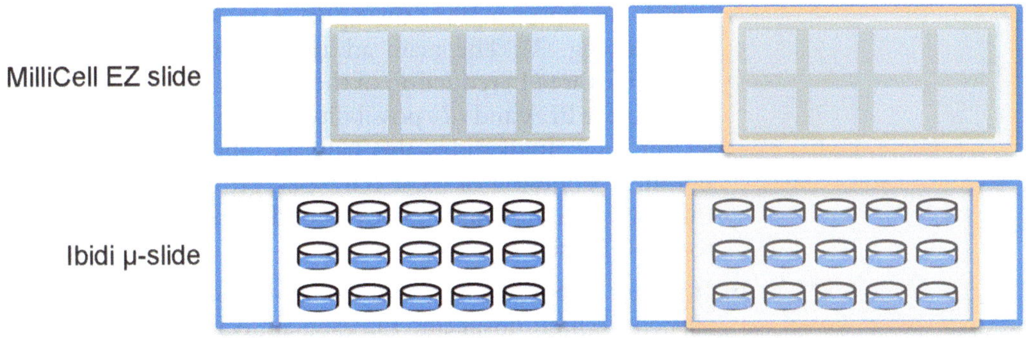

Fig. 5 Mounting of slides. Mounting medium is distributed homogeneously between the MilliCell slide and the coverslip, before sealing with nail varnish (*upper panels*). For the 15-well Ibidi slides, mounting medium is added to each well, and a coverslip is placed on top of the wells to protect samples (*lower panels*). Observation in an inverted microscope will occur through the top of the coverslip or through the bottom of the slide for the Millicell glass slide and the 15 well Ibidi µ-slide, respectively

4 Notes

1. Wash buffer A can also be prepared by dissolving 1.2 g Tris base, 8.8 g NaCl, and 0.5 mL Tween 20 in 800 mL high purity water, adjusting pH to 7.4 using HCl, and bringing the volume to 1 L with high purity water (final concentrations are 0.01 M Tris, 0.15 M NaCl, and 0.05 % Tween 20). Filter the solution through a 0.22 µm filter and store at 4 °C.
2. Wash buffer B can also be prepared by dissolving 4.24 g Tris base, 26.0 g Tris–HCl, and 5.84 g NaCl in 500 mL high purity water, adjusting pH to 7.5 using HCl, and bringing volume to 1 L with high purity water (final concentrations are 0.2 M Tris and 0.1 M NaCl). Filter the solution through a 0.22 µm filter and store at 4 °C.
3. To validate the first steps of PLA, i.e., cell fixation, cell permeabilization, and recognition of proteins of interest by the primary antibodies, it is necessary to analyze the samples of interest in parallel on a second slide by conventional immunofluorescence, using secondary antibodies tagged with a fluorophore, as indicated in Subheading 2.5.
4. In the 15-well microscope slides, the volume of fixative solution to add is 70 µL/well, whereas the minimal volume to add for primary antibodies, PLA probes, and PLA reagent solutions is 10 µL/well.
5. Before incubations with antibodies and PLA reagents, wells are removed from the MilliCell EZ slides. In order to reinforce the delimitation of the PLA reaction areas, a hydrophobic barrier is created with a grease pen. In comparison, it is not possible to

detach the bottom from the upper part of the Ibidi 15-well μ-slides (Fig. 3). The main advantages of working with the Ibidi slides are (1) reduction of the amount of cells and materials used for PLA and (2) possibility to do more assays in parallel per slide. Ibidi slides also have an "ibiTreat" surface that optimizes adhesion of most cell types. However, in our hands they gave a slightly higher background than MilliCell EZ slides for the PLA signals.

6. It is very important to prevent evaporation from the slides at all steps to limit unspecific PLA signals. A humid environment must be maintained within a heat transfer block (Fig. 4). In addition, the cover of the heat transfer block protects the fluorescently labeled oligonucleotide probes in the amplification mix from light (Subheading 3.7).
7. Mounting of slides and the way to observe samples in the confocal microscope depend on the format of the slides used for the PLA experiment (Fig. 5).
8. Several imaging software can be used to analyze PLA data, such as the DuoLink Image tool in Image J. Depending on the amount of signal, it is possible to quantify the number of individual signals (which appear as dots) [10, 11]; however, when the density of PLA signals is so high that one cannot distinguish individual spots, data are analyzed by fluorescence intensity measurement. In our recent study on [SREBP1–lamin A/C] protein complexes, PLA data were compiled as numbers of recognized pixels and as a function of their distance to the closest point of the nuclear periphery; this was done on images using a custom image recognition Image J plugin [9].

5 Conclusion

As shown in Fig. 6, the PLA protocol has allowed us to reveal distinct patterns of in situ subnuclear distribution for the protein-protein complexes [lamin A/C–emerin] and [lamin A/C–LAP2α]. While [lamin A/C–emerin] complexes appear concentrated at the nuclear periphery, [lamin A/C–LAP2α] complexes are detected throughout the nuclear interior. These observations are in accord with the dual location of lamin A/C at nuclear envelope and in the nucleoplasm and with the restricted location of emerin and LAP2α at the nuclear envelope and in the nucleoplasm, respectively [15]. These observations highlight the pertinence of using PLA to reveal where and to which extent nuclear envelope proteins form complexes with specific partners.

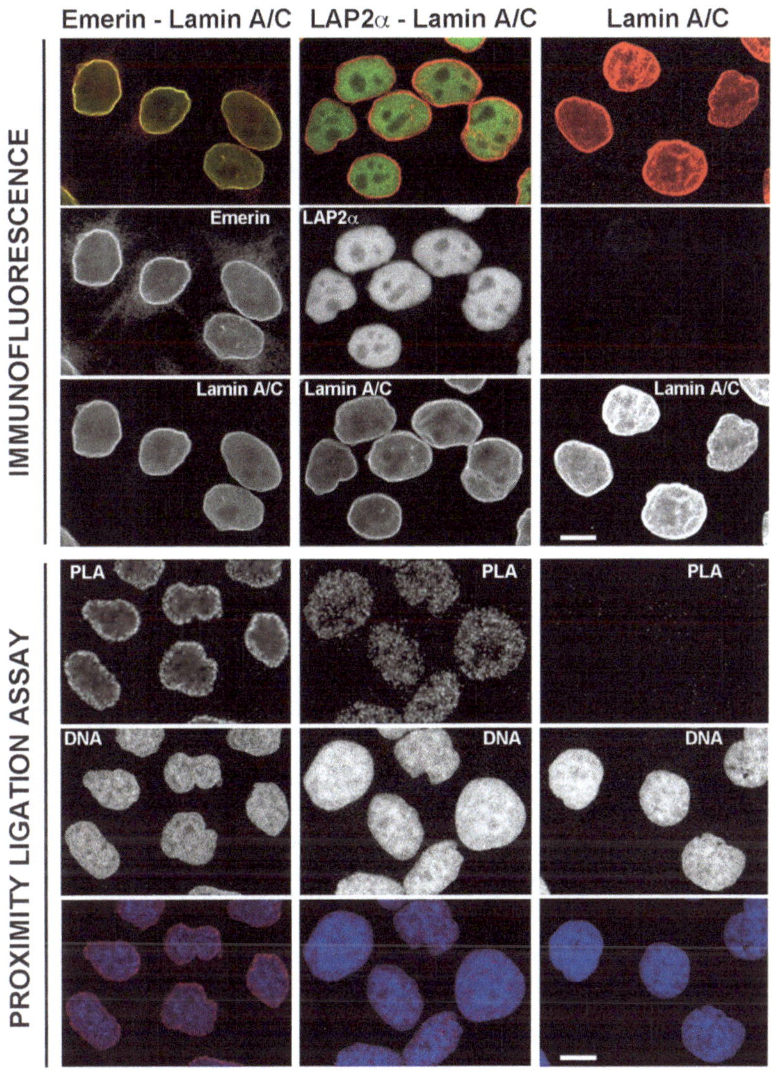

Fig. 6 Intranuclear distribution of lamin A/C, emerin, and LAP2α, with detection of their close proximities assessed by PLA in HeLa cells. *Upper part* shows the subnuclear location of lamin A/C and its two partners as revealed by conventional immunofluorescence (*left panel*, emerin in *red* and lamin A/C in *green*; *middle* and Fig. 6 (continued) right panel, LAP2α in *green* and lamin A/C in *red*). Lower part shows detection of [lamin A/C–emerin] (*left*) and [lamin A/C–LAP2α] complexes (*middle*) PLA. *Right panel*, negative control experiment using one single primary antibody (here, mouse anti-lamin A/C). Merged images show DNA (*blue*) and PLA signals (*red*). Bars, 10 μm (confocal microscopy)

Acknowledgment

This work was supported by CNRS, University Paris Diderot Paris 7, and the Association Française contre les Myopathies.

References

1. Simon DN, Wilson KL (2013) Partners and post-translational modifications of nuclear lamins. Chromosoma 122:13–31
2. Oldenburg AR, Delbarre E, Thiede B, Vigouroux C, Collas P (2014) Deregulation of Fragile X-related protein 1 by the lipodystrophic lamin A p.R482W mutation elicits a myogenic gene expression program in preadipocytes. Hum Mol Genet 23:1151–1162
3. Liu B, Ghosh S, Yang X, Zheng H, Liu X, Wang Z, Jin G, Zheng B, Kennedy BK, Suh Y, Kaeberlein M, Tryggvason K, Zhou Z (2012) Resveratrol rescues SIRT1-dependent adult stem cell decline and alleviates progeroid features in laminopathy-based progeria. Cell Metab 16:738–750
4. Liu B, Wang Z, Zhang L, Ghosh S, Zheng H, Zhou Z (2013) Depleting the methyltransferase Suv39h1 improves DNA repair and extends lifespan in a progeria mouse model. Nat Commun 4(1868):1–12
5. Krohne G (2004) Lamins. Methods Cell Biol 78:573–596
6. Shin JY, Mendez-Lopez I, Wang Y, Hays AP, Tanji K, Lefkowitch JH, Schulze PC, Worman HJ, Dauer WT (2013) Lamina-associated polypeptide-1 interacts with the muscular dystrophy protein emerin and is essential for skeletal muscle maintenance. Dev Cell 26:591–603
7. Delbarre E, Tramier M, Coppey-Moisan M, Gaillard C, Courvalin JC, Buendia B (2006) The truncated prelamin A in Hutchinson-Gilford progeria syndrome alters segregation of A-type and B-type lamin homopolymers. Hum Mol Genet 15:1113–1122
8. Ivorra C, Kubicek M, Gonzalez JM, Sanz-Gonzalez SM, Alvarez-Barrientos A, O'Connor JE, Burke B, Andres V (2006) A mechanism of AP-1 suppression through interaction of c-Fos with lamin A/C. Genes Dev 20:307–320
9. Vadrot N, Duband-Goulet I, Cabet E, Attanda W, Barateau A, Vicart P, Gerbal F, Briand N, Vigouroux C, Oldenburg AR, Lund EG, Collas P, Buendia B (2015) The p.R482W substitution in A-type lamins deregulates SREBP1 activity in Dunnigan-type familial partial lipodystrophy. Hum Mol Genet 24:2096–2109
10. Mahen R, Hattori H, Lee M, Sharma P, Jeyasekharan AD, Venkitaraman AR (2013) A-type lamins maintain the positional stability of DNA damage repair foci in mammalian nuclei. PLoS One 8:e61893
11. Barascu A, Le Chalony C, Pennarun G, Genet D, Imam N, Lopez B, Bertrand P (2012) Oxidative stress induces an ATM-independent senescence pathway through p38 MAPK-mediated lamin B1 accumulation. EMBO J 31:1080–1094
12. Fredriksson S, Gullberg M, Jarvius J, Olsson C, Pietras K, Gustafsdottir SM, Ostman A, Landegren U (2002) Protein detection using proximity-dependent DNA ligation assays. Nat Biotechnol 20:473–477
13. Söderberg O, Gullberg M, Jarvius M, Ridderstrale K, Leuchowius KJ, Jarvius J, Wester K, Hydbring P, Bahram F, Larsson LG, Landegren U (2006) Direct observation of individual endogenous protein complexes in situ by proximity ligation. Nat Methods 3:995–1000
14. Lievens S, Tavernier J (2006) Single protein complex visualization: seeing is believing. Nat Methods 3:971–972
15. Wilson KL, Foisner R (2010) Lamin-binding Proteins. Cold Spring Harb Perspect Biol 2:a000554
16. Duband-Goulet I, Woerner S, Gasparini S, Attanda W, Konde E, Tellier-Lebegue C, Craescu CT, Gombault A, Roussel P, Vadrot N, Vicart P, Östlund C, Worman HJ, Zinn-Justin S, Buendia B (2011) Subcellular localization of SREBP1 depends on its interaction with the C-terminal region of wild-type and disease related A-type lamins. Exp Cell Res 317:2800–2813

Chapter 10

Methods for Single-Cell Pulse-Chase Analysis of Nuclear Components

Marek Drozdz and David J. Vaux

Abstract

Pulse-chase methods offer powerful tools for following the evolution of a biological system over time, but are usually limited to ensemble measurements of the average behavior of very large numbers of cells. Here we describe three methods ranging from a true pulse-chase, through selective regional photoactivation, to pharmacological induction of an altered protein state, which can be applied to time-dependent studies at the single-cell level. These methods are exemplified by experimental protocols to follow region-selective nuclear envelope targeting of nascent phospholipids, a nascent nuclear lamin protein (lamin B1), and an immature lamin precursor (prelamin A).

Key words Nucleoplasmic reticulum, Pulse-chase, Single-cell assays, Super-resolution light microscopy, Secondary ion mass spectrometry, Stable isotope labeling, Photoactivatable fluorescent proteins

1 Introduction

Often in cell biology, it is necessary to compromise between analysis of the average behavior of a population of cells and less revealing assays at the single-cell level. With the emerging recognition that variation between cells in a population, whether in vitro or in vivo, may represent an important element of the nuanced and flexible biological response at the population or organismal level, this compromise has come under greater scrutiny. The problem is particularly acute for kinetic analysis of biochemical changes in subcellular morphological detail. Processes of expression, post-synthetic modification, and turnover of individual components that lend themselves to analysis at the population level with pulse tagging of a cohort of molecules and subsequent kinetic analysis do not transfer readily to single-cell assays [1, 2].

In this chapter, we describe three methods for overcoming this limitation, demonstrating these approaches with respect to components of the nuclear envelope and especially those of

Sue Shackleton et al. (eds.), *The Nuclear Envelope: Methods and Protocols*, Methods in Molecular Biology, vol. 1411, DOI 10.1007/978-1-4939-3530-7_10, © Springer Science+Business Media New York 2016

the nucleoplasmic reticulum (NR, [3, 4]). In each case, a cohort of molecules is selectively modified at a defined time, and the subsequent behavior of the marked molecules is followed over time using detection methods that image single cells at high resolution.

For our first example, we begin with the issue of following the subcellular distribution of nonprotein components over time. The ability to localize lipid components in the cell has depended largely upon subcellular fractionation (a cell population-averaging method) or by indirect marking with proteins or small molecules exhibiting specific binding to particular lipids. Here we describe a method for stable isotope labeling of fatty acyl tails of phospholipids by pulsing cells with a uniformly deuterated precursor, in this case stearic acid. Deuterated acyl chains are incorporated biosynthetically into phospholipids, enabling pulse labeling of a population synthesized at a particular time. Detection of the distribution of the heavy atom-tagged phospholipids then requires spatially resolved secondary ion mass spectrometry, using the technique of NanoSIMS [5]. In this method a raster scanned ion beam is used to ablate a surface layer of the sectioned specimen to generate atomic fragments for high-resolution mass analysis. We demonstrate this approach by showing nascent membrane phospholipids selectively incorporated into the membranes of forming nucleoplasmic reticulum channels.

In the second method, we exploit an inhibitor of proteolytic maturation to create a pulse of nascent unprocessed copies of a protein. Using an antibody that detects only the immature uncleaved form of the protein, it is then possible again to follow the delivery, distribution, and turnover of a time-tagged pool of proteins at the single-cell level. We demonstrate this by selective drug inhibition of the lamin A maturation protease ZmpSte24/FACE1 [6], which enables us to follow delivery of nascent prelamin A selectively to newly forming NR channels.

In the final example, we exploit the power of a photoactivatable fluorescent protein (e.g., PA-GFP; *see* ref. 7) as a tag that can be selectively activated and followed by time-lapse imaging. Such approaches have been used to augment selective bleaching experiments (such as FRAP and FLIP; *see* ref. 8) in the study of target protein diffusion or active trafficking. However, studying the nucleus offers the unique advantage of tagging a nascent protein pool by selective photoactivation of the cytoplasm, without direct activation of any mature preexisting copies of the protein that have already arrived in the nucleus. In this way, the delivery, distribution, and turnover of a time-tagged pool of proteins may be followed at the single-cell level. We demonstrate this method by showing delivery of nascent lamin B1 selectively to newly forming NR channels.

2 Materials

Prepare all solutions using ultrapure water. They can be stored at room temperature, unless indicated otherwise.

2.1 Cell Culture

1. Cell culture medium: Dulbecco's modified Eagle's medium, 10 % fetal calf serum, nonessential amino acids, and penicillin and streptomycin antibiotics (100 U/mL).
2. Medium composition may need to be adjusted depending on the cell type used. In addition, specific treatments of coverslips or wells may be required for adequate adhesion/spreading of particular cell types (see below).
3. Standard tissue culture supplies including, depending on the method below, 6- and 24-well tissue culture plates or 35-mm glass bottom gridded dishes (those we use are produced by MatTek Corporation).

2.2 NanoSIMS

1. 24-well cell culture plates.
2. 13-mm-diameter sterile plastic coverslips.
3. Stearic acid D35 powder [M.M. 319.69].
4. Stearic acid powder [M.M. 284.48].
5. Fatty acid-free bovine serum albumin.
6. Phosphate-buffered saline (PBS) pH 7.3.
7. 16 % paraformaldehyde, methanol free.
8. 25 % glutaraldehyde in H_2O.
9. 200 mM PIPES pH 7.4.
10. Freshly prepared osmium tetroxide: 1 % in 100 mM PIPES pH 7.4.
11. Dry ethanol.
12. Embedding capsules.
13. Epoxy resin; we are using agar 100 resin (Agar Scientific, UK) as it is less viscous and makes the infiltration and specimen handling easier.
14. Liquid nitrogen for freezing the resin blocks and detaching plastic coverslips.
15. Razor blades and glass knife.
16. 15-nm platinum-coated coverslips (or other platinum-coated coverslips compatible with the sample holder in the NanoSIMS instrument).
17. Sputter coater (e.g., Cressington 208HR high-resolution sputter coater, though others should also work) and platinum for coating.
18. Standard EM microtome, scanning electron microscope, and NanoSIMS instrument.

2.3 Structured Illumination Microscopy (OMX 3D-SIM)

1. High-precision coverslips, 22×22 mm or smaller, thickness no. 1.5H (170 μm±5 μm). These can be from a variety of sources, but must have at least equivalent optical properties and tolerances to those supplied by Marienfeld, Germany.
2. 6-well cell culture plates.
3. Phosphate-buffered saline (PBS) pH 7.3.
4. Fixative: 2.5 mL of methanol-free 16 % paraformaldehyde in 7.5 mL PBS pH 7.3.
5. 25 mM glycine solution: 28 mg glycine [M.M. 75.07] dissolved in 15 mL PBS pH 7.3.
6. 20 % Triton X-100 in PBS pH 7.3, store at 4 °C and dilute 40 times in PBS prior to use.
7. 10 % fish skin gelatine (FSG) in PBS pH 7.3. Warm up FSG stock in 50 °C water bath before preparing the 10 % solution. Use at 0.5 % for experiments.
8. Primary antibodies: goat polyclonal anti-prelamin A antibody (C-20) (Santa Cruz) and mouse monoclonal anti-lamin A/C antibody (3A6-4C11) (Active Motif). Altering antibodies should be accompanied by re-optimization of reaction parameters.
9. Secondary antibodies: polyclonal Alexa-647-conjugated donkey anti-goat antibody (Molecular Probes) and polyclonal Alexa-594-conjugated donkey anti-mouse antibody (Molecular Probes). Other secondary antibodies at these wavelengths may work, but because so many images are taken in 3D-SIM, the fluorophores must be strong and stable at the level of Alexa chemistry. It is also important to use secondary antibodies that are highly specific so that cross-reactivity does not occur. Generally donkey or goat highly cross-adsorbed is the best in this regard.
10. PBST: PBS pH 7.3, 0.1 % Tween 20.
11. DAPI (4′,6-diamidino-2-phenylindole hydrochloride) 5 mg/mL stock: 10 mg DAPI in 2 mL of dimethylformamide (DMF).
12. Mounting medium, offering good anti-fading properties, a refractive index as close to that of the coverslip as possible, and setting properties that do not lead to compression and flattening of the sample: e.g., ProLong Gold (Molecular Probes, Life Technologies Ltd, Paisley, UK) or Vectashield H-1000 (Vector Laboratories, Peterborough, UK).
13. Immersion oils with refractive indices (RI) 1.512 and/or 1.514 (depending on sample thickness) to match RI of coverslips and minimize chromatic aberrations.
14. Access to a super-resolution microscope facility.

2.4 Confocal Microscopy of Photoactivatable Fluorescent Proteins

1. 35-mm glass bottom gridded tissue culture dishes (we have used those from MatTek Corporation, Ashland, MA, USA).
2. Transfection reagent of choice.
3. Access to confocal microscope setup with 405-nm laser for photoactivation and 488-nm laser for GFP imaging.

3 Methods

All steps can be carried out at room temperature unless otherwise stated.

3.1 Cell Culture

1. Culture cells in cell culture medium. Grow cells in an incubator at 37 °C with 5 % CO_2 supply and split them when they reach about 80 % confluence.
2. Plate cells for experiments on specific media defined for each method.
3. Perform transfections according to standard protocols.

3.2 NanoSIMS

3.2.1 Preparation of Stearic Acid D35 Stock Solution

1. Dissolve 3.2 mg of stearic acid D35 in 2 mL of ethanol.
2. Complex this with 400 μL of 100 mM solution of NaOH in ethanol (this will result in the formation of a sodium salt of the stearic acid).
3. Evaporate alcohol to obtain fatty acid soaps. Evaporation can be enhanced with nitrogen gas by directing the gas from a tank to blow gently over the surface of the mixture. Be careful not to splash the solution with too strong gas stream.
4. Dissolve the fatty acid soaps in 0.5 mL of ultrapure water prewarmed to 55 °C and keep in 55 °C water bath for 10 min.
5. Dissolve 1 g of fatty acid-free bovine serum albumin in 9.5-mL DMEM and similarly warm up to 55 °C.
6. Add the dissolved fatty acid soaps to the fatty acid-free bovine serum albumin solution and then vortex mix for 10 s, followed by a further 10-min incubation at 55 °C.
7. Sterile filter the stock solution, aliquot, and keep in −20 °C. This makes a 1 mM stearic acid D35 stock solution.
8. Upon thawing to use in assays, reheat to 55 °C to dissolve the precipitate (if formed), and cool down to 37 °C before adding to cells.
9. Prepare regular stearic acid in the same way as deuterated stearic acid D35 and use it as a control treatment. Note that normal stearic acid has lower molecular mass [M.M. 284.48]; hence you will need to use only 2.8 mg to prepare 1 mM stock following the same procedure (*see* **Note 1**).

3.2.2 Stearic Acid D35 Labeling and Embedding of Cells

1. Seed mouse preadipocytes on 13-mm plastic coverslips in a 24-well plate at 50 % confluence and leave overnight (*see* **Note 2**). Depending upon the cell type(s) used, the coverslips may be pre-coated with poly-l-lysine, collagen, or more elaborate extracellular matrix preparations in order to increase cell adherence and/or spreading.
2. The next day confirm that cells are attached and look healthy, not exceeding 80 % confluence.
3. Treat cells with 10 μM stearic acid D35. Add 100 μL of 1 mM stearic acid D35 stock solution to 10-mL pre-warmed culture medium and vortex mix. Use it to replace medium in the 24-well plate (500 μL per well). Leave for 6 h in the incubator (*see* **Note 3**).
4. In the next step, aspirate medium and wash coverslips with 1 mL PBS. Discard PBS.
5. Fix cells by adding 500 μL of fixative warmed up to 37 °C. Use solution of 4 % paraformaldehyde and 1 % glutaraldehyde in 100 mM PIPES pH 7.4. Leave at room temperature for 20 min (*see* **Note 4**).
6. Change fixative to 500 μL of 2.5 % glutaraldehyde in 100 mM PIPES pH 7.4. Incubate 1 h at room temperature and then transfer to 4 °C and leave overnight (*see* **Note 5**).
7. Wash samples three times 10 min with 100 mM PIPES pH 7.4.
8. Osmicate cells with 1 % osmium tetroxide in 100 mM PIPES pH 7.4 for 1 h (*see* **Note 6**).
9. Wash samples with deionized water for 20 min.
10. Incubate samples in a graded ethanol series, starting from 50 % ethanol for 15 min, then 70 % ethanol overnight at 4 °C, then 90 % ethanol for 15 min, then 95 % ethanol for 15 min, and 100 % ethanol for 2 h with three solution changes during this time.
11. Gradually infiltrate samples with agar 100 epoxy resin. This is achieved by serial replacement of the resin solution in the coverslip-containing wells, simply by aspiration of one solution and replacement with the next higher concentration. Start with 25 % resin for 1 h, then 50 % resin for 2 h, then 75 % resin for 1 h, and 100 % resin overnight.
12. Transfer samples to fresh 100 % resin for 3 h and then repeat this step once more.
13. Resin-embed the layer of cells by carefully inverting the coverslip onto an embedding capsule filled with fresh 100 % resin. Make sure the tube is completely full and the side of coverslip with cells on it is completely immersed in the resin. Leave for 24 h at 60 °C to allow polymerization to take place.

14. Submerge blocks in liquid nitrogen and snap off the coverslip to leave the cells embedded as a monolayer in the resin (*see* **Note 7**).
15. Trim the resin block with a razor blade and glass knife so as to generate a trapezoid end containing the specimen.
16. Cut the specimen to obtain semi-thin sections of 0.5 μm using an ultramicrotome with a diamond knife, and mount them on 15-nm platinum-coated coverslips. Place on 60 °C heating block and allow to dry for a few min.

3.2.3 BSE Imaging and NanoSIMS

This procedure assumes that the researcher or a colleague helping will be familiar with the basic usage of an SEM, a NanoSIMS, and metal coating of samples. Only specific operating conditions and settings are given in the protocol for this particular application. Note that the researcher should pay particular attention to the orientation of samples throughout the procedure as this greatly enhances the ability to align BSE and NanoSIMS images in the final step:

1. Before starting backscattered electron (BSE) imaging, record areas of interest by optical microscopy.
2. Transfer sections to scanning electron microscope, and acquire BSE images with a 2-kV incident beam with a standard aperture (30 μm) and 5-mm working distance (*see* **Note 8**). The BSE signal is very sensitive to atomic number contrast, so at this low incident beam energy, biological samples contrasted with heavy metal stain (e.g., osmium, as used here) can be imaged without excess surface charging.
3. Coat sections with 5 nm of platinum in a high-resolution sputter coater (we use the Cressington 208HR) to render the surface conductive for subsequent NanoSIMS imaging.
4. In order to use the cesium ion beam to remove material from the section for $^{2}H^{-}$ and $^{1}H^{-}$ mass analysis, a small selected region of the sample must then be revealed by uncoating. Use the Cs^{+} primary beam in the NanoSIMS to remove the platinum on the surface at selected locations over regions of interest identified from the BSE image. Use a low beam dose (Cs^{+} dose of 6.5×10^{15} ions/cm^{2}) to ensure that as little as possible of the underlying biological section surface is removed before SIMS imaging (*see* **Note 9**).
5. Use small apertures (D1 = 3 or D1 = 4) for imaging single cells in order to match the primary beam size to the pixel size. Start by tuning the instrument for $^{2}H^{-}$ and $^{1}H^{-}$ to give morphological information and allow calculation of the $^{2}H/^{1}H$ ratio. Collect and process NanoSIMS images with a dwell time of 30,000 μs per pixel for a 256 × 256 pixel image. Apply a median

filter with radius of 3 pixels to the hue-saturation-intensity (HSI) encoded image.

6. Manually align the BSE and NanoSIMS images to identify regions containing structures of biological interest.
7. Using any image analysis software (we have routinely used ImageJ), construct $^{2}H/^{1}H$ ratio images from the raw hydrogen and deuterium NanoSIMS images. In the experiment described here, all fatty acyl moieties present in the cell prior to the pulse label will contain only ^{1}H; the pulse of ^{2}H-labeled acyl precursor will generate a pool of newly synthesized deuterated lipids within the cell. The ^{2}H NanoSIMS image will thus show the subcellular distribution of these newly made lipids. Constructing a $^{2}H/^{1}H$ ratio image normalizes for hydrogen abundance across the sample, and the resulting image can be interpreted to identify sites of nascent lipid incorporation (high $^{2}H/^{1}H$ ratio value) as well as sites where lipid turnover is low (low $^{2}H/^{1}H$ ratio value).
8. In the example shown (Fig. 1), a deuterated stearic acid pulse has been combined with pharmacological induction of a nucleoplasmic reticulum. The figure shows saquinavir-mediated induction of NR, in comparison with a vehicle only (DMSO)-treated control sample. The $^{2}H/^{1}H$ ratio image of the control (DMSO) sample shows that the boundary of the nucleus exhibits a uniformly low $^{2}H/^{1}H$ ratio, implying that the interphase NE is stable under these conditions, with a low rate of phospholipid turnover. In contrast, the nuclear boundary in the saquinavir-treated cell shows a discontinuous high $^{2}H/^{1}H$ ratio, implying that there are both low-turnover and high-turnover regions of NE. Moreover, there is an increased $^{2}H/^{1}H$ ratio at some NR invagination sites (white arrows), suggesting that these are structures enriched in newly made phospholipids. Interestingly, there are also NR invaginations clearly visible in the BSE image that have a very low $^{2}H/^{1}H$ ratio, implying that they are stable and have not changed composition since the start of the deuterated stearate pulse. Further work will be required to establish whether these represent baseline NR present in the nucleus before the abundance of this structure was stimulated by driving prelamin A accumulation, but the power of the approach is clear.

3.3 Structured Illumination Microscopy (OMX 3D-SIM)

1. Clean high-precision coverslips for 30 min in diluted HCl (4 mL of HCl 12 N in 250 mL H_2O) in a rack on a shaker. Rinse thoroughly in deionized water followed by a 30-min incubation in ethanol on a shaker. In a sterile environment, remove the ethanol and allow to air-dry.

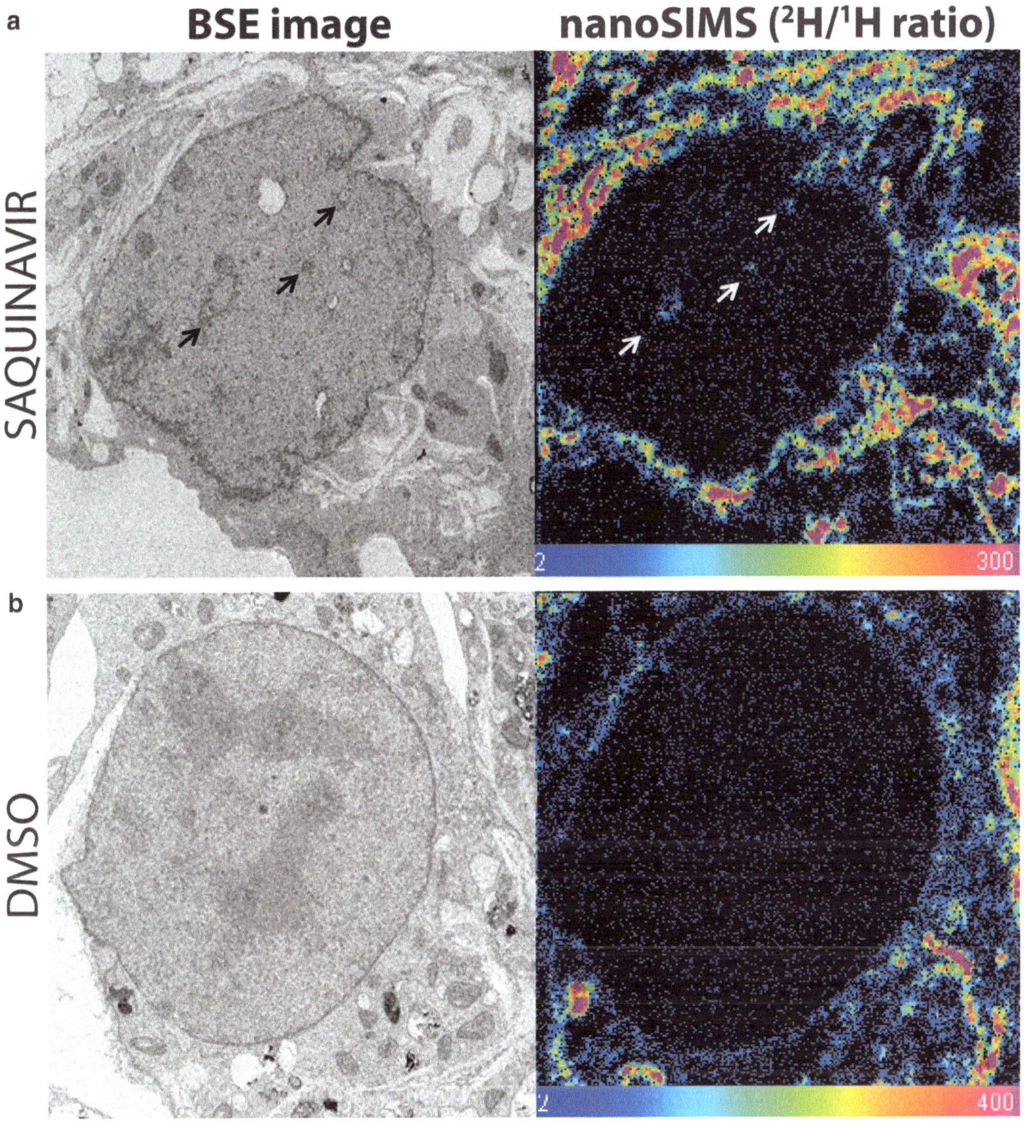

Fig. 1 Correlative backscattered electron imaging (BSE) with NanoSIMS experiments. Mouse preadipocytes were treated for 12 h with the drug saquinavir (an HIV protease inhibitor that increases occurrence of nucleoplasmic reticulum tubules) or with vehicle control (DMSO). Then cells were pulse labeled with 100 μM stearic acid D35, prepared as described, and $^2H/^1H$ ratio was measured by NanoSIMS. (**a**) Representative backscattered electron (BSE) image of a drug-treated cell with *black arrows* indicating NR tubules with a corresponding image from NanoSIMS showing enrichment of ^2H signal, also at the NR foci (indicated with *white arrows*) coming from the newly incorporated fatty acyl species. (**b**) BSE and NanoSIMS images of a control cell showing lack of NR tubules and no ^2H enrichment within the nucleus. Note highly distorted ER caused by overloading cells with stearic acid D35 (*see* **Note 3**)

2. Seed cells on sterile high-precision coverslips in a 6-well plate, and return to the incubator to allow time to adhere and spread (*see* **Note 10**). The cells should be about 80 % confluent at the time of fixation.
3. Wash cells twice with 2 mL of PBS for 5 min each time. Discard the buffer.
4. Fix cells with 4 % paraformaldehyde in PBS for 10 min (2 mL per well).
5. Aspirate fixative and add 2 mL of 25 mM glycine solution in PBS per well. Leave for 5 min.
6. Permeabilize cells by replacing the glycine solution with 2 mL of 0.5 % Triton X-100 in PBS per well. Incubate for 5 min.
7. Block cells with 2 mL of 0.5 % FSG in PBS for 30 min at room temperature or overnight at 4 °C (*see* **Note 11**).
8. Dilute primary antibodies in blocking solution according to the antibody manufacturer's recommendations for immunofluorescence microscopy (*see* **Notes 12** and **13**).
9. Place a piece of parafilm, big enough to accommodate your coverslips, on a pre-wetted paper towel. Pipette a 200-μL drop of primary antibody solution onto the parafilm. Using tweezers lift the coverslip up from the 6-well plate, and remove the excess of blocking solution by gently tapping on absorbent paper. Place the coverslip on the antibody drop, cells facing down, and incubate for 1 h under a cover (*see* **Note 14**).
10. Transfer the coverslips back to the 6-well plate, making sure that cells are facing up. Wash three times 5 min each with 2 mL of PBST.
11. Dilute appropriate fluorophore-conjugated secondary antibodies in blocking solution according to manufacturer's recommendations for immunofluorescence microscopy (*see* **Notes 15** and **16**)
12. Perform labeling with secondary antibody and then wash as described in **steps 9** and **10**, but keep in the dark from now onward.
13. Optionally, postfix by transferring coverslips to 4 % paraformaldehyde in PBS for 10 min. Wash as in **step 10.**
14. Aspirate wash buffer and add 1 mL of 2 μg/mL DAPI in PBS per well. Incubate for 5 min (*see* **Note 17**).
15. Transfer coverslips to deionized water (2 mL per well in a 6-well plate) and leave for 5 min.
16. Remove as much water as possible but without drying the cell layer (*see* **Note 18**). Either mount coverslips in 13 μL of ProLong Gold, leave in the dark overnight and only seal with

nail polish the following morning, or mount in 13 μL of Vectashield and seal with nail polish immediately (*see* **Note 19**). Other mounting agents may be used, but they should first be checked for compatibility (minimal chromatic aberration) with the super-resolution system.

17. Image slides on OMX 3D-SIM according to directions for system handling from the system manager (*see* **Notes 20** and **21**). We use the OMX V3 Blaze microscope (GE Healthcare, UK) equipped with a 60×/1.42 oil UPlanSApo objective (Olympus), 405-nm and 488-nm and 593-nm diode lasers, and sCMOS cameras (PCO AG, Germany).
18. It is important to minimize spherical aberrations by matching to the respective OTFs using immersion oil of preselected refractive indices (RIs). For imaging thick specimens, best results are typically obtained with RI 1.514 for depth adjustment in the region of optimal reconstruction a few μm into the sample.
19. Acquire 3D-SIM image stacks with five phases, three angles per image plane, and 0.125-μm *z*-distance between sections. Be careful to ensure that the full *z*-axis depth of the region of interest is included in the acquisition stack. Note that the minimum *z*-depth for effective 3D-SIM reconstruction is 1 μm.

3.4 Processing of Image Data

1. The raw data can be computationally reconstructed with SoftWoRx 6.0 (Applied Precision). Apply Wiener filter settings 0.002 and channel-specific measured optical transfer functions (OTFs) to generate super-resolution 3D image stacks for each channel. Depending on the wavelength, it is possible to obtain lateral (*x*–*y*) resolution of 100–130 nm and an axial (*z*) resolution of ~300 nm.
2. For multicolor images, images from the different color channels should be registered with alignment parameter obtained from calibration measurements with 0.2-μm-diameter TetraSpeck beads (Life Technologies) using the OMX Editor software.
3. The reconstructed super-resolution 3D images can be further analyzed using appropriate 3D-capable software, for example, Volocity (Perkin-Elmer) or Imaris (Bitplane). Such software permits free three-axis rotation and re-slicing of the volume along arbitrary planes, revealing internal 3D structure as shown (Fig. 2). While voxel-specific co-localization tools may be applied to such data, caution is required as chromatic aberration makes exact registration of multiple fluorophore channels a challenge.

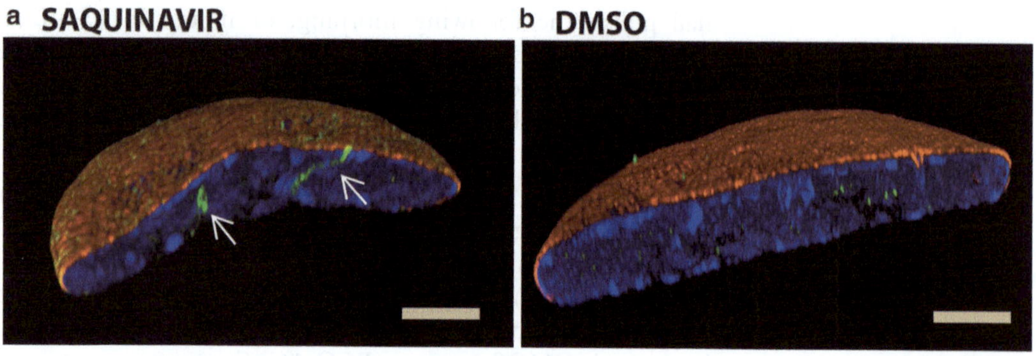

Fig. 2 Super-resolution microscopy with 3D structured illumination of the nuclear lamina in human dermal fibroblasts. Cells were exposed to saquinavir drug (**a**) or vehicle control (DMSO) (**b**) and labeled with antibodies against lamin A/C or specific for prelamin A. *Arrows* in (**a**) indicate strongly enriched prelamin A detection at the NR tubules. Lamin A/C shown in orange, prelamin A in *green*, and DAPI in *blue*

3.5 Visualizing Nuclear Import of a Protein of Interest with Photoactivatable Green Fluorescent Protein (PA-GFP)

1. Tag the cDNA for your gene of interest with cDNA encoding photoactivatable green fluorescent protein (PA-GFP) in a mammalian expression vector.
2. Seed cells in a glass bottom gridded dish (*see* **Note 22**), and transfect with a vector encoding your protein of interest tagged with PA-GFP by following the protocol specific for your transfection reagent (*see* **Note 23**). If your microscope setup does not have an environmental chamber, seed and transfect cells on appropriate coverslips for whatever live-cell imaging setup is available.
3. Allow 24–48 h for expression of the tagged protein (*see* **Note 24**).
4. Set glass bottom dishes containing transfected cells in a confocal microscope environmental control chamber maintaining 37 °C and 5 % CO_2 supply. Use high magnification (Plan Apo 63× 1.4 N.A. objective) with immersion oil.
5. Focus on cells in non-confocal transmission phase-contrast mode (*see* **Note 25**).
6. Switch to confocal mode and image cells using a laser excitation wavelength of 488 nm and an emission filter of 505 nm—long pass. Adjust laser transmission and detection gain to make sure that you are not detecting any background green fluorescence (*see* **Note 25**). If an RFP-fusion encoding plasmid has been co-transfected, it is best to do most of the initial focusing and setup in the RFP channel. If the RFP is fused to a nuclear envelope protein, then it is also easy to distinguish the cytoplasm from the nucleus.
7. In order to check that PA-GFP is adequately expressed and targeted, begin with a control photoactivation of an entire cell by scanning with the 405-nm laser at full power. Check by

imaging in the GFP channel that photoactivation has been achieved (*see* Fig. 3a). Use this step to assess the subcellular distribution of the PA-GFP-tagged protein and to ascertain the duration and intensity of 405-nm laser illumination required to give complete photoactivation.

8. Select a cytoplasmic region of interest (ROI), avoiding clipping the nucleus, and illuminate it with the 405-nm laser, using the settings determined in **step 7** in order to photoactivate PA-GFP on nascent cytoplasmic copies of the tagged protein (*see* **Note 26** and illustration of selection of an appropriate ROI in Fig. 3b). The concentration of PA-GFP molecules in the cytoplasm is low, and it is not possible to detect immediate photoactivation in the cytoplasm. Only when multiple copies of a photoactivated PA-GFP-tagged protein are trafficked to a specific subcellular location does it become possible to detect a GFP signal by confocal imaging.

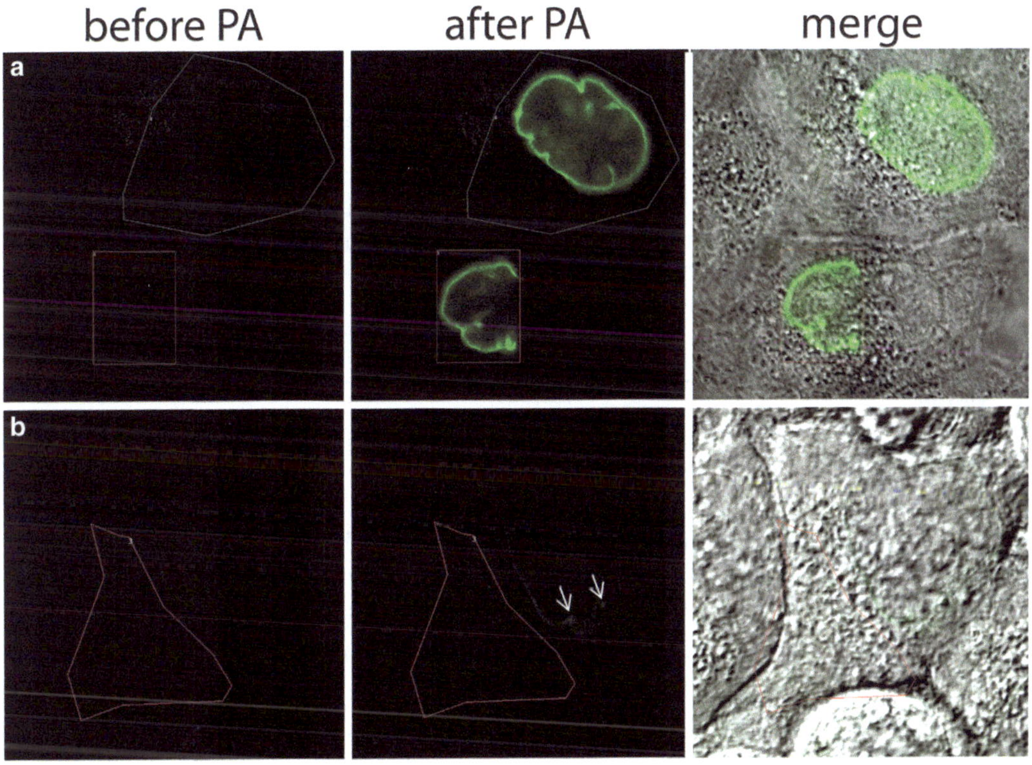

Fig. 3 Nuclear import of lamin B1 tagged with photoactivatable green fluorescent protein (PA-GFP). Hela cells were transfected with a vector encoding PA-GFP lamin B1 and live imaged 24 h later. Indicated regions of interest (ROIs) were subject to a 405-nm laser exposure. (**a**) *Top panel* shows photoactivation (PA) of lamin B1 already assembled into the nuclear lamina at the start of the experiment. The "After PA" image was taken immediately post-PA. (**b**) *Bottom panel* represents sites of incorporation of newly synthesized PA-GFP lamin B1 that has been photoactivated in the cytoplasm and then imported into the nucleus. A stronger PA-GFP signal is associated with membrane invaginations (*white arrows*). The "After PA" image was taken 15 min post-PA

9. Image PA-GFP with a 488-nm laser with settings determined in **step 6**. Acquire an image every 15–20 min for 4 h, beginning immediately after 405-nm activation, to monitor trafficking of the PA-GFP-tagged protein of interest (*see* **Notes 27** and **28**).

3.6 Data Analysis of PA-GFP-Tagged Protein Traffic

1. It may be necessary to adjust the time-lapse interval depending on the rate of your biological process of interest (*see* **Note 28**). For example, the lamin B1 targeting to NR and NE shown in Fig. 3b was obtained only 15 min after the photoactivation pulse.
2. The time-lapse image data may then be analyzed as required. In the example shown, it is the spatial distribution of the green signal from the previously photoactivated PA-GFP tags that is of interest (*see* Fig. 3b). A green GFP signal absent from the immediate post-activation image, but appearing subsequently, is indicative of trafficking of the tagged protein from the site of photoactivation in the cytoplasm to the NE and most prominently to NR, as shown here (arrows in Fig. 3b).
3. Quantitative analysis of the time series can be achieved using image processing software (we have used ImageJ and MetaMorph, although many other packages would be suitable). For example, defining an ROI that covers the subcellular destination of a tagged protein and quantitating signal intensity over time using image kinetic tools (e.g., those used to obtain FRAP recovery curves) permit estimation of delivery rate to specific sites.

4 Notes

1. Cells treated with regular stearic acid will be used for tuning the NanoSIMS instrument for background determination. It will also be a valuable control determining if there are any effects of deuterium on cell morphology.
2. We recommend preparing each treatment condition in replicates, because it is not uncommon to lose cells during embedding with resin.
3. Depending on your cell type, you may need to adjust the concentration of stearic acid used in the treatment and the incubation time. We found that increasing the concentration to 100 μM results in a very high ^{2}H signal, but also strongly affects cell morphology.
4. Remember to use personal protective equipment and do the fixation in a chemical fume hood, avoiding inhalation.
5. Cells can be left in the fixative for a couple of days at 4 °C without deterioration.

6. Osmium tetroxide is acutely toxic and must be handled with care. Always wear personal protective equipment, and do not open osmium tetroxide containers outside of a chemical fume hood. Dispose of osmium waste according to local procedures.
7. The major hazards of liquid nitrogen are associated with the properties of extreme cold and evaporation (asphyxia by exclusion of oxygen). Make sure to handle it with care.
8. The BSE signal is sensitive to atomic number contrast. Hence it is possible to image biological samples stained with heavy metals like osmium. Low-voltage images become extremely surface sensitive, because with a low-energy incident beam, the interaction volume decreases rapidly, allowing a lateral resolution of a few nanometers to be achieved in a modern SEM.
9. Maintaining the surface sensitivity for both BSE and NanoSIMS imaging is important to ensure that precisely the same material is analyzed in both techniques.
10. In general, cells do not adhere well to glass coverslips. If that is an issue with your cell type, you can pre-coat coverslips with collagen or poly-l-lysine. Non-adherent cells can be spread on concanavalin A-coated coverslips and left for 2 h before processing them.
11. Instead of using FSG, you can block cells with 3 % bovine serum albumin. The blocking agent should be chosen that yields the lowest backgrounds with your chosen antibody.
12. Depending on your instrument setup, you may be able to image in multiple channels. You can simultaneously immunolabel more than one antigen, providing you have compatible primary antibodies produced in different hosts and secondary antibodies that have minimal cross-reactivity with the heavy chains of the other species used.
13. Note that 3D-SIM requires a high signal-noise ratio (SNR) for effective super-resolution image reconstruction; it is important to select primary antibodies that yield high SNR images in the specimens of interest. You may have to optimize antibody concentration for your experiment, but from our experience, most commercial antibodies, and those listed here in particular, work fine in the 1:50–1:100 dilution range.
14. You can incubate cells with primary antibody overnight at 4 °C. However, since you are dealing with very small volumes, you will then have to limit evaporation by, for example, using a humidified chamber.
15. Make sure that you select secondary antibodies conjugated to fluorophores that precisely match the laser lines of your super-resolution instrument. Because so many images must be taken

for 3D-SIM, you want to be able to maximize the signal with the lowest excitation possible to avoid photobleaching. It is also important to select bright (i.e., high quantum yield) fluorophores and to use anti-fade agents in the mounting medium as described. In multichannel experiments it is better to select the closest emission wavelengths that can be spectrally resolved, so that the compromise required for correction of multiple wavelength-dependent (chromatic) aberrations is minimized.

16. From our experience, Alexa-conjugated commercial antibodies work fine when diluted 1:300. However, you may have to optimize the concentration for your particular experiment.
17. Do not use mounting medium containing DAPI as it will lead to high background signal and will affect the quality of your reconstructed images.
18. Use a paper towel to soak in the water excess from coverslips. You can touch the cells around the coverslip edges because you will not be imaging around the coverslip periphery, while it helps to remove most of the water. Although, be careful not to dry the cell layer completely.
19. Air bubbles are detrimental and can lead to many artifacts upon reconstruction; therefore make sure you do not trap any under the coverslip. They are best avoided by placing a needle under one side of the coverslip and slowly withdrawing it while lowering the coverslip into place.
20. Since acquiring images in super-resolution microscopy is a time-consuming process (you should estimate roughly 20–30 min per cell imaged), it is a good practice to check the antibody labeling quality on a standard microscope prior to imaging with OMX.
21. Make sure that prior to taking images of your specimen, you image TetraSpeck beads (or an equivalent multiple fluorophore-conjugated specified-sized bead) with all the colors you are going to use for your cell sample. Having several good images of TetraSpeck beads will be crucial for determination of alignment parameters for images of your specimen obtained with different color channels and in calculating the chromatic aberration for correcting images. Note that chromatic aberration should be measured at the start of every imaging session and on a long day should be imaged twice.
22. Using a grid-marked dish helps to orientate on the dish surface and makes tracking cells much easier, should you wish to remove the dish from the microscope stage and come back to it in a longer time interval.
23. Co-transfecting a vector encoding, for example, red fluorescent protein (RFP), proves to be useful in identifying trans-

fected cells without affecting PA-GFP visualization. In particular, the RFP when fused to a nuclear envelope protein can also be useful for orienting the nucleus and ER prior to photoactivation.

24. We recommend changing cell culture medium just prior to starting microscopy to remove dead cells and debris as this can obstruct imaging by interfering with light paths and also frequently produces autofluorescence.
25. You may notice that PA-GFP signal leaks even before exposure to a 405-nm laser, probably due to spontaneous photoactivation. We recommend avoiding exposure of transfected cells to very short wavelength visible or UV light, present also in daylight. Beware of cell finding using white-light phase contrast imaging; set up phase contrast through a red filter to minimize unwanted photoactivation during experimental setup.
26. If you are not getting sufficient photoactivation of GFP, you may need to consider increasing the number of iterations for 405-nm laser bleaching.
27. It is recommended to photoactivate some whole cells in a different field of view prior to live-cell imaging, in order to determine the expected maximal intensity of photoactivated GFP at the final cellular destination of the tagged protein and to adjust laser settings accordingly.
28. Prolonged exposure to laser light will cause damage to cells and cell death eventually; therefore it may be necessary to make a compromise, adjusting intervals between laser scans, illumination intensity, as well as overall experiment duration.

Acknowledgments

This work was supported by MRC grant G0801917 and BBSRC grant BB/L002159/1. The authors gratefully acknowledge assistance of the Micron BioImaging facility; Biochemistry Department; Oxford (super-resolution LM), together with Chris Grovenor and Haibo Jiang; and the Department of Materials (NanoSIMS).

References

1. Pohl S, Hasilik A (2015) Biosynthesis, targeting, and processing of lysosomal proteins: pulse-chase labeling and immune precipitation. Methods Cell Biol 126:63–83
2. Puvion-Dutilleul F, Puvion E (1995) Immunocytochemistry, autoradiography, in situ hybridization, selective stains: complementary tools for ultrastructural study of structure-function relationships in the nucleus. Applications to adenovirus-infected cells. Microsc Res Tech 31:22–43
3. Fricker M, Hollinshead M, White N, Vaux DJ (1997) Interphase nuclei of many mammalian cell types contain deep, dynamic, tubular membrane-bound invaginations of the nuclear envelope. J Cell Biol 136:531–544

4. Malhas A, Goulbourne C, Vaux DJ (2011) The nucleoplasmic reticulum: form and function. Trends Cell Biol 21:362–373
5. Jiang H, Favaro E, Goulbourne CN, Rakowska PD, Hughes GM, Ryadnov MG, Fong LG, Young SG, Ferguson DJ, Harris AL, Grovenor CR (2014) Stable isotope imaging of biological samples with high resolution secondary ion mass spectrometry and complementary techniques. Methods 68:317–324
6. Goulbourne CN, Vaux DJ (2010) HIV protease inhibitors inhibit FACE1/ZMPSTE24: a mechanism for acquired lipodystrophy in patients on highly active antiretroviral therapy? Biochem Soc Trans 38:292–296
7. Patterson GH, Lippincott-Schwartz J (2002) A photoactivatable GFP for selective photolabeling of proteins and cells. Science 297: 1873–1877
8. Ishikawa-Ankerhold HC, Ankerhold R, Drummen GP (2012) Advanced fluorescence microscopy techniques--FRAP, FLIP, FLAP, FRET and FLIM. Molecules 17: 4047–4132

Chapter 11

Analysis of Nuclear Lamina Proteins in Myoblast Differentiation by Functional Complementation

Olga Tapia and Larry Gerace

Abstract

We describe straightforward methodology for structure-function mapping of nuclear lamina proteins in myoblast differentiation, using populations of C2C12 myoblasts in which the endogenous lamina components are replaced with ectopically expressed mutant versions of the proteins. The procedure involves bulk isolation of C2C12 cell populations expressing the ectopic proteins by lentiviral transduction, followed by depletion of the endogenous proteins using siRNA, and incubation of cells under myoblast differentiation conditions. Similar methodology may be applied to mouse embryo fibroblasts or to other cell types as well, for the identification and characterization of sequences of lamina proteins involved in functions that can be measured biochemically or cytologically.

Key words Nuclear lamina, Lentivirus, siRNA, Myoblast, Functional complementation

1 Introduction

The nuclear lamina, a filamentous protein meshwork that lines the nucleoplasmic surface of the nuclear envelope (NE), comprises a polymeric array of nuclear lamins and associated transmembrane and peripheral proteins (*reviewed in* refs. 1–3). The lamina has been a topic of rapidly expanding interest over the last 20 years. In significant part, this is due to the discovery that over 15 human diseases are linked to mutations in nuclear lamins and lamina-associated proteins (which collectively may be termed "laminopathies" or "nuclear envelopathies") (*reviewed in* refs. 4–6). Whereas the majority of these diseases are caused by mutations in the gene encoding lamins A/C, similar pathologies can be caused by mutations in lamina associated transmembrane proteins such as emerin [4–6]. This suggests an intimate functional connection between these physically linked components. Most disease-causing mutations in nuclear lamina proteins affect striated muscle, but they also can target adipose tissue, bone, neurons, or multiple organ systems such as seen with Hutchinson-Gilford progeria syndrome [4–6].

Sue Shackleton et al. (eds.), *The Nuclear Envelope: Methods and Protocols*, Methods in Molecular Biology, vol. 1411, DOI 10.1007/978-1-4939-3530-7_11, © Springer Science+Business Media New York 2016

One of the major challenges in the field is to understand the molecular basis for disease caused by lamina protein mutations. Achieving this goal likely will require a detailed structure-function analysis of individual nuclear lamina proteins.

Since mutations in lamina components most commonly affect striated muscle, we have screened lamina-associated proteins for a potential role in muscle biology using cultured myoblast differentiation assays. Our work identified three new NE transmembrane proteins involved in myoblast differentiation: Lem2 [7], Net37 [8], and Net39 [9]. This work extended studies revealing that lamins A/C and emerin have a role in cultured myoblast differentiation [7, 10] and suggests that a diverse range of nuclear lamina proteins have important functions in the muscle. We have used the well-studied C2C12 murine myoblast cell line for these functional studies, as C2C12 cells readily differentiate into myotubes in culture (Fig. 1) and have been shown to closely recapitulate the myogenic differentiation program observed with primary myoblast cultures [11, 12].

In developing methods for ectopic expression of nuclear lamina proteins in C2C12 cells, we found that lentiviral methods are superior in many respects to conventional lipid-based transient transfection approaches [13]. In our hands, ectopic protein expression by lipid transfection yields highly variable cell-to-cell expression levels, and the ectopic proteins often accumulate in large cytoplasmic aggregates as well as at the NE in the majority of cells, regardless of the expression level (shown in Fig. 2a for Lem2). By contrast,

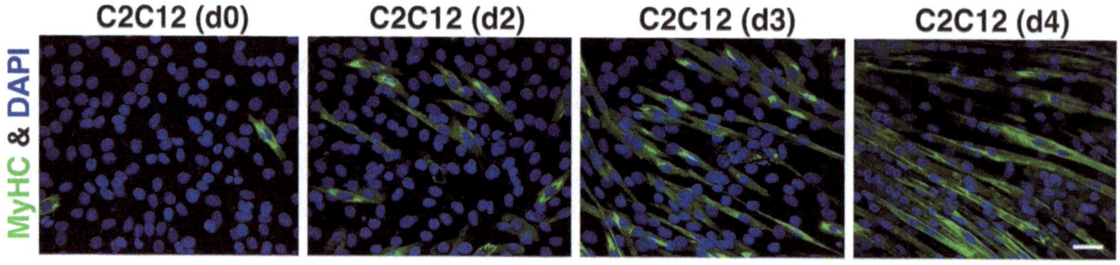

Fig. 1 The C2C12 myoblast differentiation model. Micrographs of C2C12 cells showing immunofluorescence (IF) labeling of skeletal myosin heavy chain (MyHC), a typical marker of terminally differentiated myotubes (*green*, MyHC; *blue*, DNA). MyHC is virtually absent in C2C12 cultures grown to confluency before differentiation (d0, *left*), but is highly induced in multinucleated myotubes incubated in differentiation medium for 4 days (d4, *right*). Cells stained for MyHC by IF and DAPI are used to calculate the myogenic index, defined as the percentage of nuclei present in MHC-positive cells (*see* Subheading 3.3.3). We commonly score myogenic index at d4, but in some experiments the myogenic index can increase up to day 6. Scale bar, 40 μm

Fig. 2 (continued) levels, but also is found in cytoplasmic aggregates. (**a**, *lower*) High-magnification view emphasizes the presence of V5-Lem2 in cytoplasmic aggregates. This pattern occurs in nearly all V5-positive cells, regardless of the expression level. (**b**, *upper*) Lentivirus-infected cells (*arrows*) exhibit more similar cell-to-cell expression levels of V5-Lem2. Moreover, little of the protein is found in cytoplasmic aggregates. (**b**, *lower*) Higher-magnification view shows the strong localization of V5-Lem2 at the NE in the virtual absence of cytoplasmic staining. Scale bars, 40 μm (**a** and **b**, *upper*) and 5 μm (**a** and **b**, *lower*)

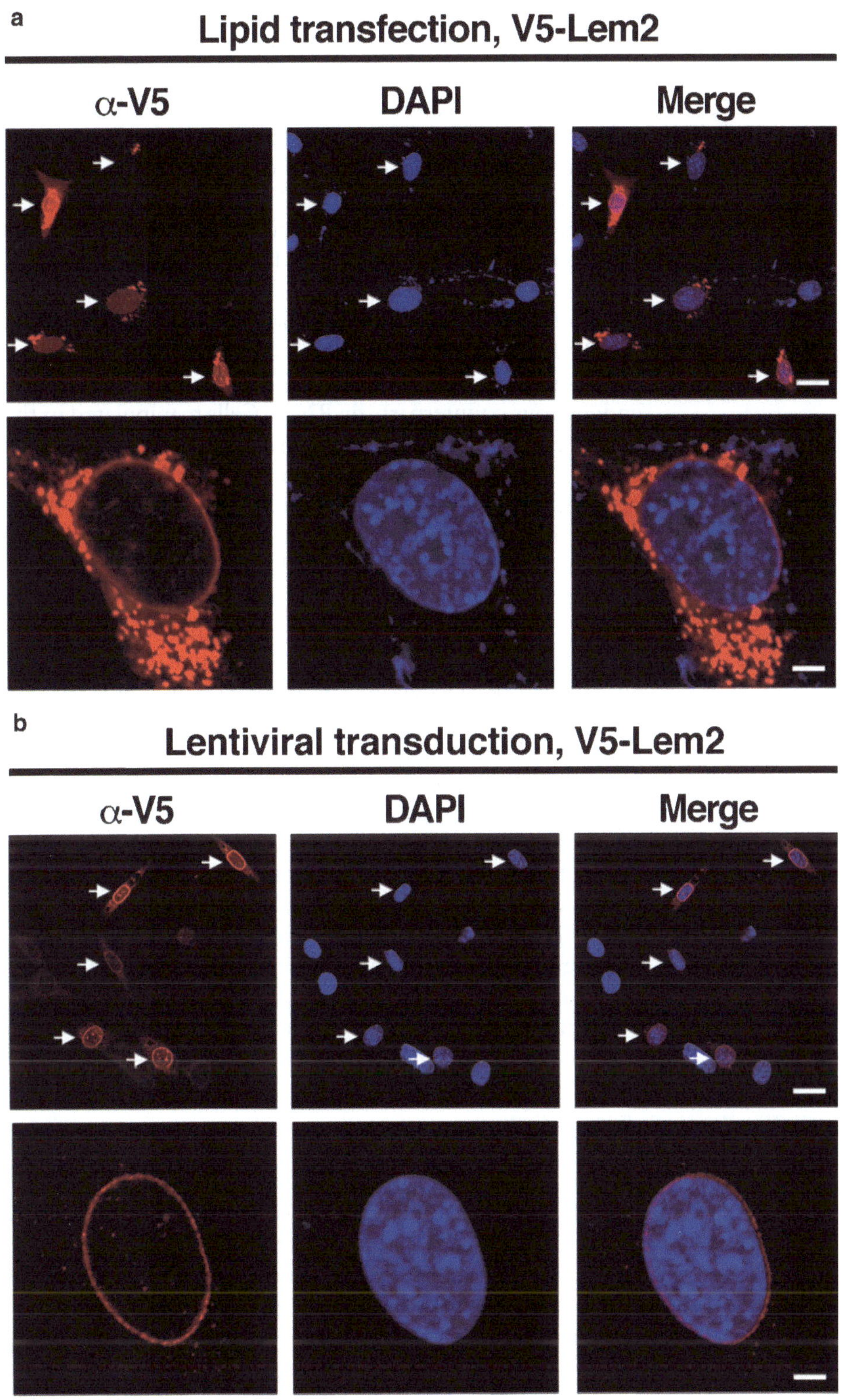

Fig. 2 Comparison of lipid-mediated transfection and lentiviral-mediated gene delivery in C2C12 cells. (**a** and **b**) Evaluation by IF (*see* Subheading 3.2.2) of C2C12 cells expressing V5-Lem2 at 24 h after transfection with Lipofectamine 2000 using the manufacturer's instructions (**a**), or at 24 h following infection with lentivirus expressing V5-Lem2 (*see* Subheading 3.2.1) (**b**). (**a**, *upper*) Transfected V5-positive cells (*arrows*) exhibit a wide cell-to-cell range in the levels of V5-Lem2. The ectopic protein is localized to the NE at various

lentiviral expression yields much more uniform cell-to-cell levels of ectopic protein expression without substantial accumulation in cytoplasmic aggregates (shown in Fig. 2b for Lem2). We have obtained similar results in mouse embryo fibroblasts (MEFs) and other cell types.

Here we describe a facile strategy for carrying out structure-function analysis of nuclear lamina proteins in myoblast differentiation using C2C12 cells. Our approach involves lentivirus-mediated expression of ectopic versions of the lamina proteins in C2C12 cells [13] and isolation of bulk-selected cell populations expressing the ectopic proteins, followed by depletion of the endogenous counterparts by RNAi. Cells manipulated in this manner are then analyzed by a myoblast differentiation protocol for the ability of the mutant (or wild-type) proteins to functionally complement the loss of the endogenous proteins (Fig. 3a). An example of this is shown for Lem2 (Fig. 3b), whose depletion strongly inhibits differentiation of C2C12 cells [7]. Using the functional complementation protocol described, differentiation is fully restored by ectopic expression of wild-type Lem2 (Fig. 3b). One technical advantage of this approach is that the rapid selection/knockdown strategy allows the analysis of mutant versions of lamina proteins that might be toxic to cells in long-term culture or that can induce compensatory changes in gene expression that may obscure primary phenotypes.

In addition to using the approaches described here to study myogenic differentiation of C2C12 cells, we have implemented similar protocols to functionally dissect nuclear lamina proteins in mouse embryo fibroblasts (MEFs) in relation to aberrant regulation of ERK signaling, which is often seen with laminopathies [5]. More generally, the methods described here should be applicable to the analysis of a diverse range of cultured cell models using various functional assays, due to the wide cell tropism of lentivirus [14] and the high efficiency of siRNA delivery to most cell types.

2 Materials

2.1 Cells

1. 293T/17 packaging cells (catalog number CRL-11268) (ATCC), a clonal derivative of the 293T cell line that constitutively expresses the simian virus 40 (SV40) large T antigen. These cells are transfected efficiently and support production of high lentivirus titers.
2. C2C12 cells (catalog number CRL-1772) (ATCC), an immortalized murine myoblast line. These cells undergo efficient differentiation into multinucleated myotubes upon growth factor withdrawal [12].
3. Growth medium: Dulbecco's Modified Eagle Medium (DMEM) [+] 4.5 g/L D-glucose, [+] 110 mg/L sodium pyruvate, [−]

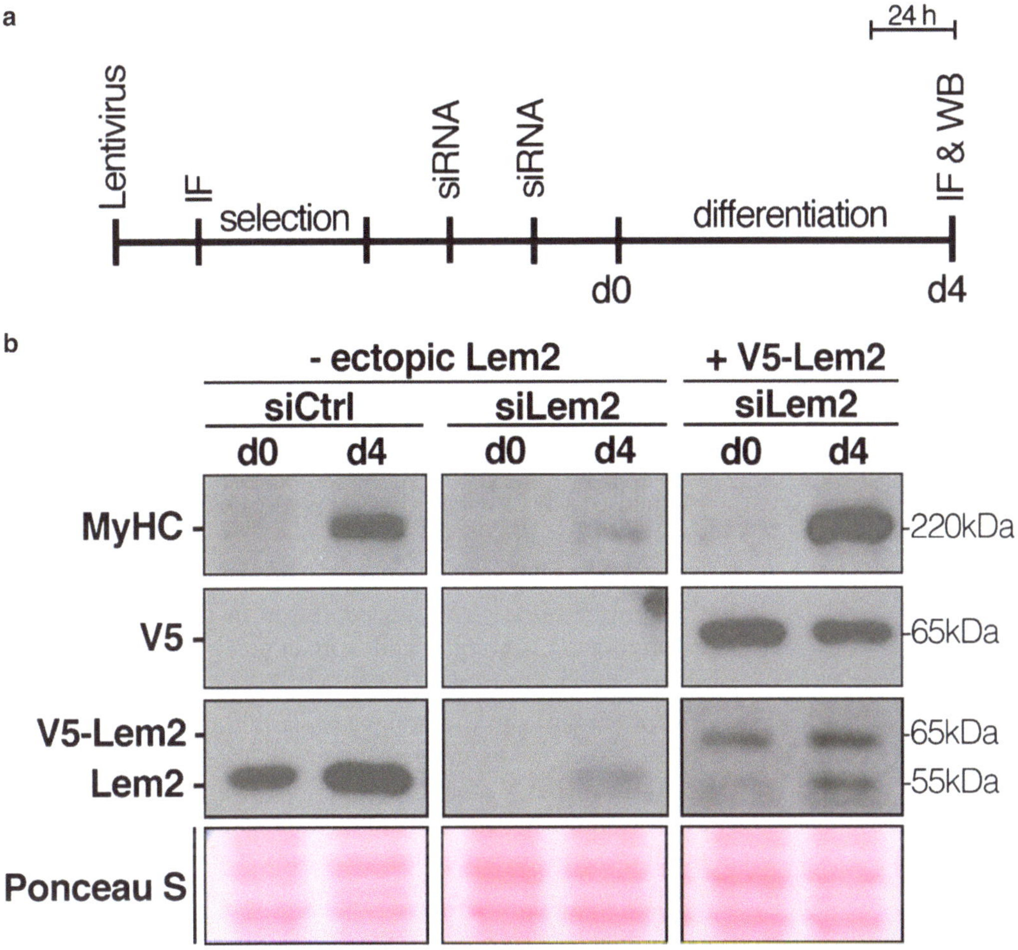

Fig. 3 Functional complementation analysis of nuclear lamina proteins in myoblast differentiation. (**a**) Schematic diagram of the experimental protocol used to analyze the effects of wild-type or mutant versions of nuclear lamina proteins on C2C12 myogenic differentiation (*see* Subheading 3.3). First, cells are transduced with lentivirus to express an ectopic epitope-tagged version of a nuclear lamina protein (or a mutant version thereof). Populations of virus-transduced cells are then selected by growth in antibiotic-containing medium. Next, cultures are subjected to two consecutive transfections with either a nontargeting control siRNA or with an siRNA targeting the 3′ or 5′ UTR of the mRNA for the endogenous nuclear lamina protein and grown to confluency (d0). Finally, after 4 days of differentiation, cells are assayed by IF and Western blotting (WB). Scale bar, 24 h. (**b**) An example of this protocol, in which expression of wild-type ectopic Lem2 is used to complement knockdown of endogenous Lem2 to support myogenic differentiation. The experiment compares non-transduced C2C12 cells ("C2C12") to C2C12 populations expressing ectopic Lem2-V5 ("C2C12-V5-Lem2"). Shown is a Western blot analysis at d0 and d4 of the differentiation protocol (*see* Subheading 3.3.4), using antibodies to detect the myogenic marker MyHC (*upper panel*), ectopically expressed Lem2 (*middle panel*), or both ectopic and endogenous Lem2 (*lower panel*). Cells transfected with nontargeting siRNA (siCtrl) or with siRNA directed against the 3′ UTR of Lem2 mRNA (siLem2) are indicated. The *upper panel* reveals strong upregulation of MyHC in d4 control cells, indicating strong myogenic differentiation. Differentiation is strongly inhibited in cells treated with siRNA to deplete endogenous Lem2 and is rescued by expression of ectopic silencing-resistant Lem2

L-glutamine supplemented with 10 % fetal bovine serum (FBS), 1 % MEM nonessential amino acids (MEM NEAA 100×), 100 U/mL penicillin, 100 μg/mL streptomycin, and 2 mM L-glutamine.

4. Differentiation medium: DMEM supplemented with 2 % donor equine serum (catalog number SH30074.02) (HyClone), 1 % MEM NEAA, 100 U/mL penicillin, 100 μg/mL streptomycin, and 2 mM L-glutamine.
5. PBS (Dulbecco's Phosphate Buffered Saline): Diluted solution prepared from stock 10× PBS in sterile H_2O.
6. Plate-coating solution: 0.15 % (w/v) gelatin in sterile H_2O.

2.2 Calcium Phosphate Transfection

1. 2× HEPES-buffered saline (HBS 2×): 50 mM HEPES, 1.5 mM Na_2HPO_4, 280 mM NaCl, 10 mM KCl, and 12 mM D-glucose. Adjust separate aliquots to pH 7.05, 7.1, and 7.15, sterilize by passing through a 0.22-μm filter, and test for optimal transfection (*see* **Note 1**).
2. Calcium chloride: 2 M stock solution in distilled water. Pass through a 0.22-μm filter and store at 4 °C.
3. Packaging plasmids for production of third-generation lentivirus, obtained from the nonprofit plasmid repository Addgene (http://www.addgene.org): plasmids encoding Gag and Pol (pMDL/pRRE; catalog number 12251), Rev (pRSV-Rev; catalog number 12253), and VSV-G (pCMV-VSV-g; catalog number 8454).
4. Lentiviral transfer vector encoding gene of interest (GOI) or GFP driven by the CMV promoter, e.g., pLenti CMV-GFP Puro (catalog number 17448) (Addgene) (*see* **Note 2**). Lentivirus transfer vectors in which the GOI is driven by other promoters also may be used.
5. 28 mm diameter syringe filter, 0.45 μm pore SFCA membrane (e.g., catalog number 431220) (Corning Incorporated).
6. Polybrene transfection reagent (e.g., catalog number TR-1003-G) (EMD Millipore).
7. Puromycin.

2.3 Immunofluorescence Staining

1. PBS: 10 mM disodium phosphate (Na_2HPO_4), 2 mM potassium phosphate dibasic (KH_2PO_4), 137 mM NaCl, and 2.7 mM KCl; pH 7.4.
2. Fixative: 4 % paraformaldehyde solution in PBS, diluted from a 16 % stock solution. Keep in dark and use within 2 h.
3. Permeabilization buffer: 0.5 % Triton X-100 in PBS.
4. Blocking buffer: 3 % (w/v) BSA (bovine serum albumin fraction V) in 0.05 % Tween-20 in PBS. Pass through a 0.22-μm filter and store at 4 °C.

5. Mounting medium containing antifading reagent, e.g., ProLong Gold Antifade Mountant with DAPI (Life Technologies).
6. Micro cover glasses, 12-mm circle no.2.
7. Primary antibodies: mouse monoclonal anti-MyHC (skeletal myosin heavy chain) (e.g., catalog number M9850-15B) (US Biologicals) and either mouse monoclonal anti-V5 tag or rabbit polyclonal anti-V5 tag.
8. Secondary antibodies: anti-mouse, Cy3-labeled; anti-mouse, Alexa Fluor 488-labeled; anti-rabbit Cy3-labeled; or anti-rabbit Alexa Fluor 488-labeled. When dual labeling is being undertaken, it is important to use secondary antibodies with minimal cross-reactivity between mouse and rabbit.

2.4 siRNA Transfection

1. OptiMEM (reduced serum medium, [+] Hepes, [+] 2.4 g/L sodium bicarbonate, [+] L-glutamine).
2. DharmaFECT-1, transfection reagent optimized for efficient and low-toxicity small RNA delivery (Dharmacon) (*see* **Note 3**).
3. 5× siRNA buffer (Dharmacon).
4. Dicer-substrate siRNA duplexes. Use siRNAs directed against the 3′ or 5′ untranslated region (UTR) of the target mRNA and as a control, a universal nontargeting siRNA (e.g., catalog number DS NC1) (IDT) (*see* **Note 4**). The transduced mRNA, which contains only the coding region of the transduced protein, will be silencing resistant. Dissolve the siRNA duplexes in siRNA buffer (prepared by diluting 5× siRNA buffer stock 1:5 in sterile H_2O) to a final concentration of 20 μM (20 pmol/μL).

2.5 Western Blotting

1. Tris-glycine electrophoresis buffer: 25 mM Tris base, 192 mM glycine (electrophoresis grade), and 0.1 % SDS; pH 8.3.
2. Transfer buffer (Towbin formulation): 25 mM Tris base, 192 mM glycine, and 20 % ethanol; pH 8.3.
3. 4–20 % Tris-Glycine Mini Protein Gels, 1.5 mm, 15 wells.
4. Nitrocellulose membranes, 0.2-μm pore size.
5. Ponceau S: 0.2 % (w/v) stock solution in 5 % acetic acid. Store in the dark at room temperature.
6. SDS sample buffer: 50 mM Tris–HCl, 2 % SDS, 0.1 % bromophenol blue, and 10 % glycerol; pH 6.8. Supplement with 100 mM dithiothreitol (DTT) and cocktail of protease inhibitors, e.g., complete mini EDTA-free (catalog number 11836170001) (Roche). If using the listed cocktail, prepare a 100× stock solution of the protease inhibitor cocktail by dissolving one tablet in 100 μL of H_2O, aliquot, and store at −20 °C.
7. Pre-stained protein ladder.

8. Washing solution: 0.05 % Tween-20 in PBS (PBS-T).
9. Blocking solution: 5 % (w/v) instant nonfat dry milk in PBS-T.
10. Primary monoclonal mouse antibodies anti-V5 tag and anti-MyHC (*see* **item** 7 in Subheading 2.3) and antibodies against nuclear lamina proteins for comparison of endogenous and ectopic protein expression levels.
11. Secondary horseradish peroxidase-conjugated antibodies for mouse and rabbit antibodies listed above.
12. Chemiluminescence detection solutions for horseradish peroxidase conjugates, e.g., SuperSignal West Pico Luminol (Thermo Scientific).
13. Autoradiography film.
14. Imaging and quantification system for chemiluminescence, e.g., BioSpectrum 810 Imaging System and VisionWorksLS Image Acquisition and Analysis Software v.7.1 (UVP).

3 Methods

Lentiviruses produced by the methods described herein, which are pseudotyped with the VSV-G protein, are tropic to a wide range of cultured cells (*see* **Note 5**). Unless otherwise stated, procedures are carried out at room temperature and under sterile conditions in BL2 (biosafety level class II) cabinets. Cells are grown in a 37 °C incubator in the presence of 5 % CO_2.

3.1 Lentiviral Production by Calcium Phosphate Transfection

1. Plate 293T cells the day before transfection by splitting an 80–90 % confluent 10-cm plate into 3 × 10-cm plates, so that cells are ~80 % confluent the day of transfection. Virus produced by each 10-cm plate is sufficient to infect 6–12 3.5-cm plates of C2C12 cells, depending on the virus titer (*see* **step 10** in Subheading 3.1).
2. The day of transfection, remove medium and add 6 mL of fresh DMEM + 10 % FBS 1 h prior to addition of the transfection mixture to the 293T cells.
3. To generate the transfection mixture for each 10-cm plate, dilute the following plasmid combinations in 350 μL H_2O in a sterile 4-mL polypropylene conical tube: 8 μg lentiviral transfer vector encoding GOI (*see* **Notes 2** and 7), 8 μg pMDL-RRE, 6.3 μg pRSV-REV, and 3 μg pCMV-VSVg. Adjust final volume to 395 μL with H_2O. Include one transfection mixture with a transfer vector encoding GFP (LV-GFP) for verification of transfection efficiency.
4. Add 55 μL of 2 M $CaCl_2$ and mix well.
5. Add the plasmid/$CaCl_2$ mixture dropwise to 450 μL HBS 2× while vortexing vigorously. The use of a proper mixing procedure

at this point is critical for success. Incubate the mixture for 15 min at room temperature.

6. Add the transfection mixture to each plate and rock the plate gently to mix. Incubate the cells overnight (16 h).

   After this step use extreme precaution while handling supernatants and materials exposed to lentivirus. While third-generation lentiviruses are self-inactivating (i.e., they can't replicate after genome integration), they can induce insertional mutagenesis. Treat all virus-contaminated tips, pipettes, plates, and discarded medium with 10 % bleach for 20 min to inactivate the virus.

7. After the overnight (16 h) incubation, check transfection efficiency in the sample producing LV-GFP using an inverted epifluorescence microscope. Approximately 80–90 % of the 293T cells should be GFP positive if the transfection is optimal.

8. Remove medium and add 6 mL of fresh DMEM + 10 % FBS (*see* **Note 6**) to each 10-cm dish. Return to the incubator for another ~36 h. Longer culture periods will lead to extensive cell death and reduction of lentivirus yield.

9. Using a 5-mL syringe, collect the lentivirus-enriched supernatants from each 10-cm dish, and filter through a 0.45-μm pore SFCA membrane filter for low protein binding (*see* **item 5** in Subheading 2.2) to remove cell debris. Keep the cleared supernatant on ice while making 0.5 mL aliquots. Store lentivirus-containing aliquots at −80 °C. Since virus titer decreases with repeated freezing and thawing, using a virus aliquot only once after thawing will promote reproducibility. When necessary, viruses can be concentrated in a Beckman 55Ti rotor at 21,000 rpm for 90 min, and the viral pellet is resuspended in 100 μL of HBS and stored at −80 °C in 10 μL aliquots.

10. Qualitatively assess the activity of the lentivirus preparation for transducing C2C12 cells using a drug selection protocol. Infect three 12-well plate wells with C2C12 cells plated the day before (~75,000 cells) with 125, 250, and 500 μL of lentivirus supernatant diluted in growth medium and supplemented with 8 μg/mL of polybrene. Adjust final culture volume to 1 mL. Follow the procedure described in **steps 3–6** in Subheading 3.2.1 and, after 48-h selection, determine the amount of lentivirus required to obtain ~60 % of puromycin (or other mammalian plasmid selection)-resistant cells. If quantitative determination of lentivirus titer is desired, follow one of the protocols described by Geraerts et al. [15].

3.2 Generation of Bulk-Transduced C2C12 Cell Populations

Analysis of specific nuclear lamina proteins in myoblast differentiation by functional complementation requires generation of cell populations that ectopically express a mutant or wild-type version of the protein of interest, ideally at an expression level near that of the endogenous protein (*see* **Note 7**). Although clones of cultured cells expressing an ectopic protein at relatively uniform levels can be isolated after various transduction methods, C2C12 cells and primary myoblasts lose differentiation capacity with extensive passaging. We have found that populations of C2C12 cells that are transduced with lentivirus to express nuclear lamina proteins and bulk selected for antibiotic resistance express similar cell-to-cell levels of ectopic proteins for 4–6 passages after selection (Fig. 2b) and that these populations are suitable for functional complementation studies in myoblasts (Fig. 3). We have used this method to successfully analyze a spectrum of deletion mutants of Lem2 to map regions required for myoblast differentiation (e.g., Fig. 3b) and for attenuation of ERK signaling.

3.2.1 Lentiviral Infection and Selection

After thawing, C2C12 cells should be cultured in growth medium (*see* **item 3** in Subheading 2.1) under subconfluent conditions (less than 50 %) and passaged every day in order to avoid spontaneous differentiation. We have observed that cells should not be used for differentiation assays after passage 12–15, as differentiation efficiency may become inconsistent. Thus, many vials of cells should be frozen at initial passages to allow the use of low-passage cells for complementation experiments (*see* **Note 8**):

1. Plate C2C12 cells at a density of 10,000–15,000 cells/cm^2 in growth medium (~150,000 cells for a 3.5-cm dish) and incubate overnight (16 h).
2. Dilute the amount of lentivirus supernatant required to transduce 40–60 % of the cells (*see* **step 10** in Subheading 3.1) in 1 mL (final volume) of growth medium supplemented with 8 μg/mL of polybrene (*see* **Note 9**).
3. Aspirate the growth medium, wash cells with PBS, and replace with lentivirus-containing growth medium. Incubate overnight (16 h).
4. The next day the plate should be ~50 % confluent. Replace the medium with fresh DMEM + 10 % FBS, and grow the infected cultures for 24–48 h to allow expression of the antibiotic resistance gene, passaging as needed to keep cells under subconfluent conditions. At the final passage, include a sterile micro coverslip at the bottom of the dish.
5. Replace growth medium with DMEM + 10 % FBS containing the selection antibiotic to all plates, including a noninfected control plate (*see* **Note 10**).
6. Allow growth of the infected cells for the selection time needed to achieve death of cells lacking the antibiotic resistance gene. This should be 48 h for cells cultured with 2 μg/mL puromycin. Verify that 100 % toxicity is achieved in the noninfected control plate.

7. Use sterile forceps to remove the micro coverslip, wash with PBS, and process the coverslip for the detection of the ectopic protein by immunofluorescence (*see* Subheading 3.2.2) to examine the expression level and subcellular localization of the ectopic protein, which should be expressed in virtually all cells.
8. Change to fresh growth medium, and expand the cell population expressing the transduced protein to a 10-cm plate to perform the assay for functional complementation of myoblast differentiation (*see* Subheading 3.3).

3.2.2 Immunofluorescence Staining

We visualize the localization of ectopic proteins in lentivirus-transduced cells using antibodies against an epitope tag fused to the N- or C-terminus. We commonly use a V5 tag, although other tags also are suitable. All steps are performed at room temperature except for the primary and secondary antibody incubation (**steps 6** and **8**):

1. Collect the glass 12-mm micro coverslip from the dish with bulk-selected lentivirus-transduced C2C12 cells, growing at ~15,000 cells/cm^2 (*see* **step** 7 in Subheading 3.2.1), and transfer to a clean 6-plate well.
2. Rinse the coverslip with PBS and add 2 mL of freshly prepared 4 % paraformaldehyde solution in PBS. Fix cells for 15 min.
3. Rinse the coverslips with PBS and permeabilize the cells with 0.5 % Triton X-100 in PBS for 10–15 min.
4. Perform three washes of 5 min with PBS followed by one wash with 3 % BSA in PBS for 15 min to limit nonspecific binding of the primary antibody.
5. Transfer the coverslip to a dry surface, remove excess moisture by blotting the edge, and add 50–100 μL of anti-V5 antibody diluted 1:100 in PBS. Incubate the coverslips for 3 h using a humid chamber (*see* **Note 11**).
6. Following the primary antibody incubation, perform three washes of 5 min with PBS.
7. Incubate the coverslip with the anti-mouse Cy3-conjugated secondary antibody diluted 1:500 in PBS, and incubate the coverslips for 45 min in a light-protected humid chamber.
8. Following the incubation time, perform three washes of 10 min with PBS.
9. Remove PBS from the coverslip by blotting the edge. Cover with 20 μL of antifade medium and remove by blotting the edge. Repeat this step, to assure complete removal of the last traces of PBS from the mounting medium.
10. Place the coverslip, cell side facing down, on a clean glass slide on top of 5–10 μL of antifade medium. Remove excess medium

with filter paper and seal with nail polish. Allow overnight (16 h) polymerization.

11. Use an epifluorescence or confocal microscope to validate ectopic protein expression and to analyze the subcellular localization pattern of ectopic proteins. Nuclear lamina proteins, when expressed at appropriate levels (*see* **Note 7**), should be strongly concentrated at the NE relative to the more peripheral ER (Fig. 2b).

3.3 Functional Complementation of Myoblast Differentiation

For functional complementation studies, we prepare bulk-selected populations of C2C12 cells expressing ectopic wild-type or mutant versions of the nuclear lamina protein. The experimental groups should include the following: (1) as a positive control, C2C12 cells transfected with nontargeting siRNAs where proper myoblast differentiation is achieved (*see* **Note 12**); (2) as a negative control, C2C12 cells treated with Dicer-substrate siRNA directed against the 5′ or 3′ UTR of the mRNA to knock down the protein that causes a defect in myoblast differentiation; and (3) the functional complementation assay, C2C12 cells stably expressing ectopic wild-type or mutant versions of the nuclear lamina protein where endogenous protein is knocked down by siRNA-mediated gene silencing. This last experimental group will reveal if ectopic mutant proteins are able to functionally rescue myogenic differentiation in cells where the endogenous protein has been depleted. Rescue of differentiation by the ectopic wild-type protein validates the specificity of the siRNA. For each experimental group, set up two plates (for analysis of initial cells and terminal differentiation, d0 and d4/d6) or 6 plates (for analysis of myoblast differentiation kinetics, d0, d1, d2, d3, d4, and d6) (*see* **Note 13**).

3.3.1 siRNA-Mediated Knockdown of Endogenous Protein

After isolating bulk-selected cells expressing GFP or a mutant or wild-type ectopic nuclear lamina protein, we typically carry out two sequential steps of siRNA treatment to deplete the endogenous protein. Our protocol for the latter is performed by "reverse" transfection, which involves addition of the siRNA solution to a trypsinized cell suspension at the time of plating. We commonly obtain 75–90 % knockdown of various nuclear lamina proteins by this method. For each experiment, we include a sample with a nontargeting control siRNA. We initially analyze two separate siRNAs against the target protein until functional complementation by the wild-type protein is established, to rule out off-target effects of individual siRNAs. The following procedure is described for reverse siRNA transfection of a 3.5-cm dish of C2C12 cells:

1. In tube A, add 10 μL of DharmaFECT-1 transfection reagent to 100 μL of OptiMEM, and in tube B, add 25–100 pmol of an appropriate siRNA (*see* **Note 14**), dissolved in siRNA buffer (*see* Subheading 2.4), to 100 μL of OptiMEM (tube B), and incubate for 5 min.

2. Add the DharmaFECT-1 solution (tube A) dropwise into the siRNA solution (tube B). Mix gently by pipetting up and down, and incubate at room temperature for 15 min.
3. Trypsinize C2C12 cell populations transduced with either GFP (the control) or the V5-tagged version of the target protein, and aliquot ~200,000 cells in 1 mL growth medium in a sterile 4-mL polypropylene conical tube.
4. Dropwise add the DharmaFECT/siRNA mixture to the cells and mix gently. After 10 min seed the cells into a 3.5-cm plate and incubate overnight (16 h) in 2 mL (final volume) DMEM + 10 % FBS.
5. The next morning, cells should be ~70 % confluent. Aspirate the medium, wash with PBS, and add 2 mL of fresh growth medium. Allow cells to grow for 8 h.
6. Perform a second siRNA reverse transfection using the same amount of siRNA as in the first transfection. Replate the cells to a 3.5-cm dish with a coverslip located at the bottom, both pre-coated with 0.15 % gelatin (*see* **Note 15**), and incubate overnight (16 h).
7. The next morning cells should be ~70 % confluent. Aspirate the medium, wash with PBS, and add 2 mL of fresh growth medium.
8. Incubate the plates for 24–48 h until the cells reach confluency. Subsequently carry out the myoblast differentiation assay (*see* Subheading 3.3.2).

3.3.2 Myoblast Differentiation Assay

1. Beginning with siRNA-transfected cells that are grown to confluency (*see* **step 8** in Subheading 3.3.1), remove the growth medium and replace with differentiation medium (*see* **item 4** in Subheading 2.1). Cells are considered to be at differentiation day 0 (d0).
2. Using sterile forceps, collect d0 coverslips, rinse with PBS, and fix with 4 % paraformaldehyde in PBS. Store at 4 °C in PBS until used for myogenic index analysis by immunofluorescence (*see* Subheading 3.3.3). Harvest the remaining cells from the 3.5-cm dish for Western blotting (*see* Subheading 3.3.4).
3. Allow cells to differentiate for 4 days (d4) (*see* **Note 16**). Under the phase-contrast microscope, elongated multinucleated myotubes should be readily visible in C2C12 cells treated with nontargeting control siRNA.
4. Collect d4 coverslips, rinse with PBS, and fix with 4 % paraformaldehyde, and process d0 and d4 coverslips for immunofluorescence staining of MyHC to determine the myogenic index of each experimental group (*see* Subheading 3.3.3). Harvest the remaining cells from the 3.5-cm dish for Western blotting (*see* Subheading 3.3.4).

3.3.3 Assessment of Myogenic Differentiation by Immunofluorescence

The myogenic index, which is a morphological parameter of differentiation, is defined as the fraction of nuclei residing in MyHC-positive cells. Coverslips collected on d0 and d4 are processed to identify cells expressing the skeletal muscle protein MyHC by immunofluorescence and to monitor nuclei by DAPI staining (*see* Subheading 3.2.2). Use an inverted epifluorescence or confocal microscope to record 3–4 micrographs (~100 cells per image) and use image analysis software (i.e., ImageJ, CellProfiler, AxioVision, Zen, etc.) for measurement of the myogenic index in the various cell populations. These values will allow evaluation of how knockdown of a protein of interest affects the differentiation capacity of cells and whether the ectopic protein functionally complements the knockdown by restoring the myogenic index to levels found in "control" cultures, i.e., C2C12 cultures treated with nontargeting control siRNA. If the myogenic index of control cultures is lower than 30 %, discard the lentivirally transduced cultures, and start the protocol with another low-passage vial of C2C12 cells.

3.3.4 Assessment of Myogenic Differentiation and Knockdown Efficiency by Western Blotting

The myogenic differentiation of C2C12 cells also can be assessed by Western blotting. Upregulation of the skeletal muscle protein MyHC at day 4 commonly is used to assess differentiation (Fig. 3b). In addition, the kinetics of differentiation can be determined by evaluating the expression of myogenin and MRF4 (Myf6), muscle-specific transcription factors that are induced at days 1–2 in the differentiation program. Finally, the knockdown of the endogenous nuclear lamina protein by siRNA (*see* Subheading 3.3.1) and the expression level of ectopic epitope-tagged protein relative to endogenous protein should be determined by Western blotting.

1. After removal of the coverslip, ~80 % of the cells remain in the plate. Wash twice with PBS and add 200 μL of SDS sample buffer (*see* **item 6** in Subheading 2.5). Scrape off the cells from the dish and transfer to a microfuge tube. Incubate the extract in a 95 °C heating block for 10 min. Centrifuge the tubes at 13,000 rpm for 10 min, and load 10 μL of each sample (~15,000 cell equivalents) on 4–20 % Tris-Glycine Mini Protein Gels. Include a lane with 7 μL of pre-stained molecular weight protein ladder.
2. Perform polyacrylamide gel electrophoresis (PAGE) to separate proteins by running the gel for ~2 h at 130 V in Tris-glycine electrophoresis buffer.
3. Transfer proteins from the polyacrylamide gel after electrophoresis to a nitrocellulose membrane using a wet transfer apparatus at 0.4A for 1.5 h.
4. Rinse the membrane with H_2O and stain the immobilized proteins in the membrane with 0.2 % Ponceau S for 2 min. Rinse membrane with H_2O twice to remove excess dye. Scan the membranes for your records.

5. Block membranes with 5 % instant nonfat dry milk blocking solution for 30 min.
6. Rinse twice with PBS-T and incubate with the primary antibodies (*see* **item 10** in Subheading 2.5) diluted 1:1000 in PBS-T during 2 h at room temperature or overnight (16 h) at 4 °C with gentle rocking.
7. Wash the membranes extensively with PBS-T, and incubate with the secondary HRP-conjugated antibodies (*see* **item 11** in Subheading 2.5) diluted 1:5000 in PBS-T for 45 min at room temperature.
8. After several washes with PBS-T, incubate the membrane with ECL substrate for 60 s, and, after removal of liquid excess, expose the membrane to X-ray film to collect an image with an automated image analyzer such as the BioSpectrum 810 Imaging System from UPV.

4 Notes

1. Although optimal transfection should occur at pH 7.05, HBS 2× may require pH optimization. Start by adjusting the pH of HBS 2× (0.2 L) to 7.0 and split into 50 mL aliquots. Titrate separate aliquots to pH 7.05, pH 7.1, and pH 7.15. Test each of the four solutions (pH 7.0–7.15) by transfecting a lentiviral transfer vector encoding GFP using the calcium phosphate protocol (*see* Subheading 3.1) into 293T cells. The HBS 2× that gives the most efficient transfection should be aliquoted and stored at −20 °C.
2. To easily subclone the GOI in the pLenti CMV-GFP Puro vector, use BamHI and SalI restriction sites. Lentivirus transfer vectors in which the GOI is driven by other promoters also can be used (*see* **Note 7**).
3. GE Dharmacon offers four specialized DharmaFECT formulations. The most suitable transfection reagents for siRNA delivery in C2C12 cells are DharmaFECT-1 and DharmaFECT-3, based on delivery of siGLO Green Transfection Indicator (Dharmacon).
4. 27-mer Dicer-substrate siRNAs can be designed using RNAi design tools available from different biotechnology companies (i.e., Integrated DNA Technologies, using the RNAi design tool at http://www.idtdna.com/Scitools/Applications/RNAi/RNAi.aspx). These siRNAs mimic endogenous Dicer-processed products and therefore are directly incorporated to the multiprotein RNA-inducing silencing complex (RISC) [16] but are sufficiently small to avoid inducing the interferon response that causes global inhibition of mRNA translation and mRNA degradation [17].

5. Lentiviral tropism is due to the ability of the viral envelope protein to interact with receptors at the target cell surface. The VSV-G envelope protein confers broad tropism over a range of species and cell types. Cronin et al. [18] describe different envelopes available and their impact on lentiviral tropism. For C2C12 cells, 1 mL of lentiviral supernatant diluted with 1 mL of growth medium typically yields ~40–50 % transduction efficiency.
6. During the production stage (*see* **step 8** in Subheading 3.1), culturing transfected 293T cells in DMEM with a lower percentage of FBS (i.e., 2–5 %) can increase the yield of lentivirus. If this variation is introduced to the protocol, adjust the FBS concentration to 10 % when diluting the lentivirus-containing supernatant with growth medium for cell infection.
7. To regulate ectopic gene expression, the use of different promoters (i.e., EF1a, PGK, or UBC) can be explored. In most cells the order of promoter strength is like CMV > EF1a > PGK > UBC. The following vectors that drive expression of the GOI under the control of the alternative promoters EF1a, PGK, and UBC, respectively, are available from the nonprofit plasmid repository Addgene (http://www.addgene.org): pEGIP (catalog number 26777), pLenti PGK GFP Puro (catalog number 19070), and Ubc.Luc.IRES.Puro (catalog number 33307). Alternatively, lentivirus transfer vectors containing inducible promoters such as the tetracycline ON system can be used to tune the expression of the ectopic protein. These vectors contain a tet-response element (TRE) upstream the promoter region that is recognized by the reverse tetracycline-controlled transactivator (rtTA). The rtTA binds to the TRE only in the presence of doxycycline to activate transcription. A single lentiviral transfer vector encoding the puromycin resistance gene, the rtTA, and the TRE3G promoter that drives a tightly regulated expression of the GOI is commercially available (Lenti-X Tet-One Inducible Expression System, Clontech Laboratories).
8. To quickly make stable populations of target cells, lentiviral transfer vectors must have a suitable antibiotic selectable marker, such as the puromycin or blasticidin resistance gene, which causes cell killing within several days (*see* **Note 10**). As an alternative to antibiotic resistance selection, lentiviral transfer vectors with a dual promoter for coexpression of the GOI and GFP can be used to isolate GFP-positive cells by fluorescence-activated cell sorting (FACS), for expansion of transduced cell populations with different levels of expression.
9. Polybrene is a cationic polymer that increases the efficiency of lentivirus infection by neutralizing negative charges and facilitating the fusion of the virus membrane and the plasma membrane of the host cell [19].

10. Antibiotic selection of any particular target cell type requires determination of the appropriate antibiotic concentration by performing a kill curve. For C2C12 cells, selection medium containing 2 μg/mL puromycin or 20 μg/mL blasticidin achieves 100 % cellular toxicity after 48 h of incubation.
11. Include the following immunofluorescence controls to validate the quality of antibodies and appropriate dilution: (1) by incubation with secondary antibody to evaluate fluorescence background signal and (2) by labeling of non-transfected cells to evaluate nonspecific binding of the primary antibody.
12. We typically analyze C2C12 cells that are lentivirally transduced with GFP in parallel as a control to validate that the lentivirus infection and selection are not deleterious to differentiation.
13. Differentiation kinetics can be measured by evaluation of the expression of early myogenic markers, myogenin and MRF4 (Myf6). Analysis involves samples each day during the course of the myoblast differentiation (d0, d1, d2, d3, and d4) to reveal the rate-limiting day.
14. Selecting the best siRNA sequences for knockdown should be empirically determined by initially comparing 3–4 different siRNAs. Even with the most potent siRNAs, the amount of siRNA needed to achieve effective knockdown will vary for different proteins, depending on the protein and mRNA stability. It is important to perform a titration experiment to determine this. Using a lower concentration of siRNA is more cost-effective and reduces the chances for off-target effects. We usually test a range of 25, 50, 75, and 100 pmol of siRNA for 200,000 cells. Whether one or two reverse transfections with siRNA are needed to achieve ~80 % knockdown also should be empirically determined. In the particular cases of Lem2 and emerin, two sequential treatments with 100 pmol siRNA on consecutive days yield 80–90 % knockdown. By contrast, maximum knockdown of Net37 is achieved with a single 50-pmol siRNA treatment.
15. For gelatin-coating plates, prepare a 1.5 % stock solution of gelatin in H_2O under sterile conditions. Dilute 1:10 in H_2O and cover each 3.5-cm dish containing a coverslip with 2 mL of 0.15 % gelatin. Incubate at 37 °C for 30 min. Aspirate and air-dry the plates at room temperature under UVA light for 30 min. Sealed plates can be stored at 4 °C.
16. When, after siRNA-mediated silencing of the nuclear lamina protein, myoblast differentiation is blocked (absence of MyHC signal by d4), further validation of C2C12 terminal differentiation by d6 should be performed. Also, differentiation kinetics should be measured by collecting samples each day (d0, d1, d2, d3, and d4) to evaluate the onset expression of early myogenic markers (i.e., myogenin and MRF4/Myf6).

References

1. Dauer WT, Worman HJ (2009) The nuclear envelope as a signaling node in development and disease. Dev Cell 17:626–638
2. Gerace L, Huber MD (2012) Nuclear lamina at the crossroads of the cytoplasm and nucleus. J Sruct Biol 177:24–31
3. Worman HJ, Schirmer EC (2015) Nuclear membrane diversity: underlying tissue-specific pathologies in disease? Curr Opin Cell Biol 34: 101–112
4. Bonne G, Quijano-Roy S (2013) Emery-Dreifuss muscular dystrophy, laminopathies, and other nuclear envelopathies. Handb Clin Neurol 113:1367–1376
5. Schreiber KH, Kennedy BK (2013) When lamins go bad: nuclear structure and disease. Cell 152:1365–1375
6. Worman HJ, Fong LG, Muchir A, Young SG (2009) Laminopathies and the long strange trip from basic cell biology to therapy. J Clin Invest 119:1825–1836
7. Huber MD, Guan T, Gerace L (2009) Overlapping functions of nuclear envelope proteins NET25 (Lem2) and emerin in regulation of extracellular signal-regulated kinase signaling in myoblast differentiation. Mol Cell Biol 29:5718–5728
8. Datta K, Guan T, Gerace L (2009) NET37, a nuclear envelope transmembrane protein with glycosidase homology, is involved in myoblast differentiation. J Biol Chem 284: 29666–29676
9. Liu GH, Guan T, Datta K, Coppinger J, Yates J 3rd, Gerace L (2009) Regulation of myoblast differentiation by the nuclear envelope protein NET39. Mol Cell Biol 29:5800–5812
10. Muchir A, Wu W, Worman HJ (2009) Reduced expression of A-type lamins and emerin activates extracellular signal-regulated kinase in cultured cells. Biochim Biophys Acta 1792: 75–81
11. Blau HM, Pavlath GK, Hardeman EC, Chiu CP, Silberstein L, Webster SG, Miller SC, Webster C (1985) Plasticity of the differentiated state. Science 230:758–766
12. Yaffe D, Saxel O (1977) Serial passaging and differentiation of myogenic cells isolated from dystrophic mouse muscle. Nature 270:725–727
13. Jackson MF, Hoversten KE, Powers JM, Trobridge GD, Rodgers BD (2013) Genetic manipulation of myoblasts and a novel primary myosatellite cell culture system: comparing and optimizing approaches. FEBS J 280:827–839
14. Cockrell AS, Kafri T (2007) Gene delivery by lentivirus vectors. Mol Biotechnol 36: 184–204
15. Geraerts M, Willems S, Baekelandt V, Debyser Z, Gijsbers R (2006) Comparison of lentiviral titration methods. BMC Biotechnol 6:34
16. Jinek M, Doudna JA (2009) A three-dimensional view of the molecular machinery of RNA interference. Nature 457:405–412
17. Dykxhoorn DM, Novina CD, Sharp PA (2003) Killing the messenger: short RNAs that silence gene expression. Nat Rev Mol Cell Biol 4:457–467
18. Cronin J, Zhang XY, Reiser J (2005) Altering the tropism of lentiviral vectors through pseudotyping (vol 5, pg 387, 2005). Curr Gene Ther 5:387–398
19. Davis HE, Rosinski M, Morgan JR, Yarmush ML (2004) Charged polymers modulate retrovirus transduction via membrane charge neutralization and virus aggregation. Biophys J 86:1234–1242

Chapter 12

Analysis of Meiotic Telomere Behavior in the Mouse

Jana Link, Ricardo Benavente, and Manfred Alsheimer

Abstract

Linear eukaryotic chromosomes are capped by the telomeres, which consist of highly repetitive nucleotide sequences bound by several telomere-specific proteins. While the general role of telomeres is to protect chromosomes from degradation and end-to-end fusion, during meiosis they are assigned with a distinct and without doubt highly fascinating function. During meiosis, telomeres attach to the nuclear envelope and mediate characteristic chromosome movements, essential for correct haploidization of the genome. Here, we provide elaborate tools to study telomeres in mammalian meiotic germ cells, which include (co-) immunofluorescence staining procedures on cell spreads and paraffin-embedded tissues. We provide detailed procedures for fluorescence labeling of telomeric DNA (Telo-FISH) to visualize telomeres at the light microscopic level, which we often use in combination with immunofluorescence staining of meiotic proteins. We also present a protocol for detection of telomeric DNA at the electron microscopic level (EM-ISH). We finally describe how meiotic telomeres can be visualized by common electron microscopic methods and how they can be analyzed at the ultrastructural level by immunogold labeling of telomere components or associated structures.

Key words Telomere, Meiosis, Nuclear envelope, Immunofluorescence analysis, FISH, EM-ISH, Electron microscopy

1 Introduction

Telomeres define the ends of linear eukaryotic chromosomes. They consist of multiple tandem repeats of short, highly conserved nucleotide sequences, such as TTAGGG in vertebrates, and a number of telomeric DNA-binding proteins. Telomeres protect chromosomes from end-to-end fusion and are pivotal for chromosome maintenance and genome stability [1]. In recent years, it turned out that during meiosis telomeres hold a very special but utmost important function. In early meiotic prophase I, telomeres move to the nuclear periphery and firmly attach to the nuclear envelope. During meiotic progression, the attached telomeres mediate the characteristic movements of meiotic chromosomes, which are required for efficient pairing, synapsis, and recombination of the homologs [2–4]. In this chapter, we give a cumulative detailed

Sue Shackleton et al. (eds.), *The Nuclear Envelope: Methods and Protocols*, Methods in Molecular Biology, vol. 1411,
DOI 10.1007/978-1-4939-3530-7_12, © Springer Science+Business Media New York 2016

overview about a set of microscopy techniques that we use to analyze mammalian meiotic telomere behavior and function. This includes immunofluorescence techniques on cell spreads and paraffin-embedded tissues as well as combinatory approaches with co-detection of telomeres by fluorescence in situ hybridization. We also present a protocol for visualization of telomeric DNA at the electron microscopic level. Lastly, we provide short methods for morphological and immunocytological investigation of telomeres by conventional electron microscopy.

2 Materials

2.1 Animals and Tissues

For standard investigation of telomeres and chromosomes in male meiosis, testis tissue is obtained from young adult, 30–40 days old wild-type mice. To analyze female meiotic prophase, ovaries are taken from female embryos (day 17–19 postfertilization) and prepared for paraffin embedding or spreading. We study time-related telomere behavior during meiotic progression on sequential testes samples from juvenile mice of increasing age (days 10–15 postpartum; *see* **Note 1**).

2.2 Meiotic Chromosome Spreads

1. Fresh mouse testis or embryonal ovary tissue.
2. Hypotonic buffer for chromosome spreads: 30 mM Tris–HCl, 17 mM Na-Citrate, 5 mM EDTA, 50 mM Sucrose, 5 mM DTT (dilute from 1 M aqueous stock solution), pH 8.2.
3. 100 mM sucrose solution in water; filter through a 0.2 μm filter and store at −20 °C until use.
4. 1 % PFA solution: dissolve 1 % (w/v) paraformaldehyde in H_2O at 58 °C; cool to room temperature (RT) and adjust to pH 9.2 with NaOH; add 0.15 % (v/v) Triton X-100 diluted from a 10 % aqueous stock solution.

2.3 Paraffin Embedding of Testes and Ovaries, Sectioning and Preparing Paraffin Samples for Immunofluorescence Microscopy

1. PBS (phosphate-buffered saline): 140 mM NaCl, 6.4 mM Na_2HPO_4, 1.4 mM KH_2PO_4, 2.6 mM KCl; adjust to pH 7.4 with NaOH.
2. 2 % PFA solution: dissolve 2 % (w/v) paraformaldehyde in PBS at 58 °C; after cooling to RT, adjust to pH 7.4 with NaOH.
3. Increasing ethanol series: 50, 60, 70, 80, 90, and 100 % (diluent: H_2O).
4. Tert-butyl alcohol (2-methyl-2-propanol).
5. Paraffin pellets (melting point 57–60 °C).
6. Glass slides (Superfrost Plus; Menzel Glas, Braunschweig, Germany).

7. Rotihistol (xylene substitute).
8. Unmasking Solution (Vector laboratories, Burlingame, CA, USA); diluted 1:100 in H_2O before use.

2.4 Components for Immunofluorescence Staining

1. PBS, pH 7.4.
2. Blocking Solution: 5 % dry milk powder, 5 % fetal calf serum (FCS), in PBS, pH 7.4. Can be stored at −20 °C; before use, remove insoluble material by centrifugation at 16,000 × *g* for 10 min.
3. 0.1 % Triton X-100 in PBS, pH 7.4.
4. PBT: 1.5 % BSA, 0.1 % Tween 20, in PBS, pH 7.4.
5. Affinity-purified polyclonal or monoclonal antibodies against proteins of interest, such as SUN1, SYCP3, lamin C2 [5, 6], or digoxigenin.
6. Appropriate fluorochrome-labeled secondary antibodies.
7. Hoechst 33258 (15 μg/mL).
8. Mounting medium: 50 % glycerol in PBS.

2.5 Telomere Fluorescence In Situ Hybridization Components

1. Telomere oligonucleotides to be used for end-labeling with digoxigenin can be obtained from a commercial source as oligonucleotides with sequences $(TAACCC)_7$ and $(GGGTTA)_7$.
2. Reagents for digoxigenin end-labeling: 100 pmol telomere oligonucleotide, Dig-ddUTP (1 mM stock solution), terminal deoxynucleotidyl transferase (TdT; 15 U/μL), 5× TdT buffer (1 M potassium cacodylate, 0.125 M Tris–HCL, 0.05 % (v/v) Triton X-100, 5 mM $CoCl_2$, pH 7.2).
3. 2× SSC: 0.3 M NaCl, 0.03 M Na-Citrate, pH 7.4.
4. Hybridization solution: 30 % formamide, 10 % dextran sulfate, 250 μg/mL *E. coli* DNA, in 2× SSC.
5. TBS: 150 mM NaCl, 10 mM Tris HCl, pH 7.4.
6. TBS/BR: 0.5 % Roche blocking reagent in TBS, pH 7.4.
7. TBST: 150 mM NaCl, 10 mM Tris–HCl, 0.05 % Tween 20, pH 7.4.
8. Mouse anti-digoxigenin antibody.
9. Appropriate fluorochrome-labeled secondary antibodies.

2.6 Electron Microscopy (EM) Components

1. Cacodylate buffer: 50 mM cacodylate, 50 mM KCl, 2.5 mM $MgCl_2$, pH 7.2.
2. 2.5 % glutaraldehyde solution: dissolve 2.5 % (v/v) glutaraldehyde in cacodylate buffer.
3. 2 % osmium tetroxide in cacodylate buffer.
4. Increasing ethanol series 30, 50, 70, 90, 96, and 100 % (diluted in H_2O).

5. Epon 812.
6. Propylene oxide/Epon 1:1.
7. Gelatine capsules.
8. Formvar-coated copper grids.
9. 2 % uranyl acetate in H_2O.
10. Reynold's lead citrate [7]; right before use, dilute 1:1 in outgassed H_2O.

2.7 EM In Situ Hybridization Components

1. PBS, pH 7.4.
2. 4 % PFA solution: dissolve 4 % (w/v) paraformaldehyde in PBS as described above; adjust to pH 7.4 with NaOH.
3. Blocking solution: 50 mM NH_4Cl in PBS, pH 7.4.
4. Increasing ethanol series: 30, 50, 70, 90, and 100 % (diluent: H_2O).
5. LR White resin, hard grain; right before use, add 0.1 g benzoyl peroxide (catalyst for LR White polymerization) to 5.05 mL of LR White resin.
6. EtOH/LR White (1:1).
7. Gelatine capsules.
8. Formvar-coated nickel grids.
9. 2× SSC (*see* Subheading 2.4).
10. RNase A (100 μg/mL in 2× SSC).
11. 0.5 mg/mL Proteinase K in 20 mM Tris–HCl, 2 mM $CaCl_2$; pH 7.4.
12. Hybridization solution, TBS, TBS/BR, TBST (*see* Subheading 2.5).
13. Mouse anti-digoxigenin antibody.
14. Appropriate gold-labeled secondary antibodies.
15. 2 % uranyl acetate in H_2O.
16. Reynold's lead citrate; right before use dilute 1:1 in outgassed H_2O.

2.8 EM Immunogold Components

1. PBS, pH 7.4.
2. 2 % PFA solution: dissolve (w/v) paraformaldehyde in PBS; adjust to pH 7.4 with NaOH.
3. Blocking solution: 50 mM NH_4Cl in PBS, pH 7.4.
4. Increasing ethanol series: 30, 50, 70, 90, 96, and 100 % (diluted in H_2O).
5. Epon 812.
6. Propylene oxide/Epon 1:1.
7. Gelatine capsules.

8. Formvar-coated copper grids.
9. Appropriate gold-labeled secondary antibodies.
10. 2 % uranyl acetate in H_2O.
11. Reynold's lead citrate; right before use dilute in outgassed H_2O.

3 Methods

3.1 Preparation of Chromosome Spreads

Chromosome spreads are essentially prepared according to a slightly adapted protocol developed by de Boer and colleagues [8].

1. For spreads of spermatocytes, remove the tunica surrounding the freshly prepared testes using fine tweezers. Transfer single tubules into a drop of hypotonic buffer on a slide; leave tubules to swell for 30–60 min. To spread oocytes, place freshly prepared ovaries in hypotonic buffer and incubate for 30–60 min.
2. Place a 20 μL droplet of the sterile 100 mM sucrose solution onto a new slide and transfer small seminiferous tubule pieces or the swollen ovaries (one per slide) from the hypotonic buffer. In this solution, wash germ cells out of the tubules/ovaries by thoroughly tearing the tissues with fine tweezers and subsequent careful resuspension with a 10 μL pipette. Immediately remove remaining larger pieces of debris before proceeding.
3. Cover a new slide with 1 % PFA solution and tilt so that the entire volume of the solution accumulates in one corner of the slide. Transfer 20 μL of the sucrose solution containing the individualized spermatocytes or oocytes in this droplet.
4. Tilt the slide in all directions to mix the solutions and allow cells to disperse evenly over the slide.
5. Place the slides in a moisture chamber and incubate in the closed chamber for 2 h to allow cells to settle down. Slightly open the lid and incubate for another 30 min. Remove the lid and dry specimen in the open moisture chamber for another 2–4 h at RT; make sure the fluid dries out entirely.
6. In case that spreads are not used immediately, wrap the dried slides in tinfoil and store at –70 °C until use.

3.2 PFA Fixation and Paraffin Embedding of Testis and Ovary Tissue Samples

1. For paraffin embedding, fix testes or ovary tissue samples in 1 % PBS-buffered formaldehyde for 3 h.
2. After washing in PBS for 1 h, dehydrate tissues in an increasing ethanol series by subsequently incubating in 50, 60, 70, 80, 90, and 100 % ethanol for at least 1 h each.
3. Immerse the tissues in tert-butyl alcohol at 25 °C for 1 h.
4. Transfer tissue samples into preheated (melted) liquid paraffin in an oven at 60 °C; incubate in the warm paraffin for another

24 h and transfer tissues into a fresh plastic mold containing melted paraffin (*see* **Note 2**).

5. While the paraffin is still liquid, adjust if necessary the position of the tissue within the paraffin using a pre-warmed needle. Once the tissue is at the desired position, remove the mold from the oven and cool the paraffin down slowly to RT. Store the paraffin blocks at 4 °C until use.

3.3 Immuno-fluorescence Staining on Chromosome Spreads of Spermatocytes and Oocytes

1. Thaw the slides with chromosome spreads at RT, keeping them still wrapped in tinfoil to avoid condensation of water on the slide surface. Once the slides have reached RT, remove the tinfoil and wash the slides three times 5 min in PBS.
2. During this time, thaw the dry milk-containing blocking solution and centrifuge 30 min at 16,000 × g and 4 °C to remove insoluble material. Use the supernatant for all subsequent blocking steps to dilute the antibodies, provided that they are compatible with this kind of blocking (*see* **Note 3**).
3. Transfer slides to a moisture chamber, cover spreads with 500 μL of the blocking solution, and block by incubating at RT for 30–60 min.
4. Apply the appropriate first primary antibodies (diluted to working concentration in blocking solution) and incubate for 1 h at RT or overnight at 4 °C (*see* **Note 4**).
5. Wash slides twice for 5 min each in PBS and re-block with blocking solution for another 30 min. To co-stain for a second structure/epitope, incubate spreads with second primary antibodies raised in a different species.
6. Wash twice 5 min in PBS and block for another 30 min in blocking solution.
7. Incubate slides with appropriate fluorescently labeled secondary antibodies for 30 min. In case of co-immunofluorescence, apply both secondary antibodies (coupled to different fluorochromes) in one solution.
8. To counterstain DNA, add in last 10 min of incubation two to three drops of Hoechst (from a 15 μg/μL stock solution) by pipetting directly onto the slides.
9. Finally, wash slides three times 5 min in PBS and mount with glycerol/PBS.

3.4 Immuno-fluorescence Staining on Paraffin Sections to Analyze Three-Dimensionally Preserved Meiotic Nuclei

1. Prepare 2–15 μm thick cross sections of paraffin-embedded testes or ovaries with a standard sliding microtome (*see* **Note 5**).
2. Transfer sections with a small brush onto a water droplet, placed on Superfrost Plus slides.
3. To remove tissue foldings, stretch the sections on top of the water droplets by gently heating to 58 °C on a heating panel.

4. As soon as the sections appear fully stretched, immediately take off the slides from the plate and remove the water underneath the paraffin section by slanting the slide and absorbing the water with a paper towel. Residual water between the paraffin section and the slide may be removed by rigorous shaking off.
5. Dry the tissue sections over night at RT.
6. Incubate the slides twice in Rotihistol for 10 min each to remove the paraffin from the tissues.
7. Transfer the slides to 100 % ethanol and allow infiltration for 10 min.
8. Rehydrate tissue samples in a decreasing ethanol series by incubating in 90, 80, 70, and 60 % ethanol, for at least 2 min each. Place slides in 50 % ethanol and incubate for 10 min to ensure complete exchange with 50 % ethanol.
9. Collect the sections in distilled water and leave in water for at least 5 min.
10. While the tissue is rehydrated, place the diluted unmasking solution in a closed container and preheat the solution to 125 °C at 1.5 bar in a table autoclave. Once the unmasking solution has reached the final temperature, release pressure of the autoclave and open the autoclave.
11. Immediately transfer the slides from water into hot unmasking solution (*see* **Note 6**), close the autoclave, and heat again until the autoclave reaches 125 °C and 1.5 bar; incubate specimen for 12–30 min depending on the needs (*see* **Note 7**).
12. When unmasking is completed, switch off the autoclave and open the valve to slowly evaporate. When the pressure is down, open the autoclave and place the slides immediately in PBS at RT (*see* **Note 8**).
13. Incubate sections in PBS containing 0.1 % Triton for 10 min.
14. Wash twice for 5 min in PBS and block samples for at least 60 min in PBT.
15. Incubate tissue sections with primary antibodies or a combination of primary antibodies for 1 h at RT or overnight at 4 °C.
16. Wash three times 5 min in PBS and incubate sections with an appropriate combination of secondary antibodies for 30 min.
17. To counterstain DNA, add Hoechst in the last 10 min of incubation.
18. Wash slides three times 5 min in PBS and mount with glycerol/PBS.

3.5 Telomere Fluorescence In Situ Hybridization in Combination with Immunofluorescence Staining

In most of our studies, Telo-FISH on meiotic cells was mainly correlated with antibody staining against synaptonemal complex components and/or proteins involved in DNA double-strand repair and recombination (see e.g. [5, 6]). To co-stain telomere DNA and meiotic proteins, we first do the in situ hybridization before proceeding to the immunofluorescence staining procedure.

1. Prepare telomere-specific probes for Telo-FISH by 3′-end-labeling of the two 42mer synthetic telomere oligonucleotides (each of the two oligonucleotides in a separate reaction): add 100 pmol oligonucleotide to 4 μL 5× TdT buffer, 1 μL Dig-ddUTP (1 mM stock) and fill to 18 μL with H_2O. Add 2 μL TdT enzyme, mix gently, and incubate at 37 °C for 20 min. Store at −20 °C until use.
2. Prepare either cell spreads or paraffin sections as described above: for cell spreads follow **steps 1–6** of Subheading 3.1; for preparing paraffin tissue sections, follow **steps 1–9** of Subheading 3.4, which includes the entire antigen unmasking procedure.
3. Wash the specimen in 2× SSC for at least 5 min.
4. Preheat the hybridization solution and both labeling reactions, all separately, to 95 °C for approx. 15 min.
5. Right before hybridization, add 2 μL of each of the two labeling reactions to 40 μL of the preheated hybridization solution, mix, and pipette onto the slides in a moisture chamber (*see* **Note 9**).
6. Close the chamber; tightly seal with Parafilm to avoid evaporation of the hot fluids.
7. Immediately place the chamber in an oven preheated to 95 °C and denature for 20 min (*see* **Note 10**).
8. Let the slides cool down and hybridize overnight at 37 °C.
9. Remove unbound telomere probes by washing twice for 10 min in 2× SSC at 37 °C.
10. Cover sections/cells with TBS/BR and incubate for 30 min at RT.
11. Apply 100 μL of diluted mouse anti-digoxigenin antibodies (1:50 in TBS/BR) for 1 h at RT.
12. Wash specimen twice in TBST for 10 min each.
13. In case of co-immunofluorescence staining for proteins on cell spread preparations, proceed with **step 3** of Subheading 3.3; in case of co-staining on tissue sections, continue with **step 11** of Subheading 3.4.

3.6 Telomere In Situ Hybridization on Electron Microscopic Samples (EM-ISH)

Analyzing meiotic telomeres on the ultrastructural level requires a sophisticated protocol that warrants accessibility for probes on structurally well-preserved electron microscopic samples. Testing and modifying different proceedings, we some years ago developed a protocol which enabled us to detect telomere DNA by in situ hybridization on structurally preserved LR-White-embedded material [9]. We here describe in detail the method of LR White embedding and the detection of telomere DNA on ultrathin sections.

1. Fix small testes tissue pieces in 4 % PBS-buffered formaldehyde for 3 h at 4 °C.
2. Wash five times 3 min each in PBS.
3. Incubate samples in 50 mM NH_4Cl (in PBS) for 15 min at RT.
4. Wash the tissue pieces five times 3 min each in H_2O.
5. Dehydrate samples in an increasing ethanol series by incubating twice in 30 % EtOH (4 °C) for 15 min each, twice in 50 %, 70 %, 90 % EtOH (−20 °C) for 30 min each, and twice in 100 % EtOH (−20 °C) for 1 h.
6. Supplement LR White resin with catalyst benzoyl peroxide (*see* **item 5** of Subheading 2.7).
7. Infiltrate samples overnight with EtOH/LR White (ratio 1:1) at 4 °C (*see* **Note 11**).
8. Transfer samples into pure LR White and infiltrate for 1 h at 4 °C.
9. Substitute with fresh LR White, incubate for another 4 h (or overnight), and repeat this step twice.
10. Transfer the tissue samples into LR White containing gelatine capsules.
11. Incubate at 40 °C for at least 3 days for complete polymerization of the resin.
12. Cut ultrathin sections of the LR-White-embedded samples on an ultramicrotome and transfer the sections to Formvar-coated nickel grids.
13. Prepare a moisture chamber and place a piece of Parafilm inside.
14. Pipette a small droplet of RNase A solution onto the Parafilm and place the nickel grid upside down on top of the droplet (*see* **Note 12**), close the chamber, and incubate at 37 °C for 1 h.
15. Wash twice in 2× SSC for 5 min each and incubate in Proteinase K solution for 8 min at RT, then wash twice in 2× SSC.
16. Place the grid on a droplet of hybridization solution containing the Dig-labeled probes (*see* **step 5** of Subheading 3.5), close the chamber, and seal with Parafilm to avoid evaporation.

17. Immediately place the chamber in a preheated oven at 93 °C for 20 min to denature the probes.
18. Hybridize overnight at 37 °C.
19. Rinse the grids in 2× SSC twice for 10 min at 37 °C to remove unbound probes.
20. To block unspecific binding, incubate samples in TBS/BR for 30 min at RT.
21. Incubate with mouse anti-digoxigenin antibodies (1:20 in TBS/BR) for 1 h at RT.
22. Wash the grids twice in TBST and incubate 1 h with appropriate gold-labeled secondary antibodies (*see* **Note 13**).
23. Wash samples twice in TBST and twice in H_2O for 10 min each.
24. For contrasting, place grids on a droplet of 2 % uranyl acetate and incubate for 4–5 min.
25. Wash two times in H_2O by shaking for 10 s in a coplin jar.
26. Counterstain with Reynold's lead citrate (diluted 1:1 in outgassed H_2O) for 4–5 min.
27. Wash twice in H_2O by shaking for 10 s in a coplin jar.
28. Dry the grids at RT before inspecting them with a transmission electron microscope.

3.7 Electron Microscopy (EM) and Immunogold EM

Standard EM and immunogold EM are well-established techniques, and protocols including minor modifications are used by the different groups worldwide. Here, we describe the protocols routinely used in our labs.

3.7.1 Standard EM

1. Small testes pieces (1 mm^3) are fixed in cacodylate buffer containing 2.5 % glutaraldehyde for 1 h. Glutaraldehyde fixation as well as the following steps are at 4 °C.
2. Wash five times for 3 min each in cacodylate buffer.
3. Fix samples in cacodylate buffer containing 2 % osmium tetroxide for 2 h.
4. Wash five times for 3 min each in H_2O.
5. Incubate the tissue pieces with 0.5 % uranyl acetate overnight.
6. Wash five times for 3 min each in H_2O.
7. Dehydrate samples in an increasing ethanol series by incubating in 30, 50, 70, 90, and 96 % EtOH for 30 min each.
8. Dehydrate samples two times in 100 % EtOH for 30 min each.
9. Incubate samples three times for 30 min each in propylene oxide at RT.
10. For embedding, infiltrate samples overnight with propylene oxide/Epon (ratio 1:1) at RT.

11. Infiltrate samples two times for 2 h each in Epon at RT.
12. Transfer the samples into Epon-containing molds.
13. For polymerization of the resin, incubate at 60 °C for 48 h.
14. Cut ultrathin sections on an ultramicrotome and transfer the sections to Formvar-coated copper grids.
15. Place the grids on a droplet of 2 % uranyl acetate for 20 min at RT.
16. Wash two times in H_2O by shaking for 10 s in a coplin jar.
17. Counterstain sections in Reynold's lead citrate for 10 min.
18. Wash in H_2O by shaking for 10 s in a coplin jar.
19. Dry the grids at RT and inspect them at the electron microscope.

3.7.2 Pre-embedding Immunogold EM

1. Cryosections (6–10 μm thick) are obtained from frozen testes pieces using a cryomicrotome.
2. Sections are placed on coverslips and air-dried.
3. Sections are fixed in PBS containing 2 % formaldehyde for 1 h at RT.
4. After washing five times 3 min each in PBS, samples are blocked in PBS containing 50 mM NH_4Cl for 15 min at RT.
5. Wash the tissue pieces five times 3 min each in H_2O.
6. Coverslips are transferred to a moisture chamber and placed on a Parafilm sheet.
7. Incubate sections with 50 μL of the desired primary antibodies for 1 h.
8. Wash three times for 10 min each.
9. Incubate sections in the moisture chamber with 50 μL of the corresponding secondary antibodies conjugated with gold particles for 1 h (*see* **Note 13**).
10. Wash three times for 10 min each.
11. Fix the sections in cacodylate buffer containing 2.5 % glutaraldehyde for 1 h. Glutaraldehyde fixation as well as the following steps are at 4 °C.
12. Wash five times for 3 min each in cacodylate buffer.
13. Refix samples in cacodylate buffer containing 2 % osmium tetroxide for 2 h.
14. Wash five times for 3 min each in H_2O.
15. Incubate with 0.5 % uranyl acetate overnight.
16. Wash five times for 3 min each in H_2O.
17. Dehydrate the tissue sections in an increasing ethanol series by incubating in 30, 50, 70, 90, and 96 % EtOH for 30 min each.
18. Dehydrate samples two times in 100 % EtOH for 30 min each.

19. Incubate samples three times for 30 min each in propylene oxide at RT.
20. For embedding, infiltrate samples overnight with propylene oxide/Epon (ratio 1:1) at RT.
21. Infiltrate samples two times for 2 h each in Epon at RT.
22. Place the coverslips on a slide covered with a tin foil and cover the section with a gelatine capsule containing 100 % Epon.
23. For polymerization of the resin, incubate at 60 °C for 48 h.
24. Detach the coverslip from tin foil.
25. Expose the gelatine capsule with the attached coverslip to liquid nitrogen to crush the glass into small pieces.
26. Remove the glass pieces from the surface of the sample with a razor blade.
27. Cut ultrathin sections on an ultramicrotome and transfer the sections to Formvar-coated copper grids.
28. Place the grids on a droplet of 2 % uranyl acetate for 20 min at RT.
29. Wash two times in H_2O by shaking for 10 s in a coplin jar.
30. Counterstain sections in Reynold's lead citrate for 10 min.
31. Wash in H_2O by shaking for 10 s in a coplin jar.
32. Dry the grids and inspect the samples at the electron microscope.

4 Notes

1. The first wave of pubertal spermatogenesis, starting approximately at 8–9 days postpartum (dpp), offers an ideal time frame for the investigation of telomere movements because the first prophase I of meiosis progresses almost synchronously within the testis of young mice. Leptotene stages are commonly enriched in mice aged 9–10 dpp, zygotene spermatocytes are mostly found in mice of approximately 11–12 dpp, whereas spermatocytes in mice older than 12 dpp are usually in the transition into pachytene or in early pachytene. Mice aged 14 dpp show enrichment of mid-pachytene spermatocytes, with meiotic progression continuing as the age of the mice increases [5, 6, 10].
2. Use warm, preheated tools to transfer the tissues in order to prevent hardening of the paraffin. Assure that the tissues are completely submerged in the liquid paraffin rather than floating on top of it. During paraffin infiltration, temperature must not exceed 60 °C; if so, tissues may become brittle and unusable.

3. In case that antibodies are not compatible with milk-containing blocking solutions, PBT (*see* Subheading 2.4) or 100 mM glycine in PBS pH 7.4 could be alternatively used for efficient blocking; in this event incubation time has to be 1 h at the least.
4. In our co-immunofluorescence studies, we used antibodies that in some cases required different conditions of incubation. While on paraffin sections, for example, antibodies against MLH1 are usually incubated overnight at 4 °C; with our antibodies against SUN1, we obtained good results when incubating for only 1 h at RT [5]. To obtain best results, conditions have to be pretested separately for each of the primary antibodies.
5. Required thickness of cross sections depends on the actual application: 4–7 μm thick sections comply with most standard immunofluorescence studies on testis tissue; for localization of telomeres and telomere-associated proteins within three-dimensionally preserved entire nuclei, sections should be thick enough to include whole nuclei (i.e., 10–15 μm).
6. At this step, the unmasking solution has approx. 100 °C. Handle with care. To avoid burning, use heatproof glassware and wear heat-resistant gloves.
7. Duration of unmasking depends on the antibody target and the thickness of the section. To get optimal results, pretest your antibodies on serial sections unmasked with increasing time and select the most promising conditions.
8. To prevent floating off the tissue sections, bubbling of the unmasking solution has to be strictly avoided during and after the unmasking procedure.
9. Compared to the ISH on tissue sections, a slightly larger volume of the hybridization solution (and thus of the labeled probes) is needed for hybridization of chromosome spreads on slides. For successful ISH of a reasonably sized central region on the slide, roughly 80–100 μL of hybridization solution is used. It may help to mark a central region on the slide using a grease pencil to be able to subject this same region to the following immunofluorescence procedure.
10. Working quickly and smoothly during this procedure greatly increases the success of the protocol. The hotter the hybridization solution containing the heated telomere probes is when it is placed onto the slides, the more efficient the hybridization reaction is.
11. Use glassware during all incubation steps with LR White resin to avoid potential plastic solvation.
12. All incubation steps starting with RNase A treatment till contrasting with 2 % uranyl acetate are performed within a moister chamber on droplets of approx. 20 μL placed on a Parafilm.

Watch out the orientation of the grids to warrant that the samples always face the solution.

13. In many cases we use goat anti-mouse antibodies coupled with 12- or 6 nm gold (Jackson ImmunoResearch). Before use, dilute the antibodies 1:7.5 in TBS/BR for EM-ISH (Subheading 3.6) or 1:5 in PBS for pre-embedding immunogold EM (Subheading 3.7.2).

References

1. Blackburn EH (2001) Switching and signaling at the telomere. Cell 106:661–673
2. Zickler D, Kleckner N (1998) The leptotene-zygotene transition of meiosis. Annu Rev Genet 32:619–697
3. Sherthan H (2001) A bouquet makes ends meet. Nat Rev Mol Cell Biol 2:621–627
4. Alsheimer M (2009) The dance floor of meiosis: evolutionary conservation of nuclear envelope attachment and dynamics of meiotic telomeres. Genome Dyn 5:81–93
5. Link J, Jahn D, Schmitt J et al (2013) The meiotic nuclear lamina regulates chromosome dynamics and promotes efficient homologous recombination in the mouse. PLoS Genet 9: e1003261
6. Link J, Leubner M, Schmitt J et al (2014) Analysis of meiosis in SUN1 deficient mice reveals a distinct role of SUN2 in mammalian meiotic LINC complex formation and function. PLoS Genet 10:e1004099
7. Reynolds ES (1963) The use of lead citrate at high pH as an electron-opaque stain in electron microscopy. J Cell Biol 17:208–212
8. de Boer E, Lhuissier FG, Heyting C (2009) Cytological analysis of interference in mouse meiosis. Methods Mol Biol 558:355–382
9. Liebe B, Alsheimer M, Höög C et al (2004) Telomere attachment, meiotic chromosome condensation, pairing, and bouquet stage duration are modified in spermatocytes lacking axial elements. Mol Biol Cell 15:827–837
10. Bellve AR, Cavicchia JC, Millette CF et al (1977) Spermatogenic cells of the prepuberal mouse. Isolation and morphological characterization. J Cell Biol 74:68–85

Part III

Nuclear Envelope Interactions with the Cytoskeleton

Chapter 13

Identification and Validation of Putative Nesprin Variants

Flavia Autore, Catherine M. Shanahan, and Qiuping Zhang

Abstract

Nesprins are a family of multi-isomeric scaffolding proteins that were originally identified at the nuclear envelope (NE), where they bind to lamin A/C, emerin, and SUN-domain containing proteins, to form the LInker of Nucleoskeleton-and-Cytoskeleton (LINC) complex that connects the NE to the actin cytoskeleton. However, nesprin genes also give rise to a variety of tissue-specific variants of different sizes with potential roles beyond the NE. These variants are generated through alternative initiation, termination, and splicing, which makes nesprin biology very complex to study due to the difficulty in generating specific antibodies and/or short interfering RNAs (siRNA) to particular isoforms. In order to distinguish genuine nesprin variants and eliminate confusion with degradation products of larger nesprin isoforms, in this chapter we discuss methods including 5′ and 3′ Rapid Amplification of cDNA Ends (RACE) and RT-PCR in combination with EST database searching, for identifying and validating putative nesprin isoforms. This information is essential to allow a better understanding of nesprin functions in different cell types.

Key words Nesprins, Alternative splicing, EST database, RACE, Multi-tissue RT-PCR

1 Introduction

Nesprins comprise a family of multi-isomeric scaffolding proteins that were originally identified as components of the nuclear envelope (NE) [1]. There are four mammalian genes (*SYNE 1–4*)-encoding nesprin proteins. Nesprin-1, -2, and -3 are ubiquitously expressed in human tissues, while nesprin-4 is mainly expressed in highly polarized epithelial cells [2–5]. All four nesprin proteins contain a C-terminal Klarsicht/ANC-1/Syne Homology (KASH) domain, which mediates the interaction with SUN domain containing proteins (SUN1 and SUN2) at the nuclear membrane, forming the *LI*nker of the *N*ucleoskeleton and the *C*ytoskeleton (LINC) complex [2–4]. The nesprins reside in the outer nuclear membrane (ONM) where they interact with cytoskeletal filaments. SUN 1/2 reside at the inner nuclear membrane (INM) and bind to lamin A/C directly via their nucleoplasmic domain. Therefore, the LINC complex provides a physical connection between the nucleus and the cytoskeleton [5, 6]. Nesprins-1 and -2 are also

Sue Shackleton et al. (eds.), *The Nuclear Envelope: Methods and Protocols*, Methods in Molecular Biology, vol. 1411,
DOI 10.1007/978-1-4939-3530-7_13, © Springer Science+Business Media New York 2016

able to bind to lamin A/C and emerin at the INM [1, 7]. Mutations in the genes encoding all of these proteins contribute to Emery–Dreifuss muscular dystrophy (EDMD) and dilated cardiomyopathy (DCM) [8, 9], indicating that nuclear-cystoskeletal connection is fundamental to muscle cell function.

There are sequence diversities at the N-terminus of nesprins, allowing them to bind to different components of the cytoskeleton: actin filaments, microtubules (MT), and intermediate filaments (IF). Nesprin-1 and nesprin-2 are defined as giant proteins due to their molecular weights of ~1 MDa and ~800 kDa, respectively. Both are comprised of N-terminal paired Calponin Homology domains (CHD) that bind to F-actin, a C-terminal KASH domain, and a central Spectrin Repeat (SR)-containing rod domain, mediating protein–protein interactions [10, 11]. Nesprin-3 and nesprin-4 are characterized by a much less extended rod domain, which reflects their lower molecular weights of 110 and 42 kDa, respectively [11]. Despite lacking the CHDs, they are both able to bind cytoskeleton filaments: nesprin-3 binds IFs through a plectin-binding domain, whereas nesprin-4 interacts with kinesin-1, a motor protein associated with MT networks [12, 13].

Nesprin-1 and -2 are encoded by 146 and 115 exons, respectively, which give rise to a large number of different variants [14]. Most variants have overlapping domains, indicating that the expression of nesprin variants is a highly flexible and regulated process. This flexibility is likely an adaptive response that cells require to meet the requirements of different cellular cues or stimuli. In particular in muscle tissue, nesprins not only switch isoform expression of the same gene, but also change expression between the different nesprin genes. For example, during human muscle development, expression level of nesprin-1 giant increases during early myogenesis *in vitro*, while nesprin-2 partially replaces nesprin-1 at the NE. Ultimately, the shorter nesprin KASH isoforms become the dominant variants during the transition from immature to mature muscle fibers *in vivo* [15].

We documented that nesprin-1 and -2 give rise to a vast array of tissue-specific isoforms of different sizes [14] including (1) isoforms lacking the KASH domain (14 for nesprin-1 and 5 for nesprin-2); (2) isoforms lacking the CHDs (16 for nesprin-1 and 12 for nesprin-2) [16], and (3) SR-only isoforms for nesprin-1 [17]. The isoform repertoire reflects emerging new functions of nesprins beyond the NE. The isoforms are generated by alternate promoter usage, pre-mRNA 3′ end processing, and alternative splicing (including or excluding cassette exons) [14, 16, 18]. For example, nesprin-2α1 and -2α2 variants are generated via different initiation within 5′ untranslated region (UTR) (Fig. 1a) [1, 7]. Alternative splicing is another mechanism by which exons are removed from the mature nesprin mRNA. For example, nesprin-1ΔKASH is generated through the removal of cassette exon 145 of

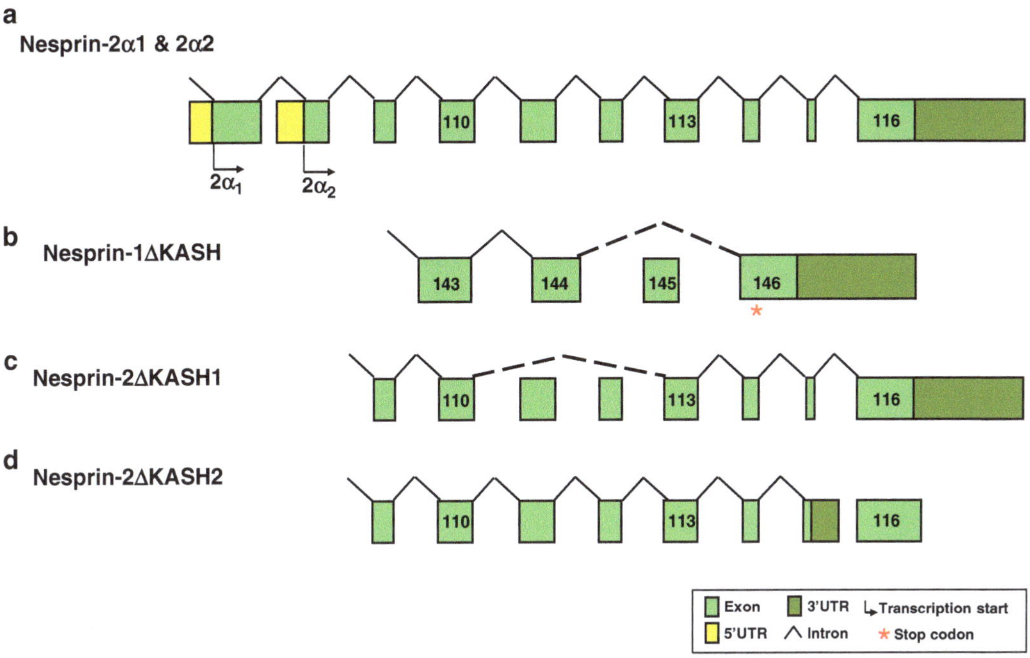

Fig. 1 Generation of nesprin-2α1 and 2α2, as well as nesprin-1 and -2 ΔKASH variants. Nesprin-2α1 and -2α2 variants were generated via the potential combinations of 5′ and 3′ UTRs (**a**). Nesprin-1ΔKASH and nesprin-2ΔKASH1 were generated through the removal of cassette exon 145 (**b**) and exons 111–112 (**c**), respectively, via the indicated splicing events (splicing shown in *dash lines*). Nesprin-2 ΔKASH2 was generated through utilization of an alternative 3′ UTR juxtaposed to exon 115 (**d**)

nesprin-1, resulting in a stop codon at exon 146 and deletion of the KASH domain (Fig. 1b). Similarly, nesprin-2ΔKASH1 is generated through the removal of exons 111–112 (Fig. 1c). In contrast, nesprin-2ΔKASH2 is generated via different termination point through utilization of an alternative 3′ UTR juxtaposed to exon 115 via a retained intron, to produce an alternative variant (Fig. 1d). These alternative transcript generation mechanisms have also been observed experimentally in other studies of nesprin 1 variants [19].

All the nesprin isoforms published so far were predicted mainly using the NCBI EST (database for "Expressed Sequence Tag") [20], but only a small subset has been verified experimentally. The variety of sizes and isoforms that can be generated make nesprin biology very complex due to the difficulty in generating (1) specific antibodies and (2) short interfering RNAs (siRNA) [21]:

(1) The use of specific antibodies is limited to a small subset of isoforms, because the antibodies available detect not only the real products of alternative splicing (of which there may be many), but often degradation products that are commonly present during the manipulation of proteins of large molecular weight [14, 15, 22]. (2) In addition, the most common event for alternate isoform

formation is the use of alternative 5′ and 3′ UTRs, with common coding exons, which complicates the exclusive knockdown of a particular isoform as it is likely to interfere with the expression of other variants. In the literature, the siRNAs currently used for knocking down small isoforms also hamper the expression of the giant nesprins due to this lack of specifically targeted exon regions [14]. Moreover, the efficiency of knockdown can only be detected at the RNA level by qPCR using oligonucleotides designed specifically for each of the siRNA-targeted exons, because the protein expression level is usually not verifiable due to lack of antibody specificity as described above in (1) [14].

We have employed multiple techniques to identify and verify transcripts encoding novel nesprin variants. In some cases, these analyses have enabled us to identify variants that do have some sequence specificity at the protein level, which enables the generation of more selective antibody tools. Therefore, the identification and verification of novel transcripts is essential for a complete understanding of nesprin biology [1, 7, 14]. In this chapter, we discuss the techniques we have used including: (1) 5′ and 3′ Rapid Amplification of cDNA Ends (RACE) for identifying cDNA ends that represent novel 5′ and/or 3′ UTRs, in combination with the EST database analysis, in which there is a vast collection of partial sequences of cDNA clones available; (2) RT-PCR from multi-tissue cDNAs using alternatively combined 5′ and 3′ UTRs for identifying and validating putative isoforms [1, 7, 16, 23].

2 Materials

A range of gene-specific primers for amplification of the region of interest of the cDNA using RACE (*see* **Note 1** and Fig. 2).

Marathon®-Ready cDNA from a tissue of interest (Clontech).
Advantage® 2 PCR kit (Clontech).
GC-RICH PCR system (Roche Applied Science).
First Strand cDNA (OriGene).
PCR-grade water.
DreamTag Green PCR master mix (Fisher Scientific).
pGEM-T Easy vector (Promega).
Eppendorf Mastercycler Nexus (Eppendorf).

3 Methods

3.1 Bioinformatics Analyses

Published nesprin-1 and nesprin-2 UTRs were identified via screening the NCBI expressed sequence tag (EST) database (http://www.ncbi.nlm.nih.gov/dbEST/) [20], which contains partial sequences of cDNA clones ranging from ~200 to ~700 bp. To screen for novel nesprin UTRs in the EST database:

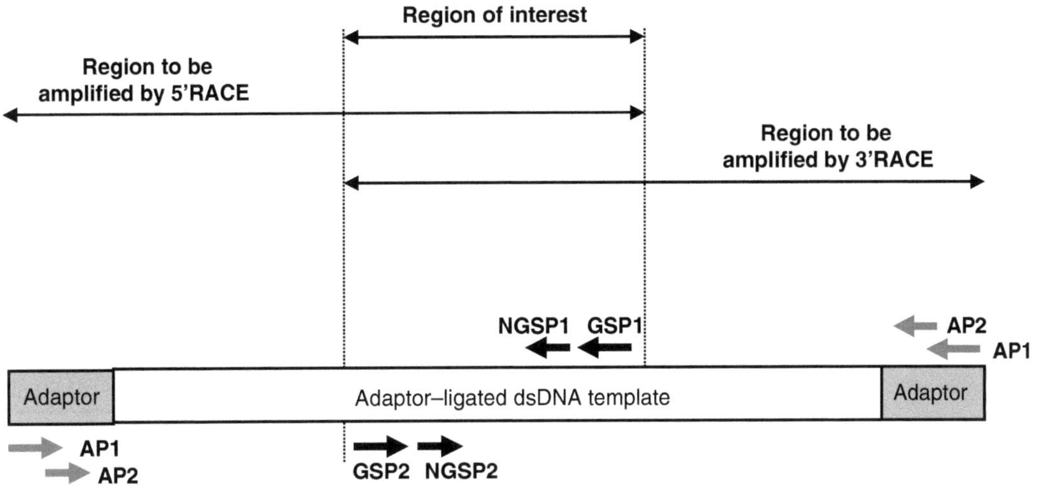

Fig. 2 Illustration of the template and primers used in Marathon RACE reactions. AP1 and AP2: adaptor primers 1 and 2; GSP and NGSP: gene-specific primers, GSP1 and NGSP1 for 5′RACE, GSP2 and NGSP2 for 3′RACE

1. Submit consecutive nesprin cDNA sequences of 1 kb with 500 bp overlapping, covering the entire gene length, to the EST database [14].
2. Select the 5′ UTRs containing an identifiable and viable Kozak sequence near the first start codon for verification.
3. Filter the detected 3′ UTRs with polyadq software (http://rulai.cshl.org/tools/polyadq/polyadq_form.html) [24] (*see* **Note 2**).

3.2 Rapld Amplification of cDNA Ends (RACE)

RACE is a suitable method to obtain full-length novel isoforms and only requires knowledge of a small internal region of the cDNA sequence as well as both an adaptor-specific primer and a gene-specific primer (GSP) [25]. The presence of a known specific adaptor sequence at the 5′ or 3′ end of the cDNA is essential for the libraries being used in RACE. Furthermore, a second PCR reaction (named nested PCR) is often required, by using the PCR product from the initial run to produce a more specific amplification product. GSP (including nested GSP) design is an important step to obtain efficient and specific PCR amplifications (*see* **Note 1**).

1. Design GSP Primers for 5′ or 3′ RACE against the region of interest on the gene, as illustrated in Fig. 2 and in the Marathon-ready cDNA user manual (Clontech).
2. Perform an initial round of PCR reactions using Advantage 2 polymerase and the Marathon-Ready cDNA (Clontech) according to manufacturer's protocol, also briefly described in Table 1, using PCR cycling parameters as (an example) follows:

   94 °C for 1 min

   30 cycles:

Table 1
RACE PCR recipe (50 μl)

RACE ready cDNA	5 μl
10× Advantage 2 PCR Buffer	5 μl
dNTP mix (10 mM)	1 μl
Adaptor primer (10 μM)	1 μl
GS primer (10 μM)	1 μl
Advantage 2 Polymerase	1 μl
Nuclease-free water	to 50 μl

- 94 °C 30 s (denaturation)
- 68 °C 4 min (anneal/extension)

68 °C for 10 min (final extension)

3. Dilute the initial PCR product (1:50) with PCR grade water, and use 5 μl from the dilution for nested PCR reactions with a nested adapter primer and a nested GSP (NGSP) (*see* **Note 3**). Use PCR cycling parameters as follows:

   94 °C for 1 min

   20 cycles:

   - 94 °C 30 s
   - 68 °C 4 min

   68 °C 10 min

4. Run the nested PCR products on an agarose gel, excise the desired fragments from the gel and clone into pGEM-T Easy vector (Promega).
5. Subject each cloned fragment to sequencing. Align the obtained sequences against the appropriate NCBI species genome to identify novel cDNA ends, which will be either upstream or downstream of the cDNA region of interest.

3.3 Multi-tissue RT-PCR

To characterize the RACE PCR products and validate expression of the putative isoform at RNA level, RT-PCR is required. To amplify short transcripts, design the 5′ and 3′ GSPs to either existing or newly identified regions of 5′ and 3′ UTRs identified from the EST database analysis and/or RACE above. For larger transcripts, design a 5′ GSP or 3′ GSP to the newly identified regions of 5′ UTR or 3′ UTR, and a 3′ GSP or 5′ GSP against the nearest coding exon, or primers in a range of exons (*see* **Note 4**) [14].

1. Perform RT-PCR on a multiple tissue cDNA panel (Origene) using GC-RICH PCR system (Roche Applied Science) according

Table 2
GC-rich PCR master mix 1 (35 μl)

First strand cDNA	50 ng
dNTP mix (10 mM)	2 μl
Forward primer (5 μM)	2 μl
Reverse primer (5 μM)	2 μl
5 M GC RICH resolution solution	5 μl
PCR grade water	to 35 μl

Table 3
GC-rich PCR master mix 2 (15 μl)

PCR grade water	4 μl
5× GC Rich PCR buffer	10 μl
GC Rich enzyme mix	1 μl

to manufacturer's protocol (Tables 2 and 3), also amplifying housekeeping genes as controls. Use PCR cycling parameters as follows:

95 °C for 3 min

30–40 cycles:

- 95 °C 30 s
- 48–55 °C 1 min (depending on the required annealing temperature for different primers)
- 68 °C 1–3 min (depending on the length of putative isoforms)

68 °C for 7 min

2. Analyze the PCR products by agarose gel electrophoresis. Excise and clone the PCR product into a pGEM-T Easy vector (Promega). Subject clones to DNA sequencing for further validation.

3.4 Alternative Splicing of Cassette Exons

Putative cassette exons of nesprins can be identified using EST databases. The resulting alternative splicing events could either occur in multiple variants or be specific for a single variant, which are also likely to be activated in a tissue-specific manner [14]. These post-transcriptional events would further increase the complexity of nesprin variants [18].

1. To verify the occurrence of alternative splicing events, design-specific primers that target the exons located on either side of the identified cassette exons via the EST database analysis.

Table 4
RT-PCR recipe (50 μl)

cDNA	50 ng
2× *Taq* master mix[a]	25 μl
Forward primer (10 μM)	1 μl
Reverse primer (10 μM)	1 μl
Nuclease-free water	to 50 μl

[a]DreamTag Green PCR master mix contains *Taq* polymerase, dNTP, and reaction buffer at 2× concentration

2. Use the designed primers to perform PCR screening of cDNA libraries from different tissues of interest (*see* **Note 5**) [14].
3. Perform all PCR reactions as described in Subheading 3.3 using GC-RICH PCR system, or as in Table 4, using the following PCR cycling parameters:

   95 °C 1 min

   30 cycles:

   - 95 °C 30 s
   - 55 °C 1 min (depending on the required annealing temperature for different primers)
   - 72 °C 1 min per kb

   72 °C 10 min
4. Purify and clone the PCR products, then subject to DNA sequence for validation, as described in Subheading 3.3 (*see* **Note 6**).

4 Notes

1. General rules for designing primers are applied. Efficient RACE primers must be specific for their target sequences in order to form stable duplexes without forming secondary structures such as hairpin loops, as it may interfere with polymerization. For advice on RACE Primer design, refer to the Marathon-ready cDNA user manual (Clontech). Adaptors required for RACE are ligated and ready for use in the Marathon-Ready cDNA kit.
2. Only the 3′ UTRs containing a poly (A) tail or at least one poly (A) site downstream of the initial open reading frame (ORF) termination are selected for further experiments.
3. A nested PCR reaction is performed using the PCR product from the initial run for a more specific amplification product. For nested RACE, the nested GSP (NGSP) is designed to

anneal with one to ten nucleotides downstream of the GSP used in the first round of amplification (for 5′ RACE) or upstream of the GSP (for 3′ RACE).

4. For validation of 5′ UTRs, the forward primer is designed within the 5′ UTR and the reverse primer within the second coding exon to control for genomic contamination. For validation of 3′ UTRs, the forward primer is designed within the second to last coding exon and the reverse primer within the 3′ UTR to also control for genomic contamination.
5. For cassette exons (either included and excluded), forward and reverse primers should be separated by at least one exon-intron boundary to control for genomic contamination.
6. There are some examples for cassettes exons described for nesprins. When PCR amplification was carried out across nesprin-1 exon 93 (one of the cassettes detected via EST database searching), two PCR products were obtained from two different cDNA libraries. One PCR product showed the presence of the exon 93 while the other showed absence of the exon, which encodes a unique 47 amino acid peptide sequence. Nesprin-2 also showed its propensity of tissue-specific alternative splicing. For example, regarding exon 107, another cassette detected by EST, two PCR products were detected in U2OS cells, either containing or missing exon 107, which encodes a unique 23 amino acids. Only isoforms containing exon 107 were expressed in vascular smooth muscle cells [14].

Acknowledgment

This work was funded by a British Heart Foundation (BHF) project grant to QPZ (PG/11/58/29004) and programme grant to CMS (RG/11/14/29056).

References

1. Zhang Q, Skepper JN, Yang F, Davies JD, Hegyi L, Roberts RG, Weissberg PL, Ellis JA, Shanahan CM (2001) Nesprins: a novel family of spectrin-repeat-containing proteins that localize to the nuclear membrane in multiple tissues. J Cell Sci 114:4485–4498.
2. Sosa BA, Kutay U, Schwartz TU (2013) Structural insights into LINC complexes. Curr Opin Struct Biol 23(2):285–291.
3. Sosa BA, Rothballer A, Kutay U, Schwartz TU (2012) LINC complexes form by binding of three KASH peptides to domain interfaces of trimeric SUN proteins. Cell 149:1035–1047.
4. Wang W, Shi Z, Jiao S, Chen C, Wang H, Liu G, Wang Q, Zhao Y, Greene MI, Zhou Z (2012) Structural insights into SUN-KASH complexes across the nuclear envelope. Cell Res 22:1440–1452.
5. Crisp M, Liu Q, Roux K, Rattner JB, Shanahan C, Burke B, Stahl PD, Hodzic D (2006) Coupling of the nucleus and cytoplasm: role of the LINC complex. J Cell Biol 172: 41–53.
6. Meinke P, Nguyen TD, Wehnert MS (2011) The LINC complex and human disease. Biochem Soc Trans 39:1693–1697.

7. Zhang Q, Ragnauth CD, Skepper JN, Worth NF, Warren DT, Roberts RG, Weissberg PL, Ellis JA, Shanahan CM (2005) Nesprin-2 is a multi-isomeric protein that binds lamin and emerin at the nuclear envelope and forms a subcellular network in skeletal muscle. J Cell Sci 118:673–687.
8. Zhang Q, Bethmann C, Worth NF, Davies JD, Wasner C, Feuer A, Ragnauth CD, Yi Q, Mellad JA, Warren DT, Wheeler MA, Ellis JA, Skepper JN, Vorgerd M, Schlotter-Weigel B, Weissberg PL, Roberts RG, Wehnert M, Shanahan CM (2007) Nesprin-1 and -2 are involved in the pathogenesis of Emery Dreifuss muscular dystrophy and are critical for nuclear envelope integrity. Hum Mol Genet 16:2816–2833.
9. Puckelwartz MJ, Kessler EJ, Kim G, Dewitt MM, Zhang Y, Earley JU, Depreux FF, Holaska J, Mewborn SK, Pytel P, McNally EM (2010) Nesprin-1 mutations in human and murine cardiomyopathy. J Mol Cell Cardiol 48(4):600–608
10. Autore F, Pfuhl M, Quan XP, Williams A, Roberts RG, Shanahan CM, Fraternali F (2013) Large-scale modelling of the divergent spectrin repeats in nesprins: giant modular proteins. PLoS One 8(5):e63633.
11. Simpson JG, Roberts RG (2008) Patterns of evolutionary conservation in the nesprin genes highlight probable functionally important protein domains and isoforms. Biochem Soc Trans 36:1359–1367.
12. Morgan JT, Pfeiffer ER, Thirkill TL, Kumar P, Peng G, Fridolfsson HN, Douglas GC, Starr DA, Barakat AI (2011) Nesprin-3 regulates endothelial cell morphology, perinuclear cytoskeletal architecture, and flow-induced polarization. Mol Biol Cell 22:4324–4334.
13. Roux KJ, Crisp ML, Liu Q, Kim D, Kozlov S, Stewart CL, Burke B (2009) Nesprin 4 is an outer nuclear membrane protein that can induce kinesin-mediated cell polarization. Proc Natl Acad Sci U S A 106:2194–2199.
14. Rajgor D, Mellad JA, Autore F, Zhang Q, Shanahan CM (2012) Multiple novel nesprin-1 and nesprin-2 variants act as versatile tissue-specific intracellular scaffolds. PLoS One 7(7): e40098
15. Randles KN, Lam LT, Sewry CA, Puckelwartz M, Furling D, Wehnert M, McNally EM, Morris GE (2010) Nesprins, but not sun proteins, switch isoforms at the nuclear envelope during muscle development. Dev Dyn 239:998–1009.
16. Rajgor D, Shanahan CM (2013) Nesprins: from the nuclear envelope and beyond. Expert Rev Mol Med 15:e5.
17. Rajgor D, Mellad JA, Soong D, Rattner JB, Fritzler MJ, Shanahan CM (2014) Mammalian microtubule P-body dynamics are mediated by nesprin-1. J Cell Biol 205(4):457–475
18. Duong NT, Morris GE, le Lam T, Zhang Q, Sewry CA, Shanahan CM, Holt I (2014) Nesprins: tissue-specific expression of epsilon and other short isoforms. PLoS One 9(4):e94380.
19. Djinovic-Carugo K, Gautel M, Ylänne J, Young P (2002) The spectrin repeat: a structural platform for cytoskeletal protein assemblies. FEBS Lett 513:119–123
20. Boguski MS, Lowe TM, Tolstoshev CM (1993) dbEST—database for "expressed sequence". Nat Genet 4(4):332–333
21. King SJ, Nowak K, Suryavanshi N, Holt I, Shanahan CM, Ridley AJ (2014) Nesprin-1 and nesprin-2 regulate endothelial cell shape and migration. Cytoskeleton 71(7):423–434
22. le Lam T, Böhm SV, Roberts RG, Morris GE (2011) Nesprin-2 epsilon: a novel nesprin isoform expressed in human ovary and Ntera-2 cells. Biochem Biophys Res Commun 412(2):291–295
23. Zhang Q, Ragnauth C, Greener MJ, Shanahan CM, Roberts RG (2002) The nesprins are giant actin-binding proteins, orthologous to Drosophila melanogaster muscle protein MSP-300. Genomics 80:473–481.
24. Tabaska JE, Zhang MQ (1999) Detection of polyadenylation signals in human DNA sequence. Gene 231(1-2):77–86
25. Yeku O, Frohman MA (2011) Rapid amplification of cDNA ends (RACE). Methods Mol Biol 703:107–122

Chapter 14

Detection of Diverse and High Molecular Weight Nesprin-1 and Nesprin-2 Isoforms Using Western Blotting

James Carthew and Iakowos Karakesisoglou

Abstract

Heavily utilized in cell and molecular biology, western blotting is considered a crucial technique for the detection and quantification of proteins within complex mixtures. In particular, the detection of members of the nesprin (nuclear envelope spectrin repeat protein) family has proven difficult to analyze due to their substantial isoform diversity, molecular weight variation, and the sheer size of both nesprin-1 and nesprin-2 giant protein variants (>800 kDa). Nesprin isoforms contain distinct domain signatures, perform differential cytoskeletal associations, occupy different subcellular compartments, and vary in their tissue expression profiles. This structural and functional variance highlights the need to distinguish between the full range of proteins within the nesprin protein family, allowing for greater understanding of their specific roles in cell biology and disease. Herein, we describe a western blotting protocol modified for the detection of low to high molecular weight (50–1000 kDa) nesprin proteins.

Key words LINC complex, Nesprins, Nuclear envelope, Polyvinylidene difloride, SDS-PAGE, SYNE, Western blotting

1 Introduction

The LINC (Linker of Nucleoskeleton and Cytoskeleton) complex is a structural bridge spanning the entire nuclear envelope of eukaryotic cells, serving as a functional connection between the nuclear interior and the cytoskeleton [1, 2]. At its core are KASH- and SUN-domain protein associations. Proteins that are members of the vertebrate KASH-domain family are known as nesprins (nuclear envelope spectrin repeat proteins). Nesprins are encoded by the *SYNE* genes and display a plethora of isoforms, which are structurally and functionally diverse. These isoforms are generated via alternative initiation, termination, and splicing. As a consequence, nesprins vary drastically in their modular organization, including in their spectrin repeat copy number, ABD (actin-binding domain), PBD (plectin-binding domain), and KASH (Klarsicht, ANC-1, and *SYNE* homology) domain compositions [3].

Sue Shackleton et al. (eds.), *The Nuclear Envelope: Methods and Protocols*, Methods in Molecular Biology, vol. 1411, DOI 10.1007/978-1-4939-3530-7_14, © Springer Science+Business Media New York 2016

The ability to detect and distinguish between this wide range of isoforms is crucial to the advancement of nesprin research. With key roles in directed cell migration, nuclear structure, ciliogenesis, DNA damage repair, and cellular signalling pathways, nesprins exhibit a high degree of functional diversity [3]. Not surprisingly, nesprin knockout mice exhibit severe neuromuscular and cardiac disease pathologies [3–5], high-frequency hearing loss [6], and memory defects [7]. This phenotypic diversity suggests that advanced analytical methods are required to comprehensively investigate the roles of distinct nesprin isoforms and paralogues within the cell.

Western blot analysis relies on the separation of proteins by molecular weight via sodium dodecyl sulphate polyacrylamide gel electrophoresis (SDS-PAGE) [8–11] and subsequent protein transfer to a membrane. Existing western blot protocols accurately detect proteins between 10 and 250 kDa; however, proteins with broader molecular weight ranges, such as nesprins (between 20 and 1000 kDa), require additional adaptations. These include the use of polyacrylamide gels fortified with agarose [12] and gradient gels ranging from 3 to 15 % acrylamide [13–15], in order to aid in their acquisition. Protein transfer can be achieved via a variety of techniques including diffusion [9], vacuum-assisted solvent flow [16], and electrophoretic transfer [11]. For the transfer of larger molecular weight proteins (>250 kDa), adapted electrophoretic transfer methods have been developed, one of which will be described within this protocol. Using the method described herein, it has been possible to accurately resolve nesprin proteins ranging from 20 to 1000 kDa. This was achieved through the adaptation of a gradient polyacrylamide gel system, which allows for direct comparisons to be made between nesprin isoform variants.

2 Materials

All solutions described should be prepared using purified deionized water and analytical grade reagents unless otherwise stated. Prepare and store all reagents at room temperature (unless indicated otherwise). When disposing of waste materials and solutions, follow your respective guidelines and COSHH (control of substances hazardous to health) forms as to their safe disposal.

2.1 Cell/Tissue Lysis

1. Sterile filtered, cell culture grade PBS.
2. Protein lysis buffer (RIPA): 50 mM Tris–HCl, pH 7.5, 150 mM NaCl, 0.1 % SDS (sodium dodecyl sulphate), 1 % Nonidet P-40, 0.5 % sodium-deoxycholate, and protease inhibitors (*see* **Note 1**).

3. Protease inhibitor cocktail (PIC, 100×): Once made into solution, aliquots should be stored at −20 °C.
4. Cell scrapers.
5. 1 mL syringes.
6. 24G × ¾ in. disposable hypodermic needles.
7. Laemmli sample buffer (5×): 0.25 M Tris–HCl, pH 6.8, 20 % glycerol, 4 % SDS, 1.43 M β-mercaptoethanol and 0.2 % bromophenol blue (*see* **Note 2**).

2.2 SDS-PAGE

1. Gel Caster: (Hoefer Inc.) Dual gel caster for mini vertical units.
2. Gel electrophoresis running chamber: (Hoefer Inc.) SE260 Mighty Small II. Deluxe Mini Vertical Electrophoresis Unit with 1.5 mm thick spacers (*see* **Note 3**).
3. 10.6 cm × 10.1 cm glass plate: (Web Scientific).
4. Aluminium back plate: (Web Scientific).
5. 25 mL Gradient mixer: (VWR).
6. Peristaltic pump.
7. 3 % Polyacrylamide resolving gel stock: For 100 mL, combine 25 mL of 1.5 M Tris–HCl, pH 8.8 with 10 mL [30 %] acrylamide (*see* **Note 4**). Add a further 1 mL of [10 %] SDS and top up with 64 mL H_2O. Stock solution should be stored at 4 °C for maximum 1 month.
8. 15 % Polyacrylamide resolving gel stock: For 100 mL, combine 25 mL of 1.5 M Tris–HCl, pH 8.8 with 50 mL [30 %] acrylamide (*see* **Note 4**). Add a further 1 mL [10 %] SDS and top up with 24 mL H_2O. Store the stock solution at 4 °C for 1 month.
9. 4 % Polyacrylamide stacking gel stock: For 100 mL, combine 20 mL of 0.5 M Tris–HCl, pH 6.8 with 13.3 mL [30 %] acrylamide (*see* **Note 4**). Add a further 1 mL [10 %] SDS and top up with 65.6 mL H_2O. Store the stock solution at 4 °C for 1 month.
10. SDS gel running buffer (10×): 0.25 M Tris–HCl, pH 8.3, 2 M glycine, 1 % SDS.
11. Ammonium persulfate (APS): 10 % solution in H_2O. Leave one aliquot for use and store the rest at −20 °C.
12. *N,N,N′,N′*-tetramethylethylenediamine (TEMED). Store at 4 °C.

2.3 Immunoblotting

1. Nesprin-2 affinity-purified C-terminal antibodies (pAbK1; [17, 18]).
2. Anti-Rabbit POD (Horseradish Peroxidase) secondary antibodies.

3. Prestained protein ladder.
4. Western blot transfer buffer: 25 mM Tris-Base, 190 mM glycine, 0.1 % SDS, 10 % ethanol.
5. Western blot transfer tank: e.g. Electroblotting unit BTV100 (VWR).
6. Whatman paper.
7. Polyvinylidene difloride (PVDF) membrane: Immobilon-P 0.45 μm pore size (*see* **Note 5**).
8. Wet transfer compression cassette (VWR).
9. Fiber pads (VWR).
10. Blocking solution: 5 % skim milk powder in PBS.
11. Tween-20 washing buffers: Two buffers are utilized containing 0.1 and 0.3 % Tween-20 in PBS.
12. Enhanced chemiluminescence (ECL) detection solution: For 20 mL, add 17.7 mL H_2O followed by 2 mL of 1 M Tris–HCl, pH 8.5. Add a further 200 μL of 250 mM luminol (3-Aminophthalhydrazide) and 89 μL of 90 mM *p*-Coumaric acid. Just before usage, add 6.1 μL of 30 % H_2O_2. Luminol and *p*-Coumaric acid stock solutions are prepared in dimethyl sulfoxide (DMSO), stored at ambient temperature and protected from light.
13. High performance chemiluminescence films.

3 Methods

All methods described below should be undertaken at room temperature unless otherwise specified.

3.1 Sample Preparation: From Cell Culture to Cell Lysates

1. For a T-75 culture flask at 70 % cell confluency: prepare 500 μL ice-cold RIPA buffer, combine with 5 μL of 100× concentrated PIC (protease inhibitor cocktail), and then leave on ice in an Eppendorf tube.
2. Take the T-75 flask from the cell incubator, aspirate the media, and wash the cells three times for 1 min each with ice-cold cell culture grade PBS.
3. Perform all subsequent processes on ice to prevent autolysis.
4. Remove the PBS and add the ice-cold RIPA buffer containing PIC. Evenly distribute the solution across the surface while using a cell scraper to dislodge cells (*see* **Note 6**).
5. Dislodge all cells from the flask surface and transfer the lysate into a pre-cooled Eppendorf tube and leave on ice for 15 min.

6. Perform a mechanical breakdown of the lysate during the 15 min incubation on ice using a sterile 1 mL syringe fitted with a 24G × ¾ in. disposable hypodermic needle. Syringe sample 15 times to ensure lysate homogenization, efficient protein extraction, and genomic DNA shearing (*see* **Note 7**).
7. Centrifuge the resulting lysate at 4 °C for 15 min at 12,000 × *g* to pellet the remaining cell debris.
8. Carefully remove the supernatant and place into a new Eppendorf tube (*see* **Note 8**).
9. Add 120 μL of 5× concentrated Laemmli sample buffer to the extracted supernatant and then incubate at 99 °C for 4 min (*see* **Note 9**).
10. After incubation, store samples at −20 °C (*see* **Note 10**).
11. Equal loading of cell lysates can be obtained by protein quantitation assays (e.g. BCA, Lowry) before Laemmli sample buffer addition, or by our preferred method of assessing Coomassie blue staining after SDS-PAGE (*see* **Note 11**).

3.2 Sodium Dodecyl Sulphate Polyacrylamide Gel Electrophoresis (SDS-PAGE)

1. Assemble the gel casting equipment using the 10.6 cm × 10.1 cm glass and aluminium back plates and the spacers. Add water to ensure that no leakage occurs across the bottom of the caster (*see* **Note 12**).
2. For large molecular weight proteins such as nesprins, utilize a gradient gel system (Fig. 1), to produce 3–15 % acrylamide gradient gels. Alternatively, commercially available precast gradient gels can be used.
3. Mix 5.5 mL of 15 % polyacrylamide stock with 18.3 μL of 10 % APS and 9.7 μL TEMED in a 50 mL conical flask, and insert solution into column A. Once the solution is inserted into column A, mix 5.5 mL of 3 % polyacrylamide stock with 18.3 μL 10 % APS and 9.7 μL TEMED in a 50 mL conical flask, then load the solution into column B (*see* **Note 13**). Elevate the gradient maker and then activate the magnetic stirrers and the peristaltic pump (50 rpm). Open valve 1 first, followed by valve 2. As the 3 % acrylamide is pulled from column B to column A, the 15 and 3 % acrylamide solutions mix, forming a gradual decrease in % acrylamide from the base of the gel to the top (*see* **Note 14**). Ensure enough space is allowed for the gel comb.
4. Gently pour 2 mL isobutanol over the gel to ensure a level gel surface is formed (*see* **Note 15**).
5. Pour off the isobutanol once the resolving gel is set (ca. 30 min). In a separate conical tube, combine 3 mL of 4 % polyacrylamide stock with 30 μL 10 % APS and 16 μL TEMED to

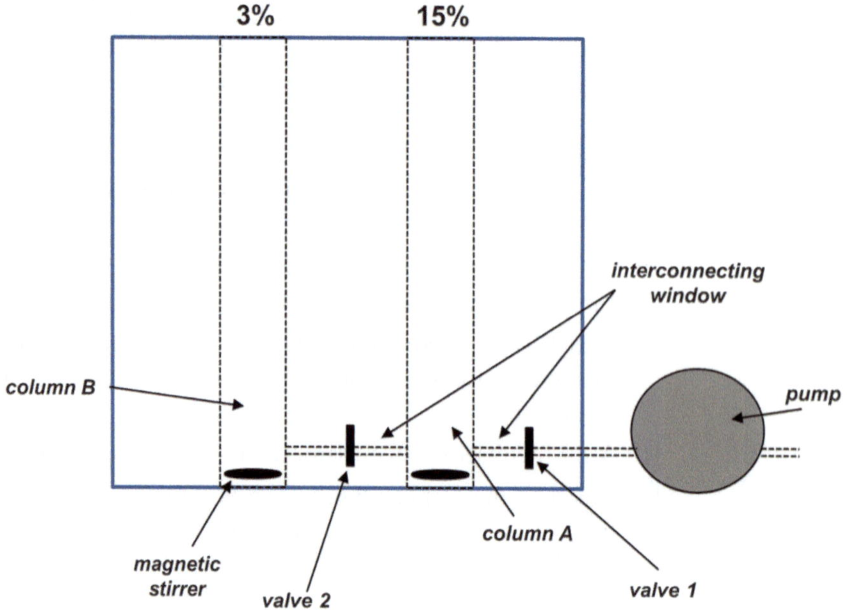

Fig. 1 Schematic depicting the self-casting gradient mixer system. Both the 15 and 3 % acrylamide solutions mix within column A as the pump draws liquid to the gel caster. This produces an acrylamide concentration gradient of 15 % at the base of the gel, which decreases to 3 % at the top

prepare the stacking gel solution. Pour the stacking gel solution onto the resolving gel and add the comb (*see* **Note 16**).

6. Thaw the sample aliquots produced in Subheading 3.1 and load equal amounts of protein alongside a standard protein ladder.
7. After loading, electrophorese the samples at a constant voltage of 80–100 V for 20 min. Increase the voltage to 120 V and electrophorese the samples until the dye front has reached the base of the gel (*see* **Note 17**).
8. Immediately following electrophoresis, carefully separate the glass plates using a spatula or similar tool, leaving one side of the gel still attached to one glass plate. Transfer the gel to a separate container and submerge in transfer buffer for 1 min (*see* **Note 18**).
9. Cut 4 sections of Whatman paper and 1 section of PVDF membrane to the same size as the gel. Before transfer, activate the PVDF membrane by immersing for 15 s in methanol, followed by a 2 min wash in distilled H_2O and a final 5 min wash in transfer buffer.

3.3 Tank (Wet) Western Blot Electro-Transfer

1. Following protein separation via SDS-PAGE, wet electrotransfer is conducted. An example of the final transfer set up using a compression cassette and fiber pads is shown in Fig. 2.

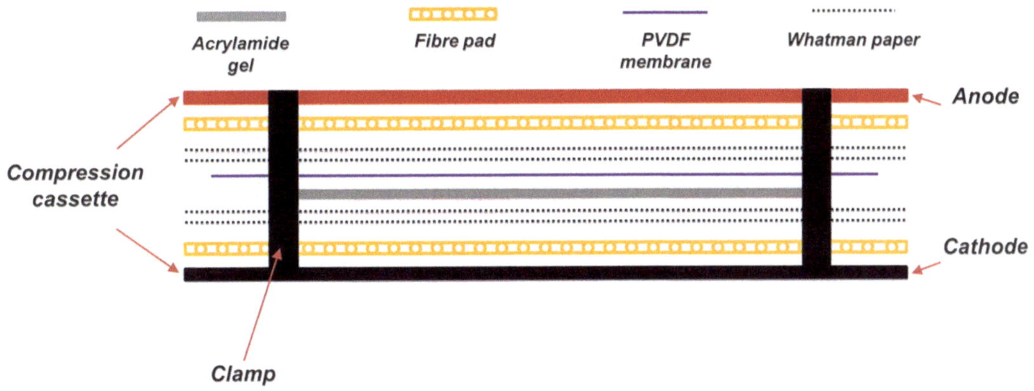

Fig. 2 Schematic indicating the Whatman paper, acrylamide gel, and PVDF membrane "sandwich" assembly in the compression cassette. The assembly order is as follows; (1) 1 fiber pad, (2) 2 pieces of transfer buffer-soaked Whatman paper, (3) acrylamide gel, (4) PVDF membrane, (5) 2 pieces of transfer buffer-soaked Whatman paper, (6) 1 fiber pad

2. Place 1 fiber pad followed by 2 Whatman paper layers, soaked with western blot transfer buffer, in the compression cassette. Then gently place the acrylamide gel across the surface of the Whatman paper (*see* **Note 18**).
3. Carefully place the PVDF membrane across the surface of the gel (*see* **Note 19**).
4. Add a further 2 Whatman paper layers soaked in transfer buffer, and the final fiber pad (*see* **Note 20**).
5. Place the compression cassette into an electrophoretic transfer tank filled with 2 L western blot transfer buffer.
6. For lower molecular weight proteins (<100 kDa), transfer can be conducted for 2 h at a constant voltage of 50 V. For large molecular weight proteins (>250 kDa), transfer is conducted at a constant voltage of 10 V for a minimum of 18 h, followed by a further 2 h at a constant voltage of 45 V (*see* **Note 21**).

3.4 Membrane Blocking and Immuno-Blotting

1. Following transfer, disassemble the compression cassette and label the prestained molecular weight standards and gel positioning on the PVDF membrane (*see* **Note 22**).
2. Wash the PVDF membrane in methanol for 10 s, then sandwich in Whatman paper and leave to dry for 15 min (*see* **Note 23**).
3. Incubate the membrane in blocking solution for 30 min at ambient temperature or overnight at 4 °C to reduce unspecific antibody binding to the membrane.
4. Rinse the membrane two times for 15 s each time in PBS + 0.1 % Tween-20, followed by a single 5 min wash in PBS + 0.1 % Tween-20.

5. Add primary antibody, diluted in blocking solution, to the membrane. For lower molecular weight proteins of <120 kDa, antibodies are incubated for 1 h at room temperature. However, for larger proteins of >120 kDa, overnight primary antibody incubation at 4 °C using a rocking shaker is required.
6. Wash the membrane three times with PBS + 0.1 % Tween-20, 10 min each time.
7. Dilute the secondary antibody in blocking buffer and incubate for 1 h at room temperature using a rocking shaker.
8. Wash the membrane initially three times with PBS + 0.3 % Tween-20, for 5 min each time.
9. Wash an additional three times using PBS + 0.1 % Tween-20, for 5 min each time (*see* **Note 24**).
10. Remove the washing solution from the membrane and carefully drain excess liquid with a paper towel (*see* **Note 25**).
11. Place the freshly made ECL detection solution onto the membrane for 1 min. Remove the excess solution and leave a thin film of ECL solution lining the surface of the membrane.
12. Wrap the membrane carefully within Saran wrap, place it into a standard film cassette, and overlay blot with a photographic film (*see* **Note 26**). Films should be developed at a range of exposures to ensure adequate documentation.
13. Label the molecular weights of the reference standard on the film.
14. To ensure equal loading has been achieved, standard western blots against GAPDH, β-actin, or β-tubulin are initially performed. At least 2–3 loading control proteins should be assessed as proteome variability between cell lines, and varying cell culture conditions can lead to differential expression between these standards.
15. Protein smearing is an issue which occurs during western blotting and often results from incomplete DNA shearing and lysate homogenization during Subheading 3.1, **step 6** in this protocol. Figure 3 displays representative images of variable protein smearing states in which panel (a) displays extreme smearing and protein overloading, while panel (b) displays reduced smearing and unequal total protein loading. The protein smearing effects seen in Fig. 3b can be removed through re-syringing the sample and adjusting the total protein content (Fig. 3c).
16. See pan-nesprin-2 western blot examples in Fig. 4. Panel (a) displays the simultaneous detection of both giant (~800 kDa) and low molecular weight (~25 kDa) nesprin-2 isoforms. The same gel developed at a lower exposure time can be seen in panel (b), highlighting how lower nesprin isoforms vary in their expression relative to nesprin-2 giant.

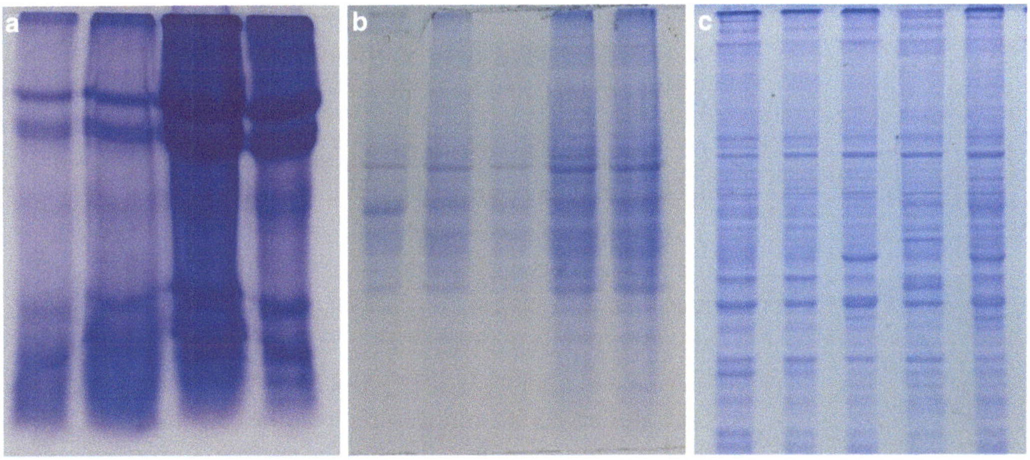

Fig. 3 Coomassie blue-stained SDS-PAGE gels, displaying varying protein lysate qualities and quantities. Panel (**a**) displays protein overloading and lysates containing cell debris. Panel (**b**) displays less drastic smearing, recoverable through re-syringing and centrifugation. Panel (**c**) displays optimal experimental conditions

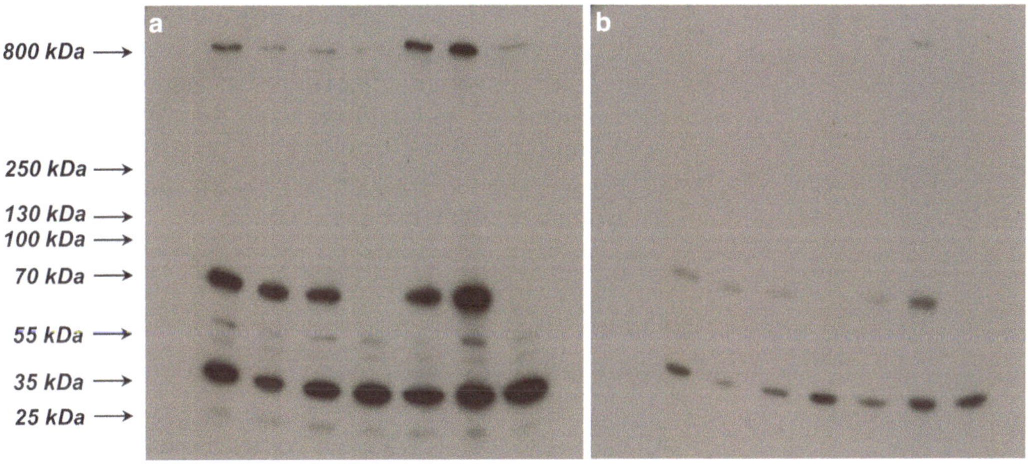

Fig. 4 Nesprin 2 western blot analysis of various cell lines using polyacrylamide gradient gels, displaying both 2 min (**a**) and 30 s (**b**) exposure times. The western blot was executed utilising the pAbK1 anti-nesprin-2 affinity-purified C-terminal polyclonal antibody [17, 18]. Nesprin isoforms between 25 and 800 kDa are attained, displaying dissimilar expression patterns. Our detection method therefore provides increased accuracy when assessing isoform variability across broad molecular weight ranges. The lanes represent samples from the following cell lines: HeLa (cervical cancer), SW620 and SW480 (colon adenocarcinoma), MDA-MB-231 and MCF7 (breast cancer epithelia), HaCaT (immortalized keratinocytes), and COS7 (African green monkey fibroblast-like kidney cells)

4 Notes

1. Once prepared, the RIPA (RadioImmuno-Precipitation Assay) buffer must be protected from light. Leave one 10 mL aliquot on ice for immediate use, and store remaining aliquots at –20 °C.

2. One aliquot should remain at ambient temperature for immediate use and remaining aliquots should be stored at −20 °C. SDS precipitates at 4 °C, so Laemmli buffer must always be warmed prior use.
3. 1.5 mm spacers are preferred over thinner spacers for the detection of large molecular weight proteins.
4. Acrylamide solution used contains 30 % (w/v) acrylamide, 0.8 % (w/v) bis-acrylamide stock (37.5:1).
5. For the specific detection of nesprin proteins, PVDF membranes are preferred over nitrocellulose due to their increased protein-binding affinity and re-probing potential.
6. Care should be taken not to remove the cell culture flask from the ice while scraping cells from the surface. Ensure all cells are removed from the surface of the cell culture flask as if too few cells are dislodged, the overall protein concentration may be too low for successful detection. Roughly 2 min of thorough scraping is enough to achieve complete removal of cells. As a rough estimate, cell debris should be seen within the lysis buffer when tilted at 45°.
7. Draw the lysis solution into the syringe before the needle tip is attached. Further care should be taken during the first few syringing-steps; the high viscosity of the protein solution can result in high pressure, causing the hypodermic needle to fall off the syringe, resulting in sample loss.
8. Care should be taken not to disturb the pellet. If a distinct pellet is not seen, re-centrifuge the sample to ensure complete separation.
9. Seal the Eppendorf tubes securely with a locking clip during boiling to prevent sample loss.
10. Continual freeze/thaw cycles will compromise lysate quality; ensure enough sample aliquots are produced at this stage in order to avoid this.
11. Coomassie blue staining provides an early indication of total protein quantity and lysate quality, as cellular debris, unsheared genomic DNA, and pellet fragments can be detected as smears. If smears do appear, samples can be re-syringed and/or sonicated prior centrifugation at 12,000 × g for 15 min.
12. Spacers, glass, and aluminium back plates should be thoroughly cleaned with ethanol and dried before use. Residual dust and dirt can cause gels to stick to the plates, making it difficult to remove the gels for transferring. Use flawless plates and ensure proper alignment.
13. When mixing APS and TEMED with the acrylamide stocks, ensure the solution is mixed thoroughly so that a uniformed gel is produced. If bubbles appear, gently tap the solution to remove them.

14. Stirrer and peristaltic pump speeds should be kept to a minimum. Increasing both speeds can cause the two acrylamide solutions to mix too quickly, compromising the quality of the gradient gel.
15. Ensure the isobutanol is evenly distributed across the surface of the gel.
16. Mix the solution quickly as the mixture sets rapidly. As the comb is inserted, take care that no bubbles form across the bottom of the wells as this can result in unequal electrophoretic migration of the proteins within the resolving gel.
17. The lower voltage used initially helps to ensure all proteins within the sample collect together at the bottom of the wells, allowing proper protein separation to occur in the resolving gel.
18. Care must be taken not to distort or break the acrylamide gel.
19. It is best to correctly position the PVDF membrane onto the gel on the first attempt.
20. It is important that no bubbles form between the gel and PVDF membrane, as this will impair protein transfer. To avoid this, use a 10 mL pipette to gently roll over the Whatman paper surface.
21. The tank can overheat during overnight transfers, resulting in distorted gel morphology, and hence, an abnormal western blot pattern (*see* Fig. 4, nesprin-2 low molecular weight isoform "smiling" effects). To counteract this, the blotting procedure should be performed in a cold room at 4 °C.
22. Noting ladder marker positioning allows accurate molecular weights to be assigned to each western blot signal after the procedure is completed.
23. At this stage, the membrane can be left for several days within the Whatman paper at 4 °C without further processing.
24. If the western blot analysis yields high background, increase the washing times and the PBS + 0.3 % Tween-20 washing steps.
25. It is important to ensure that the membrane is not completely dried at this stage. Only the excess washing solution should be removed from the membrane.
26. The Saran wrap helps ensure that the membrane doesn't dry out during the final western blot stages, as some proteins will require film exposures of up to 30–45 min.

Acknowledgments

This work was supported by Breast Cancer Now. With thanks to Dr. Martin Goldberg for critically reading the contents of this chapter and valuable discussions.

References

1. Stewart-Hutchinson PJ, Hale CM, Wirtz D et al (2008) Structural requirements for the assembly of LINC complexes and their function in cellular mechanical stiffness. Exp Cell Res 314:1892–1905
2. Crisp M, Liu Q, Roux K et al (2006) Coupling of the nucleus and cytoplasm: role of the LINC complex. J Cell Biol 172:41–53
3. Cartwright S, Karakesisoglou I (2014) Nesprins in health and disease. Semi Cell Dev Biol 29:169–179
4. Zhang X, Xu R, Zhu B et al (2007) Syne-1 and Syne-2 play crucial roles in myonuclear anchorage and motor neuron intervention. Development 134:901–908
5. Puckelwartz MJ, Kessier E, Zhang Y et al (2009) Disruption of nesprin-1 produces an Emery Dreifuss muscular dystrophy-like phenotype in mice. Hum Mol Genet 18:607–620
6. Horn HF, Brownstein Z, Lenz DR et al (2013) The LINC complex is essential for hearing. J Clin Invest 123:740–750
7. Zhang X, Lei K, Yuan X et al (2009) SUN1/2 and Syne/Nesprin-1/2 complexes connect centrosome to the nucleus during neurogenesis and neuronal migration in mice. Neuron 64:173–187
8. Kurien BT, Scofield RH (2006) Western blotting. Methods 38:283–293
9. Renart J, Reiser J, Stark GR (1979) Transfer of proteins from gels to diazobenzyloxymethyl-paper and detection with antisera; a method for studying antibody specificity and antigen structure. Proc Natl Acad Sci U S A 76:3116–3120
10. Gershoni JM, Palade GE (1983) Protein blotting: principles and applications. Anal Biochem 131:1–15
11. Towbin H, Staehelin T, Gordon J (1979) Electrophoretic transfer of proteins from polyacrylamide gels to nitrocellulose sheets; procedure and some applications. Proc Natl Acad Sci U S A 76:4350–4354
12. Tatsumi R, Hattori A (1995) Detection of giant myofibrillar proteins connectin and nebulin by electrophoresis in 2 % polyacrylamide slab gels strengthened with agarose. Anal Biochem 224:28–31
13. Walker JM (1984) Gradient SDS polyacrylamide gel electrophoresis. Methods Mol Biol 1:57–61
14. Casas-Terradellas E, Garcia-Gonzalo FR, Hadjebi O et al (2006) Simultaneous electrophoretic analysis of proteins of very high and low molecular weights using low-percentage acrylamide gel and a gradient SDS-PAGE gel. Electrophoresis 27:3935–3938
15. Bustamante JJ, Garcia M, Gonzalez L et al (2005) Separation of proteins with a high molecular mass difference of 2 kDa utilising preparative double-inverted gradient polyacrylamide gel electrophoresis under non-reducing conditions: application to the isolation of 24 kDa human growth hormone. Electrophoresis 26:4389–4395
16. Peferoen M, Huybrechts R, De Loof A (1982) Vacuum-blotting: a new simple and efficient transfer of proteins from sodium dodecyl sulfate-polyacrylamide gels to nitrocellulose. FEBS Lett 145:369–372
17. Lu W, Schneider M, Neumann S et al (2012) Nesprin interchain associations control nuclear size. Cell Mol Life Sci 69:3493–3509
18. Lüke Y, Zaim H, Karakesisoglou I et al (2008) Nesprin-2 giant (NUANCE) maintains nuclear envelope architecture and composition in skin. J Cell Sci 121:1887–1898

Chapter 15

The Use of Polyacrylamide Hydrogels to Study the Effects of Matrix Stiffness on Nuclear Envelope Properties

Rose-Marie Minaisah, Susan Cox, and Derek T. Warren

Abstract

Matrix-derived mechanical cues influence cell proliferation, motility, and differentiation. Recent findings clearly demonstrate that the nuclear envelope (NE) adapts and remodels in response to mechanical signals, including matrix stiffness, yet a plethora of studies have been performed on tissue culture plastic or glass that have a similar stiffness to cortical bone. Using methods that allow modulation of matrix stiffness will provide further insight into the role of the NE in physiological conditions and the impact of changes in stiffness observed during ageing and disease on cellular function. In this chapter, we describe the polyacrylamide hydrogel system, which allows fabrication of hydrogels with variable stiffness to better mimic the environment experienced by cells in most tissues of the body.

Key words Mechanotransduction, Hydrogels, Extracellular matrix and stiffness

1 Introduction

Signals derived from the extracellular microenvironment regulate many cellular processes including proliferation, migration, and differentiation. Our understanding of how cells sense and transmit these signals has increased rapidly in recent years and we are now beginning to appreciate the role of biophysical signalling, in addition to biochemical pathways that regulate these processes. Matrix stiffness has emerged as a major regulator of cellular behavior and the stiffness of the microenvironment transmits "outside-in" forces to cells. This process is dependent on ECM adhesions that convey force between the ECM and cytoskeleton [1]. Cells respond to outside-in signals by exerting actomyosin-based contractile forces on the matrix (inside-out forces) that increase cell stiffness and scale with ECM stiffness [2]. Rho/ROCK signalling is rapidly activated at ECM adhesions in response to matrix stiffness to augment actomyosin activity, via actin polymerization and myosin light chain phosphorylation, and increase cell stiffness [3, 4].

Sue Shackleton et al. (eds.), *The Nuclear Envelope: Methods and Protocols*, Methods in Molecular Biology, vol. 1411,
DOI 10.1007/978-1-4939-3530-7_15, © Springer Science+Business Media New York 2016

The nuclear envelope (NE) consists of an outer nuclear membrane (ONM) and an inner nuclear membrane (INM) with both membranes joined at nuclear pores [5]. Importantly, the NE has emerged as a major regulator of cytoskeletal organization and differentiation [6]. Underlying the INM is an intermediate filament (IF) protein meshwork, called the nuclear lamina, which is composed of lamins (A/C and B) and lamina-associated proteins [5]. The nuclear lamina is indirectly coupled to the cytoskeleton via the LInker of Nucleoskeleton and Cytoskeleton (LINC) complex that spans the NE [7, 8]. This complex is comprised of nesprin and SUN family members. Giant nesprin-1/2 isoforms reside on the ONM and bind F-actin via a pair of N-terminal Calponin Homology (CH) domains [9]. Complex stability is maintained via interactions between the C-terminal Klarsicht, Anc-1, Syne-1 Homology (KASH) domain of the nesprins, and the SUN domain of SUN1 and SUN2 in the perinuclear space [7, 8]. SUN1/2 span the INM and interact directly with lamins A/C, physically coupling the actin cytoskeleton to the nuclear lamina [8]. Thus, the plasma membrane, actin cytoskeleton, and nucleus function as a single mechanically coupled system. Recent studies have highlighted the importance of this network in matrix-derived mechanosignalling and lamin A levels scale with matrix stiffness in mesenchymal stem cells (MSCs) [10, 11]. Lamin A also modulates actin dynamics to control the nuclear availability of the serum response factor (SRF)-coactivator myocardin-related transcription factor-A (MRTF-A) to regulate SRF-mediated transcription [12]. Importantly, the LINC complex has recently been shown to directly transmit biophysical mechanical signals across the NE and mechanical stimulation of nesprin-1 giant on the ONM increases lamin A coupling to the INM [13].

The Young's modulus, E (a measure of material stiffness), of different tissues within the body demonstrates considerable variation, from very soft (fat 0.1–1 kPa), to stiff (muscle10–20 kPa), to extremely stiff (cortical bone 10–20 GPa) [14]. Moreover, aging and disease augment tissue stiffness. For example, healthy aorta has a Young's modulus of between 10 and 20 kPa, whereas atherosclerotic plaques contain a stiffened fibrous cap (E= 60–250 kPa) and aortas from spontaneously hypertensive rats display a two- to four-fold increase in stiffness compared to age-matched controls [15–18]. A plethora of studies have been performed on tissue culture plastic (E= 10 GPa) and glass (E= 55 GPa) that possess a similar Young's modulus to cortical bone (E= 10–20 GPa). However, following recent advances in our understanding of how the NE adapts and remodels in response to matrix-derived mechanical cues, methods that allow manipulation of these signals will provide a better understanding of NE function in physiological and pathological conditions. One strategy to regulate ECM stiffness is via the utilization of hydrogels that possess defined Young's moduli.

Polyacrylamide (PA) hydrogels have a number of advantages including: (1) no specialized equipment is required for fabrication, (2) they possess constant surface chemistry despite altered mechanical properties, and (3) only ECM molecules covalently linked to the hydrogel can serve as ligands for cell attachment [14, 19]. In this chapter, we will discuss the process of PA hydrogel fabrication, a multistep process that includes coverslip activation, hydrogel polymerization, and hydrogel functionalization.

2 Materials

(3-Aminopropyl)-triethoxysilane (APES).
0.5 % Glutaraldehyde.
Ammonium persulphate (APS), 10 % dissolved in water.
N,N,N,N′ tetramethyl-ethylenediamine (TEMED).
Sulfo-SANPH, 1 mg/mL in water.
2.5 % Bis-acrylamide.
40 % Acrylamide.
Phosphate buffered saline (PBS).
Earle's balanced salt solution.
30 mm thickness No. 1 coverslip.
1 mm thick glass slide.
6-Well culture plate.
Collagen-1 (or ECM component of choice).
A trans-illuminator or UV lamp that irradiates at 300–460 nm. We use a 25 W UVP trans-illuminator.

3 Method

All steps are performed at room temperature, unless otherwise stated.

3.1 Coverslip Activation

1. In a fume hood, cover the top surface of a 30 mm coverslip with APES for 5 min.
2. Rinse three times with 2 mL distilled water, making sure that the top and bottom surfaces of the coverslip are well-washed (*see* **Note 1**).
3. Immerse coverslips in 0.5 % glutaraldehyde solution and incubate for 30–60 min.
4. Rinse three times with distilled water and air dry. In our hands, activated coverslips remain functional for 1 week and can be stored at room temperature.

3.2 Hydrogel Preparation

1. Mix together the appropriate ratios of acrylamide and bis-acrylamide in distilled water or PBS. Typical ratios that we use, and their corresponding Young's moduli (determined by

Table 1
Stiffness of hydrogel mixtures measured by atomic force microscopy

Hydrogel stiffness (kPa)	Acrylamide (final %)	Bis-acrylamide (final %)
1.879 ± 0.2835	5	0.1
11.78 ± 1.270	7.5	0.15

atomic force microscopy), are shown in Table 1 (*see* **Note 2**). The Young's moduli of other acrylamide/bis-acrylamide ratios can be found elsewhere [14] (*see* **Note 3**).

2. To initiate hydrogel polymerization, add 1:100 APS and 1:1000 TEMED to the aliquot of stock solution and mix well (*see* **Note 4**).
3. Place 50 μL of the mix in the centre of a glass slide. Place the activated side of the coverslip into the solution, gently lowering the coverslip and being careful not to create air bubbles. Incubate for 5–10 min, using any leftover mix to assess polymerization state (*see* **Note 5**).
4. Remove the glass slide and place the coverslip, hydrogel side up, into the bottom of a 6-well plate or 35 mm petri dish (*see* **Note 6**).
5. Wash hydrogels with distilled water to remove any unpolymerized mixture. Hydrogels can be stored in distilled water or PBS at 4 °C for 1 week.

3.3 Functionalizing Hydrogels for Tissue Culture

Cell attachment to the hydrogels requires an ECM coating. We use collagen; however, you can use any ECM component, depending on cell type and integrin pathways of interest. The ECM component is covalently cross-linked to the hydrogel, and to achieve this, we use sulpho-SANPH, a photo-activated protein cross-linker using the following method.

1. Remove the distilled water/PBS from the hydrogel.
2. Add 500–1000 μL of sulfo-SANPH, making sure the hydrogel is fully immersed to allow complete coverage (*see* **Note 7**).
3. Expose to 365 nm UV light for 5–10 min. The hydrogel will be coated in dark red sulfo-SANPH. Repeat if necessary (*see* **Note 8**).
4. Remove the sulfo-SANPH and wash the hydrogel 3× in sterile PBS (*see* **Note 9**).
5. Immerse the coverslip in 0.1–0.3 mg/mL collagen or ECM component of your choice. Incubate at 4 °C for between 1 and 4 h (*see* **Note 10**).

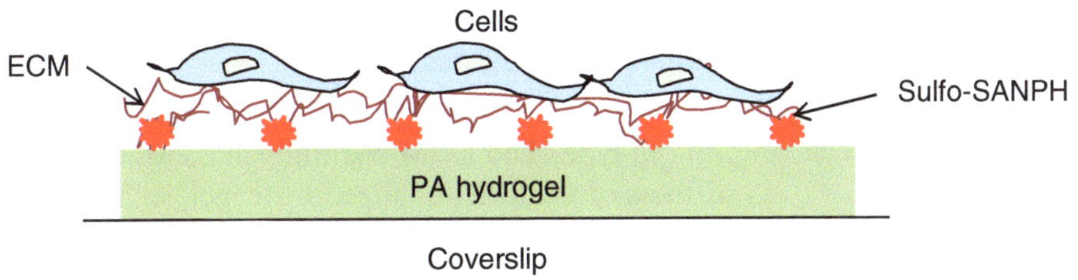

Fig. 1 Schematic diagram representing the final functionalized hydrogel set-up with cells attached

6. Wash once in PBS and then incubate in prewarmed Earle's balanced salt solution (or equivalent) for 10–30 min at 37 °C to allow the hydrogel to warm up (*see* **Note 11**).
7. Plate cells onto the hydrogel in their standard medium (*see* **Note 12**). A schematic of the final functionalized hydrogel set-up is shown in Fig. 1.
8. Cells can be trypsinized and maintained on hydrogels for weeks.

4 Notes

1. It is important to wash well as any remaining APES will react with glutaraldehyde in subsequent steps leaving a brown material that could interfere with downstream experiments.
2. For mapping of cellular traction forces, fluorescent beads may be added to the hydrogel.
3. We use hydrogels that mimic the physiological and pathological stiffness experienced by VSMCs and fibroblasts. The choice of hydrogel stiffness is cell type-dependent and should approximate conditions experienced in vivo.
4. For consistency between batches of hydrogels, we prepare a 50 mL stock solution and perform atomic force microscopy (AFM) to confirm the Young's modulus of each stock. Solutions can be stored at 4 °C for several months.
5. Extended polymerization times can make it difficult to remove the coverslip from the slide.
6. Trying to lift the coverslip directly off the slide often results in snapping of the coverslip. To overcome this, we first slide the coverslip and then lift.
7. As sulfo-SANPH is light-sensitive, it should be made fresh and kept in the dark until required.
8. UV exposure makes the sulfo-SANPH change to a dark red colour. If this does not happen, then cross-linking has failed and the step needs to be repeated.

9. Perform all subsequent steps in a tissue culture hood to limit the chance of contamination.
10. We incubate for 1 h but this may need to be increased, depending on cell type. The ECM coating will contribute to the overall stiffness of the functionalized hydrogel, so it is important to keep the concentration of ECM component and the incubation time consistent. It is also important to confirm the functionalized hydrogel stiffness to confirm the ECM coat is not altering overall stiffness.
11. We find warming the hydrogel yields better cell attachment/viability.
12. Seeding number is cell size- and assay-dependent. For example, we aim for a confluency of between 60 and 70 % for Western blotting, whereas for immunofluorescence we aim for 20–30 % confluency so that the cells are not touching one another. In our hands, confluent vascular smooth muscle cell (VSMC) and fibroblast cultures detach from the hydrogel.

Acknowledgments

This work is supported by a British Heart Foundation (BHF) Intermediate Basic Science Research Fellowship awarded to D.W. (FS/11/53/29020).

References

1. Ankam S, Suryana M, Chan LY, Moe AA, Teo BK, Law JB, Sheetz MP, Low HY, Yim EK (2013) Substrate topography and size determine the fate of human embryonic stem cells to neuronal or glial lineage. Acta Biomater 9(1):4535–4545. doi:10.1016/j.actbio.2012.08.018
2. Etienne J, Fouchard J, Mitrossilis D, Bufi N, Durand-Smet P, Asnacios A (2015) Cells as liquid motors: mechanosensitivity emerges from collective dynamics of actomyosin cortex. Proc Natl Acad Sci U S A 112(9):2740–2745. doi:10.1073/pnas.1417113112
3. Beningo KA, Hamao K, Dembo M, Wang YL, Hosoya H (2006) Traction forces of fibroblasts are regulated by the Rho-dependent kinase but not by the myosin light chain kinase. Arch Biochem Biophys 456(2):224–231. doi:10.1016/j.abb.2006.09.025
4. Huang X, Yang N, Fiore VF, Barker TH, Sun Y, Morris SW, Ding Q, Thannickal VJ, Zhou Y (2012) Matrix stiffness-induced myofibroblast differentiation is mediated by intrinsic mechanotransduction. Am J Respir Cell Mol Biol 47(3):340–348. doi:10.1165/rcmb.2012-0050OC
5. Burke B, Ellenberg J (2002) Remodelling the walls of the nucleus. Nat Rev Mol Cell Biol 3(7):487–497. doi:10.1038/nrm860
6. Mellad JA, Warren DT, Shanahan CM (2011) Nesprins LINC the nucleus and cytoskeleton. Curr Opin Cell Biol 23(1):47–54. doi:10.1016/j.ceb.2010.11.006
7. Crisp M, Liu Q, Roux K, Rattner JB, Shanahan C, Burke B, Stahl PD, Hodzic D (2006) Coupling of the nucleus and cytoplasm: role of the LINC complex. J Cell Biol 172(1):41–53. doi:10.1083/jcb.200509124
8. Haque F, Lloyd DJ, Smallwood DT, Dent CL, Shanahan CM, Fry AM, Trembath RC, Shackleton S (2006) SUN1 interacts with nuclear lamin A and cytoplasmic nesprins to provide a physical connection between the nuclear lamina and the cytoskeleton. Mol Cell Biol 26(10):3738–3751. doi:10.1128/MCB.26.10.3738-3751.2006

9. Zhang Q, Ragnauth CD, Skepper JN, Worth NF, Warren DT, Roberts RG, Weissberg PL, Ellis JA, Shanahan CM (2005) Nesprin-2 is a multi-isomeric protein that binds lamin and emerin at the nuclear envelope and forms a subcellular network in skeletal muscle. J Cell Sci 118(Pt 4):673–687. doi:10.1242/jcs.01642
10. Swift J, Ivanovska IL, Buxboim A, Harada T, Dingal PC, Pinter J, Pajerowski JD, Spinler KR, Shin JW, Tewari M, Rehfeldt F, Speicher DW, Discher DE (2013) Nuclear lamin-A scales with tissue stiffness and enhances matrix-directed differentiation. Science 341(6149):1240104. doi:10.1126/science.1240104
11. Buxboim A, Swift J, Irianto J, Spinler KR, Dingal PC, Athirasala A, Kao YR, Cho S, Harada T, Shin JW, Discher DE (2014) Matrix elasticity regulates lamin-A, C phosphorylation and turnover with feedback to actomyosin. Curr Biol 24(16):1909–1917. doi:10.1016/j.cub.2014.07.001
12. Ho CY, Jaalouk DE, Vartiainen MK, Lammerding J (2013) Lamin A/C and emerin regulate MKL1-SRF activity by modulating actin dynamics. Nature 497(7450):507–511. doi:10.1038/nature12105
13. Guilluy C, Osborne LD, Van Landeghem L, Sharek L, Superfine R, Garcia-Mata R, Burridge K (2014) Isolated nuclei adapt to force and reveal a mechanotransduction pathway in the nucleus. Nat Cell Biol 16(4):376–381. doi:10.1038/ncb2927
14. Tse JR, Engler AJ (2010) Preparation of hydrogel substrates with tunable mechanical properties. Curr Protoc Cell Biol. Chapter 10:Unit 10 16. Doi:10.1002/0471143030.cb1016s47
15. Sehgel NL, Sun Z, Hong Z, Hunter WC, Hill MA, Vatner DE, Vatner SF, Meininger GA (2015) Augmented vascular smooth muscle cell stiffness and adhesion when hypertension is superimposed on aging. Hypertension 65(2):370–377. doi:10.1161/HYPERTENSIONAHA.114.04456
16. Sehgel NL, Zhu Y, Sun Z, Trzeciakowski JP, Hong Z, Hunter WC, Vatner DE, Meininger GA, Vatner SF (2013) Increased vascular smooth muscle cell stiffness: a novel mechanism for aortic stiffness in hypertension. Am J Physiol Heart Circ Physiol 305(9):H1281–H1287. doi:10.1152/ajpheart.00232.2013
17. Tracqui P, Broisat A, Toczek J, Mesnier N, Ohayon J, Riou L (2011) Mapping elasticity moduli of atherosclerotic plaque in situ via atomic force microscopy. J Struct Biol 174(1):115–123. doi:10.1016/j.jsb.2011.01.010
18. Hayenga HN, Trache A, Trzeciakowski J, Humphrey JD (2011) Regional atherosclerotic plaque properties in ApoE–/– mice quantified by atomic force, immunofluorescence, and light microscopy. J Vasc Res 48(6):495–504. doi:10.1159/000329586
19. Cretu A, Castagnino P, Assoian R (2010) Studying the effects of matrix stiffness on cellular function using acrylamide-based hydrogels. J Vis Exp. (42). Doi: 10.3791/2089

Chapter 16

Cell Microharpooning to Study Nucleo-Cytoskeletal Coupling

Gregory Fedorchak and Jan Lammerding

Abstract

To evaluate the intracellular force transmission between the nucleus and cytoskeleton, we optimized a single cell-based assay that involves the manipulation of living, adherent cells with a fine glass microneedle and a microscope-mounted micromanipulator. The user inserts the microneedle into the cytoplasm and then, using a custom-programmable computer script, pulls the needle laterally toward the cell periphery. Normalized cross-correlation is applied to recorded time-lapse image sequences to determine average displacements within predefined regions of the nucleus and the cytoskeleton. These regional displacements, together with calculations of nuclear elongation, nuclear centroid translocation, and nuclear shape changes, enable quantitative assessments of nucleo-cytoskeletal coupling in both normal and disease conditions and provide an improved understanding of the role of specific nuclear envelope proteins in intracellular force propagation.

Key words Force transmission, Mechanotransduction, LINC complex, Cell signaling, Cell mechanics, Nesprin, SUN proteins, Lamins

1 Introduction

Connecting the nucleus to the cytoskeleton is important for a number of cellular processes, such as nuclear positioning, cell migration, cellular differentiation, chromosome movements, and mechanotransduction signaling [1]. However, the specific molecular connectors that maintain nuclear shape and position under various mechanical stimuli are just beginning to emerge. One important component is the linker of nucleoskeleton and cytoskeleton (LINC) complex, comprised of SUN and nesprin proteins, which span the inner and outer nuclear membranes, respectively. In 2006, Crisp and colleagues first identified the LINC complex and characterized the consequences of LINC complex disruption [2].

Electronic supplementary material: The online version of this chapter (doi:10.1007/978-1-4939-3530-7_16) contains supplementary material, which is available to authorized users.

Sue Shackleton et al. (eds.), *The Nuclear Envelope: Methods and Protocols*, Methods in Molecular Biology, vol. 1411,
DOI 10.1007/978-1-4939-3530-7_16, © Springer Science+Business Media New York 2016

This foundational work has led to the recent discovery of various human pathologies related to mutations in LINC complex proteins, including Emery-Dreifuss muscular dystrophy, dilated cardiomyopathy, and various types of cancer, prompting a search for additional LINC complex components and regulators [3–5]. Despite these efforts, the precise mechanisms underlying these diseases are still not well understood. This may be due in part to the limited arsenal of tools to study intracellular force coupling within living cells.

The microharpoon assay—a technique which harpoons the cytoskeleton of single adherent living cells with a glass microneedle and exerts a pulling force on the cytoskeleton and nucleus while monitoring the induced nuclear displacement and deformation—is well equipped to study the force-transmitting properties of nuclear envelope proteins. The first applications of this technique in the late 1990s demonstrated that the cell is mechanically interconnected, with forces being transmitted from the cytoskeleton to the nucleus, where they can induce large deformation [6]. In the same seminal study, Maniotis et al. also used a glass micropipette and a micromanipulator to pull on extracellular matrix (ECM)-coated microbeads attached to the cell surface. However, shortcomings of the technique included endocytosis of the microbeads and difficulty in getting a single microbead per cell positioned at the appropriate distance from the nucleus.

Other techniques to probe nuclear envelope mechanics exist, each with their own merits and limitations. Force spectroscopy with optical tweezers offers great precision, but struggles to generate forces required to substantially affect the nucleus. A recent study found that forces of several nanonewtons are necessary to induce nuclear deformation and translocation, much larger than the forces generated by individual kinesin and dynein motor forces (≈2–7 pN), which collectively drive nuclear positioning [7, 8], and larger than the forces obtained with optical tweezers (typically up to 100 nN). It should be noted that force propagation in some cells (e.g., beating cardiac myocytes and migrating cells) is sufficient to cause visible deformation of the cell nucleus, rendering nanonewton force magnitudes generated by the microharpoon biologically relevant. Other, magnetic bead-based approaches can generate larger pulling or twisting forces, but in these assays it is difficult to control the number and localization of the paramagnetic beads [9, 10]. The use of isolated nuclei enables direct probing of the nucleus [11], yet it may conceal important cytoskeletal effects, and the isolation procedure may cause damage to the nuclear exterior and/or affect chromatin organization based on the exact buffer conditions [12, 13]. A recent technique combines microneedle manipulation with micropipette aspiration to "directly" apply a mechanical load on the nuclear surface of intact cells. Rapid nuclear relaxation dynamics are then monitored upon release of the nucleus from the pipette [7].

The microharpoon assay described herein advances the approach pioneered by Maniotis and colleagues [6] and subsequently refined by our group [14, 15] and addresses major technical difficulties—such as the risk of permanently damaging the cell membrane during the pull—by minimizing vibrations and abrogating the need for potentially cytotoxic dyes to track cytoskeletal movements. The use of computer automation allows for precise control of the micromanipulator, making the pull of the needle more consistent and reproducible. Lastly, sophisticated image processing helps to extract maximal information from the time-lapse image sequences. Future applications may help to identify new protein–protein interactions at the nuclear envelope, uncover potential functional overlap and redundancies among nuclear envelope proteins (e.g., nesprin-1, nesprin-2, and nesprin-3), and characterize the effects of disease-causing mutations in nuclear envelope proteins on intracellular force transmission.

2 Materials

2.1 Image Acquisition and Analysis

1. Inverted epifluorescence microscope (*see* **Note 1** and Fig. **1**).
2. 40× air objective with phase contrast (*see* **Note 1**).
3. Lower magnification (5× and/or 10×) long working distance objective(s).
4. Microscope-mounted digital camera: CCD or CMOS.
5. Image acquisition software: e.g., MATLAB (*see* **Note 2**).

2.2 Micro-manipulation

1. Motorized microscope-mounted micromanipulator with micropipette holder: e.g., InjectMan NI 2 (Eppendorf).
2. Borosilicate glass tubing: OD: 1.0 mm, ID: 0.78 mm; 10 cm length (Sutter).
3. Micropipette puller: e.g., P-97 Model (Sutter).

2.3 Cell Culture and Labeling Reagents

1. Cell permeable DNA stain, such as Hoechst 33342 (Invitrogen) (*see* **Note 3**).
2. Dulbecco's modified Eagle medium (DMEM).
3. Phenol red-free DMEM with 25 mM HEPES.
4. Fetal bovine serum (FBS) (Aleken Biologicals).
5. Penicillin/Streptomycin (P/S) (Life Technologies).
6. Dulbecco's Phosphate Buffered Saline (PBS).
7. 35 mm glass bottom culture dishes (FluoroDish, World Precision Instruments, Inc.).
8. Fibronectin (EMD Millipore) (*see* **Note 4**).

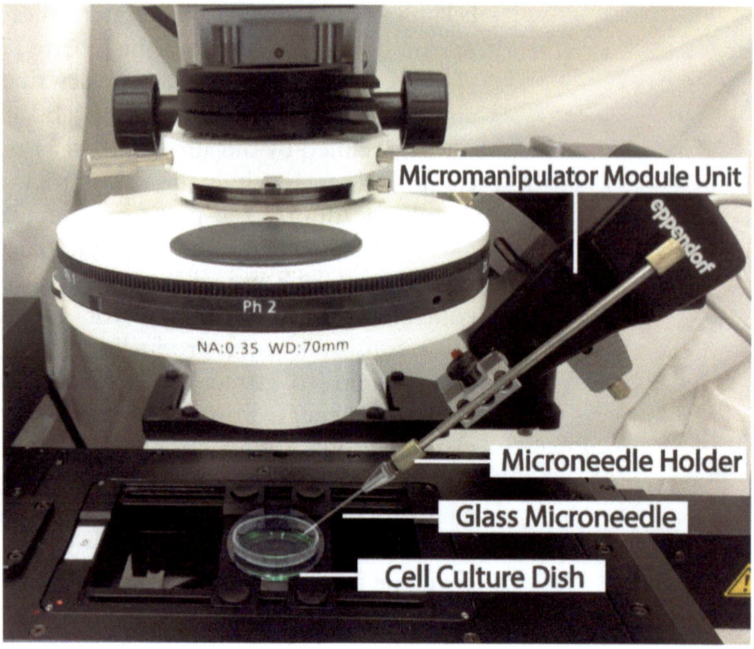

Fig. 1 Micromanipulator microscope setup. An inverted fluorescence microscope equipped with a long working distance condenser to accommodate the motorized micromanipulation module unit, e.g., InjectMan NI 2 (Eppendorf). The glass microneedle is held within a micropipette holder that inserts the microneedle into the cell culture dish axially at 45°. The micromanipulator controller and its connection to the computer are not visible in this image

3 Methods

The protocol has been developed for studying mouse embryonic fibroblasts (MEFs); however, it is easily adaptable for many other adherent cell types. The protocol has been used effectively on a variety of mouse and human cell types (e.g., human fibroblasts, NIH 3T3, MDA-MB-231, etc.), including wild-type cells and cells lacking specific nuclear envelope proteins.

3.1 Cell Culture Preparation

1. Prepare 35 mm glass bottom cell culture dishes for cell seeding. For many cell lines, it is sufficient to use cell culture-treated glass. Otherwise, incubate the dish with a low concentration of fibronectin (0.5 μg/mL) in Phosphate Buffered Saline (PBS) or other suitable cell-adhesion protein for 2 h at 37 °C (*see* **Note 4**). Wash dishes 2× with PBS.
2. Detach cells with 0.05 % trypsin. Seed 2 mL of cell suspension in growth medium (DMEM w/o sodium pyruvate, 10 % FBS, 1 % P/S) in 35 mm glass bottom cell culture dish to achieve a subconfluent density of ≈70 % (*see* **Note 5**). The number of cells should be optimized for the specific cell type used (e.g., for MEFs, the density will be approximately 80,000 cells/mL).

3. Incubate cells at 37 °C (5 % CO_2) for a minimum of 3 h to allow spreading (*see* **Note 6**). Immediately before the microharpooning procedure, add Hoechst 33342 nuclear dye (1 μg/mL final concentration) to the growth medium and incubate cells for 10 min in a 37 °C incubator.
4. Wash the cells 1× in PBS or growth medium for 30 s at room temperature to remove the residual Hoechst and then add phenol red-free growth medium (DMEM w/o sodium pyruvate and with 25 mM HEPES, 10 % FBS, 1 % P/S) to the cells for imaging.

3.2 Microscope Set-Up, Microharpooning, and Image Acquisition

1. Set up the inverted epifluorescence microscope for phase contrast and fluorescence (Hoechst/DAPI filter cube) imaging and initiate the image acquisition software.
2. Pull the borosilicate glass tubing with a micropipette puller to generate fine microneedles with tip diameters of ≈1 μm (*see* **Note 7** and Fig. **2**). This can be done in advance or immediately preceding the experiment.
3. Load the needle into the needle holder and fasten into the module unit at a 45° angle (*see* Fig. **1**). Start by positioning the needle above the liquid level of the dish (*see* **Note 8**). Slowly lower the needle down toward the cells (−*z* direction) using the micromanipulator controls, until the cells are in focus and the needle is visible at the desired magnification (*see* **Notes 9–11**). For the subsequent experiments, the needle is typically maintained in a central position within the field of view; cells are selected by moving the microscope stage. Adjustments are then made by moving the needle.
4. Select a well-spread, isolated, and healthy-looking cell (*see* **Note 12**) and acquire 1–2 sets of images prior to needle insertion, each set consisting of one phase contrast image and one image of the fluorescent Hoechst DNA stain. Use the same reflector/filter-cube for both contrast and fluorescence image acquisition to avoid vibrations caused by rotating the filter block turret (*see* **Note 13**).
5. Using a micromanipulator, carefully insert the microneedle into the cytoplasm of a cell at fixed distance (5 μm) away from the nuclear periphery. Insert the needle axially to minimize membrane damage. Use a piece of reference tape on the screen to mark out the distance between the edge of the nucleus and needle insertion site (*see* **Note 14**). Once the needle locally indents the plasma membrane, which is visible in phase contrast mode, insert the needle up to an additional ≈2 μm into the cytoplasm and then halt the needle. The needle should be deep enough so that it "catches" cytoskeletal structures and will not simply slide over the surface or detach from the mem-

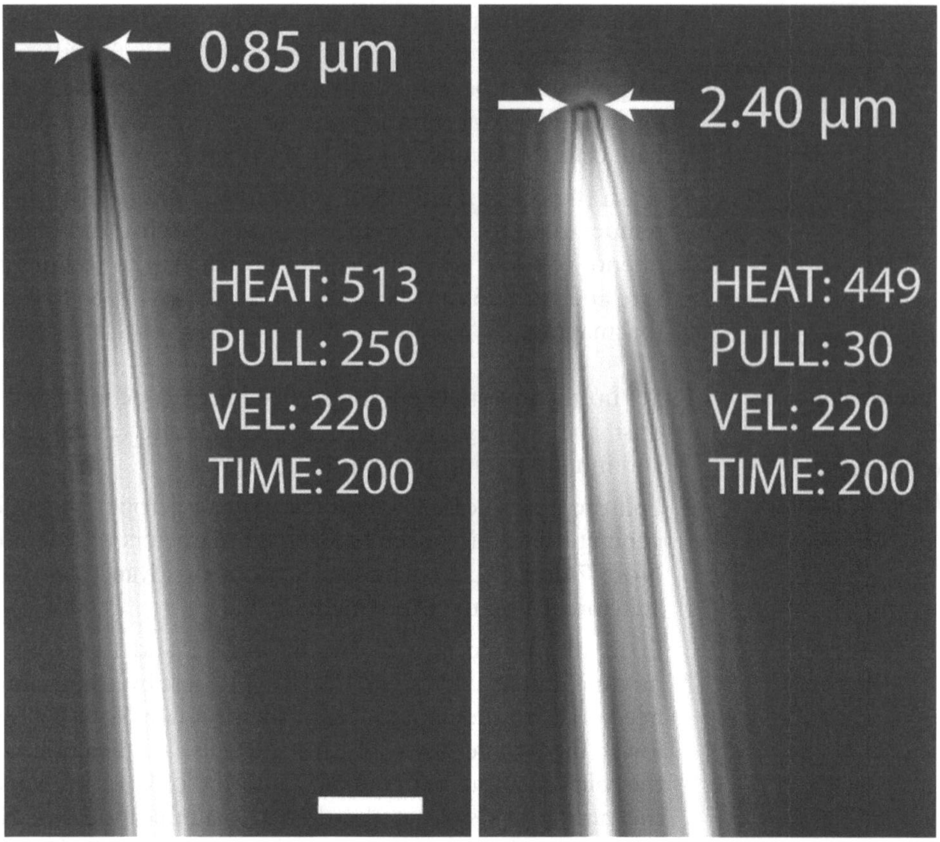

Fig. 2 Microneedle tip dimensions and pull parameters. A Sutter P-97 micropipette puller was used to create microharpoons from borosilicate glass rods. A tip diameter of ≈1 μm (*left*) or smaller is well suited for the microharpoon assay. Different tip diameters and taper lengths result from changing the pull parameters (*right*, *lower* HEAT and PULL). Note that a wider tip (*right*) is less ideal for the microharpoon assay as it could damage the cell membrane. Scale bar: 10 μm

brane during the pull, but not too deep, causing unwanted premature deformations in the regions of interest prior to executing the pull. Inserting the needle too deep may also damage the cell (*see* **Note 13**) and potentially cause collision between the needle and glass-bottom dish.

6. Initiate the microneedle manipulation sequence to move the microneedle away from the nucleus towards the cell periphery at a user-specified speed and distance (*see* **Note 15**). This is done with a computer connected to the micromanipulator (e.g., using a USB-interface in combination with a custom-written MATLAB script available from the Lammerding laboratory upon request) or by using a programmable micromanipulator. Simultaneously, collect time-lapse images (brightfield and fluorescence) throughout the pull (*see* **Note 16**, Movie 1 and Fig. **3**) and after the

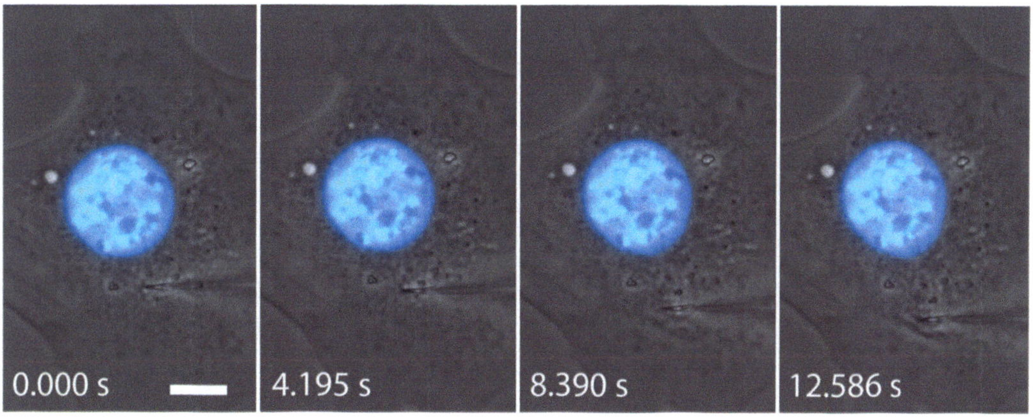

Fig. 3 Representative time-lapse image sequence showing nuclear deformation during manipulation with the microharpoon. The frames are taken approximately 4 s apart and follow the 10 μm translocation of the microharpoon. Scale bar: 10 μm. The corresponding movie is available in the Supplemental materials (*see* Movie 1)

needle has reached its final position. Collect images at fixed time intervals every 2–5 s.

7. Using the computer or the micromanipulator control panel to control the micromanipulator, remove the needle by rapidly retracting it away from the cell in the axial direction (20 μm/s works well).
8. Acquire 1–2 final sets of images following removal of the microneedle from the cytoskeleton to check for cell damage and viability (*see* **Note 17**).

3.3 Image Processing and Analysis

The following steps should be tailored to the specific needs of the user. While we chose a MATLAB-based approach, alternative approaches for image processing and analysis exist, many of which are publically available (*see* **Note 2**). Custom-written MATLAB scripts are available from the Lammerding laboratory upon request.

1. Using custom analysis software, generate displacement maps by tracking phase contrast features in the cytoplasm and fluorescently labelled features in the nucleus (*see* Movie 2 and Fig. **4**). One approach divides the image into a grid of regularly spaced small image regions ($\approx 2 \times 2$ μm^2) and applies a normalized cross-correlation algorithm to each small image region in subsequent image frames (*see* **Note 18**). This approach enables the tracking of each region-center from frame-to-frame, which is then used to calculate the displacement of each image region. The collection of displacement vectors forms a displacement map of the cell for each time-point, from which average displacements within regions of interest can be computed.

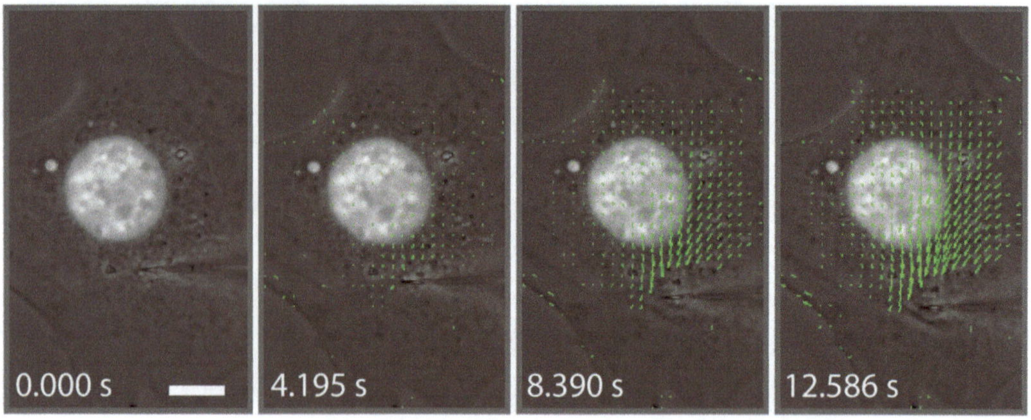

Fig. 4 Representative time-lapse image sequence illustrating the computed cytoskeletal strain map during micromanipulation. As the microharpoon pulls on the cytoskeleton, forces and deformations (indicated by the growing displacement vectors) are transmitted throughout the cell. The displacement vectors are drawn 2× their actual size. Similar displacement maps are computed for nuclear displacements. Scale bar: 10 μm. The corresponding movie is available in the Supplemental materials (*see* Movie 2)

2. In order to quantify nucleo-cytoskeletal coupling, compute induced intracellular displacements within four discrete, 2 μm × 2 μm regions: (a) near the microneedle manipulation site (cyto near); (b) an area inside the nucleus near the microneedle insertion site (nuc near); (c) an area on the opposite side of the nucleus (nuc far); and (d) the cytoskeleton on the other side of the nucleus (cyto far) (*see* **Note 19** and Fig. 5a). Compare displacement differences between a "pre-pull" frame, showing the harpooned cell immediately prior to pull initiation, and a "post-pull" frame of the cell just before harpoon removal.
3. To analyze the nuclear shape and deformation, apply thresholding and smoothing to the fluorescence image of the nucleus. From this binary image, nuclear area and shape changes can be tracked over time. Fitting an ellipsoid to the binary image can be used to extract additional parameters such as effective major and minor radius, eccentricity, etc. Nuclear strain along the axis of force application is calculated by dividing the nuclear elongation ($\Delta L = L - L_0$) by the initial length, L_0, where L is the final length of the nucleus (in the microneedle pull direction) at the end of the strain application. Nuclear centroid movements can also be quantified (*see* **Note 20**).
4. Interpret results. Figure 5c schematically illustrates how to identify nucleo-cytoskeletal coupling defects based on the plot of nuclear and cytoskeletal displacements. Figure 5a, b show expected results for wild-type cells and cells with a force transmission defect, respectively (*see* **Note 19**).

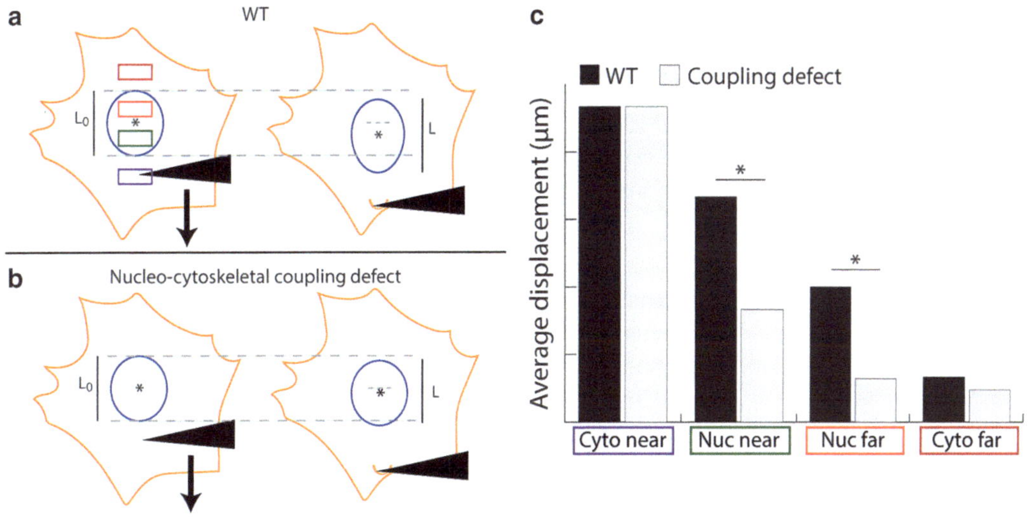

Fig. 5 Schematic overview of microharpoon assay and sample results comparing wild-type cells with those possessing a nucleo-cytoskeletal coupling defect. (**a**) Wild-type cell before and after the microharpoon pull. The four discrete regions used to evaluate nucleo-cytoskeletal coupling are shown in the colored boxes: "cyto near" in *purple*, "nuc near" in *green*, "nuc far" in *orange,* and "cyto far" in *red.* (**b**) Expected results for a cell with a force transmission defect. Note the decrease in nuclear strain along the axis of force application, calculated by dividing the nuclear elongation ($\Delta L = L - L_0$) by the initial length, L_0, where L is the final length of the nucleus (in the microneedle pull direction) at the end of strain application. Also note the decrease in nuclear centroid displacement. (**c**) Hypothetical plot of expected results comparing the cells in (**a**) and (**b**). The "nuc near" and "nuc far" regions typically provide the most relevant information about nucleo-cytoskeletal coupling, whereas the "cyto far" region is often influenced by high noise. Displacements in the "cyto near" region should show comparable results for both cell types, as these measurements reflect the applied cytoskeletal strain

3.4 Optional Experimental Variations

1. Selectively disrupting the various cytoskeletal systems (e.g., using nocodazole for microtubules, cytochalasin D for actin, etc.) may help better define the mechanism of force transmission for a given protein of interest (e.g., novel LINC complex candidates).
2. In the future, combining the microharpoon assay with 3-D confocal microscopy may provide a more detailed view of cellular force transmission and may reveal additional phenomena (e.g., 3-D nuclear shape changes, differences between apical and basal protein distribution, and response to force).

4 Notes

1. The experimental procedure requires a fluorescence microscope equipped with a long working distance condenser to accommodate the micromanipulation unit (*see* Fig. **1**), a filter block for Hoechst, and a high-magnification air objective (e.g.,

Zeiss EC Plan-Neofluar 40×/0.75 Ph2 M27 with 1.6× Tubelens Optovar).

2. Other programming platforms may be used for the analysis. For MATLAB, the Mathworks online FileExchange has some well-documented and helpful digital image correlation (DIC) programs, such as the following: http://www.mathworks.com/matlabcentral/fileexchange/43073-improved-digital-image-correlation--dic.
3. Alternatives to Hoechst 33342 may be used to stain the nucleus, such as the dye SYTO 59. This minor-groove binding molecule has been shown to work well for live-cell imaging with limited cytotoxicity. Dyes with longer excitation wavelengths are advantageous as this will minimize potential phototoxicity. While we and others have successfully used fluorescently labeled histones (e.g., GFP-histone H2B) to visualize nuclear deformations, a recent study found that expression of GFP-histone H1.1 altered nuclear mechanics [7].
4. The concentration and type of ECM protein can substantially affect cell morphology, spreading, and cytoskeletal organization, which is likely to influence intracellular force transmission. Therefore, it is crucial that the cells are not allowed to spread too thin, which increases the risk of inserting the needle through the cells and/or ripping the cytoskeleton with the needle. This can be achieved by using only low concentrations of ECM molecules for coating the cell culture surface, or by micropatterning by microcontact printing. Incubation with cell adhesion protein can go longer than 2 h or overnight at 4 °C.
5. It is important to minimize cell–cell contacts in order to reduce variability in the assay. We found that a cell confluency of 50–70 % provides a good balance between having a large number of cells to choose from and maintaining cells in a "happy" state while keeping the number of cell–cell contacts low. Substrate patterning approaches such as direct printing using deep UVs [16] or microcontact printing can help to achieve this.
6. The incubation time required for sufficient cell spreading may vary depending on cell type. We have performed the microharpoon assay at various time-points (e.g., 3, 6, 24, or 48 h post seeding) and concluded that results at the 3, 6, and 24 h time-points are essentially identical. At 48 h, the cells display greater displacements in the "nuc far" and "cyto far" regions, potentially due to increased cytoskeletal tension. We therefore recommend performing the experiments between 3 and 24 h after seeding the cells.
7. Using a P-97 micropipette puller (Sutter), we had success with the following pull parameters: HEAT: 513, PULL: 250, VEL:

220, TIME: 200. Please note that the exact settings will vary depending on the machine and its configuration. Generally, increasing the HEAT and PULL parameters results in a smaller microneedle diameter and a longer taper. We recommend experimenting with varying the parameters until a reproducible, long, and sharp needle shape can be produced (*see* Fig. **2**).

8. The starting position of the microneedle should be just above the level of medium. This will be helpful in the case where a cell or piece of debris sticks to the needle. Surface tension at the medium–air interface can be used to shear off the debris upon rapid removal of the needle in the + *z* direction.

9. To position the needle at the experimental onset, start with a low magnification (e.g., 5× or 10×) long working distance objective and bring the needle into focus at the center of the viewing field. Lower the focal plane using the coarse adjustment knob of the microscope and slowly lower the needle into focus. Repeat this process until the cells appear in focus with the needle slightly above. As the needle is drawn closer to the cells, change the objective to the final magnification (40× or higher). The need to change objectives (e.g., in order to efficiently locate the microneedle) favors use of an air objective over an immersion objective.

10. The same needle may be used for multiple cells and multiple dishes; however, it should be changed in the case of damage or debris stuck to the needle (*see* **Note 7**).

11. Once the cells are in view at 40×, change the controls to "Fine" and "Axial" (for the Eppendorf InjectMan NI 2 micromanipulator) to enable precise control over the microneedle movement in the axial direction.

12. Try to select interphase cells that resemble a sunny-side-up egg, with the nucleus as close to the center as possible. Cells with nuclei close to the cell periphery often behave inconsistently in response to microneedle manipulation. For the analysis, it is also necessary to have a minimum of 5 μm of visible cytoplasm beyond the far side of the nucleus. Once again, substrate patterning approaches may help to increase throughput and reduce experimental variability.

13. The formation of visible, expanding lacerations in the cell membrane and cytoskeleton caused by microneedle manipulation was a major challenge during optimization of the technique. Upon imaging the microneedle pull using a high-speed digital camera, we noticed the presence of vibrations induced by rotation of the high-speed filter turret. This issue could be prevented by acquiring all of the images using a single filter cube (for the Hoechst fluorescence signal). An alternative approach would be to use an external excitation filter wheel or

LED illumination in combination with multi-band filter cube. We confirmed that cytoskeletal displacements calculated based on phase contrast images closely match results obtained in previous approaches that tracked GFP- or mCherry-actin and GFP-vimentin as fiducial markers of cytoskeletal displacements. Other reasons for cytoskeletal ripping may include excessive cell spreading, improper microneedle tip shape/size, or vibration of the microscope stage, for example, caused by the camera cooling fan or the lack of an anti-vibration platform.

14. It is helpful to position a piece of tape on the computer screen displaying the live camera image in order to mark the targeted needle insertion site. The length of the tape should correspond to an actual length of 5 μm in the microscope sample, which can be easily determined by image calibration. In our settings, this corresponds to tape of about 10 mm length. During the experiments, the stage is moved so that the bottom edge of the nucleus aligns with the top end of the tape. The microharpoon is then inserted even with the bottom end of the tape.
15. The micromanipulator can be interfaced with MATLAB through the computer USB terminal in order to achieve automated and highly reproducible needle translocations. For MEFs, moving the needle a total distance of 10 μm at 1 μm/s provides sufficient nuclear deformations that can be analyzed using our algorithm and compared between cell types. We found that human cells (e.g., human skin fibroblasts) are more rigid than mouse cells and require greater forces to achieve detectable nuclear deformations and displacements. This can be accomplished by moving the needle a further distance at a greater speed (e.g., 30 μm at 5 μm/s). Under some conditions, the cytoskeletal tension in the cell will resist the motion of the needle, causing a slight bend in the glass needle. Therefore, the tip of the needle does not always travel the full distance specified by the software.
16. It is essential to have one set of images with the needle inserted prior to the start of the pull, and another set of images at the final position of the needle translocation. In order to calculate profiles of nuclear strain rate and nuclear centroid speed, one should acquire 3–5 frames during the pull, itself, taken at intervals of about 3 s. It can be advantageous to synchronize the micromanipulator and the image acquisition software so that the needle will always be at the same location for a given frame (assuming all other parameters are consistent). However, this step is not essential if the analysis is only based on the comparison between the initial and final frames of the image sequences.
17. After the pull, check for nuclear retraction and major rips in the cytoplasm (*see* **Note 13**). Damaged cells must be excluded

from the analysis. Additionally, a live/dead assay or membrane damage assay may be performed.

18. Our cross-correlation windows are 2 μm × 2 μm corresponding to 20 × 20 pixels with a center-to-center distance of 0.5 μm or 5 pixels.
19. The displacements in the "cyto near" site should all be very similar since the same microneedle displacement is applied to all cells. Comparing the "nuc far" and especially the "nuc near" displacements between modified (e.g., mutant or knockdown) and control cells gives the best indication of nucleo-cytoskeletal coupling. Reduced displacements compared to control cells suggest that forces are not transmitted very well across the nuclear envelope. Displacements in the "cyto far" region become more difficult to interpret as they are often quite small (on the order of 1–2 pixels), approaching the detection limit. For MEFs, our "cyto near" and "cyto far" regions are both 5 μm from the nuclear membrane. In our software, the user selects regions of interest and the program calculates displacements within 2 μm × 2 μm (20 pixel × 20 pixel) windows. This is repeated 3× using partially overlapping windows, and the results are averaged for each cell. Statistical analysis is then performed on about 15–30 cells for each condition, with data collected from at least three independent experiments.
20. The analysis described here allows one to determine rates of nuclear deformation and translation. One may also consider examining the rate and extent of nuclear retraction following needle removal. These measurements reflect the relative magnitude of elastic versus viscous resistance in nuclear mechanics [7]. Other image processing approaches may be used to estimate the ratio of nuclear to cytoplasmic stiffness [6]. The microharpoon assay is best used in conjunction with other approaches such as membrane strain, micropipette aspiration, cell migration, and perfusion experiments [17] to help elucidate physical/mechanical consequences of certain mutations, etc.

Acknowledgments

This work was supported by National Institutes of Health awards [R01 HL082792 and R01 NS59348]; a Department of Defense Breast Cancer Idea Award [BC102152]; a National Science Foundation CAREER award to J Lammerding [CBET-1254846]; and a Pilot Project Award by the Cornell Center on the Microenvironment & Metastasis through Award Number U54CA143876 from the National Cancer Institute, as well as a NSF graduate research fellowship to G Fedorchak [2014163403].

References

1. Chang W, Worman HJ, Gundersen GG (2015) Accessorizing and anchoring the LINC complex for multifunctionality. J Cell Biol 208(1):11–22
2. Crisp M, Liu Q, Roux K, Rattner JB, Shanahan C, Burke B, Stahl PD, Hodzic D (2006) Coupling of the nucleus and cytoplasm: role of the LINC complex. J Cell Biol 172(1):41–53. doi:10.1083/jcb.200509124
3. Luxton GG, Starr DA (2014) KASHing up with the nucleus: novel functional roles of KASH proteins at the cytoplasmic surface of the nucleus. Curr Opin Cell Biol 28:69–75
4. Mejat A, Decostre V, Li J, Renou L, Kesari A, Hantai D, Stewart CL, Xiao X, Hoffman E, Bonne G, Misteli T (2009) Lamin A/C-mediated neuromuscular junction defects in Emery-Dreifuss muscular dystrophy. J Cell Biol 184(1):31–44. doi:10.1083/jcb.200811035
5. Méjat A (2010) LINC complexes in health and disease. Nucleus 1(1):40–52
6. Maniotis AJ, Chen CS, Ingber DE (1997) Demonstration of mechanical connections between integrins, cytoskeletal filaments, and nucleoplasm that stabilize nuclear structure. Proc Natl Acad Sci U S A 94(3):849–854
7. Neelam S, Chancellor TJ, Li Y, Nickerson JA, Roux KJ, Dickinson RB, Lele TP (2015) Direct force probe reveals the mechanics of nuclear homeostasis in the mammalian cell. Proc Natl Acad Sci U S A 112(18):5720–5725
8. Tanenbaum ME, Akhmanova A, Medema R (2011) Bi-directional transport of the nucleus by dynein and kinesin-1. Commun Integr Biol 4(1):21–25
9. Iyer KV, Pulford S, Mogilner A, Shivashankar G (2012) Mechanical activation of cells induces chromatin remodeling preceding MKL nuclear transport. Biophys J 103(7):1416–1428
10. Poh Y-C, Shevtsov SP, Chowdhury F, Wu DC, Na S, Dundr M, Wang N (2012) Dynamic force-induced direct dissociation of protein complexes in a nuclear body in living cells. Nat Commun 3:866
11. Guilluy C, Osborne LD, Van Landeghem L, Sharek L, Superfine R, Garcia-Mata R, Burridge K (2014) Isolated nuclei adapt to force and reveal a mechanotransduction pathway in the nucleus. Nat Cell Biol 16(4): 376–381
12. Lammerding J, Dahl KN, Discher DE, Kamm RD (2007) Nuclear mechanics and methods. Methods Cell Biol 83:269–294. doi:10.1016/S0091-679X(07)83011-1
13. Dahl KN, Engler AJ, Pajerowski JD, Discher DE (2005) Power-law rheology of isolated nuclei with deformation mapping of nuclear substructures. Biophys J 89(4):2855–2864
14. Lombardi ML, Jaalouk DE, Shanahan CM, Burke B, Roux KJ, Lammerding J (2011) The interaction between nesprins and sun proteins at the nuclear envelope is critical for force transmission between the nucleus and cytoskeleton. J Biol Chem 286(30):26743–26753. doi:10.1074/jbc.M111.233700
15. Zwerger M, Jaalouk DE, Lombardi ML, Isermann P, Mauermann M, Dialynas G, Herrmann H, Wallrath LL, Lammerding J (2013) Myopathic lamin mutations impair nuclear stability in cells and tissue and disrupt nucleo-cytoskeletal coupling. Hum Mol Genet 22(12):2335–2349. doi:10.1093/hmg/ddt079
16. Azioune A, Carpi N, Tseng Q, Thery M, Piel M (2010) Protein micropatterns: a direct printing protocol using deep UVs. Methods Cell Biol 97:133–146
17. Isermann P, Davidson PM, Sliz JD, Lammerding J (2012) Assays to measure nuclear mechanics in interphase cells. Curr Protoc Cell Biol. Chapter 22:Unit 22.16. Doi: 10.1002/0471143030.cb2216s56

Chapter 17

Wound-Healing Assays to Study Mechanisms of Nuclear Movement in Fibroblasts and Myoblasts

Wakam Chang, Susumu Antoku, and Gregg G. Gundersen

Abstract

The rearward positioning of the nucleus is a characteristic feature of most migrating cells. Studies using wounded monolayers of fibroblasts and myoblasts have shown that this positioning is actively established before migration by the coupling of dorsal actin cables to the nuclear envelope through nesprin-2G and SUN2 linker of nucleoskeleton and cytoskeleton (LINC) complexes. During nuclear movement, these LINC complexes cluster along the actin cables to form adhesive structures known as transmembrane actin-associated nuclear (TAN) lines. Here we described experimental procedures for studying nuclear movement and TAN lines using wounded monolayers of fibroblasts and myoblasts, the acquisition of data using immunofluorescence microscopy and live-cell imaging, and methods for data analysis and quantification.

Key words LINC complex, SUN protein, Nesprin, TAN line, Retrograde actin flow, Nuclear lamina, Nuclear positioning, Centrosome orientation, Cell migration

Abbreviations

CH	Calponin homology
DAPI	4′,6-Diamidino-2-phenylindole
GFP	Green fluorescent protein
KASH	Klarsicht, Anc1 and Syne homology
LINC	Linker of nucleoskeleton and cytoskeleton
LPA	Lysophosphatidic acid
MEFs	Mouse embryo fibroblasts
miniN2G	Mini-nesprin-2G
MT	Microtubule
MRCK	Myotonic dystrophy-related, Cdc42-binding kinase
SUN	Sad1 and Unc83
TAN lines	Transmembrane actin-associated nuclear lines

Sue Shackleton et al. (eds.), *The Nuclear Envelope: Methods and Protocols*, Methods in Molecular Biology, vol. 1411,
DOI 10.1007/978-1-4939-3530-7_17, © Springer Science+Business Media New York 2016

1 Introduction

The nucleus maintains its proper position in most cells and undergoes active movement during various cellular and developmental processes, including migration, fertilization, cell division, and differentiation in animals cells [1]. Abnormal nuclear positioning is associated with diseases such as muscular dystrophy, cardiomyopathy [2], and lissencephaly [3]. The nucleus is moved and positioned by forces generated by the actin and microtubules networks [1]. In most cases this movement is mediated by the linker of nucleoskeleton and cytoskeleton (LINC) complex, which is composed of inner nuclear membrane SUN (*S*ad1 and *Un*c83 homology) domain proteins and outer nuclear membrane KASH (*K*larsicht, *A*nc1 and *S*yne *h*omology) domain proteins [4]. The SUN domain projects into the perinuclear space where it interacts with the KASH domain, establishing a protein complex that spans the two nuclear membranes. KASH proteins (termed nesprins in mammals) interact with cytoskeletal elements in the cytoplasm through direct and indirect means whereas SUN proteins bind to the nuclear lamina and other inner nuclear membrane proteins [5].

Various systems have been developed to study nuclear movement and positioning, including fertilized eggs [6], embryonic *C. elegans* hypodermal cells [7], *Drosophila* ommatidia [8, 9], neuronal progenitors and migrating neurons [10], developing myotubes [11, 12], junction formation in epithelial clusters [13–15], and wounded monolayers of fibroblasts and myoblasts polarizing for migration [16–22]. The wounded monolayer systems have been particularly useful for understanding the molecular mechanisms of nuclear movement and offer the ease of manipulating tissue culture cells. Additionally, with serum-starved wounds, nuclear movement can be synchronously triggered by adding serum or the serum factor lysophosphatidic acid (LPA). The latter induces movement of the nucleus toward the cell rear without inducing membrane protrusion or migration, thus allowing nuclear movement to be studied separately from cell migration [5, 18]. LPA-stimulated rearward nuclear movement generates cell polarity by orienting the centrosome, and this occurs through two Cdc42 pathways: an actin/myosin pathway that moves the nucleus and a MT/dynein pathway that maintains the centrosome in the cell center [18, 23]. Inhibition of the actin/myosin pathway impairs nuclear movement so both the nucleus and centrosome stay near the centroid of the cell. On the other hand, inhibition of the MT/dynein pathway disrupts centration of the centrosome, resulting in its rearward movement with the nucleus [18]. The mechanism of actin-dependent nuclear movement has been extensively studied [17–22]. Stimulated by Cdc42, myotonic dystrophy-related, Cdc42-binding kinase (MRCK) activates myosin IIA and IIB by phosphorylating their regulatory light chains [18, 20]. Active myosin IIA is responsible for the formation of dorsal actin cables and their movement, whereas myosin IIB interacts with emerin, a nuclear envelope protein, to regulate the directionality of actin movement [20]. Moving dorsal actin cables are captured by the LINC complex proteins nesprin-2G (the giant isoform of nesprin-2) via its N-terminal CH domains and by interacting with the formin FHOD1, while SUN2 mediates the anchorage of nesprin-2G via its C-terminal KASH domain [19, 22]. Nesprin-2G, FHOD1, and SUN2 accumulate along dorsal actin cables to form transmembrane actin-associated nuclear (TAN) lines [19]. TAN lines are anchored to the nuclear lamina through SUN2 interaction with A-type lamins [21]. The inner nuclear membrane proteins, Samp1 and emerin, are also found in TAN lines and contribute to their function [17, 20].

Nuclear movement in fibroblasts was first studied using NIH3T3 fibroblasts [18], mouse embryo fibroblasts (MEFs), and human fibroblasts [21]. More recently it was found that undifferentiated C2C12 myoblasts utilize a similar mechanism for nuclear movement [5]. In all of these studies, modified wound-healing assays were used. In brief, a confluent monolayer of cells is serum starved to deprive the cell of polarization stimuli. The monolayer is then wounded, and LPA is added to stimulate nuclear movement. Cells are then fixed, permeabilized, and stained with a combination of DAPI, phalloidin, and anti-centrosome/tubulin antibodies to visualize the nucleus, dorsal actin cables, and the centrosome. Fluorescence images of cells at the wound edge are acquired and subjected to analysis of centrosome orientation and nuclear and centrosomal positions. Data analysis and quantification can be done manually using *Metamorph* (Molecular Devices, Sunnyvale, CA) or *ImageJ* (NIH, Bethesda, MD) but are quite time-consuming, so we developed custom software, *Cell Plot*, to quantify nuclear and centrosomal positions in fixed cell images, which automates the analysis [20]. Live-cell imaging using phase contrast or DIC microscopy is useful to study movement of the nucleus, because it minimizes photodamage. However, to study the dynamics of actin flow and the movement of TAN lines, fluorescent live-cell microscopy is necessary. This is usually accomplished by co-expressing a fluorescent protein-tagged actin probe (e.g., LifeAct-mCherry) and GFP-mini-nesprin-2G (GFP-miniN2G), an engineered form of nesprin-2G which lacks the spectrin repeats 3–54 but contains the actin-binding CH domains and the KASH domain and rescues nuclear movement in cells depleted of nesprin-2G [19]. The velocity and directionality of TAN lines and nuclear movement are then measured from kymographs prepared from the movies. Here, we describe these methods in detail.

2 Materials

NIH3T3 fibroblasts are cultured in Dulbecco's Modified Eagle Medium (DMEM) with 10 % calf serum. Other cell lines including mouse C2C12 myoblasts, MEFs, and primary human fibroblasts are cultured in DMEM containing 10 % fetal bovine serum [5].

2.1 General

1. Serum-free medium: DMEM with 10 mM HEPES (pH 7.4) (*see* **Note 1**).
2. Growth medium: Serum-free medium with 10 % calf serum.
3. NIH3T3 fibroblasts, MEFs, human primary fibroblasts, or C2C12 mouse myoblasts: cultured in growth medium.
4. 0.25 % Trypsin-0.9 mM EDTA in Hank's Balanced Salt Solution (HBSS).
5. LPA (Avanti Polar Lipids, Alabaster, AL): 1 mM solution in 100 mM NaCl, 1 % fatty acid-free Bovine Serum Albumin, 10 mM HEPES, pH 7.4 (*see* **Note 2**).
6. Microscope slides and acid-washed coverslips (22 mm × 22 mm) (*see* **Note 3**).
7. Glass-bottom imaging dishes (*see* **Note 3**).
8. siRNA transfection reagent, e.g., Lipofectamine RNAiMax (Life Technologies).
9. siRNA: 20 μM stock solution in RNase-free water.
10. Micromanipulator for microinjection (*see* **Note 4**).
11. Recording medium: HBSS supplemented with essential and non-essential MEM amino acids, 2.5 g/L glucose, 2 mM L-glutamine, 1 mM sodium pyruvate, and 20 mM HEPES (pH 7.4).

2.2 Nuclear Movement and TAN Line Detection in Fixed Cells

1. Fixation buffer: 4 % paraformaldehyde in phosphate buffered saline (PBS).
2. Blocking/permeabilization buffer: 0.3 % Triton X-100, 5 % normal goat serum or 5 % BSA in PBS.
3. Staining regents: Alexa-488 phalloidin, DAPI, anti-tubulin, and/or anti-pericentrin antibodies. We use rat-monoclonal anti-tyrosinated-tubulin (YL1/2, the European Collection of Animal Cell Cultures, Salisbury, UK) and/or mouse monoclonal anti-pericentrin antibodies (BD Biosciences, San Jose, CA) (*see* **Note 5**).
4. Anti-fade mounting and sealing medium.
5. GFP-miniN2G plasmid [19]: 5 mg/mL in HKCl buffer (140 mM KCl, 10 mM HEPES, pH.7.4) (*see* **Note 6**).
6. Epifluorescence microscope equipped with high NA oil immersion 60× objective and fluorescence filter cubes for three or four color imaging.

2.3 Nuclear Movement (Live-Cell Phase Contrast Microscopy)

1. Phase contrast microscope with temperature control and motorized stage (*see* **Notes 7** and **8**).

2.4 Actin Flow and TAN Line Movement (Live-Cell Fluorescent Microscopy)

1. GFP-miniN2G and LifeAct-mCherry plasmids [19]: 5 mg/mL GFP-miniN2G plasmids and 25 mg/mL LifeAct-mCherry plasmids in HKCl buffer (*see* **Note 6**).
2. Epifluorescence microscope with temperature control and motorized stage (*see* **Note 9**).

2.5 Data Analysis

1. *ImageJ* software.
2. *Cell Plot* (WC, Gundersen Laboratory: http://www.columbia.edu/~wc2383/software.html).

3 Methods

The standard protocol that has been developed to study nuclear movement in NIH3T3 fibroblasts [24] can also be applied to human fibroblasts, MEFs, and myoblasts. With each of these systems, proper optimization of the starting cell density before serum starvation and the length of serum starvation is required to obtain a confluent, but not overcrowded, monolayer of cells. For instance, for NIH3T3 fibroblasts and C2C12 myoblasts, the optimal starting cell density for serum starvation is ~80 and ~40 %, respectively. Length of starvation time is also critical. Without sufficient starvation, cells at the wounded monolayer edge can move their nuclei or even migrate in the absence of LPA or serum. A useful criterion to confirm proper serum starvation is to stain cells for filamentous actin. In well-starved cells, there should almost be no visible actin cables. Starvation times vary from 2 days for NIH3T3 fibroblasts (*see* **Note 10**) to 4 days in C2C12 cells. Microinjection of purified proteins and plasmids is an efficient way to introduce proteins into cells at the wound edges. This technique and RNAi-mediated depletion of proteins are invaluable in deciphering the molecular pathway of nuclear movement.

3.1 Preparation of Cells with siRNA Knockdown

1. Transfect NIH3T3 fibroblasts with siRNAs of interest (40 nM) using siRNA transfection reagent according to the manufacturer's instruction (*see* **Note 11**). Transfected cells are plated onto a 22-mm square acid-washed coverslip in a 35-mm dish (for immunostaining) or on a glass-bottom coverslip dish (for live-cell imaging). Initial cell density is ~20 % so that it reaches ~80 % in 2 days. For assaying nuclear position in fixed cells, prepare two coverslips for each siRNA: one for LPA stimulation and the other without LPA stimulation (this also serves as

a check on the serum starvation, since incompletely starved cells will move their nuclei without added LPA).

2. Incubate cells for 2 days.
3. Remove medium from the dish and briefly wash three times with serum-free medium.
4. Add 2 mL of serum-free medium to each dish and incubate cells for 2 days.
5. Wound the monolayers with a small pipette tip or a jeweler's screwdriver and incubate cells for at least 30 min to allow them to recover (*see* **Note 12**).

3.2 Expression of Proteins in Wound Edge Cells by Microinjection

1. Prepare the plasmid expressing the protein of interest in HKCl buffer, back-load this into a glass micropipette, and inject into the nuclei of serum-starved cells at the wound edge using a micromanipulator (*see* **Notes 4, 6, 13, 14**).
2. Incubate the cells for 1–4 h, depending on the construct and the desired level of expression, to allow the cells to express the protein of interest.

3.3 Assay of Nuclear Position in Fixed Cells

1. Add 10 μM LPA to the serum-starved wounded monolayer of cells. Reserve one dish without LPA stimulation to serve as a negative control.
2. Incubate cells for 2 h.
3. Briefly wash the coverslips with PBS and fix the cells with 4 % PFA at room temperature for 10–20 min.
4. Wash the coverslips three times with PBS, each for 5 min.
5. Permeabilize with 0.5 % Triton X-100 in PBS for 5 min and block with blocking buffer at room temperature for 30 min.
6. Follow standard immunostaining procedure and mount the coverslips on pre-cleaned glass slides. DAPI and anti-tubulin and/or anti-pericentrin antibodies should be used to visualize the nucleus and the centrosome (*see* **Note 5**).
7. Mount the slides onto a microscope and acquire images of cells at the wound edges. Avoid imaging isolated cells and only image cells within a continuous wound edge.

3.4 TAN Line Detection (Fixed Cells)

1. Prepare serum-starved cells expressing GFP-miniN2G or use untransfected cells if endogenous TAN lines are to be detected (*see* **Note 15**).
2. Add 10 μM LPA to the dishes and incubate cells for 1 h (*see* **Note 16**).
3. Fix, permeabilize, and block the coverslips as described in Subheading 3.3, **steps 3–5**.

4. Follow standard immunostaining procedure and mount the coverslips to slides. Use DAPI, phalloidin, and anti-GFP (or anti-nesprin-2G) antibody staining to visualize the nucleus and TAN lines.
5. Mount the slides onto a microscope and acquire images of injected cells. TAN lines will be found on the dorsal surface of the cells (*see* **Note 17**).

3.5 Nuclear Movement (Live-Cell Phase Contrast Microscopy)

1. Add conditioned recording medium to a wounded monolayer of serum-starved cells on glass-bottom imaging dishes (*see* **Note 18**).
2. Mount dishes on a phase contrast microscope with temperature control (37 °C) (*see* **Notes 19** and **20**).
3. Locate cells at the wound edge and record their positions for multi-position acquisition using microscope control software.
4. Add 20 μM LPA to induce nuclear movement.
5. Acquire time-lapse images for each position for at least 2 h at an acquisition rate of 5 min/frame.

3.6 Actin Flow and TAN Line Movement (Live-Cell Fluorescent Microscopy)

1. Add conditioned recording medium to a wounded monolayer of serum-starved cells in a coverslip dish that were previously microinjected with GFP-miniN2G and LifeAct-mCherry (*see* **Note 18**).
2. Mount the dish onto an epifluorescence microscope with temperature control (37 °C) (*see* **Notes 19** and **20**).
3. Locate expressing cells at the wound edge and record their positions for multi-position acquisition using microscope control software.
4. Add 20 μM LPA to induce nuclear movement.
5. Acquire two-channel image stacks for each position for at least 2 h. An acquisition rate of 3–5 min/frame is sufficient to capture the movement of TAN lines. TAN lines are on the dorsal side of the nuclei so include the top of nuclei in the range of the *Z* stacks. Retrograde actin cables originate from the leading edge, so the bottom of the cells should also be included if retrograde actin movement is of interest (*see* **Note 16**).

3.7 Data Analysis: Centrosome Orientation

Analysis of centrosome orientation has been previously described in detail [24, 25] and is summarized here. This measurement can be done with an image program or judged by eye.

1. Using the drawing tool in the imaging program, draw lines from the center of the nucleus to the two sides of the cell's leading edge where it contacts neighboring cells (Fig. 1).

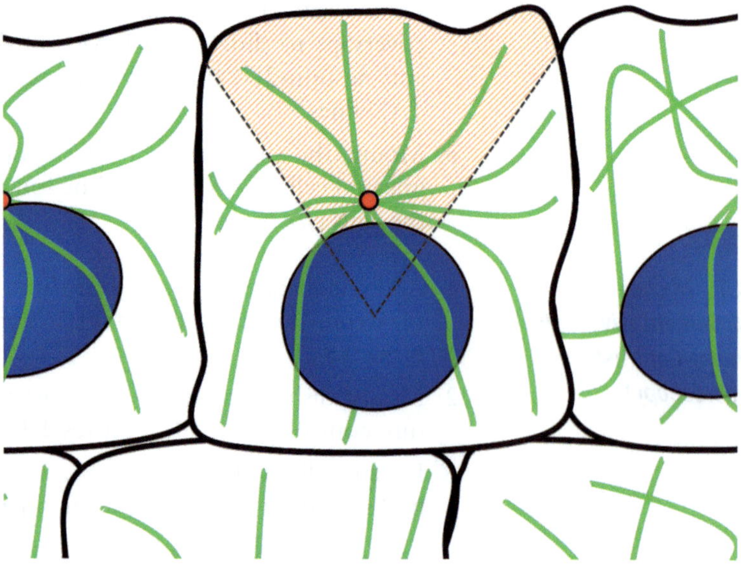

Fig. 1 The centrosome is considered "oriented" if and only if it is located within the shaded area (excluding the nucleus) and is not on the nucleus

2. The centrosome is considered "oriented" if: (1) it lies between these two lines and the leading edge and (2) it is not located on top of the nucleus (*see* **Notes 21** and **22**).
3. For each experimental condition at least 30 cells should be counted. In the absence of LPA and serum ~33 % of cells have oriented centrosome because of random position of the centrosome. The number increases to 60–70 % of the cells after induction with serum or LPA.

3.8 Data Analysis: Analyzing Nuclear and Centrosome Positions with ImageJ

Nuclear and centrosome positions can be manually measured using *ImageJ*, *Metamorph*, or similar software. We have developed *Cell Plot* freeware (Fig. 2) to semi-automate these measurements. Here we include a brief description of how this is done manually with *ImageJ*, because it is helpful to know the nature of these measurements to understand results from *Cell Plot*.

1. Open *ImageJ* and make sure that both "Area" and "Centroid" are selected in the "Set Measurements" dialog.
2. Open the images of interest in *ImageJ*. Merge the channels if necessary.
3. Select a cell at the wound edge and rotate the image so that the leading edge of the cell is parallel to bottom of the image window and the wound is at the bottom.
4. Use the freehand selection tool to draw the boundary of the cell creating a region of interest (ROI) and click menu item

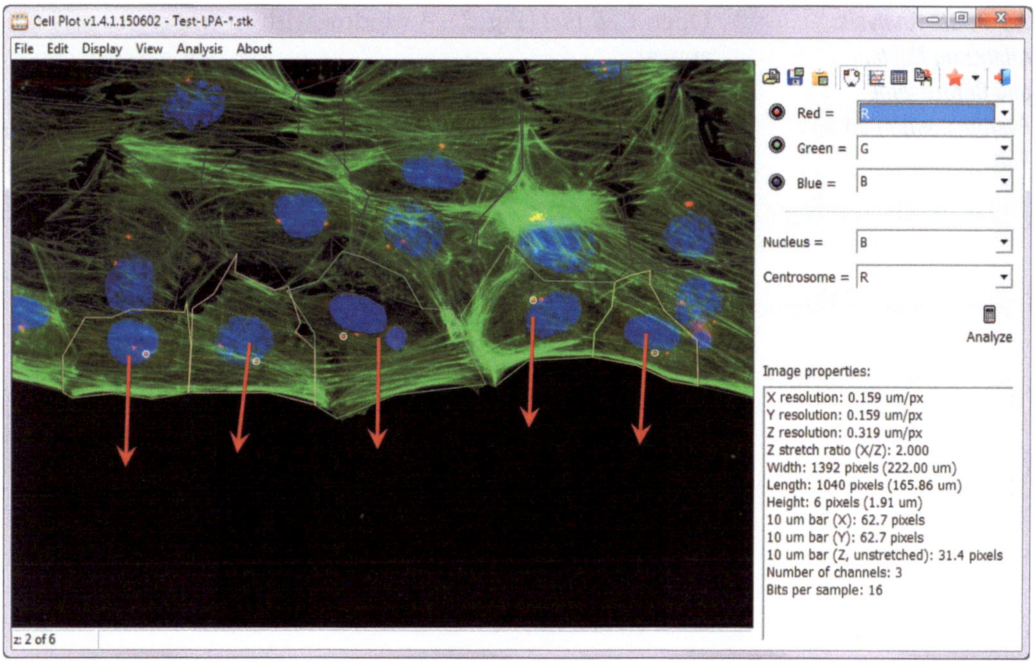

Fig. 2 The user interface of *Cell Plot*. After analysis is finished, identified cells are marked by *gray* boundaries. Double-clicking on a cell highlights it and includes it in the output results. When a highlighted cell is selected, its boundary can be modified by moving the points that define the boundary. The direction of wound edge (*red arrows*) and centrosome position (*white circles*) can also be adjusted with a mouse

"Measure (Ctrl + M)." The area of the ROI and the (x, y) position of the centroid ROI should be logged.

5. Use the oval selection to draw a circle to represent the nucleus and click "Measure (Ctrl + M)."
6. Use the point selection to mark the centrosome and "Measure (Ctrl + M)."
7. Repeat **steps 2–6** for every cell to be analyzed. We typically analyze 30–40 cells for each condition.
8. Export the data to a spreadsheet program such as *Microsoft Excel.*
9. Calculate the relative position of the nucleus and centrosome by subtracting the Y position of the cell centroid from their Y positions (in pixels).
10. Calculate the average radius of the cell from its area: $r = \sqrt{\frac{\text{Area}}{\pi}}$ (in pixels).
11. Normalize the relative position of the nucleus and centrosome to the radius (no unit, expressed as %). By this measurement, positive values represent positioning toward the leading edge, and negative values represent rearward location.

3.9 Data Analysis: Analyzing Nuclear and Centrosome Positions with Cell Plot

1. Open *Cell Plot* (Fig. 2). A window with an empty image should appear.
2. Drag and drop image files into *Cell Plot* or use the "Load images…" function to browse for the images. *Cell Plot* currently accepts TIFF stack files (*.tif files from *ImageJ* or *.stk files from *MetaMorph*). All frames in the files are analyzed together. *See* **Note 23** for file name requirements.
3. Select the "nucleus" and "centrosome" channels using the drop-down lists on the right panel.
4. Click "Analyze."
5. Double click on a cell to include/exclude it from analysis.
6. Use the mouse to correct the cell boundary (polygons), the direction or wound edge (red arrows), and the position of centrosomes (white circles) (*see* **Note 24**).
7. Use PageUp and PageDown to move between frames.
8. Click "Export to Excel…" to open *Microsoft Excel* with all results entered.

4 Notes

1. 1× Penicillin and streptomycin can be added to growth medium and serum-free medium.
2. It is important to use fatty acid-free BSA (for example, Sigma-Aldrich #A6006), as regular BSA contains LPA.
3. Coverslips are treated with 2 M HCl for 5 min and washed with running tap water for 15 min, ddH_2O for 5 min (three times), and 95 % ethanol for 5 min before drying. Glass-bottom imaging dishes should also be acid washed.
4. Microinjection is preferred for introducing plasmid DNA into cells over liposome transfection because the former gives immediate expression, better control of the protein expression level, and less trauma to cells.
5. Antibody to a junctional/plasma membrane protein (e.g., β-catenin) or phalloidin staining of actin may be used to stain the cell boundary. Efficient depletion of the protein of interest should be confirmed by immunofluorescence staining and western blotting with an appropriate antibody.
6. DNA concentration needs to be optimized for each plasmid. Plasmid DNA is diluted in HKCl buffer, and aggregates in the solution are removed by centrifugation ($\geq$15,000 × *g*, 30 min).
7. A microscope with motorized *x–y* stage is preferred because it allows recording several samples/positions sequentially. With our Proscan II *x–y* stage (Prior), we are able to record more

than 50 positions at a rate of 5 min per frame with autofocus on. Autofocus or a focus-maintaining system is necessary for live-cell phase contrast microscopy.

8. A dry objective is preferable for phase contrast microscopy. We use a Nikon 40× ELWD Plan objective (NA 0.6) and a CoolSNAP HQ CCD camera on a Nikon TE300 microscope. In some circumstance a 20× objective gives acceptable resolution.
9. For live-cell fluorescence microscopy, a 60× or 100× Plan Apo oil objective should be used to provide optimal resolution. Photodamage from the fluorescence illumination can lead to imaging artifacts and cell toxicity; to limit photodamage, use neutral density filters, limit exposure times, and/or reduce the number of focal planes and positions.
10. Starvation time may vary for NIH3T3 fibroblasts from different sources.
11. When using liposome transfection, we use a "reverse transfection" protocol. In this protocol, trypsinized cells and medium are added to pre-plated transfection complexes.
12. The monolayer can be wounded multiple times to increase the number of the cells at the wound edges.
13. Only inject cells that are part of a continuous wound edge. Microinjection should take less than 15 min to reduce cell damage due to pH changes in low CO_2 environment outside of the incubator.
14. When injecting multiple coverslips, it does not affect the results if the monolayers are wounded all at once since without serum or LPA addition, there will be no response from the cells if they have been properly serum starved.
15. The expression of GFP-miniN2G enhances detection of TAN lines compared to immunofluorescence staining of endogenous N2G. TAN lines can be detected in 30–40 % of GFP-miniN2G expressing cells, while endogenous TAN lines are detectable in ~15 % cells at the wound edge. Note that the number of cells exhibiting TAN lines (30–40 % of wound edge cells) is similar to the percentage of cells that actively move their nucleus [18, 19].
16. It is important to note that TAN lines are transient structures and are optimally detected 30–60 min after LPA stimulation of NIH3T3 fibroblasts and C2C12 myoblasts.
17. Nuclear membrane folding may also cause linear nesprin-2 signals so we only consider linear nesprin-2G colocalizing with actin cables as TAN lines. An additional criterion is the colocalization of SUN2 with the nesprin-2G and actin filaments.

18. Using recording medium reduces background fluorescence relative to DMEM. Care is needed when changing the serum-free medium of serum-starved cells as application of fresh medium can lead to detachment of the cells. If medium must be replaced, use conditioned serum-free medium prepared by harvesting medium from separate batches of serum-starved cells.
19. Pour enough mineral oil on top of the medium to reduce evaporation of water if a humidity chamber is not available.
20. Live-cell imaging may also be done at 34–35 °C. Cells are in general healthier, and it also reduces focal drift on microscopes without a focus controlling system.
21. For centrosomes located on the boundary of the shaded area (*see* Fig. 1), consider half of them as oriented.
22. In some cells, the two centrosomes are separated and can be distinguished. In this case, use the MT channel to determine which one is the dominant MT organization center.
23. Multichannel images should be saved as multiple single-channel TIFF files, and they should be named as "filename-G.tif", "filename-R.tif", "filename-B.tif", and "filename-I.tif". For example, when a file named "3T3-LPA-B.tif" is opened, *Cell Plot* looks for "3T3-LPA-G.tif", "3T3-LPA-R.tif", and "3T3-LPA-I.tif" and opens them all at once.
24. The cell boundary can be modified by moving the points with a mouse. Alternatively, a boundary can be directly drawn by holding the "Alt" key and moving the mouse.

Acknowledgments

The methods described in this paper were developed with support by NIH grants R01GM099481 and R01AR068636.

References

1. Gundersen GG, Worman HJ (2013) Nuclear positioning. Cell 152:1376–1389
2. Folker ES, Baylies MK (2013) Nuclear positioning in muscle development and disease. Front Physiol 4:363
3. Morris NR, Efimov VP, Xiang X (1998) Nuclear migration, nucleokinesis and lissencephaly. Trends Cell Biol 8:467–470
4. Starr DA, Fridolfsson HN (2010) Interactions between nuclei and the cytoskeleton are mediated by SUN-KASH nuclear-envelope bridges. Annu Rev Cell Dev Biol 26:421–444
5. Chang W, Antoku S, Ostlund C, Worman HJ, Gundersen GG (2015) Linker of nucleoskeleton and cytoskeleton (LINC) complex-mediated actin-dependent nuclear positioning orients centrosomes in migrating myoblasts. Nucleus 6:77–88
6. Reinsch S, Gonczy P (1998) Mechanisms of nuclear positioning. J Cell Sci 111(Pt 16):2283–2295
7. Starr DA, Han M (2003) ANChors away: an actin based mechanism of nuclear positioning. J Cell Sci 116:211–216

8. Mosley-Bishop KL, Li Q, Patterson L, Fischer JA (1999) Molecular analysis of the klarsicht gene and its role in nuclear migration within differentiating cells of the Drosophila eye. Curr Biol 9:1211–1220
9. Kracklauer MP, Banks SM, Xie X, Wu Y, Fischer JA (2007) Drosophila klaroid encodes a SUN domain protein required for Klarsicht localization to the nuclear envelope and nuclear migration in the eye. Fly (Austin) 1:75–85
10. Vallee RB, Seale GE, Tsai JW (2009) Emerging roles for myosin II and cytoplasmic dynein in migrating neurons and growth cones. Trends Cell Biol 19:347–355
11. Cadot B, Gache V, Vasyutina E, Falcone S, Birchmeier C, Gomes ER (2012) Nuclear movement during myotube formation is microtubule and dynein dependent and is regulated by Cdc42, Par6 and Par3. EMBO Rep 13:741–749
12. Wilson MH, Holzbaur EL (2015) Nesprins anchor kinesin-1 motors to the nucleus to drive nuclear distribution in muscle cells. Development 142:218–228
13. Stewart RM, Zubek AE, Rosowski KA, Schreiner SM, Horsley V, King MC (2015) Nuclear-cytoskeletal linkages facilitate cross talk between the nucleus and intercellular adhesions. J Cell Biol 209:403–418
14. Desai RA, Gao L, Raghavan S, Liu WF, Chen CS (2009) Cell polarity triggered by cell-cell adhesion via E-cadherin. J Cell Sci 122: 905–911
15. Dupin I, Sakamoto Y, Etienne-Manneville S (2011) Cytoplasmic intermediate filaments mediate actin-driven positioning of the nucleus. J Cell Sci 124:865–872
16. Luxton GW, Gomes ER, Folker ES, Worman HJ, Gundersen GG (2011) TAN lines: a novel nuclear envelope structure involved in nuclear positioning. Nucleus 2:173–181
17. Borrego-Pinto J, Jegou T, Osorio DS, Aurade F, Gorjanacz M, Koch B, Mattaj IW, Gomes ER (2012) Samp1 is a component of TAN lines and is required for nuclear movement. J Cell Sci 125:1099–1105
18. Gomes ER, Jani S, Gundersen GG (2005) Nuclear movement regulated by Cdc42, MRCK, myosin, and actin flow establishes MTOC polarization in migrating cells. Cell 121:451–463
19. Luxton GW, Gomes ER, Folker ES, Vintinner E, Gundersen GG (2010) Linear arrays of nuclear envelope proteins harness retrograde actin flow for nuclear movement. Science 329:956–959
20. Chang W, Folker ES, Worman HJ, Gundersen GG (2013) Emerin organizes actin flow for nuclear movement and centrosome orientation in migrating fibroblasts. Mol Biol Cell 24: 3869–3880
21. Folker ES, Ostlund C, Luxton GW, Worman HJ, Gundersen GG (2011) Lamin A variants that cause striated muscle disease are defective in anchoring transmembrane actin-associated nuclear lines for nuclear movement. Proc Natl Acad Sci U S A 108:131–136
22. Kutscheidt S, Zhu R, Antoku S, Luxton GW, Stagljar I, Fackler OT, Gundersen GG (2014) FHOD1 interaction with nesprin-2G mediates TAN line formation and nuclear movement. Nat Cell Biol 16(7):708–715
23. Schmoranzer J, Fawcett JP, Segura M, Tan S, Vallee RB, Pawson T, Gundersen GG (2009) Par3 and dynein associate to regulate local microtubule dynamics and centrosome orientation during migration. Curr Biol 19: 1065–1074
24. Gomes ER, Gundersen GG (2006) Real-time centrosome reorientation during fibroblast migration. Methods Enzymol 406:579–592
25. Palazzo AF, Joseph HL, Chen YJ, Dujardin DL, Alberts AS, Pfister KK, Vallee RB, Gundersen GG (2001) Cdc42, dynein, and dynactin regulate MTOC reorientation independent of Rho-regulated microtubule stabilization. Curr Biol 11:1536–1541

Chapter 18

Methods for Assessing Nuclear Rotation and Nuclear Positioning in Developing Skeletal Muscle Cells

Meredith H. Wilson, Matthew G. Bray, and Erika L.F. Holzbaur

Abstract

Skeletal muscle cells are large syncytia, containing hundreds of nuclei positioned regularly along the length of the fiber. During development, nuclei are actively distributed throughout the myotube by the microtubule motor proteins, kinesin-1, and cytoplasmic dynein. Nuclear movement consists of translocation along the long axis of the cell concurrent with three-dimensional rotation of nuclei. In this chapter we describe methods for quantitatively assessing the speed of nuclear rotation in cultured myotubes using live-cell imaging techniques coupled with rigid body kinematic analyses. Additionally, we provide protocols for analyzing nuclear distribution in myotubes.

Key words Skeletal muscle, Nuclear positioning, Nuclear rotation, Angular velocity, Rigid body kinematics

1 Introduction

Early observation of myogenesis in primary chick and rat muscle cells revealed that nuclei in myotubes are highly mobile [1–3]. In addition to making linear excursions through the cytoplasm, nuclei displayed prominent rotational dynamics. Rotation was observed in more than one plane, and nuclei were observed to rotate in both clockwise and counterclockwise directions; individual nuclei could also change their direction of rotation over time. Linear translocation of nuclei in cultured myotubes was observed at rates of up to 18 μm/h, with an average rate of ~6–9 μm/h [3], but only rough estimates of rotation rates were possible because this early work was imaged with phase contrast microscopy and thus only provided information in a single focal plane.

Advances in microscopy now allow the imaging of nuclear dynamics in three-dimensional space. Using confocal microscopy to obtain XY information in multiple *Z*-planes, we can visualize the entire nucleus as it rotates [4]. Rotational dynamics are most clearly observed by imaging nuclei stained with fluorescent Hoechst dye,

Sue Shackleton et al. (eds.), *The Nuclear Envelope: Methods and Protocols*, Methods in Molecular Biology, vol. 1411,
DOI 10.1007/978-1-4939-3530-7_18, © Springer Science+Business Media New York 2016

which binds DNA and brightly labels dense areas of heterochromatin. These chromocenters are distributed throughout the nucleus and retain their positions relative to one another as a nucleus translocates and rotates over short observation periods (minutes). Image analysis software can be used to track the movement of these chromocenters in three-dimensional space over time. Using rigid body kinematic analyses, we have developed protocols to quantitatively assess nuclear rotation [4].

Here, we describe methods for culturing C2C12 mouse myotubes and imaging myonuclear rotation. We include protocols for tracking chromocenters and provide a MATLAB-based algorithm that we developed to calculate the orientation and angular velocity of a nucleus in motion. In this algorithm, the nucleus is analyzed as a rigid body; it is assumed that the chromocenters retain the same relative orientation within the nucleus as it undergoes motion. We have used these methods to show that nuclear rotation is abolished in the absence of microtubules and severely compromised following siRNA-mediated reduction in kinesin-1 motor expression [4].

We have also shown that nuclear dynamics are necessary for proper distribution of nuclei throughout the developing myotube. Inhibition of rotation and translocation results in aggregation of nuclei [4, 5], which is correlated with muscle dysfunction in mouse and fly muscles [6, 7]. Additionally, mispositioned nuclei are found in patients with centronuclear myopathies and Emery-Dreifuss muscular dystrophy [8, 9]. In recent years, significant progress has been made in understanding the mechanisms driving nuclear movement and distribution in developing muscle cells. In these studies, analyzing and graphically representing the distribution of nuclei in individual myotubes and in populations of myotubes following experimental perturbation is essential. Therefore, as an additional part of this chapter, we detail our methods for imaging whole myotubes, quantifying the position of the nuclei along the long axis of these cells, and effectively displaying distribution data for myotube populations. We have used these methods to show that loss of kinesin-1 motors from the nuclear envelope results in abnormal aggregation of nuclei at the midline of myotubes and that nesprins act as motor cargo adaptors for myonuclei [4, 5].

Though we describe the culture and analysis of differentiated C2C12 myotubes, an immortalized mouse muscle cell line [10, 11], we have found that the methods for analyzing nuclear rotation and distribution described here are also valid for primary myotube cultures. Furthermore, we propose that these methods may be effectively applied in the future to the analysis of nuclear rotation in other cell systems.

2 Materials

2.1 C2C12 Myotube Cell Culture

1. Mouse C2C12 myoblasts (CRL_1772) obtained from the American Type Culture Collection (ATCC) (*see* **Note 1**).
2. Dulbecco's modified Eagle's medium (DMEM).
3. Fetal bovine serum (FBS).
4. Horse serum.
5. Stable L-glutamine supplement.
6. Growth medium: DMEM supplemented with 10 % (v/v) FBS and 2 mM L-glutamine supplement.
7. Differentiation medium: DMEM supplemented with 10 % horse serum and 2 mM L-glutamine supplement.
8. Trypsin solution: 0.25 % trypsin.
9. Glass-bottom dishes, 35 or 50 mm, e.g., FluoroDish (World Precision Instruments).
10. ACLAR embedding film (Ted Pella, Inc.) cut into squares that fit into the wells of 12-well culture dishes.
11. Collagen coating solution: Rat tail collagen, type 1 diluted to 50 μg/ml in 0.02 N acetic acid.
12. Dulbecco's phosphate buffered saline (DPBS).
13. Incubator at 37 °C with 5 % CO_2.
14. Inverted light microscope.
15. Benchtop centrifuge.

2.2 Cell Transfection Reagents

1. Lipofectamine 2000 transfection reagent (Life Technologies, 11668-027).
2. Unsupplemented DMEM.
3. Mammalian expression construct for EGFP.

2.3 Live-Cell Microscopy for Nuclear Rotation Analysis

1. Spinning disk confocal fluorescence microscope equipped with a motorized stage, 60× and 100× oil-immersion objectives, EMCCD camera, Volocity 3D Image Analysis Software (Improvision), and, if available, an automated focusing system for drift correction (e.g., Perkin Elmer UltraVIEW VoX spinning disk confocal on a Nikon Ti microscope equipped with a perfect focus system, 60×/1.49 NA and 100×/1.49 NA oil-immersion apochromatic objectives (Nikon), and a Hamamatsu EMCCD C9100 50 camera).
2. Microscope environmental chamber maintained at 37 °C and 5 % CO_2 (CO_2 control is optional for imaging less than ~5 h).

3. Imaging media: phenol-red-free DMEM, high glucose with 25 mM HEPES, supplemented with 10 % horse serum and 2 mM L-glutamine supplement.
4. Live-cell Hoechst dye (Hoechst 33342, trihydrochloride, trihydrate, FluoroPure grade) (0.5 μg/ml final concentration).
5. Mineral oil.
6. Volocity 3D tracking software protocol described below.
7. MATLAB nuclear rotation analysis algorithm included below.

2.4 Cell Fixation and Immunofluorescence

1. Phosphate-buffered saline (PBS): 1.3 M NaCl, 70 mM $Na_2HPO_4{\cdot}2H_2O$, and 30 mM $NaH_2PO_4{\cdot}H_2O$ in ddH_2O. Adjust pH to 7.2.
2. 4 % Paraformaldehyde (PFA) in PBS, warmed to 37 °C.
3. 0.1 % Triton X-100 in PBS.
4. Bovine serum albumin (BSA).
5. Goat serum.
6. Blocking solution: 1× PBS supplemented with 1 % BSA, 5 % goat serum (filtered and stored at 4 °C).
7. Anti-alpha-actinin primary antibody (clone EA-53, mouse, Sigma-Aldrich), 1:500 dilution.
8. Anti-alpha-tubulin primary antibody (YL1/2, rat, AbD Serotec), 1:500 dilution.
9. Goat anti-mouse IgG (H+L) secondary antibody, Alexa Fluor 488 conjugate (Life Technologies, A-11001), 1:500 dilution.
10. Goat anti-rat IgG (H+L) Secondary antibody, Alexa Fluor 594 conjugate (Life Technologies, A-11007), 1:500 dilution.
11. Hoechst dye (Hoechst 33342, see above), 1:500 dilution.
12. Antifade mounting media, e.g., ProLong Gold Antifade (Life Technologies, P36930).
13. 40×22 mm (#1.5) glass coverslips.

2.5 Fixed Cell Microscopy for Nuclear Distribution Analysis

1. Spinning disk confocal fluorescence microscope equipped with a motorized stage, 40× oil-immersion objective, EMCCD camera and Volocity 3D Image Analysis Software (e.g., Perkin Elmer UltraVIEW VoX spinning disk confocal on a Nikon Ti microscope equipped with a motorized stage, 40×/1.30 NA oil-immersion apochromatic objective (Nikon), Hamamatsu EMCCD C9100-50 camera, and Volocity 3D Image Analysis Software (Improvision/Perkin Elmer)).
2. Microsoft Excel.
3. GraphPad Prism (GraphPad software).

3 Methods

3.1 C2C12 Myotube Culture Protocols

1. Culture proliferating C2C12 myoblasts in growth medium: DMEM + 10 % FBS + 2 mM L-glutamine supplement. Myoblasts may be cultured on plastic culture dishes. To split, wash cells once with DPBS, add 0.25 % trypsin and incubate until cells release from the dish, add growth medium, and spin cells at 1000 rpm for 2 min in a 15 ml conical tube to pellet. Gently resuspend cells in 1 ml of fresh growth medium before diluting and plating cells on 10 cm culture dishes in additional growth medium (*see* **Note 2**).
2. For myotube cultures, myoblasts must be plated in growth medium on collagen-coated glass-bottom dishes (for live-cell imaging or immunofluorescence) or on collagen-coated ACLAR plastic embedding film coverslips in 12-well plates (for immunofluorescence) (*see* **Notes 3** and **4**).
3. To coat plates with a thin layer of collagen:

   Dilute stock collagen solution to 50 μg/ml using 0.02 N acetic acid.

   Add diluted collagen solution to dishes as follows:

   1–2 ml to the glass center of a 35 mm dish.

   2–3 ml to the glass center of a 50 mm dish.

   1 ml/well on a 12-well dish containing ACLAR coverslips (*see* **Note 4**).

   Incubate at room temperature for 1 h in the hood.

   Aspirate the remaining solution.

   Rinse well with DPBS to remove the acid.

   Air-dry plates in the hood, exposed to UV for 10 min.

   Plates may be used immediately or stored at 4 °C under sterile conditions.

   We have stored plates for up to 3 weeks without adverse effects on myotube growth or morphology.
4. To induce differentiation of myoblasts to myotubes, switch to differentiation medium when myoblasts are ~70 % confluent. Cells will continue to divide for a time after switching to differentiation medium, and if they are too confluent when put into differentiation medium, they will likely overgrow and die rather than differentiate.
5. Differentiation medium should be replaced daily after induction to replenish nutrients, remove cell debris, and maintain pH. Typically, cells will begin to fuse ~2–3 days after induction of differentiation. Multinucleated myotubes will increase in size and maturity over the next 5 days. In more mature cultures, myofibril twitching can be observed as well as spontaneous contraction of whole myotubes. As cultures get older,

>7 days post-differentiation, significant branching of myotubes can be observed, sometimes creating large networks of fused myotubes. At this stage, the health of the culture begins to decline compromising further analyses of nuclear dynamics and distribution. Typically, analysis of myotubes is performed 6–7 days post-induction of differentiation.

3.2 C2C12 Transfection (Optional, *See* Note 5)

1. Transfect differentiating myotubes with EGFP plasmids using Lipofectamine 2000 reagents according to the manufacturer's protocol. Transfections have been effective on days 4 and 5 following induction of differentiation (48–72 h prior to fixation or live-cell analysis).

3.3 Live-Cell Imaging for Analysis of Nuclear Rotation

1. Aspirate differentiation medium from the glass-bottom dish and rinse myotubes with pre-warmed (37 °C) phenol-red-free imaging media.
2. Add pre-warmed imaging medium containing 0.5 μg/ml Hoechst dye to dish, overlay the medium with warm mineral oil, and place the dish in the microscope environmental chamber to incubate for 20 min prior to the start of imaging (*see* **Note 6**).
3. Locate myotube nuclei to image. Nuclei in myotubes can be distinguished from myoblast nuclei by their shape and pattern of heterochromatin. Nuclei in myoblasts are typically flatter and wider and have more numerous and smaller chromocenters. In comparison, nuclei in myotubes are more spherical and have a greater z-depth and fewer, larger chromocenters. One can often distinguish myotube nuclei within the same myotube by looking for a line of nuclei sharing this appearance. *See* Fig. 1. Additionally, if myotubes have been transfected with EGFP, this signal can be used to identify myotubes and associated nuclei (*see* **Note 5**).
4. Using a 60× or 100× objective, capture images of nuclei over time using the following guidelines:
   (a) Obtain z-series encompassing the entire depth of the myotube (~15–40 μm) with a 0.5 μm step size (*see* **Note 7**).
   (b) Take images at a rate of 1 z-series per minute for at least 15 min (16 timepoints). *See* Fig. 2a.
   (c) Adjust exposure time, gain, and offset settings to obtain quality images while minimizing myotube/nuclei exposure to laser light. If EGFP is used to identify myotubes, it is best to only take one z-series in the 488 channel at the start or end of the time series (*see* **Note 8**).
   (d) If available, a perfect focus system on the microscope will help to minimize axial focus fluctuations, thereby improving the quality of the data.

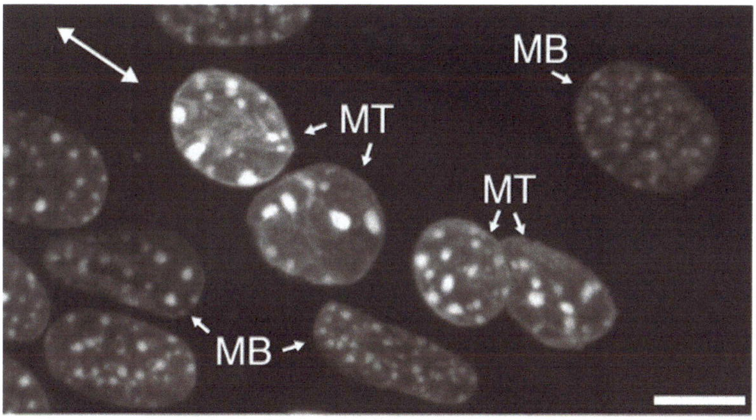

Fig. 1 Examples of myoblast and myotube nuclei labeled with Hoechst dye. Hoechst dye binds DNA and brightly labels dense areas of heterochromatin (chromocenters). The nucleus in a mononucleated myoblast (MB) is typically flatter and wider and has smaller and more numerous chromocenters. The nuclei in myotubes (MT) are usually more spherical; have fewer, larger chromocenters; and are often found in a line along the long axis of the myotube (*two-headed arrow*). Scale = 10 μm

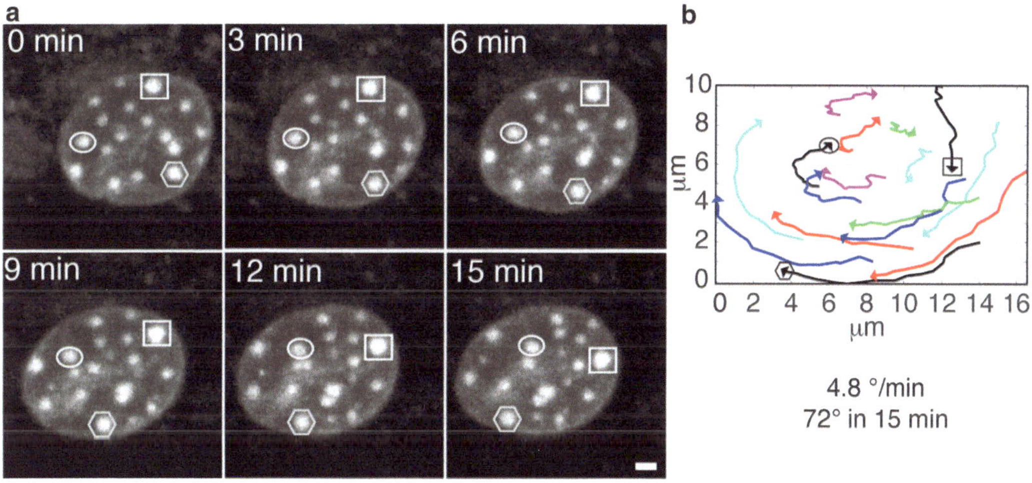

Fig. 2 Example of nuclear rotation and chromocenter tracking. (**a**) DNA was labeled with Hoechst dye, and a nucleus within a myotube was imaged at a frame rate of 1 frame per minute for 15 min. Maximum projections of confocal z-stacks are shown over time. Three of the chromocenters are highlighted to aid in visualization of rotation. Scale bar: 2 μm. (**b**) Chromocenters were tracked in *X*, *Y*, and *Z* over time using Volocity 3D Image Analysis Software, and the resulting 2D projection of the tracks is shown. Black tracks correspond to the highlighted chromocenters in panel (**a**). The nucleus rotates while translocating to the left. Applying the algorithm included in the *NucleusAngularVec.m* MATLAB script, the mean total angular velocity of this nucleus is 4.8°/min. The nucleus rotates 72° in 15 min

5. Continue imaging nuclei in additional myotubes on this plate as needed. Switch to a new plate of cells after 2–3 h or if the health of cells begins to decline.

3.4 Nuclear Rotation Analysis

In this analysis, the chromocenters within a nucleus are first tracked in *XYZ* over time in Volocity 3D Image Analysis Software. The resulting tracking data is used to calculate the orientation and angular velocity of rotation using rigid body kinematic analyses performed in MATLAB. The nucleus is analyzed as a rigid body where it is assumed that the chromocenters retain the same relative orientation in the nucleus as it undergoes motion. A minimum of three chromocenters (spatial points) are required for the calculation where each track contains the time-dependent coordinates of a single chromocenter at each time point. More than three points can be used and in general will result in more accurate results for the spatial orientation. The shape of the nucleus also deforms to some degree during its course of travel, and the tracking of chromocenters has a finite error. However, the rigid body analysis performed here is still applicable, and the error introduced by these factors can be averaged out through the use of several tracking points.

1. Open the file containing the time series images of nuclei rotating in Volocity 3D Image Analysis Software.
2. Open the first captured time series and crop the images in XY to include only one nucleus of interest. Make sure to advance through the time points and verify that the cropped region of interest (ROI) is large enough to encompass all linear movement of the nucleus over time (*see* **Note 9**).
3. In the Volocity Measurements View, set up the following protocol to identify and track the bright Hoechst dye-labeled chromocenters in *XYZ* over time:
   (a) Find Objects—Define the Threshold (~3–5 %) and the Minimum Object Size (~0.18–0.2 μm^3).
   (b) Remove noise from objects—medium filter.
   (c) Separate touching objects—0.5 μm^3.

   The software will identify chromocenters in the ROI, both those in the nucleus of interest and those in any other nuclei present in the ROI.

   Advance through each frame of the time series to verify that the program is correctly identifying chromocenters. Adjust the threshold and minimum object size as needed to achieve optimal identification.

   While a minimum of three chromocenters is necessary for angular velocity calculations, additional points in the nucleus will improve the rotation data. We have found that the largest chromocenters are tracked most accurately in *XYZ* space.

   (d) Track objects—choose the following parameters: shortest path, ignore new objects, set the maximum distance between objects. Verify by eye that the tracks are correct by playing the time series through. Check that the software is able to track each individual chromocenter continuously from the first frame to the last frame. Alter the parameters above as necessary to achieve accurate tracks (*see* **Note 10** and Fig. 2).

4. Even with significant optimization, it may not be possible to achieve complete, correct tracks for all chromocenters in a nucleus. If this is the case, in the measurements window, filter the track data by time span and eliminate data for all tracks shorter than the duration of the time series (highlight a track in the measurements window and press delete). Additionally, remove any data for chromocenters/tracks that are not in the nucleus of interest. Sort the data by Track ID in the Measurement Table.
5. In the Measurements tab, choose Columns. In the opened window, define which measurements to include in the data file. Include Object ID, Track ID, Timepoint, Relative Time (s), Centroid X (μm), Centroid Y (μm), and Centroid Z (μm). Other data may also be included in the data file but are not necessary for angular velocity analysis.
6. In the Measurements tab, choose Make Measurement Item. Name the Measurement Item as desired; all of the data will now appear in a spreadsheet in the Volocity library. Additionally, export this table as a .CSV file.
7. Open the MATLAB script entitled *ChromocenterDataParse.m*. When prompted enter the name of the .CSV file containing the tracking data. This script parses the Volocity data into MATLAB. Refer to **Note 11** for source code listing.
8. To calculate the total angular speed, run *NucleusAngularVec.m*. Refer to **Note 12** for source code listing and a detailed explanation of the algorithm.
9. Repeat **steps 1–8** for each nucleus to be analyzed.

3.5 Cell Fixation and Immunostaining for Analysis of Nuclear Distribution

1. Pre warm 4 % PFA in PBS to 37 °C.
2. Aspirate differentiation medium from the 12-well plate with ACLAR coverslips, and rinse the myotubes with warm DPBS.
3. Add warm 4 % PFA to fix the cells and incubate at room temperature 10 min.
4. Aspirate the 4 % PFA and wash the coverslips with PBS (3 × 5 min) at room temperature.
5. Permeabilize the cells by incubating with 0.1 % Triton X-100 in PBS for 5 min.
6. Wash the cells 3 × 5 min with PBS.
7. Incubate fixed myotubes for 1 h in blocking solution at room temperature.

8. Incubate myotubes for 2 h with the anti-alpha-actinin antibody (1:500) and anti-alpha-tubulin primary antibody (1:500) in blocking solution at room temperature (*see* **Note 13**).
9. Wash 3 × 10 min with PBS.
10. Incubate myotubes for 1–2 h with Alexa 488 conjugated anti-mouse secondary antibodies (1:500), Alexa 594 conjugated anti-rat secondary antibodies (1:500), and Hoechst dye (0.5 μg/ml) in blocking solution at room temperature.
11. Wash 3 × 10 min with PBS.
12. Mount ACLAR coverslips either on a glass slide or on a 22 × 40 mm glass coverslip (#1.5) with antifade media (*see* **Note 14**).
13. Let antifade cure overnight at room temperature before imaging.

3.6 Imaging Myotubes for Analysis of Nuclear Distribution

1. Before starting to obtain images, calibrate the motorized stage on the microscope.
2. Using a 40× objective, capture tiled images of large regions of fixed immunostained myotubes with Volocity using the following guidelines:
   (a) Define the region of interest (ROI) to be captured on the grid in the XY Stage view. Typically a region of ~1000 μm × 1000 μm is appropriate. Volocity will acquire the minimum number of fields to completely cover the ROI. A ROI ~1000 μm² is typically achieved by a 4 × 4 grid of fields or tiles (16 tiles) with 10 % overlap at 40× (*see* Fig. 3 and **Note 15**).
   (b) Find the approximate center focal plane of all of the myotubes in the current field of view. Set this position as zero. Set the image acquisition parameters to obtain a z-series for each field of view, starting ~20 μm below and moving to ~20 μm above this center point, with a 1–2 μm z-step size. The size of this z-stack should encompass the entire depth of all of the myotubes in the field of view (typically ~40 μm).
   (c) Set the software to obtain images in the appropriate channels for Hoechst dye, alpha-actinin/Alexa 488, and tubulin/Alexa 594 and adjust exposure settings as needed to obtain quality images.
   (d) Set the software to save all raw tiles. You can choose to either stitch the tiles together immediately after obtaining the images or to stitch at a later time. Stitching large regions can take significant amounts of time, so offline stitching is recommended. In Volocity V6.0 or higher, select Stitch Images in the Tools menu; in the pop-up win-

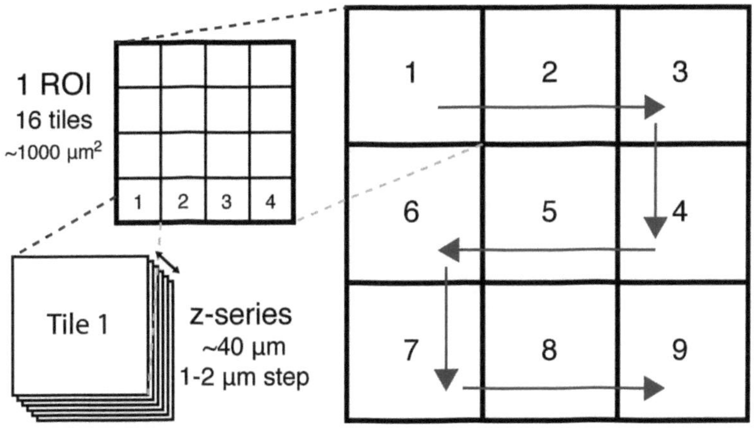

Fig. 3 Guideline for imaging large regions of myotubes. Using a 40× objective, capture a ROI of ~1000 μm^2. This is typically composed of 16 tiles with 10 % overlap. Each of these tiles is composed of a z-stack encompassing a depth of ~40 μm. To avoid imaging the same region more than once, use the grid pattern in the Volocity XY stage view to define ROIs, taking care to move across and down the coverslip in a sequential pattern

dow, choose "Create a Stitched Image" and click to option "Correct for Brightness."

3. Use the grid on the XY stage view to define the next ROI. To avoid imaging the same region twice, it is advised to define ROIs in a sequential pattern. If the myotube cultures have differentiated and fused well, 9–10 ROIs are typically sufficient to obtain nuclear distribution data for 50–60 myotubes.

3.7 Analysis of Nuclear Distribution

1. Create a maximum projection of a stitched ROI in Volocity. Include the Hoechst dye, alpha-actinin, and tubulin channels in the image (*see* Fig. 4a and **Note 16**).
2. In the following analysis, only include myotubes that lie fully within the stitched ROI. Do not include branched myotubes in the analysis.
3. In the Volocity Measurements View, starting at the bottom and/or left end of a given myotube, as shown by the alpha-actinin staining, use the measurement tools to determine the *X,Y* coordinates of the myotube end (e1) (Fig. 4a, b). These coordinates can either be determined in pixels or preferably in microns if the measurement tools are calibrated to the objective.
4. Determine the *X,Y* coordinates of the centroid of each nucleus in the myotube, moving sequentially from the left of the myotube to the right (or the bottom to top, if the myotube is oriented vertically). Assign sequential numbers to each nucleus (Fig. 4a, b).

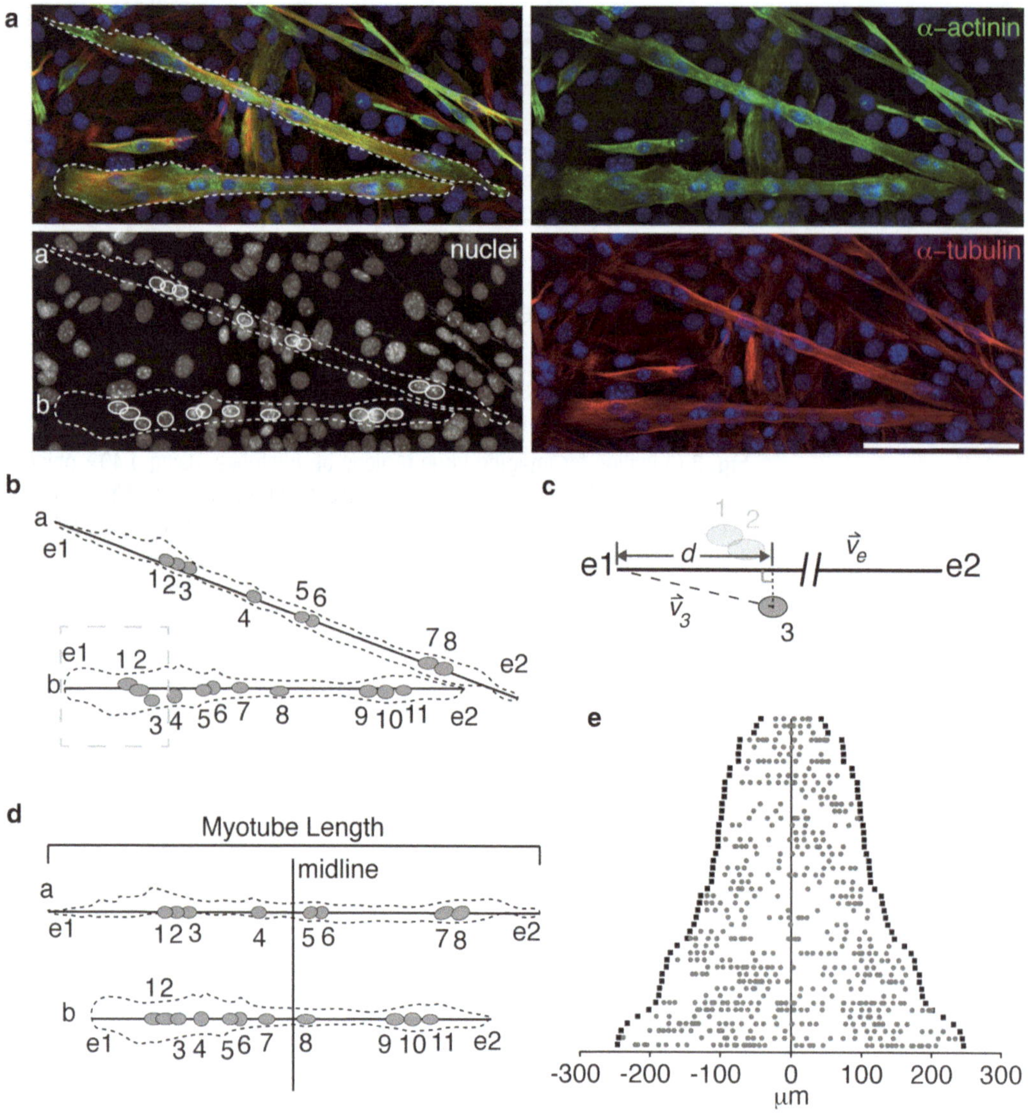

Fig. 4 Myotube nuclear distribution analysis. (**a**) A small region of a stitched ROI shows C2C12 myotubes and surrounding myoblasts. Myotubes were immunostained α-actinin (*green*) and α-tubulin (*red*), and DNA was stained with Hoechst dye (*blue*). Image is a maximum projection of confocal z-sections. This region includes two complete myotubes suitable for nuclear distribution analysis (**a** and **b**, outlined for clarity; nuclei within the myotubes are *circled*). This region also includes numerous myoblasts and additional myotubes extending out of the shown region. Scale = 50 μm. (**b**) Illustration of the two myotubes in (**a**); nuclei in each myotube are numbered, and a line segment between the ends (e1 and e2) of the myotubes is shown. (**c**) Depiction of the boxed area in (**b**) illustrating the vectors used to calculate the distance, d, between e1 and the centroid of nucleus 3 when projected onto the line segment extending between myotube ends. (**d**) Illustration of the myotubes following projection of all nuclei onto the line segment between e1 and e2. Myotubes are aligned at the midline of the cells. (**e**) Example of a nuclear distribution plot for a population of untreated myotubes. Each line on the *y*-axis represents an individual myotube, organized according to length. The ends of the myotube are marked with a *dark square*; data points represent individual nuclei

5. Determine the X,Y coordinates of the right (or top) end of the myotube (e2).
6. Repeat this process for all of the myotubes in each stitched ROI.
7. Using the X,Y coordinates for each myotube end, determine the length of each myotube using the distance equation:

$$d = \sqrt{\left((X_2 - X_1)^2 + (Y_2 - Y_1)^2\right)}$$

8. Using e1 and e2 to define a vector, $\vec{v}_e$, project the X,Y coordinates of the centroid of each nucleus along a perpendicular path onto the line segment (Fig. 4c). Assuming $\vec{v}_3$ is the vector from e1 to the centroid of a nucleus labeled 3, the distance d from e1 to projected point on the line is $d = \left\| \frac{\| \vec{v}_e \cdot \vec{v}_3 \| \vec{v}_e}{\| \vec{v}_e \|^2} \right\|$.

   After this projection, all of the centroids of each nucleus lie along the line segment extending between myotube ends (Fig. 4d).
9. Determine the distance between adjacent nuclei on this line segment.
10. To visually represent the position of nuclei in individual myotubes within a population of myotubes, organize your data to make a plot of nuclear distribution using GraphPad Prism software (*see* Fig. 4e).
    (a) For each myotube, set the midpoint between e1 and e2 to zero and determine the distance of each nucleus (and myotube ends) relative to this midpoint, with nuclei to the left of the midline being assigned negative values and the nuclei to the right of the midline assigned positive values.
    (b) Number the myotubes in a population according to length, setting the longest myotube equal to one, second longest equal to 2, and so on.
    (c) Create an X,Y graph in Prism in which the number of the myotube (Y value) is assigned to all points in the myotube, including e1, e2, and each nucleus (X values).
    - In Prism make the following selections:
      - Make New Data Table (+ Graph).
      - Choose X,Y graph and Enter and plot a single Y value for each point.
    - In the table created, enter the distance data for myotube ends and centroid of each nucleus, as calculated in 10a, into the X column. Group this data by myotube, and organize the myotubes in the column according to myotube length, starting with the longest myotube.

- In the first Y column, enter the number of the myotube, as determined in 10b, for all of the X values corresponding to myotube ends (i.e., the first and last data point in each myotube).
- In the second Y column, enter the number of the myotube, as determined in 10b, for all of the X values corresponding to nuclei.
- In the graph that is produced, change the color or shape of values from the first Y column to delineate that they are myotube ends.

4 Notes

1. Primary mouse myotube cultures can also be analyzed using these protocols. For further information on generating primary cultures, we refer the reader to the following references [12, 13].
2. To maintain proliferating C2C12 myoblasts, cells must be split regularly at sub-confluency, before the cells commit to differentiation. As the cells become increasingly confluent, they will begin to elongate and align with one another, thereby decreasing the myoblast population capable of further proliferation. Even with strict splitting habits, the differentiation potential of these cells typically decreases with passage number; therefore, it is best to thaw a new vial of myoblasts after ~10 passages or when you start to notice a decline in myotube formation. Upon obtaining a vial of C2C12 myoblasts, it is advisable to initially expand the culture and freeze down stock vials of low-passage number cells at high density.
3. Plates and ACLAR coverslips can also be coated with other substances, including gelatin or laminin [14], a fibroblast substratum [15], or Matrigel [16].
4. ACLAR plastic embedding film is used in place of glass coverslips because it was noted that the myotubes readily pull off of the glass surface when they begin to spontaneously contract. However, it has been our experience that C2C12 cells grow well on one side of the ACLAR film but pull into abnormal piles of cells when cultured on the opposite side. We have not been able to distinguish which is the "correct" side visually; therefore, it is important to test pieces of ACLAR in both orientations by growing cells on each side and keeping careful track of the "correct" side of the film. It will be necessary to test each new sheet of ACLAR in this manner. Sterilize scissors prior to cutting ACLAR and work in the hood.
5. Although it is not necessary, transfecting the cells with DNA constructs for EGFP or other cytosolic or plasma membrane-

bound fluorescent protein makes it easier to distinguish myotube boundaries, myotube branching, and relative position of nuclei within the live myotubes. In the absence of such a marker, it is important to become adept at recognizing nuclei present within myotubes compared with nuclei in mononucleated myocytes, which can often lie beneath myotubes on the culture plate. However, if it is necessary to know where the nuclei lie within the myotube and/or that all nuclei being analyzed are in the same myotube, expression of EGFP will aid in delineating the boundaries of the myotube.

6. The use of HEPES buffered medium will help maintain pH in the absence of the 5 % CO2 atmosphere. We have imaged myotube cultures up to ~5 h without noticing alterations in myotube health or changes in nuclear dynamics. Overlaying the imaging media with mineral oil reduces evaporation, thereby preventing changes in osmolarity. Note that it is difficult to sufficiently remove the mineral oil following imaging if it is necessary to fix and stain these myotubes for further analysis. Alternatively, the environmental chamber may be humidified during imaging.
7. We have used z-step sizes as small as 0.2 μm. We have found that step sizes smaller than 0.5 μm only marginally improve precision of chromocenter tracking in the z-dimension. The 0.5 μm z-step size minimizes exposure time while still allowing for reliable chromocenter tracking and subsequent assessment of angular velocity.
8. When the myotubes are exposed to excessive laser light, they may begin to contract forcefully, which can cause the myotube to detach from the plate. This is typically rare during a 15 min time series but becomes more likely as the length of time series increases and when EGFP is used to identify myotubes.
9. It is not necessary to crop the images for the analysis, but by eliminating the image data around a specific nucleus, we have noticed that Volocity more effectively finds chromocenters in successive frames and creates valid tracks.
10. Initially, try using the Volocity command "estimate maximum distance between objects automatically." This will provide a guideline for defining this distance parameter manually in subsequent analysis. If a number is manually entered, the software will report the same tracks in repeated analyses of the same chromocenter; however, if it estimates automatically, the tracking data will be slightly different in repeated analyses.
11. ChromocenterDataParse.m source code

```
% ChromocenterDataParse.m
% Centroid for each chromocenter and time
for each track is read into
```

```
% xCent yCent zCent time matrices
%
% Open CSV file containing track information
[filename,           pathname]=uigetfile('*.
csv','Select a CSV file');
fid=fopen(fullfile(pathname,filename));
tline=fgetl(fid); %First line is a space
tline=fgetl(fid); %Second line are labels
labelread=textscan(tline,'%s','Delimi
ter',',');
labels=labelread{1}; %Cell array of column
labels
clear xCent yCent zCent time tracks dataCnt
trackCnt=0;  dataCnt=0;  tracks=0;
tline=fgetl(fid);
while(tline~=-1)
    cnt=1;    cflag=0;
    while(tline)
        label=strtok(labels{cnt},'"');
        data tline]=strtok(tline,',');
        if( strcmp(label, 'Type') ) %Don't
        read Track Objects
            if(strcmp(data,'"Track"'))
                tline=fgetl(fid);
                cflag=1;
                break;
            end;
        end
        if( strcmp(label, 'Rel. Time (s)'))
            RelTime=str2double(strtok
            (data,'"'));
        end
        if( strcmp(label, 'Centroid X (µm)') )
            xc=str2double(strtok(data,'"'));
        end
        if( strcmp(label, 'Centroid Y (µm)') )
            yc=str2double(strtok(data,'"'));
        end
        if( strcmp(label, 'Centroid Z (µm)') )
            zc=str2double(strtok(data,'"'));
        end
        if( strcmp(label, 'Track ID') )
            trackID=str2double(strtok
            (data,'"'));
        end
        cnt=cnt+1;
    end
```

```
    if(cflag) continue; end;
    tCID=find(tracks==trackID);
    if(tCID)
        dataCnt(tCID)=dataCnt(tCID)+1;
    else
        trackCnt=trackCnt+1;
        tCID=trackCnt;
        tracks(tCID)=trackID;
        dataCnt(tCID)=1;
    end
    time(tCID,dataCnt(tCID))=RelTime;
    xCent(tCID,dataCnt(tCID))=xc;
    yCent(tCID,dataCnt(tCID))=yc;
    zCent(tCID,dataCnt(tCID))=zc;
    tline=fgetl(fid);
end
fclose(fid);
```

12. NucleusAngularVec.m source code

```
%NucleusAngularVec.m
%This algorithm calculates the euler angles at each time step
%from the centroid data loaded by the dat-aParse routine.
%It outputs angular velocity
%
%It uses xCent,yCent,zCent and time from Chromocenter DataParse.m
Nt=size(xCent,2);% number of time points
Np=size(xCent,1);% number of track points
for t=2:Nt
    %Calculate Centriods
    v1c=[0;0;0]; v2c=[0;0;0];
    for ii=1:Np
        v1c=v1c+[xCent(ii,t-1);yCent(ii,
        t-1);zCent(ii,t-1)];
        v2c=v2c+[xCent(ii,t);yCent(ii,t)
        ;zCent(ii,t)];
    end
    v1c=v1c/Np; v2c=v2c/Np;
    %Calculate Correlation matrix
    c=zeros(3,3);
    for ii=1:Np
        vec1=[xCent(ii,t-1);yCent(ii,
        t-1);zCent(ii,t-1)]-v1c;
        vec2=[xCent(ii,t);yCent(ii,t);zC
        ent(ii,t)]-v2c;
        c=c+vec2*vec1';
    end
```

```
        c=c/Np;
        %Calculate Rotation Matrix
        [u,w,v]=svd(c);
        R=u*[1 0 0;0 1 0; 0 0 det(u*v')]*v';
        %Get Euler Angles from Rotation matrix
        theta=-asin(R(3,1));
        psi=atan2(R(3,2)/cos(theta),R(3,3)/
        cos(theta));
        phi=atan2(R(2,1)/cos(theta),R(1,1)/
        cos(theta));
        %Get angular velocity
        disp(R)
        disp([theta psi phi])
        dt=time(1,t)-time(1,t-1);
        %Angular velocity of each Euler
        component
        wv(t-1,:)=([theta psi phi])/dt;
        %Resultant angular velocity (angular
        speed)
        wc(t-1)=norm([theta psi phi])/dt;
        %Translation vector of nucleus from
        previous time setp
        %Not used here - can be used to deter-
        mine linear velocity,
        %acceleration, etc…
        translation(t-1,:)=v2c-R*v1c;
    end
    str=sprintf('Time Average Angular velocity
    (total)    %0.5    g    degrees    per
    second',mean(wc*180/pi));
    disp(str);
    str=sprintf('Time Average Angular velocity
    (pitch)    %0.5    g    degrees    per
    second',mean(wv(:,1)*180/pi));
    disp(str);
    str=sprintf('Time Average Angular velocity
    (yaw)    %0.5    g    degrees    per
    second',mean(wv(:,2)*180/pi));
    disp(str);
    str=sprintf('Time Average Angular velocity
    (roll)    %0.5    g    degrees    per
    second',mean(wv(:,3)*180/pi));
    disp(str);
    figure
    hold off
    plot(time(1,1:(Nt-1)),wc*180/
    pi,'k','Linewidth',3)
    hold on
```

```
plot(time(1,1:(Nt-1)),wv(:,1)*180/
pi,'r:','Linewidth',1)
plot(time(1,1:(Nt-1)),wv(:,2)*180/
pi,'b:','Linewidth',1)
plot(time(1,1:(Nt-1)),wv(:,3)*180/
pi,'g:','Linewidth',1)
legend('Total Angular Velocity','Pitch
Angular Velocity','Yaw Angular
Velocity','Roll Angular Velocity');
xlabel('time (seconds)');
ylabel('Angular Velocity (degrees per
second)');
```

To begin the tracking analysis, we sample a set of *Np* points (*XYZ* centroid of chromocenters) on the nucleus in three-dimensional space at a set of discrete times $t(n)$. The set of points at one time epoch n are recorded as column vectors. This data is retrieved from the Volocity 3D Image Analysis Software (*see* Subheading 3.4).

$$\mathbf{x}(n)=\begin{bmatrix} x_1(n) \\ x_2(n) \\ \vdots \\ x_{N_p}(n) \end{bmatrix},\ \mathbf{y}(n)=\begin{bmatrix} y_1(n) \\ y_2(n) \\ \vdots \\ x_{N_p}(n) \end{bmatrix},\ \mathbf{z}(n)=\begin{bmatrix} z_1(n) \\ z_2(n) \\ \vdots \\ z_{N_p}(n) \end{bmatrix}$$

The centroid of the set of points is then calculated through averaging. Note this is the centroid of the set of chromocenters of a single nucleus. Subheading 3.4 refers to finding the centroid of single chromocenter.

$$x_c(n)=\frac{1}{N_p}\sum_{m=1}^{N_p}x_m(n),\ y_c(n)=\frac{1}{N_p}\sum_{m=1}^{N_p}y_m(n),\ z_c(n)=\frac{1}{N_p}\sum_{m=1}^{N_p}z_m(n)$$

Next, a vector **vm** from each point on the body to the centroid of the set of points at times n and $n-1$ is calculated. The vector **vm** is then multiplied by its transpose from the previous time step and summed over all points to calculate the correlation matrix, **C** [17].

$$\mathbf{v}_m(n)=\begin{bmatrix} x_m(n)-x_c(n) \\ y_m(n)-y_c(n) \\ z_m(n)-z_c(n) \end{bmatrix},\ \mathbf{v}_m(n-1)=\begin{bmatrix} x_m(n-1)-x_c(n-1) \\ y_m(n-1)-y_c(n-1) \\ z_m(n-1)-z_c(n-1) \end{bmatrix},\ \mathbf{C}=\frac{1}{N_p}\sum_{m=1}^{N_p}\mathbf{v}_m(n)\mathbf{v}_m(n-1)^T$$

The 3×3 correlation matrix can then be used to calculate the rotation matrix between the set of points at times n and $n-1$ using singular value decomposition. The singular value decomposition of the correlation matrix $\mathbf{C}$ results in $\mathbf{C} = \mathbf{U\Sigma V}^*$. Following the singular value decomposition, the rotation matrix can be calculated as

$$\mathbf{R} = \mathbf{U}\begin{bmatrix} 1 & 0 & 0 \\ 0 & 1 & 0 \\ 0 & 0 & \left|\mathbf{UV}^*\right| \end{bmatrix}\mathbf{V}^*$$

Using the components of the rotation matrix the Euler angles can be calculated. We adopted the convention where

$$\mathbf{R_X}(\psi) = \begin{bmatrix} 1 & 0 & 0 \\ 0 & \cos(\psi) & -\sin(\psi) \\ 0 & \sin(\psi) & \cos(\psi) \end{bmatrix}$$

$$\mathbf{R_Y}(\theta) = \begin{bmatrix} \cos(\theta) & 0 & \sin(\theta) \\ 0 & 1 & 0 \\ -\sin(\theta) & 0 & \cos(\theta) \end{bmatrix}$$

$$\mathbf{R_Z}(\phi) = \begin{bmatrix} \cos(\phi) & -\sin(\phi) & 0 \\ \sin(\phi) & \cos(\phi) & 0 \\ 0 & 0 & 1 \end{bmatrix}$$

$$\mathbf{R} = \begin{bmatrix} R_{11} & R_{12} & R_{13} \\ R_{21} & R_{22} & R_{23} \\ R_{31} & R_{32} & R_{33} \end{bmatrix} = \mathbf{R_Z}(\phi)\mathbf{R_Y}(\theta)\mathbf{R_X}(\psi)$$

Solving for the Euler angle θ, ψ, and ϕ, we get

$$\theta = -\sin^{-1}(R_{31})$$

$$\text{if } R_{31} \neq \pm 1 \text{ then } \psi = \text{atan2}\left(\frac{R_{32}}{\cos\theta}, \frac{R_{33}}{\cos\theta}\right) \text{ else } \psi = \text{atan2}(R_{32}, R_{33})$$

$$\text{if } R_{31} \neq \pm 1 \text{ then } \phi = \text{atan2}\left(\frac{R_{21}}{\cos\theta}, \frac{R_{11}}{\cos\theta}\right) \text{ else } \phi = 0$$

The angular velocity is then calculated as the time rate of change of the Euler angles, and the total angular speed is calculated from the resultant vector:

$$\omega = \frac{[\theta, \psi, \phi]}{\Delta t} \; and \; \omega_t = |\omega|$$

To calculate the total angular speed, a MATLAB script was developed which incorporates the equations as described in the previous steps. First, line 6 of the MATLAB script ChromocenterDataParse.m is modified so that it opens the

CSV output from Volocity and it is run. This script parses the Volocity data into MATLAB. Next NucleusAngularVec.m is run to calculate the total angular speed.

13. Immunostaining for alpha-actinin is an effective way to identify the ends of the myotubes in the culture dish. Phase contrast images would also be appropriate, if this is an option on the microscope. Immunostaining for beta-tubulin aids in assessing whether a given nucleus is within the myotube.
14. It is possible to image the myotubes through the ACLAR plastic film, but we have had more success mounting the ACLAR pieces on glass coverslips and imaging through the glass instead. Adjustable stage adaptors accommodate 40 mm glass coverslips.
15. If the myotubes have been transfected with a tagged protein of interest, but the numbers of myotubes expressing the transgene are small and it is only these transfected myotubes that are to be part of the nuclear distribution analysis, it may be better to define smaller ROIs that encompass only a single transfected myotube (or a few if they are close together). In this situation, taking large, unbiased ROIs may result in substantial unusable data if they lack any myotubes expressing the transgene. In this case, move systematically through the coverslip to obtain images of all of the expressing myotubes present on the coverslip or until the predefined number of myotubes has been obtained.
16. Although the distribution of nuclei is analyzed in the maximum projection image, it is important to refer to the original images in the z-series when there is question as to whether a given nucleus is within the myotube of interest. As discussed previously, often myoblasts will be below myotubes, and in a maximum projection, it may not be immediately clear that the myoblast nucleus is in a separate cell. Along with using the shape and chromocenter pattern in a nucleus to assess whether it is within the myotube, it is often helpful to move through the z-series plane-by-plane or view the stack in an *X,Z* or *Y,Z* orientation. Myoblast nuclei will almost always appear in a *Z* plane below the myotube nuclei and will likely not be in a plane with significant alpha-actinin signal. Furthermore, the absence of signal for alpha-actinin or beta-tubulin where the nucleus resides also typically indicates the nucleus of interest is within the myotube.

Acknowledgments

This work was supported by the National Institutes of Health (P01 GM087253 to E.L.F.H., T32 GM-07229, and T32 AR-053461 to M.H.W.) and the American Heart Association (#13PRE16090007 to M.H.W.).

References

1. Capers CR (1960) Multinucleation of skeletal muscle in vitro. J Biophys Biochem Cytol 7:559–566
2. Cooper WG, Konigsberg IR (1961) Dynamics of myogenesis in vitro. Anat Rec 140:195–205
3. Englander LL, Rubin LL (1987) Acetylcholine receptor clustering and nuclear movement in muscle fibers in culture. J Cell Biol 104:87–95
4. Wilson MH, Holzbaur EL (2012) Opposing microtubule motors drive robust nuclear dynamics in developing muscle cells. J Cell Sci 125:4158–4169
5. Wilson MH, Holzbaur EL (2015) Nesprins anchor kinesin-1 motors to the nucleus to drive nuclear distribution in muscle cells. Development 142:218–228
6. Metzger T, Gache V, Xu M, Cadot B, Folker ES, Richardson BE, Gomes ER, Baylies MK (2012) MAP and kinesin-dependent nuclear positioning is required for skeletal muscle function. Nature 484:120–124
7. Wang Z, Cui J, Wong WM, Li X, Xue W, Lin R, Wang J, Wang P, Tanner JA, Cheah KS et al (2013) Kif5b controls the localization of myofibril components for their assembly and linkage to the myotendinous junctions. Development 140:617–626
8. Romero NB (2010) Centronuclear myopathies: a widening concept. Neuromuscul Disord 20:223–228
9. Zhang Q, Bethmann C, Worth NF, Davies JD, Wasner C, Feuer A, Ragnauth CD, Yi Q, Mellad JA, Warren DT et al (2007) Nesprin-1 and -2 are involved in the pathogenesis of Emery Dreifuss muscular dystrophy and are critical for nuclear envelope integrity. Hum Mol Genet 16:2816–2833
10. Yaffe D, Saxel O (1977) Serial passaging and differentiation of myogenic cells isolated from dystrophic mouse muscle. Nature 270:725–727
11. Blau HM, Chiu CP, Webster C (1983) Cytoplasmic activation of human nuclear genes in stable heterocaryons. Cell 32:1171–1180
12. Shefer G, Yablonka-Reuveni Z (2005) Isolation and culture of skeletal muscle myofibers as a means to analyze satellite cells. Methods Mol Biol 290:281–304
13. Danoviz ME, Yablonka-Reuveni Z (2012) Skeletal muscle satellite cells: background and methods for isolation and analysis in a primary culture system. Methods Mol Biol 798:21–52
14. Kummer TT, Misgeld T, Lichtman JW, Sanes JR (2004) Nerve-independent formation of a topologically complex postsynaptic apparatus. J Cell Biol 164:1077–1087
15. Cooper ST, Maxwell AL, Kizana E, Ghoddusi M, Hardeman EC, Alexander IE, Allen DG, North KN (2004) C2C12 co-culture on a fibroblast substratum enables sustained survival of contractile, highly differentiated myotubes with peripheral nuclei and adult fast myosin expression. Cell Motil Cytoskeleton 58:200–211
16. Langen RC, Schols AM, Kelders MC, Wouters EF, Janssen-Heininger YM (2003) Enhanced myogenic differentiation by extracellular matrix is regulated at the early stages of myogenesis. In Vitro Cell Dev Biol Anim 39:163–169
17. Challis JH (1995) A procedure for determining rigid body transformation parameters. J Biomech 28:733–737

Chapter 19

Imaging Approaches to Investigate Myonuclear Positioning in *Drosophila*

Mafalda Azevedo, Victoria K. Schulman, Eric Folker, Mridula Balakrishnan, and Mary Baylies

Abstract

In the skeletal muscle, nuclei are positioned at the periphery of each myofiber and are evenly distributed along its length. Improper positioning of myonuclei has been correlated with muscle disease and decreased muscle function. Several mechanisms required for regulating nuclear position have been identified using the fruit fly, *Drosophila melanogaster*. The conservation of the myofiber between the fly and vertebrates, the availability of advanced genetic tools, and the ability to visualize dynamic processes using fluorescent proteins in vivo makes the fly an excellent system to study myonuclear positioning. This chapter describes time-lapse and fixed imaging methodologies using both the *Drosophila* embryo and the larva to investigate mechanisms of myonuclear positioning.

Key words *Drosophila*, Muscle, Nuclear movement, Embryo, Larvae

1 Introduction

The skeletal muscle provides an important system in which to study mechanisms of myonuclear positioning. Myofibers, the cellular units of the skeletal muscle, are multinucleated and position their myonuclei to reside above the sarcomeres at the cell periphery and along the length of the fiber to maximize internuclear distance. Moreover, mispositioned myonuclei correlate with muscle disease [1], and recent data suggest that mispositioned myonuclei may cause muscle weakness [2–4], illustrating the functional importance of proper positioning.

To identify the cellular mechanisms of myonuclear movement in vivo, we have utilized the musculature of *Drosophila* embryos and larvae [2–6]. *Drosophila* muscles share the conserved myofiber structure found in vertebrates: *Drosophila* myofibers are multinucleated, maintain particular sizes and shapes, attach to particular tendon cells, become innervated by specific motoneurons, build sarcomeres, and are essential for locomotion (reviewed in [7, 8]).

Sue Shackleton et al. (eds.), *The Nuclear Envelope: Methods and Protocols*, Methods in Molecular Biology, vol. 1411, DOI 10.1007/978-1-4939-3530-7_19, © Springer Science+Business Media New York 2016

The *Drosophila* system also offers unparalleled genetic approaches to manipulate genes of interest within the whole organism or specifically within muscle tissue. In addition to their powerful genetics, *Drosophila* embryos are also amenable to drug treatments commonly used to manipulate cell biological processes in culture, providing additional opportunities to probe the cellular machinery responsible for myonuclear movement [6]. Lastly *Drosophila* embryos are amenable to time-lapse microscopy without perturbing development such that cell biological processes, including myonuclear dynamics, can be assessed [5].

There are 30 individual muscles per abdominal hemisegment of the *Drosophila* embryo (Fig. 1). Formation of these individual body wall myofibers depends on the specification and fusion of two myoblast cell types: founder cells (FCs) and fusion-competent myoblasts (FCMs). Each FC contains the information necessary to direct the formation of a specific muscle. FCs can be identified by the expression of identity genes, such as the transcriptional regulators *even-skipped* and *apterous* [9–11]. The combination of identity genes and chromatin regulators expressed by a particular FC regulates the final morphology of the specific muscle. In contrast, FCMs are more naïve cells. Upon fusion to an FC, FCMs become reprogrammed to the specific developmental program of the FC, as evidenced by the observation that each newly incorporated FCM nucleus begins to express the same combination of identity genes as the FC [7]. Myoblast fusion is an iterative process; depending on the particular muscle, body wall muscles in *Drosophila* embryos contain between 2 and 25 myonuclei. As myoblast fusion concludes, the polarized

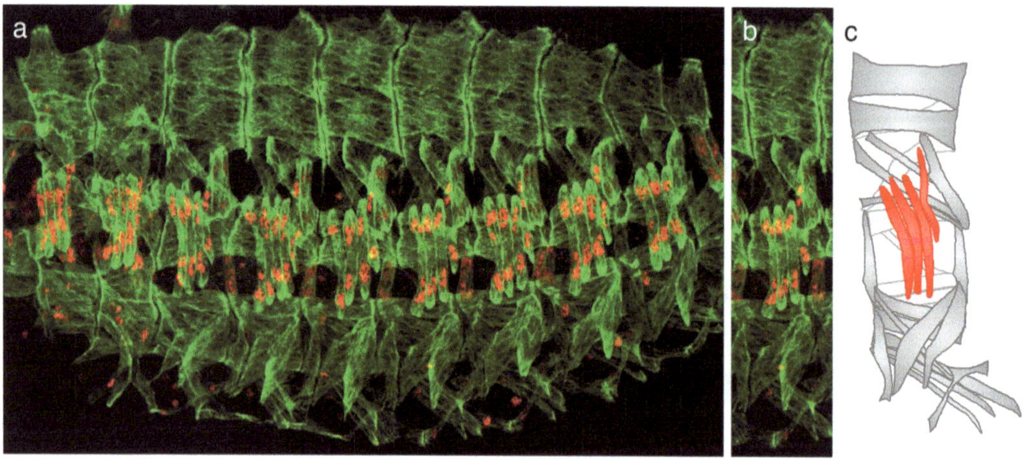

Fig. 1 The *Drosophila* embryonic musculature. The *Drosophila* embryo is divided in hemisegments, each containing 30 muscles. (**A**) Stage 16 embryo showing the musculature in green (Tropomyosin) and the nuclei of the lateral transverse (LT) muscles in red (DsRed). (**B**) Image of a single hemisegment. (**C**) Schematic representation of all the muscles present in one hemisegment; the LT muscles are highlighted in *red*. For additional detail *see* Dobi et al. [7]

syncytial myotube extends processes toward specific tendon cells and forms stable myotendinous junctions. Specific motoneurons innervate each muscle, leading to coordinated contraction at the final stage of embryogenesis [8, 12].

Although mispositioned nuclei are observed in all muscles when certain genes are disrupted, myonuclear positioning has been best characterized in the lateral transverse (LT) muscles of the developing embryo. We have shown that LT myonuclei undergo not one but a series of movements to maximize internuclear distance in vivo (Fig. 2; [2–5]). Post-fusion (stage 14), the myonuclei are located in a single group within the myotube. Over the course of development, these nuclei segregate into smaller groups that follow characteristic migration patterns (stages 15, 16) before evenly distributing throughout the myofiber at the end of embryonic stage 17 [2, 8]. During larval development, in which the muscles grow up to 40-fold [13], the myonuclei in the individual muscles maintain regular spacing along the muscle fiber. These myonuclei are positioned above the sarcomeres along the periphery of the muscle fiber (Fig. 3; [2]).

To date, there are several mechanisms required for the different movements of myonuclei in *Drosophila*. These mechanisms require microtubules; microtubule-associated proteins, such as ensconsin; the microtubule motor proteins, kinesin and dynein; and motor protein adaptors, such as kinesin light chain, dynein light chain, p150/Glued, cytoplasmic linker protein-190 (CLIP190), sunday driver (Syd or JNK-interacting protein 3 (JIP3)), and JNK signaling [2–5]. We refer the reader to Schulman et al. [8] in which we discuss our current thinking about the different mechanisms at work during each stage of myonuclear movement. To identify these mechanisms, we have developed various techniques to define and quantify myonuclear position in *Drosophila* [2–5]. This chapter describes these methodologies in both fixed and live *Drosophila* embryonic and larval preparations.

2 Materials

2.1 Imaging of Fixed Embryonic Muscles

2.1.1 Embryo Collection and Fixation (Fig. 4)

1. Embryo laying pot: 100 ml plastic beaker punched with holes to allow air exchange and prevent condensation. To make holes use a hot syringe needle with gauge (Sub-Q 26G5/8).
2. Embryo collection plates (apple juice plates): Microwave 1500 ml of ddH_2O, 50 g of granulated sugar, and 45 g of agar until a homogeneous solution is formed. Add 500 ml of cold apple juice, cool solution to 65 °C, and add 40 ml of 10 % Tegosept in 100 % ethanol. Pour solution into petri dishes (60 × 15 mm); make approximately 200 plates. Secure these plates to the laying pots with rubber bands.

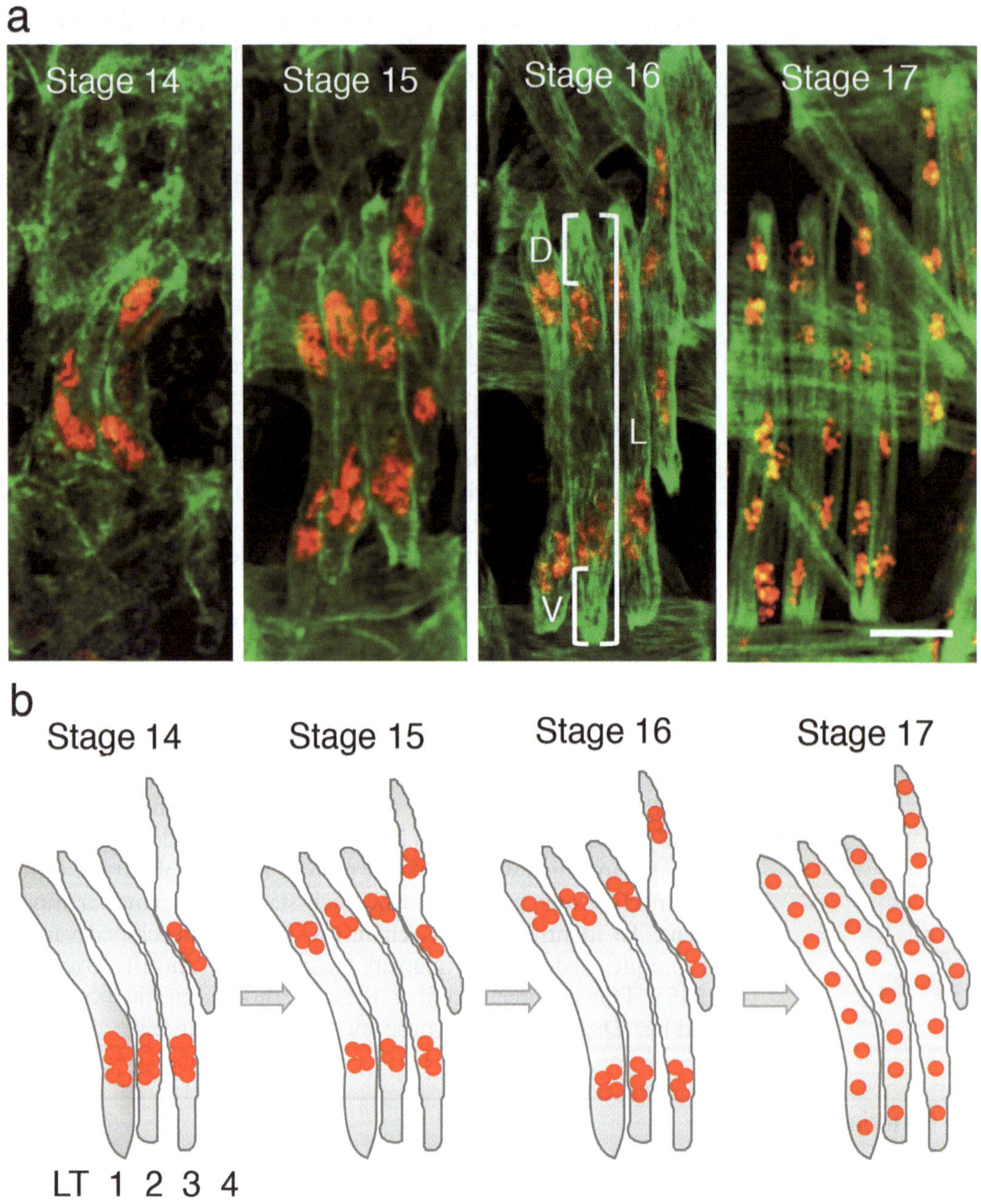

Fig. 2 Myonuclear positioning in the LT muscles during embryonic development. (**A**) Myonuclear position from stage 14 to stage 17. In the images of stages 14–16, the LT muscles are stained with Tropomyosin (*green*) and DsRed (*red*) to show the muscle structure and the nuclei, respectively. In the stage 17 image, the LT muscles are labeled with *Tropomyosin-GFP* (*green*) and the nuclei are labeled with *apterousME::NLSdsRed* (*red*) (*see* Table 2). At stage 16, image measurements are taken: D represents the dorsal measurement, V represents the ventral measurement, and L represents the total length of the muscle. Scale bar, 10 μm. (**B**) Schematic representation of myonuclear movements from stage 14 to stage 17 as shown in panel (**A**). LT muscles 1–3 have 6–8 nuclei each, whereas LT4 has 4–6 nuclei

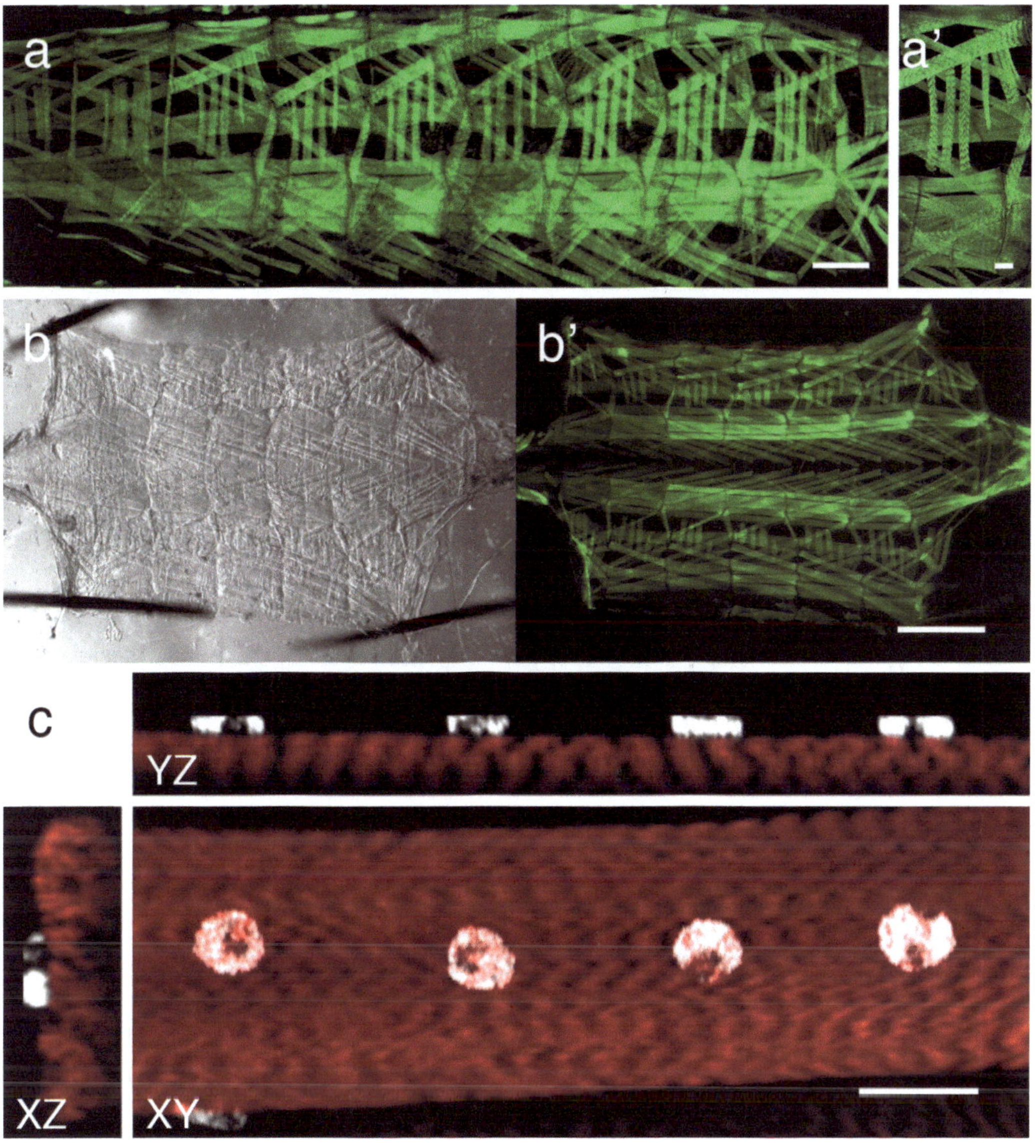

Fig. 3 Larval musculature and myonuclear positioning in a 3rd instar larva. (**A**) Image of a live *Drosophila* 3rd instar larva expressing *Tropomyosin-GFP*, a GFP gene trap that labels all muscles. Note the repeated muscle pattern in each hemisegment. Scale bar, 100 μm. (**A**′) Higher magnification of one hemisegment. Scale bar, 25 μm. (**B**, **B**′) Dissected flat prep of a larva carrying *Tropomyosin-GFP* with all the muscles exposed. Scale bar, 50 μm. (**B**) Bright field image. (**B**′) Wide field fluorescence image. (**C**) Orthogonal view of a ventral longitudinal 4 muscle (VL4). *Red*, actin; *white*, nuclei. Scale bar, 25 μm

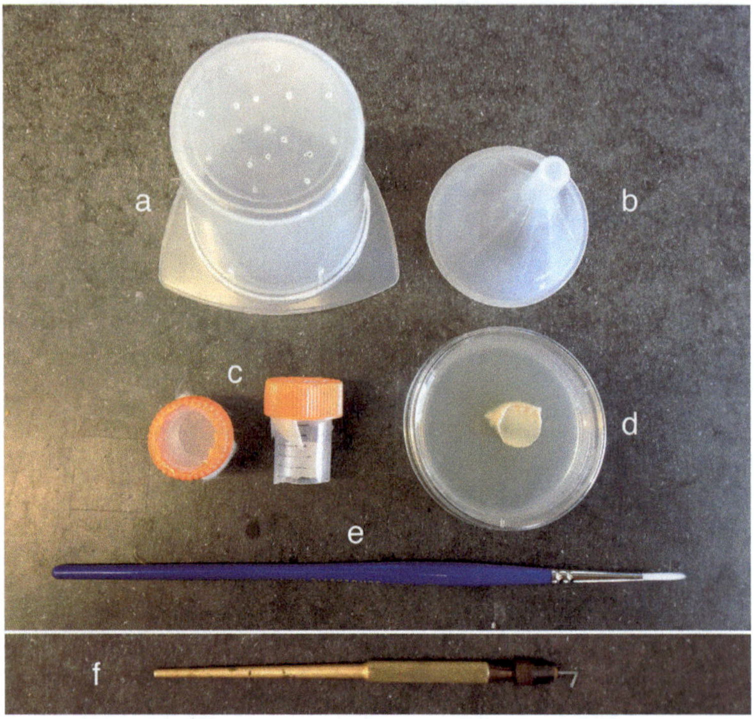

Fig. 4 Materials used for embryo collection. (**A**) Embryo laying pot perforated on the *bottom*. (**B**) Small funnel. (**C**) Collection baskets: *top* and *side* views. (**D**) Yeasted apple juice plate. (**E**) Paintbrush. (**F**) Embryo hook

3. Yeast paste: Distill water plus dry baker's yeast and stir to make a paste. Store at 4 °C.
4. Paintbrush (size 0).
5. Collection baskets: Cut the top off of a 15 ml Falcon tube at approximately the 12 ml line and cut out the center of the tube cap.
6. Nylon membrane (pore size, 140 μm) screwed onto the collection basket with the cap.
7. 3 % sodium hypochlorite (50 % bleach).
8. Small funnel.
9. Eppendorf tubes (1.5 ml).
10. Heptane.
11. Methanol.
12. PFA: 4 % EM grade paraformaldehyde diluted in PBS (phosphate-buffered saline).
13. Platform shaker.
14. Vortex with 1.5 ml Eppendorf tube adaptor.

15. PFA/heptane waste receptacle.
16. Methanol/heptane waste receptacle.
17. Glass Pasteur pipets.

2.1.2 Embryo Staining

1. 3 % Triton X-100: For 200 ml, add 6 ml of Triton X-100 and 194 ml ddH_2O.
2. 10× PBS (phosphate-buffered saline; 0.01 M KH_2PO_4, 0.1 M Na_2HPO_4, 1.37 M NaCl, 0.027 M KCl, pH 7.0).
3. PBT (phosphate-buffered saline containing 0.3 % Triton): For 100 ml, add 10 ml 10× PBS, 10 ml 3 % Triton X-100, and 80 ml ddH_2O.
4. PBT-BSA (phosphate-buffered saline containing 0.3 % Triton and 0.1 % BSA): For 100 ml, add 10 ml 10× PBS, 10 ml 3 % Triton X-100, 1 ml 10 % BSA, and 79 ml ddH_2O.
5. Primary and secondary antibodies of interest. Refer to Table 1 for a suggested list of antibodies and probes routinely used to label *Drosophila* muscle and nuclei.
6. Three-dimensional rotating mixer such as a Nutator.
7. Eppendorf tubes (0.5 ml).
8. Glass Pasteur pipets.

2.1.3 Embryo Mounting

1. Microscope slides.
2. Scotch tape.
3. Cover slips: size 22 × 40 mm with thickness 0.16–0.19 mm.
4. Mounting medium, such as ProLong Gold antifade reagent or Vectashield.
5. Razor blades.
6. Metal washers (any hardware store).

2.1.4 Embryo Imaging

All images in the lab are acquired on a Leica SP5 laser-scanning confocal microscope equipped with a 20× 0.7NA HC PL Apochromat oil objective, 63× 1.4NA HCX PL Apochromat oil objective, and LAS AF 2.2 software. Equivalent microscopes and software may be used.

2.2 Time-Lapse Imaging of Embryonic Muscles

2.2.1 Embryo Collection and Mounting (Fig. 5)

Materials described in Subheading 2.1.1, plus:

1. Plain-tipped applicator (wooden stick).
2. Halocarbon oil 700.
3. Air-permeable Teflon membrane mounted on a circular Perspex frame (designed by E. Wieschaus; [14, 15]).
4. Cover slips: size 22 × 40 mm with thickness 0.16–0.19 mm.

Table 1
Antibodies and probes used to probe *Drosophila* myonuclear positioning[a]

Primary antibodies[b]	Expression detected	Company, reference #
Muscle markers		
Tropomyosin	Actin cytoskeleton	Abcam, ab50567
Myosin heavy chain (MHC)	Myosin, actin motor protein	DSHB, EB165-s
Nuclear markers		
Lamin Dm0	Nuclear envelope	DSHB, ADL67.10
Lamin C	Nuclear envelope	DSHB, LC28.26
Specific muscle nuclear markers		
Even-skipped	DA1	[20]
Vestigial	DA1, DA2, DA3, LL1, VLs1-4	[21]
Others		
DsRed	Red fluorescent protein	Clontech, 632496
GFP	Green fluorescent protein; useful to identify the present of fluorescent balancers (e.g., DGy, CTG)	Clontech, 632381 Invitrogen, A-11120
α-Tubulin	Microtubules	Sigma, T9026
Secondary antibodies	**Expression detected**	**Company, reference #**
Alexa Fluor 488 conjugated		Life Technologies
Alexa Fluor 555 conjugated		Life Technologies
Alexa Fluor 647 conjugated		Life Technologies
Probes antibodies	**Expression detected**	**Company, reference #**
Alexa Fluor 546 conjugated phalloidin	F-actin	Life Technologies, A22283
Hoechst	DNA	Invitrogen, 33342

[a]We refer the reader to several papers for reagents that probe specific proteins involved in myonuclear positioning [2–6, 22]
[b]*See* **Notes 1** and **2**

2.2.2 Embryo Imaging

Mounted embryos are imaged on a Leica SP5 laser-scanning confocal microscope equipped with a 63× 1.4NA HCX PL Apochromat oil objective and LAS AF 2.2 software (for imaging nuclear clusters or individual nuclei [5]) or on a Zeiss AxioPlan2 epifluorescence imaging system with light microscopy with a 40× Plan-NeoFluar 0.75NA objective and AxioVision Rel 4.8 software (for myonuclear movement in the whole embryo; [2]). Equivalent microscopes and software may be used.

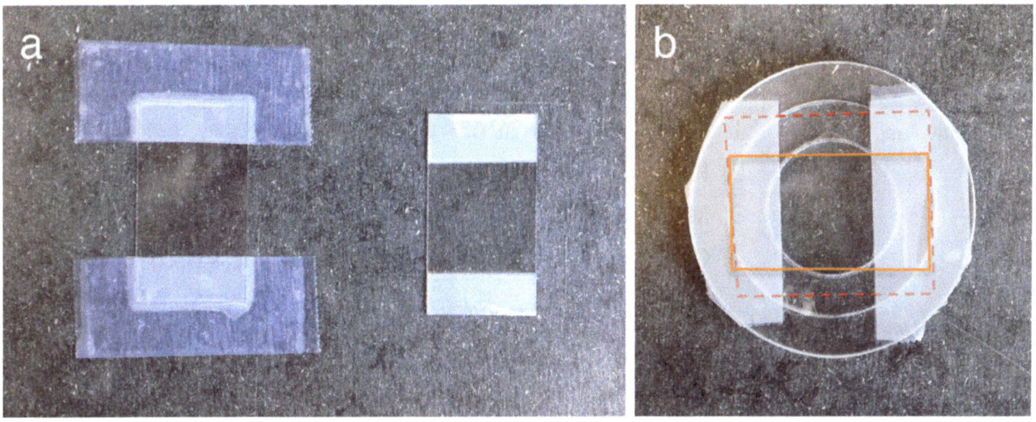

Fig. 5 Materials used for mounting embryos. (**A**) Cover slip with a Scotch tape bridge used to mount fixed embryos for imaging. The image shows a cover slip with two pieces of tape on both ends, before the excess is removed (*left*) and after (*right*). (**B**) Frame used to mount embryos for time-lapse imaging. The image shows an air-permeable Teflon membrane (*red dashed line box*) mounted on a circular Perspex frame. Two pieces of tape provide support for a cover slip bridge where the embryos are placed (*center of the frame*). The orange solid line box represents the cover slip; it is placed on top of both sides of the tape as shown in the image

2.3 Imaging of Formalin-Fixed Larval Muscles

2.3.1 Larvae Collection

1. Laying pot as described in Embryo Collection and Fixation from Imaging of Fixed Embryonic Muscles section (*see* Fig. 4).
2. Lightly yeasted apple juice plates (*see* Subheading 2.1.1, **step 2**).
3. 15 % sucrose solution: sucrose diluted in water.
4. Embryo hook or straight forceps (tip dimensions, 0.1 × 0.06 mm).
5. Dissection microscope equipped with fluorescence.
6. Paintbrush.

2.3.2 Larvae Dissection and Fixation

1. Dissection pins (diameter, 0.1 mm).
2. Two pairs of straight forceps (tip dimensions, 0.1 × 0.06 mm).
3. Dissection plate: petri dish (150 × 15 mm) coated with a thick layer of Sylgard.
4. Vannas spring scissors (sharp tips; non-serrated; cutting edge, 3 mm; tip diameter, 0.05 mm).
5. Cold HL3.1 buffer: 70 mM NaCl, 5 mM KCl, 0.2 mM $CaCl_2$, 20 mM $MgCl_2$, 10 mM $NaHCO_3$, 5 mM trehalose, 115 mM sucrose, and 5 mM HEPES; pH 7.3 [16].
6. 1× PBS.
7. Formalin solution, 10 %.
8. Glass Pasteur pipets.

2.3.3 Larvae Staining and Mounting

As in Subheadings 2.1.2 and 2.1.3, plus:

1. Straight forceps (tip dimensions, 0.1 × 0.06 mm).

2. Kimwipes tissues.
3. Dissection scope.

2.3.4 Heat-Fixed Larvae

1. Water (distilled) at 65 °C.
2. Halocarbon oil or 50 % glycerol.
3. Microscope slides.
4. Cover slips: size 22 × 40 mm with thickness 0.16–0.19 mm.

2.3.5 Larval Imaging

Images are acquired using the 40× 1.25–0.75NA HXC PL Apochromat oil objective of a Leica SP5 laser-scanning confocal microscope and the LAS AF 2.2 software. Equivalent microscopes and software may be used.

2.4 Fly Stocks

The *Drosophila* community has developed a rich collection of genetic reagents to probe various cellular processes, including myonuclear positioning. In addition to classic mutations, there is a wide range of available fly stocks that allow tissue-specific manipulation of a gene of interest. Spatial and temporal manipulation is accomplished using the bipartite GAL4/UAS system [17]. Reagents useful for the study of myonuclear positioning are listed in Table 2. New reagents are constantly being developed, particularly with the development of CRISPR genome-editing approaches in *Drosophila*. We refer the reader to FlyBase (http://flybase.org/), a community resource that provides a comprehensive, searchable compendium of genes, mutations, expression patterns, available stocks, and references.

3 Methods

3.1 Imaging of Fixed Embryonic Muscles

3.1.1 Embryo Collection

1. Set up a laying pot with 60 females/female virgins and 30 males at 22 °C (room temperature), 25 °C, or 29 °C, depending on the experiment (*see* **Note 3**).
2. Label and change the apple juice plates on laying pots, keeping the old ones for embryo collection.
3. Pour enough 50 % bleach to cover embryos on each plate and remove dead flies with a paintbrush. Dechorionate embryos for 4 min at room temperature. During this period, label 1.5 ml Eppendorf tubes and add 500 μl heptane to each.
4. Gently disturb the embryos on the plate with a paintbrush and transfer them to a collection basket using a small funnel. Rinse the plate with water to collect any remaining embryos and transfer them to the collection basket. Rinse the embryos in the collection basket liberally with additional distilled water to remove any traces of bleach.
5. Move the embryos to the Eppendorf tube containing the heptane with the paintbrush.

Table 2
***Drosophila* stocks useful for the study of myonuclear positioning**

Fly stock	Use	Reference
GAL4 lines	Available from stock centers/*Drosophila* labs	
Twist-GAL4	Somatic mesoderm expression from gastrulation to stage 14	[23]
Mef2-GAL4	Somatic mesoderm expression from stage 7 and throughout larval and adult development	[24]
24B-GAL4	Somatic mesoderm expression from stage 7	[17]
MHC-GAL4	Somatic muscle expression—stage 13 throughout larval and adult development	[25]
G7-GAL4	Muscle expression in larvae	[26]
Alpha-GAL4	Muscle expression in larvae	K. Broadie lab, Vanderbilt University
ApterousME-GAL4	LT muscle-specific expression in embryo	[27]
5053-GAL4	VL1 muscle-specific expression in embryo	[28]
Rp298-GAL4	Founder cell-/myotube-specific expression in embryo/pupa	[29]
sns-GAL4	Fusion competent myoblast in embryo/pupa	[30]
UAS lines	Available for gene of interest at *Drosophila* stock centers (i.e., Bloomington or Vienna stock centers) or from individual *Drosophila* labs	
UAS-RNAi	Examples: UAS-DhcRNAi, UAS-SydRNAi	[3, 4]
UAS-fluorescent reporter lines targeted to nuclei or other organelle	Examples: UAS-cytoplasmic GFP, UAS-nuclear localized GFP; UAS-lamin GFP labels nuclear envelope; UAS-EB1-GFP labels microtubule plus ends	[3, 5]
Fluorescent reporter lines		
apterousME::NLSdsRed	Labels all myonuclei in the LT muscles	[2, 27]
Tropomyosin-GFP	GFP gene trap in tropomyosin. Labels all muscles	*Drosophila* FLYTRAP collection
Zasp66-GFP	GFP gene trap in Zasp66. Labels all muscles	*Drosophila* FLYTRAP collection
Gene mutants	Available for gene of interest at stock centers or from *Drosophila* labs: (http://flybase.org/)	

3.1.2 Embryo Fixation

1. Thaw an aliquot of 4 % PFA.
2. Add 500 μl of PFA to the Eppendorf tube containing the heptane and embryos.
3. Using a platform shaker set at 250 rpm, incubate the embryos for 20 min at room temperature.

4. Remove both the PFA and heptane phases, leaving the embryos in the Eppendorf tube.
5. Add 500 μl of fresh heptane and 500 μl of 100 % methanol (for an alternative fixation method to label fixed embryos with phalloidin, *see* **Note 4**).
6. Vortex for 1 min at room temperature. Embryos that have been devitellinized will fall to the bottom of the tube, whereas those that remain at the interface or float are not devitellinized and should be discarded.
7. Remove all liquid.
8. Rinse embryos in methanol three times. To rinse: add solution, invert tube, let settle, remove, and repeat.
9. Add fresh 100 % methanol for storage at −20 °C. Embryos can be stored and used reliably for up to 6 months.

3.1.3 Embryo Staining

1. Transfer embryos to a 0.5 ml Eppendorf tube and remove excess methanol.
2. Rinse three times with PBT-BSA.
3. Add fresh PBT-BSA and incubate for 30 min on Nutator at room temperature. BSA serves to reduce nonspecific binding of antibodies.
4. Add primary antibodies diluted in PBT-BSA (Table 1). As an example, use antibodies directed to tropomyosin or myosin heavy chain to visualize muscles and DsRed to label the nuclei of the lateral transverse (LT) muscles expressing the *apterousME::NLSdsRed* (Figs. 1 and 2). Incubate primary antibodies with the embryos on the Nutator for 1 h at room temperature or overnight at 4 °C.
5. Remove primary antibodies and store at 4 °C (most can be reused up to three times).
6. Rinse embryos with PBT-BSA three times.
7. Wash embryos with PBT-BSA three times for 10 min at room temperature on the Nutator.
8. Remove PBT-BSA and add secondary antibodies (Table 1), diluted in PBT-BSA, to the embryos and incubate on the Nutator for 1 h at room temperature. As fluorescent antibodies are light sensitive, cover them with aluminum foil. From this point forward, keep them covered as much as possible until the staining protocol is finished.
9. Remove and discard secondary antibodies.
10. Rinse embryos with PBT (no BSA) three times.
11. Wash embryos with PBT three times for 10 min on the Nutator at room temperature.

3.1.4 Embryo Mounting

1. While washing the embryos (**step 12** above), prepare slides for mounting. To mount embryos, first make a "tape bridge," comprised of two superimposed pieces of Scotch tape, on two opposite sides of the cover slip (Fig. 5a).
2. Label the slide with date, genotype or cross, and the antibodies used.
3. Place a metal washer on the slide and rest the cover slip on the washer with the tape bridge facing up.
4. Cut 1 cm off of the end of a p200 pipet tip for each sample so that embryos can be pipetted without damage.
5. Remove PBT from the last wash (**step 12** above).
6. Add 80 μl of ProLong Gold antifade reagent to embryos using the truncated p200 tip (ProLong Gold is extremely viscous) and slowly mix them together.
7. Transfer embryos to the cover slip. Remove any visible bubbles.
8. Place a new cover slip over the tape bridge and line up both cover slips to form a sandwich.
9. Store overnight in the dark until the ProLong Gold sets.
10. Remove the washer and tape the cover slip sandwiches to the slide. These slides are ready to be imaged.
11. Slides can be stored at −20 °C for up to 6 months.

3.1.5 Confocal Microscopy

While the data indicate that similar movements and mechanisms are active in other *Drosophila* muscle subsets, our lab has focused on studying myonuclear movements in a subset of muscles, the lateral transverse muscles (LT). Hence, the methodology for confocal imaging described below is focused on this subset.

1. On the confocal microscope, identify an embryo at the correct stage (stages 14–16) positioned with its lateral side facing up using the low-magnification objective. Staging is done based on overall embryo shape, the intensity of DsRed and tropomyosin signal when used, gut morphology, and the morphology of the trachea [2, 3, 18]. At stage 16, a control LT muscle shows two clusters of nuclei (3–4 nuclei in each cluster) near each longitudinal pole of the muscle.
2. After finding a correctly staged embryo for imaging, move to a higher magnification objective, such as the 63× objective, and increase the zoom to 1.5×. On the computer, position the embryo in a horizontal position with four hemisegments visible in the imaging field. Choose hemisegments 3–7 for imaging (*see* **Note 5**).
3. To set the *Z*-stack, begin by selecting the first slice in which the LT muscles are visible, and after moving through the muscle,

stop and select the last slice in which these muscles are no longer visible. The recommended step size is 0.5 μm. The total size of the stack will vary between embryos, but the step size should be kept constant. The gain for each channel should be adjusted according to the intensity of the corresponding signal.

3.1.6 Image Analysis

Multiple software packages, including Volocity Visualization Software 6.1.1 (Improvision), Imaris (Bitplane), and Adobe Photoshop CS4, are used for image processing. ImageJ (NIH) software is used for measurements. With the appropriate plug-in (*bioformats_package*), however, files from Leica or Zeiss confocal microscopes can be opened directly in ImageJ, and all rendering, processing, and analysis can be done within ImageJ. Therefore, the instructions below will refer to ImageJ software:

1. Open your files by dragging them into ImageJ toolbar.
2. For the nuclear positioning analysis, select the extended focus view to visualize whole muscles. Select: Image > Stacks > Z Project and select Standard Deviation as the projection type (*see* **Note 6**). Adjust the number of slices to be included in the final projection.
3. Merge channels by selecting Image > Color and Merge Channels.
4. Brightness and contrast, as well as other settings, can be adjusted by selecting Adjust under the Image menu (Image > Adjust > Brightness/Contrast).
5. For analysis, position the image with the anterior side of the muscle set to the left and dorsal to the top of the viewing field. The image can be rotated by selecting Transform under the Image menu (Image > Transform > Rotate).
6. For measuring nuclear positioning, turn on the segmented line option. To do this, go to the ImageJ toolbar, right-click on the line option (represented as a straight diagonal line), and select Segmented Line.
7. Three different measurements are made for each muscle (Fig. 2a): Dorsal (D)—Distance from the dorsal end of each muscle to the nearest edge of the closest nucleus; Ventral (V)—Distance from the ventral end of each muscle to the nearest edge of the closest nucleus; and Length (L)—Length of each muscle following the curvature of the muscle. The edges of the nuclei and ends of the myofibers are defined by a clear change from signal to background within the relevant viewing channel. Each LT muscle is considered individually in all cases. To register each measurement, hit the M key (a new window will pop up when this is assessed for the first time).
8. For the analysis, a total of three measurements (dorsal, ventral, and length) per muscle, from four muscles per hemisegment in

four hemisegments per embryo, are measured in each of ten embryos per genotype.

9. Copy all measurements to an Excel sheet. To calculate the dorsal and ventral distances, represent each value as a percent of the total muscle length to account for variation in muscle size. This will provide a quantification of the proximity of each cluster to each end of the muscle, which can be compared to control values. To calculate the nuclear spread, use the measurement values in the following formula: $\frac{\text{Length} - \text{Dorsal} - \text{Ventral}}{\text{Length}}$. This quantifies the distance that both clusters have traveled relative to each other and to the muscle length.
10. For statistical analysis, use either Excel or an advanced statistical software, such as SPSS (Statistical Package for the Social Sciences, IBM) or GraphPad Prism. The standard statistical test used is the Student's *t*-test.

3.2 Time-Lapse Imaging of Embryonic Muscles

3.2.1 Embryo Collection and Mounting

1. Follow **steps 1–4** from Subheading 3.1.1.
2. Place the collection basket on a small petri dish and fill it up with water so that the embryos are floating.
3. Dip a wooden stick into halocarbon oil and use the halocarbon oil on the end of the stick to collect the floating embryos from the collection basket. Then transfer them to an air-permeable Teflon membrane mounted on a circular Perspex frame (Fig. 5b). Place two pieces of tape on each side of the embryos to create a bridge that will prevent embryo damage. Place a cover slip on top of both pieces of tape.

3.2.2 Confocal Microscopy

1. For time-lapse imaging of whole embryos, *Z*-stacks are taken with optical sections set at 3 μm on the Zeiss AxioPlan2. To set the stack, the first slice is set when the LT nuclei are in view in multiple hemisegments. The final slice is set below the nuclei. A frame rate of 20 s is used. Only a single Z-section with the greatest number of nuclei in focus is selected for each time point. Images are processed using ImageJ and compiled into a video using QuickTime [2] (*see* **Note** 7).
2. For time-lapse imaging of myonuclear clusters and individual myonuclei, *Z*-stacks of two hemisegments are imaged simultaneously, using a high-magnification objective, such as the 63× 1.4 NA HCX PL Apochromat oil objective (*see* Subheading 2.2.2) and increasing the zoom to 2–2.5×. Set the stack starting slightly above the nuclei and finishing slightly below the nuclei. Set the step size to 0.5 μm and the time interval to be between 20 s and 3 min [5] (*see* **Note** 7).

3. Depending on the types of nuclear movement analyzed, collect data from 10 h (whole embryo) to 60 min (individual nuclear movements). Record a minimum of five time-lapse series per genotype.

3.2.3 Image Analysis

1. Quantification of myonuclear movements and myonuclear morphology changes are completed manually using measurement functions in ImageJ (NIH) as described in Subheading 3.1.6.
2. *Nuclear translocation speed:* The velocity at which myonuclear clusters in the LT muscles separate during stage 15 is assessed by measuring the distance between the dorsal and ventral clusters of nuclei at $t=0$ h and at $t=1$ h and determining the increased distance between the two clusters. By tracking individual nuclei as they moved relative to a fixed position in the embryo over the course of 10 min, the speed of individual myonuclei is determined. One hundred individual myonuclei from five different embryos are assayed for each genotype.
3. *Nuclear rotations:* The number of LT myonuclei that rotate is determined by examining the time-lapse videos. A myonucleus that rotates, stops, and then rotates again is counted as one rotating myonucleus. Two hundred and fifty myonuclei are normally examined.
4. *Nuclear translocation directionality:* The leading edge of a myonucleus is defined as the edge of the myonucleus furthest in the direction of translocation. A myonucleus is judged to have changed direction if it persistently moves a distance of at least one nuclear radius in the opposite direction from its previous direction of translocation. Two hundred and fifty myonuclei are examined.
5. *Nuclear shape changes:* The aspect ratios of moving myonuclei are determined by dividing the length of the dorsal-ventral axis of a myonucleus by the length of the anterior-posterior axis of a myonucleus using ImageJ software (NIH). Aspect ratios are considered to change if a difference of ≥ 0.7 was observed in these measurements. Two hundred and fifty myonuclei are examined with at least ~80 % exhibiting the changes in noted behavior.

3.3 Imaging of Formalin-Fixed Larval Muscles

3.3.1 Larval Collection (Fig. 4)

1. Set up a laying pot with 60 females/female virgins and 30 males at 22 °C (*see* **Note 3**).
2. Label and change the apple juice plates on laying pots, keeping the old ones for embryo collection.
3. Day 0: Under the dissection scope, select stage 16/17 embryos and transfer them to a half-yeasted plate with the help of an embryo hook. Arrange the embryos in a line next to a very thin layer of yeast. This will make it easier to find the 1st instar larvae on the next day.

4. Day 1: Within 24 h, transfer the hatched larvae to a vial with food and store the vial at 25 °C for an additional 3 days or 22 °C for 4 days.
5. On day 4 or day 5 (depending on the temperature), larvae will be 3rd instar larvae. Pour room temperature 15 % sucrose solution in the vial. Gently disturb the food at the bottom of the vial with a spatula. The larvae will float and can be collected using a paintbrush. In the event that some larvae are stuck in the food, release by gentle agitation. Transfer the larvae to an apple juice plate.
6. If larvae expressing a fluorescent protein that was not previously visible in the embryo are desired, select the appropriate larvae under the fluorescence scope before starting dissecting.
7. Prepare separate Eppendorf tubes with PBS for each genotype/cross and keep them in ice.

3.3.2 Larval Dissection and Fixation with Formalin

1. Before starting the dissections, gather at least six dissection pins on the dissection plate.
2. Transfer one larva to the dissection plate and add a drop of cold HL3.1 buffer using a glass pipet. This "relaxing buffer" will slow larval movements.
3. Orient the larva with its dorsal side up (identified by visualization of the main trunks of the trachea) and pin both the anterior and posterior ends of the larva to the plate. Specifically, after securing one end, use the other pin to stretch the larva to its normal length, being careful to not overstretch it, before securing the other end.
4. Using the scissors, make small incisions perpendicular to the length of the larva close to each pin (cutting both trunks of the trachea). Then, cut along the larval body, ideally between the two main trunks of the trachea, to expose the inner organs.
5. With forceps, gently remove the gut, trachea, and neurons, being careful to not damage muscles.
6. Pin each corner of the larval body to the plate, making it flat. Once more, stretch out the tissue, but avoid overstretching. The flatter the larva during preparation, the easier the image acquisition later.
7. Remove the HL3.1 buffer, add a few drops of 10 % formalin, and incubate for 20 min at room temperature.
8. Rinse with cold PBS three times.
9. Add fresh PBS and remove pins.
10. Transfer to an Eppendorf tube with cold PBS.
11. Store at 4 °C for a maximum of 1 week (for optimal staining results). Do not store dissected larvae at −20 °C as this will damage the structure of the muscle.

3.3.3 Larval Staining and Mounting

1. Staining larvae is similar to staining embryos (*see* Subheading 3.1.3), except when using fluorescence-conjugated probes such as phalloidin or Hoechst. If these types of probes are used, there is only one incubation of 1 h at room temperature followed by PBT washes. If antibodies are also used, incubate with primary antibodies first, as in the embryo staining protocol, followed by incubation with appropriate secondary antibodies and the aforementioned probes.
2. After completing the staining, transfer a maximum of six larvae to a microscope slide.
3. Under a dissection scope, position larvae with the head at the top and flatten them by exposing the muscles facing up. Spread out the larvae evenly on the slide. While doing this, do not let larvae dry out. Add more PBT if necessary.
4. Remove any excess PBT by absorbing it with a Kimwipes.
5. Place a drop of ProLong Gold antifade reagent on top of each larva and remove any visible bubbles.
6. Cover with a cover slip and let it dry at room temperature in the dark. Note that there is no bridging of the cover slips for the mounting of larvae.
7. Slides can be stored at −20 °C for up to 6 months.

3.3.4 Confocal Microscopy

1. For formalin-fixed larval muscles, we image the ventral lateral muscles 3 and 4 (VL3 and VL4) because they are easily observed and are more resistant to damage during dissection. For simplicity, it is recommended to focus on VL4, with its single row of myonuclei, rather than VL3, which contains two rows of nuclei.
2. A VL4 muscle should fit in the imaging field on the microscope computer. Set the *Z*-stack, starting a little above the muscle and finishing just prior to the slice when other muscles first become apparent. Set the Z-step size to 0.5 μm. Image six muscles from each of five larvae per genotype.

3.3.5 Image Analysis

As for embryo image analysis (*see* Subheading 3.1.6), different software packages can be used. All the instructions herein will refer to ImageJ software (NIH):

1. Follow **steps 1–4** from Confocal Microscopy in the Imaging of Fixed Embryonic Muscles (*see* Subheading 3.1.6).
2. For assessing myonuclear position in larvae, different types of measurements are made: muscle length, internuclear distance (Fig. 6a, b), nearest neighbor (Fig. 6c), largest gap (Fig. 6d), and myonuclear number.
3. *Muscle length*: Use the straight line option on the ImageJ toolbar and hit the M key to open a new window with the values.

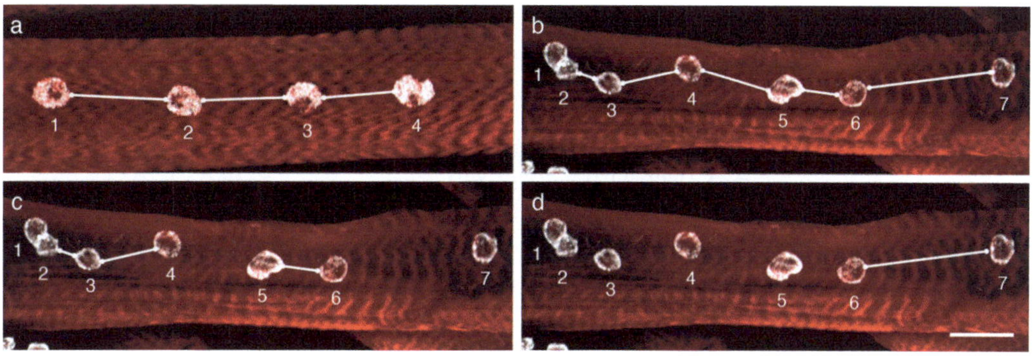

Fig. 6 Myonuclear positioning measurements in larval muscles. In a control VL4 muscle (**A**) and in a VL4 muscle with disrupted myonuclear positioning (**B–D**), white lines represent different measurements made to determine myonuclear positioning defects. (**A**) Internuclear distance. Nuclei are approximately equidistant. (**B**) Internuclear distance. The distance between nuclei is variable. (**C**) Nearest neighbor. Two different scenarios: the nearest neighbor of nucleus 5 is nucleus 6 and vice-versa; the nearest neighbor of nucleus 4 is nucleus 3 but the opposite is not the case. Nucleus 2 is the nearest neighbor to nucleus 3. (**D**) Largest gap. In this muscle, the largest gap between two nuclei is the distance between nuclei 6 and 7. Scale bar, 25 μm

4. *Internuclear distance*: Measure the distance between every nucleus. In control muscles or muscles without nuclear clumping, the resulting standard deviation will be very small compared to the measurements from muscles with nuclear clumping.
5. *Nearest neighbor*: Measure the distance to the nearest edge of the closest nucleus from each nucleus. Assume distance equals 0 when two nuclei are touching. The closer the nuclei, the lower the final average will be.
6. *Largest gap*: Measure the longest distance between two nuclei within a given muscle. For more severe defects in myonuclear positioning, this distance is bigger, whereas in muscles with control-like nuclear distribution, this value is smaller.
7. *Number of nuclei per muscle*: A control VL4 muscle contains 7–9 nuclei.
8. Copy all measurements to an Excel sheet. To calculate each of the parameters mentioned above, always consider the length of the muscle and express values as a percentage of muscle length.
9. For statistical analysis, either use Excel or a more advanced statistical software, such as SPSS (Statistical Package for the Social Sciences, IBM) or GraphPad Prism. The standard statistical test used is the Student's *t*-test.

3.4 Heat-Fixed Larvae

An alternative method of fixing larvae for microscopy is heat fixation (adapted from [19]). For this approach, use larvae at any developmental stage. Since there is no staining with this protocol, larvae expressing fluorescent markers in both the muscles and the nuclei are ideal (*see* Table 2):

1. Pick a larva using forceps. Be careful not to damage muscles.
2. Dip the larva in water at 65 °C for 1–2 s.
3. Mount 5–6 larvae in halocarbon oil or 50 % glycerol in a microscope slide.
4. Cover with a cover slip.

4 Notes

1. For antibodies with high levels of nonspecific background staining, preabsorption using fixed control embryos is recommended. For example, some antibodies listed in Table 1, such as anti-Eve, are preabsorbed to obtain better results [6].
2. Due to the formation of the embryonic cuticle at stage 17, antibodies can only penetrate the embryo from the initial stages of embryogenesis until stage 16. For visualizing myonuclear position as well as muscle structure at stage 17, fluorescently tagged proteins that label the myonuclei and the muscles must be examined (*see* Table 2).
3. Although the ideal number of flies in a laying pot is 60 females/female virgins and 30 males, laying pots can be set up with as few as 25 females/female virgins and 15 males. Using a large number of flies in a laying pot is important when collecting large amounts of embryos.
4. For labeling fixed embryos with phalloidin, an alternative fixation method in which methanol is substituted with ethanol should be used. After fixing in heptane and PFA, add 100 % ethanol, vortex for 45 s, discard the ethanol, add 100 % ethanol, vortex for 45 s, collect embryos to a new tube, and store in fresh 90 % ethanol at −20 °C.
5. When imaging embryos, it is recommended to focus on hemisegments 3–7 because these hemisegments are developmentally more similar in age and are less prone to variations in muscle structure.
6. When rendering a *Z*-stack in ImageJ, a number of projection types are available (average intensity, maximum intensity, minimum intensity, sum of slices, standard deviation, and median). The type selected depends on what is being measured. Maximum intensity is suggested if the user is measuring that particular signal. We use standard deviation as it renders an excellent image of the sarcomeric structure in the muscle; however, we are measuring distances between the nuclei.
7. While acquiring time-lapse images, check every 10 min to adjust focus as embryos will move and the nuclei will shift out of focus. However, note that adjusting this focus means restart-

ing the time-lapse sequence. This cannot be done within a time-lapse. After a few adjustments, the focal plane will essentially stabilize and it can be left for hours to get the desired time-lapse sequence.

Acknowledgments

We thank the members of the Baylies Lab for advice, particularly, Krista Dobi, Jonathan Rosen, and Stefanie Windner for providing images for Fig. 3A (Dobi) and 2A Stages 14/15 and Stage 17 (Rosen and Windner, respectively). We also acknowledge our funding agencies, SFRH/BD/52041/2012 (Azevedo), and MDA and NIH NIAMS RO1-068128 (Baylies).

References

1. Romero NB (2010) Centronuclear myopathies: a widening concept. Neuromuscul Disord 20(4):223–228
2. Metzger T et al (2012) MAP and kinesin-dependent nuclear positioning is required for skeletal muscle function. Nature 484(7392): 120–124
3. Folker ES, Schulman VK, Baylies MK (2012) Muscle length and myonuclear position are independently regulated by distinct Dynein pathways. Development 139(20):3827–3837
4. Schulman VK et al (2014) Syd/JIP3 and JNK signaling are required for myonuclear positioning and muscle function. PLoS Genet 10(12):e1004880
5. Folker ES, Schulman VK, Baylies MK (2014) Translocating myonuclei have distinct leading and lagging edges that require kinesin and dynein. Development 141(2):355–366
6. Schulman VK, Folker ES, Baylies MK (2013) A method for reversible drug delivery to internal tissues of Drosophila embryos. Fly (Austin) 7(3):193–203
7. Dobi KC, Schulman VK, Baylies MK (2015) Specification of the somatic musculature in Drosophila. Wiley Interdiscip Rev Dev Biol 4(4):357–375
8. Schulman VK, Dobi KC, Baylies MK (2015) Morphogenesis of the somatic musculature in Drosophila melanogaster. Wiley Interdiscip Rev Dev Biol 4(4):313–334
9. Bourgouin C, Lundgren SE, Thomas JB (1992) Apterous is a Drosophila LIM domain gene required for the development of a subset of embryonic muscles. Neuron 9:549–561
10. Fujioka M et al (2005) Embryonic even skipped-dependent muscle and heart cell fates are required for normal adult activity, heart function, and lifespan. Circ Res 97(11):1108–1114
11. Knirr S, Frasch M (2001) Molecular integration of inductive and mesoderm-intrinsic inputs governs even-skipped enhancer activity in a subset of pericardial and dorsal muscle progenitors. Dev Biol 238(1):13–26
12. Landgraf M, Thor S (2006) Development of Drosophila motoneurons: specification and morphology. Semin Cell Dev Biol 17(1):3–11
13. Demontis F, Perrimon N (2009) Integration of Insulin receptor/Foxo signaling and dMyc activity during muscle growth regulates body size in Drosophila. Development 136(6): 983–993
14. Kaltschmidt JA et al (2000) Rotation and asymmetry of the mitotic spindle direct asymmetric cell division in the developing central nervous system. Nat Cell Biol 2(1):7–12
15. Haseloff J, Dormand EL, Brand AH (1999) Live imaging with green fluorescent protein. Methods Mol Biol 122:241–259
16. Brent JR, Werner KM, McCabe BD (2009) Drosophila larval NMJ dissection. J Vis Exp (24), e1107
17. Brand AH, Perrimon N (1993) Targeted gene expression as a means of altering cell fates and generating dominant phenotypes. Development 118(2):401–415
18. Beckett K, Baylies MK (2007) 3D analysis of founder cell and fusion competent myoblast arrangements outlines a new model of myoblast fusion. Dev Biol 309(1):113–125

19. Schnorrer F et al (2010) Systematic genetic analysis of muscle morphogenesis and function in Drosophila. Nature 464(7286):287–291
20. Frasch M et al (1987) Characterization and localization of the even-skipped protein of Drosophila. EMBO J 6(3):749–759
21. Williams JA, Bell JB, Carroll SB (1991) Control of Drosophila wing and haltere development by the nuclear vestigial gene product. Genes Dev 5(12B):2481–2495
22. Elhanany-Tamir H et al (2012) Organelle positioning in muscles requires cooperation between two KASH proteins and microtubules. J Cell Biol 198(5):833–846
23. Baylies MK, Bate M (1996) Twist: a myogenic switch in Drosophila. Science 272(5267): 1481–1484
24. Ranganayakulu G et al (1998) Divergent roles for NK-2 class homeobox genes in cardiogenesis in flies and mice. Development 125(16):3037–3048
25. Chen EH, Olson EN (2001) Antisocial, an intracellular adaptor protein, is required for myoblast fusion in Drosophila. Dev Cell 1(5):705–715
26. Zhang YQ et al (2001) Drosophila fragile X-related gene regulates the MAP1B homolog Futsch to control synaptic structure and function. Cell 107(5):591–603
27. Richardson BE et al (2007) SCAR/WAVE and Arp2/3 are crucial for cytoskeletal remodeling at the site of myoblast fusion. Development 134(24):4357–4367
28. Ritzenthaler S, Suzuki E, Chiba A (2000) Postsynaptic filopodia in muscle cells interact with innervating motoneuron axons. Nat Neurosci 3(10):1012–1017
29. Dutta D et al (2002) Real-time imaging of morphogenetic movements in Drosophila using Gal4-UAS-driven expression of GFP fused to the actin-binding domain of moesin. Genesis 34(1-2):146–151
30. Kocherlakota KS et al (2008) Analysis of the cell adhesion molecule sticks-and-stones reveals multiple redundant functional domains, protein-interaction motifs and phosphorylated tyrosines that direct myoblast fusion in Drosophila melanogaster. Genetics 178(3): 1371–1383

Part IV

Nuclear Envelope-Chromatin Interactions

Chapter 20

Mapping Nuclear Lamin-Genome Interactions by Chromatin Immunoprecipitation of Nuclear Lamins

Anja R. Oldenburg and Philippe Collas

Abstract

The nuclear lamina is a meshwork of A- and B-type lamins which interact with chromatin and regulate many nuclear functions. Recent studies have reported the discovery of chromatin domains interacting with nuclear lamins by chromatin immunoprecipitation (ChIP) of lamin A/C or B1 and identification of the associated DNA sequences by microarray or high-throughput sequencing. ChIP has been used over many years to get a snapshot of interactions between DNA and proteins in cells, including modified histones, transcription factors, chromatin remodelers, and recently, structural proteins such as nuclear lamins. We describe here the procedure we have established in our laboratory for ChIP of lamin A/C and lamin B1 from human cultured cells. The protocol is compatible with polymerase chain reaction and high-throughput sequencing analysis of the co-immunoprecipitated DNA.

Key words Chromatin, Chromatin immunoprecipitation, Nuclear lamina, Lamin A/C, Lamin B, Antibody, Sonication

1 Introduction

In eukaryotic cells, the nuclear envelope separates the nuclear genome from the cytoplasm. The nuclear envelope consists of an outer and inner nuclear membrane perforated by nuclear pore complexes and, interfacing the inner membrane and chromatin, the nuclear lamina, a meshwork of type V intermediate filaments called lamins [1]. Nuclear lamins are implicated in the regulation of nuclear functions such as DNA replication, transcription, and chromatin organization. Lamins consist of A-type lamins (lamins A and C, commonly designated lamin A/C, which are splice variants of the *LMNA* gene) and of B-type lamins (lamins B1 and B2, products of the *LMNB1* and *LMNB2* genes). While B-type lamins are expressed in essentially all cell types, A-type lamins are not expressed in preimplantation embryos, undifferentiated embryonic stem cells, or induced pluripotent stem cells. Variations in B-type lamin level and distribution have been linked to aging and senescence [2–4].

Sue Shackleton et al. (eds.), *The Nuclear Envelope: Methods and Protocols*, Methods in Molecular Biology, vol. 1411, DOI 10.1007/978-1-4939-3530-7_20, © Springer Science+Business Media New York 2016

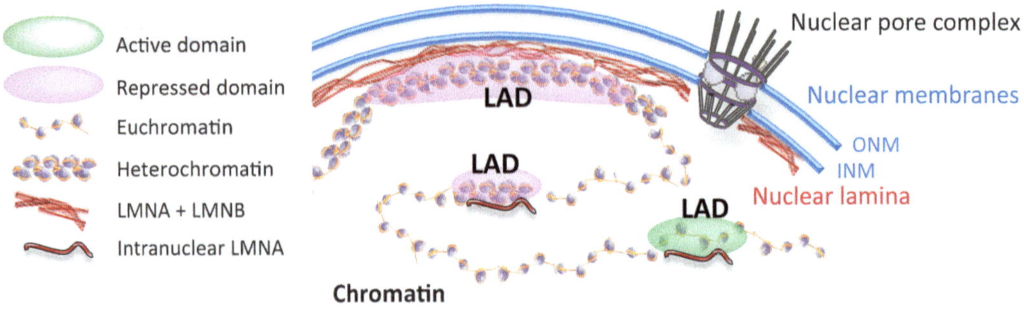

Fig. 1 Nuclear lamins interact with chromatin. At the nuclear periphery, A- and B-type lamins associate with chromatin domains (LADs) that are largely heterochromatic. In the nuclear interior, A-type lamins are also able to interact with chromatin. LADs may contain both inactive and active genic regions

Both A- and B-type lamins have been shown to interact with chromatin (Fig. 1) through so-called lamina-associated domains (LADs), of typically 0.1–10 megabases (Mb) in length [5–7]. LADs have initially been identified using DamID (DNA adenine methyltransferase [Dam] identification), a proximity assay based on 6-methyl-adenosine tagging of DNA sequences interacting with lamins, and identification of these sequences by hybridization to microarrays or by high-throughput sequencing [5–7]. LADs are generally gene poor and heterochromatic, and genes within LADs tend to be silent [8].

LADs have also been identified by chromatin immunoprecipitation (ChIP) of lamin A/C, coupled to array hybridization [9] or high-throughput sequencing (ChIP–seq) [10, 11], or by ChIP–seq analysis of lamin B1 [3, 4, 11]. The overall width of LADs reflects overall broad genome coverage by A- or B-type lamins (~30 % of the genome). Lamin ChIP–seq signals reflect overall low-level occupancy [3, 7, 10]. To enable a reliable detection of genomic regions associated with nuclear lamins, we have disclosed an algorithm, enriched domain detector (EDD), tailored for the detection of broad peaks of enrichment [10]. An advantage of EDD over other existing broad peak calling algorithms is its sensitivity to the width of enriched domains rather than to strength of enrichment at a given site and its robustness against local variations in enrichment levels.

Recent work from our laboratory suggests that LADs are not necessarily restricted to the nuclear periphery. A-type lamins are not only found at the nuclear periphery but also exist as a nucleoplasmic pool [12, 13] which is also susceptible of interacting with chromatin [11]. Accordingly, we define here "LADs" as *lamin*-associated domains, in order not to necessarily imply a localization at the nuclear periphery as implied in the original definition of LAD [11]. Recent findings show that LADs harbor distinct properties, including heterochromatic regions and euchromatic nuclease-accessible domains containing active genes and regulatory elements [11].

ChIP has arguably been adopted as the method of choice to obtain a snapshot of protein–DNA interactions in cells. ChIP has been used for mapping the genomic localization of histones, transcription factors, chromatin remodelers, and structural proteins such as nuclear lamins. In the ChIP assay [14], chromatin is fractionated into ~200–500 base pair (bp) fragments, by sonication in buffer containing an ionic detergent, or by enzymatic digestion with micrococcal nuclease (MNase), which digests the linker DNA between nucleosomes. Chromatin fragmentation by sonication is a priori less selective for "open" (MNase accessible) regions and has been until recently [11] exclusively used for the mapping of LADs. A lamin ChIP protocol based on chromatin fragmented by partial digestion with MNase is reported by Duband-Goulet in this volume. We describe here a protocol for ChIP analysis of lamin A/C and lamin B1 from chromatin prepared by sonication as it is carried out in our laboratory.

2 Materials

2.1 Laboratory Equipment

1. Magnetic racks suited for 1.5- and 15-mL tubes.
2. Filtered pipette tips (10, 200 and 1000 μL).
3. 1.5-mL centrifuge tubes.
4. 15-mL conical tubes.
5. Bioruptor® bath sonicator (Diagenode) or any other bath sonicator suitable for ChIP.
6. Rotator (rotating wheel) with adjustable speed, placed in a cold room.
7. Roller platform.
8. Table top centrifuge with a swing-out rotor for 1.5-mL tubes.
9. Centrifuge with a swing-out rotor suited for 15-mL conical tubes.
10. Mini centrifuge.
11. Vortex.
12. Thermomixer.
13. Heating block.
14. Agarose gel electrophoresis apparatus.
15. Qubit spectrophotometer.
16. Thermal cycler with real-time capacity.

2.2 Reagents

1. Ice bath.
2. MilliQ deionized water.
3. 36.5 % Formaldehyde solution.

4. Dynabeads® Protein A or G. Beads should be well suspended before pipetting.
5. Primary antibody for ChIP (*see* **Note 1**).
6. 5 M NaCl.
7. 400 mM EGTA.
8. 500 mM EDTA.
9. 1 M Tris–HCl, pH 7.5.
10. 1 M Tris–HCl, pH 8.0.
11. Glycine: 2.5 M stock solution in PBS.
12. Acrylamide carrier.
13. Proteinase K: 20 mg/mL solution in MilliQ water.
14. SDS: 10 % solution in MilliQ water.
15. Protease inhibitor mix (100×).
16. PMSF: 100 mM stock (100×) solution in 100 % ethanol.
17. Phosphate-buffered saline (PBS).
18. Phenol–chloroform–isoamyl alcohol (25:24:1).
19. Chloroform–isoamyl alcohol (24:1).
20. 500 μg/mL RNAse A.
21. 3 M NaAc (pH 5.2).
22. IQ SYBR® Green (Bio-Rad).

2.3 Buffers

1. RIPA "zero-SDS" buffer: 10 mM Tris–HCl, pH 8.0, 140 mM NaCl, 1 mM EDTA, 0.5 mM EGTA, 0.1 % (wt/vol) sodium deoxycholate, 1 % (vol/vol) Triton X-100, protease inhibitor mix (1:100 dilution from stock), 1 mM PMSF (1:100 dilution from stock). Add protease inhibitor mix and PMSF immediately before use.
2. RIPA lysis buffer: to a RIPA zero-SDS solution, add SDS (from the 10 % stock) to 1 % final concentration immediately before use.
3. RIPA buffer: to a RIPA zero-SDS solution, add SDS (from a 10 % stock) to 0.1 % final concentration immediately before use.
4. Elution buffer: 20 mM Tris–HCl, pH 7.5, 50 mM NaCl, 5 mM EDTA, 1 % SDS. Add SDS (from the 10 % stock) immediately before use.

3 Methods

3.1 Coupling of Antibody to Magnetic Dynabeads

This procedure consists in preparing a slurry of Dynabeads® Protein A (*see* **Note 2**) and coupling ChIP antibodies to the Dynabeads.

1. Vortex the bottle of Dynabeads stock for 1 min before taking out beads. It is important that the beads are well suspended before use.

2. Calculate the amount of beads needed based on the required number of ChIPs. For one lamin ChIP in one 15-mL tube, use 42 μL of suspended beads from **step 1**.
3. Add 1 mL RIPA buffer and vortex. Pipette until solution is homogeneous.
4. Incubate on magnet for 1 min and remove the buffer. To ensure beads trapped in the lid are captured, snap spin the tube in a centrifuge before placing the tube on the magnet.
5. Repeat **steps 3** and **4**.
6. Resuspend the beads in 2 mL RIPA buffer.
7. To each 15-mL tube, add 10 μg lamin A/C or lamin B1 ChIP antibody. This amount is for the antibodies we are using but may vary for different antibodies.
8. Place tubes for 3 h to overnight on a rotator at 25 rpm in a cold room (*see* **Note 2**).

3.2 Cross-Linking

1. Harvest cells according to your lab protocol for the cell type of interest.
2. Count cells and sediment them at 300 × *g* for 7 min at room temperature.
3. Resuspend up to 10^7 cells in 2.5 mL PBS. Transfer cells into a 15-mL tube.
4. Add 70 μL formaldehyde from the 36.5 % stock, mix, and incubate for 10 min at room temperature on a roller.
5. Add 135 μL of the 2.5 M glycine stock (0.13 M final concentration) and incubate 5 min at room temperature on a roller.
6. Centrifuge at 470 × *g* for 8 min at 4 °C using a swing-out rotor. Discard the formaldehyde waste appropriately in a fume hood.
7. Resuspend cells in 2 mL ice-cold PBS.
8. Centrifuge at 470 × *g* for 10 min at 4 °C and discard the supernatant.
9. Repeat **steps** 7 and **8** once. Pellets of cross-linked cells can be snap frozen in liquid nitrogen and stored at −80 °C for 1 month without significant loss of quality.

3.3 Cell Lysis, Chromatin Fragmentation, and Chromatin Dilution

1. Resuspend the cell pellet from Subheading 3.2, **step 9**, in 300 μL RIPA lysis buffer on ice.
2. Vortex and allow lysis to occur for 10 min on ice. Pipette the sample if some occurs.
3. Place the tube into a Bioruptor® sonicator according to the manufacturer's instructions and fragment chromatin for 4 × 10 min using a 30 s on/off program at high power (*see* **Note 3**) Snap spin the tube in the sonication breaks.

4. Centrifuge at 10,000×*g* for 10 min at 4 °C in a swing-out rotor.
5. Transfer the supernatant ("chromatin"; ~300 μL) into one clean 15-mL tube; take care of avoiding the top layer and the bottom fraction (debris). Chromatin must be kept on ice at this stage or snap frozen if it is to be used later.
6. Take out 10 μL of chromatin, and place it into a clean 1.5-mL tube for fragment size analysis.
7. Dilute the rest of the chromatin sample from **step 5** (~290 μL) 1:10 in RIPA zero-SDS buffer on ice; this should yield ~2.9 mL of diluted chromatin. Chromatin dilution in RIPA zero-SDS buffer lowers the SDS concentration to 0.1 %, which is tolerable for most ChIP antibodies.
8. Take out 100 μL of diluted chromatin from **step 7**, and place it into a clean 1.5-mL tube marked *Input chromatin* and place at 4 °C for up to 24 h. Freeze at −20 °C if stored longer. This sample will be used later for cross-link reversal and DNA purification together with the ChIP DNA sample.

3.4 Analysis of Sonicated Chromatin Fragment Size

We determine chromatin fragment length by agarose gel electrophoresis before using chromatin for ChIP. We aim for a size range of ~200–400 base pairs.

1. Add 1 μL RNase A (from the 500 μg/mL stock) to the 10 μL chromatin sample from **step 6**, Subheading 3.3, and incubate at 37 °C for 20 min.
2. Add 190 μL elution buffer.
3. Add 5 μL Proteinase K from a tenfold dilution of the 20 mg/mL stock, and incubate at 68 °C, 500 rpm for 1 h on a Thermomixer.
4. Purify the DNA (*see* Subheading 3.6, starting from **step 2**).
5. Dissolve the DNA in 10 μL MilliQ H_2O.
6. Resolve DNA in a 1.5 % agarose gel with suitable size markers to determine fragment size (*see* **Note 3**).

3.5 Immunoprecipitation and Washes

1. Snap spin the tube(s) containing the antibody-Dynabead® complexes from **step 8**, Subheading 3.1, in a centrifuge to bring down any solution trapped in the lid.
2. Place tube in a chilled magnetic rack (on ice) for 1 min.
3. Remove and discard the supernatant while tube is on magnet.
4. To this tube, add the entire diluted chromatin preparation from **step 8**, Subheading 3.3.
5. Incubate overnight at 25 rpm on a rotator in a cold room.
6. Snap spin tubes in a centrifuge to bring down any solution trapped in the lid.

7. Place tubes in a chilled magnet on ice for 2 min.
8. Remove and discard the supernatant while tubes are on magnet, being careful to avoid the Dynabeads®.
9. Add 2 mL of RIPA buffer, mix with pipette, and incubate 4 min at 25 rpm on a rotator in a cold room.
10. Repeat **steps 6–9** once.
11. Snap spin tubes in a centrifuge to bring down any solution trapped in the lid.
12. Place tubes in a chilled magnet on ice for 2 min.
13. Remove and discard the supernatant while tubes are on magnet, being careful to avoid the Dynabeads®.
14. Add 1 mL of RIPA buffer, mix with pipette, transfer to a clean 1.5-mL tube, and incubate 4 min at 25 rpm on a rotator in a cold room.
15. Repeat **steps 6–8** and put tubes on ice.

3.6 RNase Treatment, Cross-Link Reversal, and DNA Elution

1. To the ChIP samples from **step 15**, Subheading 3.5, add 260 μL elution buffer and 10 μL RNase A (*see* **Note 4**).
2. To the "Input" chromatin sample (100 μL from **step 8**, Subheading 3.3), add 190 μL elution buffer and 10 μL RNase A.
3. Incubate ChIP and "Input" samples at 37 °C for 20 min at 1300 rpm on a Thermomixer.
4. Add 2 μL Proteinase K (from the 20 mg/mL stock) to each tube, mix, and incubate for another 20 min at 37 °C at 1300 rpm on a Thermomixer.
5. Set the Thermomixer temperature to 68 °C, and incubate for 6 h to overnight at 1300 rpm on a Thermomixer.
6. Snap spin tubes, place tubes on magnet for 1 min, and transfer the ChIP and "Input" eluates to separate clean 1.5-mL tubes. At this stage, both ChIP and "Input" samples are ready for DNA purification.

3.7 DNA Purification

1. Add 200 μL elution buffer to both ChIP and "Input" samples; total volume is now 500 μL.
2. Add 500 μL (i.e., 1× volume) phenol–chloroform–isoamyl alcohol, premixed at 24:24:1 volume ratios, mix by manually inverting the tubes several times, and centrifuge at 15,000 × *g* for 5 min at room temperature.
3. Carefully collect 450 μL of the upper phase, and transfer into a clean 1.5-mL tube.
4. Add 450 μL chloroform–isoamyl alcohol, premixed at a 24:1 volume ratio, manually invert the tubes, and centrifuge at 15,000 × *g* for 5 min.

5. Carefully collect 400 μL of the upper phase, and transfer into a clean 1.5-mL tube.
6. Add 40 μL (i.e., 0.1× volume) 3 M Na-acetate, 10 μL acrylamide carrier (0.25 % final concentration), and 1 mL (i.e., 2.5× volume) ice-cold 96 % ethanol.
7. Put samples at −80 °C for 2 h to overnight to precipitate the DNA.
8. Centrifuge at 20,000 ×*g* for 15 min at 4 °C.
9. Air-dry the DNA for 15 min at room temperature by leaving the tube open.
10. Dissolve the DNA in 20 μL MQ H_2O (*see* **Note 5**). Leave the tube at room temperature to for several hours to overnight to enable complete solubilization of the DNA.
11. Freeze the DNA at −20 °C. DNA can be stored frozen for months.

3.8 Real-Time Polymerase Chain Reaction (PCR) Setup

1. Prepare a master mix for real-time PCR analysis, and aliquot the following, for individual 25 μL qPCR reactions and for all ChIP and "Input" samples:
   (a) MilliQ water: 3.25 μL.
   (b) SYBR Green Master Mix (2× stock): 6.25 μL.
   (c) Forward primer (20 μM stock): 0.25 μL.
   (d) Reverse primer (20 μM stock): 0.25 μL.
   (e) DNA template: 2.5 μL.
2. Prepare a standard curve with purified genomic DNA, using a wide range of DNA concentrations (we recommend 0.005–20 ng/μL) to cover the range of your ChIP DNA samples. Use 5 μL DNA as template in each PCR. Establish one standard curve for each primer pair and for each real-time PCR plate.
3. Set up a 40-cycle real-time PCR program.
4. Acquire the data using your real-time PCR data acquisition program.
5. Calculate the amount of DNA in each sample using the standard curve.
6. Export the data into Excel spreadsheets.
7. Determine the amount of precipitated DNA relative to input as ([Amount of ChIP DNA]/[Amount of input DNA]) × 100. We analyze at least three independent ChIPs, each in duplicate qPCRs, and express the data as percent (±SD) of input (Fig. 2).

3.9 Setup for ChiP–Seq

For ChIP–seq analysis, a sequencing library should be prepared according to your laboratory or sequencing facility protocol. We sequence our samples after Illumina ChIP–seq library preparation starting from 12 ng ChIP DNA.

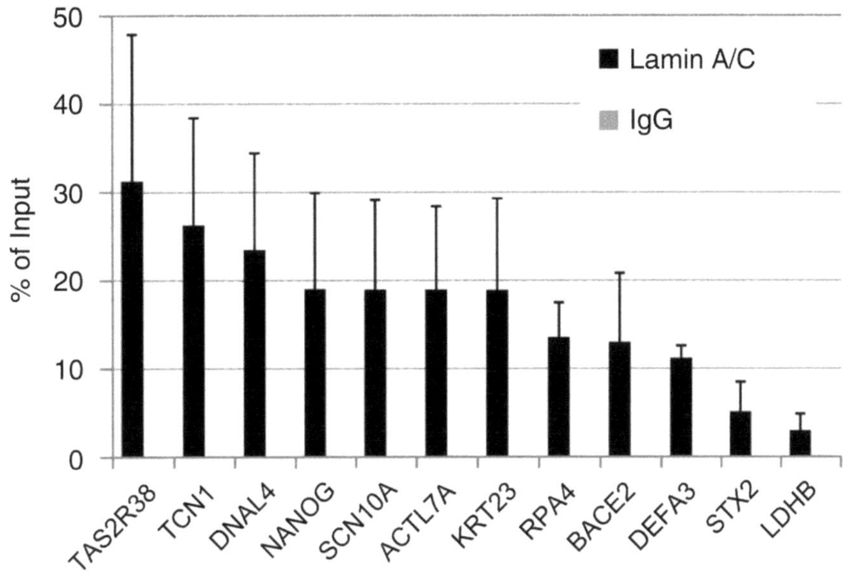

Fig. 2 ChIP-qPCR representation of association of lamin A/C with designated genomic sites in cultured primary human skin fibroblasts. Data show the mean ± SD percent precipitated DNA relative to "Input," from duplicate independent ChIP experiments. Data for an unspecific mouse IgG used for ChIP are also shown

4 Notes

1. We have used the following antibodies: anti-lamin A/C (Santa Cruz sc-7292x; mouse monoclonal); anti-lamin B1 antibody (Abcam ab16048; rabbit polyclonal). For negative control ChIPs, we use unspecific mouse- or rabbit-purified IgGs according to the species of the lamin antibody. Other antibodies may be used but should be tested for their ability to immunoprecipitate the target proteins and associated DNA under ChIP conditions.
2. Dynabeads® Protein A and G are used depending on origin and isotype of the primary antibody. The antibody-bead incubation step should be carried out during the cross-linking (*see* Subheading 3.2), cell lysis, and chromatin preparation (*see* Subheading 3.3) steps. If necessary, it can be prolonged until all chromatin samples are ready for ChIP.
3. Sonication should produce chromatin fragments of ~200–400 base pairs. The sonication regime indicated is suitable for several types of cells but should be optimized for each cell type for the immunoprecipitation is carried out. Make sure that the sonication bath is cooled.
4. For PCR analysis of the ChIP DNA, RNAse treatment is not necessary. It is however required for analysis by high-throughput sequencing.

5. For qPCR analysis, this can be adjusted according to the number of target regions examined. If whole-genome amplification is necessary before qPCR (for instance, to examine a large number of loci), dissolve in 10 μL MilliQ H_2O and use everything in the amplification reaction. For high-throughput sequencing, measure DNA concentration with a Qubit spectrophotometer, and adjust as required by your sequencing facility.

Acknowledgments

Our work is supported by the Norwegian Research Council, the University of Oslo and the Norwegian Center for Stem Cell Research (P.C.), and by South East Health Norway (A.R.O.).

References

1. Burke B, Stewart CL (2013) The nuclear lamins: flexibility in function. Nat Rev Mol Cell Biol 14:13–24
2. Li Z, Zhu Y, Zhai Y, Castroagudin R, Bao Y, White TE, Glavy JS (2013) Werner complex deficiency in cells disrupts the nuclear pore complex and the distribution of lamin B1. Biochim Biophys Acta 1833:3338–3345
3. Sadaie M, Salama R, Carroll T, Tomimatsu K, Chandra T, Young AR, Narita M, Perez-Mancera PA, Bennett DC, Chong H, Kimura H, Narita M (2013) Redistribution of the Lamin B1 genomic binding profile affects rearrangement of heterochromatic domains and SAHF formation during senescence. Genes Dev 27:1800–1808
4. Shah PP, Donahue G, Otte GL, Capell BC, Nelson DM, Cao K, Aggarwala V, Cruickshanks HA, Rai TS, McBryan T, Gregory BD, Adams PD, Berger SL (2013) Lamin B1 depletion in senescent cells triggers large-scale changes in gene expression and the chromatin landscape. Genes Dev 27:1787–1799
5. Guelen L, Pagie L, Brasset E, Meuleman W, Faza MB, Talhout W, Eussen BH, de Klein A, Wessels L, de Laat W, van Steensel B (2008) Domain organization of human chromosomes revealed by mapping of nuclear lamina interactions. Nature 453:948–951
6. Kind J, Pagie L, Ortabozkoyun H, Boyle S, de Vries SS, Janssen H, Amendola M, Nolen LD, Bickmore WA, van Steensel B (2013) Single-cell dynamics of genome-nuclear lamina interactions. Cell 153:178–192
7. Meuleman W, Peric-Hupkes D, Kind J, Beaudry JB, Pagie L, Kellis M, Reinders M, Wessels L, van Steensel B (2013) Constitutive nuclear lamina-genome interactions are highly conserved and associated with A/T-rich sequence. Genome Res 23:270–280
8. Collas P, Lund EG, Oldenburg AR (2014) Closing the (nuclear) envelope on the genome: how nuclear lamins interact with promoters and modulate gene expression. BioEssays 36:75–83
9. Lund EG, Oldenburg A, Delbarre E, Freberg C, Duband-Goulet I, Eskeland R, Buendia B, Collas P (2013) Lamin A/C-promoter interactions specify chromatin state-dependent transcription outcomes. Genome Res 23:1580–1589
10. Lund EG, Oldenburg A, Collas P (2014) Enriched domain detector: a program for detection of wide genomic enrichment domains robust against local variations. Nucleic Acids Res 42:e92
11. Lund EG, Duband-Goulet I, Oldenburg A, Buendia B, Collas P (2015) Distinct features of lamin A-interacting chromatin domains mapped by ChIP-sequencing from sonicated or micrococcal nuclease-digested chromatin. Nucleus 6:30–38
12. Dechat T, Gesson K, Foisner R (2010) Lamina-independent lamins in the nuclear interior serve important functions. Cold Spring Harb Symp Quant Biol 75:533–543
13. Kolb T, Maass K, Hergt M, Aebi U, Herrmann H (2011) Lamin A and lamin C form homodimers and coexist in higher complex forms both in the nucleoplasmic fraction and in the lamina of cultured human cells. Nucleus 2:425–433
14. Collas P (2009) The state-of-the-art of chromatin immunoprecipitation. Methods Mol Biol 567:1–25

Chapter 21

Lamin ChIP from Chromatin Prepared by Micrococcal Nuclease Digestion

Isabelle Duband-Goulet

Abstract

It is now clearly demonstrated that nuclear lamins interact with the genomic DNA and largely contribute to its three-dimensional organization and transcriptional regulation. Emergence of genome-wide mapping techniques such as DamID technology or chromatin immunoprecipitation (ChIP) followed by array hybridization or high-throughput sequencing has allowed the mapping of large lamin-interacting genomic areas called lamina-associated domains. These cover up to 40 % of the genome and are preferentially located in transcriptionally silent heterochromatin at the nuclear periphery. We recently showed that the use of enzymatic rather than physical fragmentation of chromatin in ChIP experiments uncovers new chromatin compartments with features of euchromatin that interacts with A-type lamins. We describe here a detailed ChIP procedure to covalently cross-link protein–DNA, fragment the chromatin fibers by micrococcal nuclease digestion, and solubilize the lamin network with a short sonication pulse prior to immunoprecipitating the lamin-DNA complexes using specific antibodies. Enriched DNA fragments from the lamin-binding sites are then purified as suitable samples for qPCR analysis or high-throughput sequencing.

Key words Chromatin, Micrococcal nuclease digestion, ChIP, Lamins A/C, Lamin B1, Lamina-associated domain

1 Introduction

In eukaryotic cells, the DNA undergoes several levels of compaction. At the first level, DNA is wrapped around an octamer of core histone proteins to form the nucleosome core particle. Nucleosome core particles are connected to each other by short stretches of linker DNA and form nucleosome arrays that are further compacted in higher order structures [1–3]. These structures have long been described as either compact areas of heterochromatin generally considered as the transcriptionally silent part of the genome or less dense areas of euchromatin considered to be transcriptionally active. In the last decade, genome-wide mapping techniques including DamID and chromatin immunoprecipitation (ChIP) have provided new insights into chromosomal organization, giving rise to new

Sue Shackleton et al. (eds.), *The Nuclear Envelope: Methods and Protocols*, Methods in Molecular Biology, vol. 1411, DOI 10.1007/978-1-4939-3530-7_21, © Springer Science+Business Media New York 2016

emerging principles. Not only intra- and interchromosomal contacts but also interactions of specific loci with nuclear scaffolds such as lamins or other transmembrane proteins of the nuclear envelope contribute to the global folding of chromatin [4].

Lamins are nuclear intermediate A- and B-type filaments that associate to form a meshwork called the nuclear lamina, localized between chromatin and the inner nuclear membrane. A- and B-type lamins differ in their expression in that B-type lamins are ubiquitous, whereas A-type lamins are developmentally regulated [5, 6]. A- and B-type lamins also differ in their posttranslational processing as lamin A rapidly loses its C-terminal farnesylation after synthesis while B-type lamins remain farnesylated [7], conferring different anchoring properties to the nuclear envelope to A- and B-type lamins. This is illustrated by the fact that A-type lamins can also be localized outside the nuclear lamina, in the nuclear interior, as opposed to B-type lamins which are always tethered to the nuclear periphery [8].

Nuclear lamin polymers act as anchoring platform where interactions between chromatin and A- and B-type lamins occur through large genomic areas called lamina-associated domains (LADs), which are 0.1–1 megabases long and cover up to 40 % of the genome [9–12]. LADs are more likely positioned in transcriptionally silent heterochromatin, with poor gene content, and enriched at the nuclear periphery. However, A-type lamins have recently been reported to interact with subregions of promoters with various transcriptional features, and these interactions are not necessarily positioned at the nuclear periphery [13], suggesting that LADs may be distributed all over the nucleus (consistent with nucleoplasmic localization of A-type lamins) and associated with loci of various transcriptional status. By comparing enzymatic and physical fragmentation of the chromatin fibers in ChIP protocols, we were recently able to reveal that different chromatin compartments with distinct gene contents, histone modification patterns and histone variants enriched in either hetero- or euchromatin can interact with A-type lamins with little overlap with domains interacting with B-type lamins [14]. This new class of LADs reflects the capacity of lamins to accommodate different chromatin environments.

Micrococcal nuclease (MNase) has both endonuclease and exonuclease activity that results in the digestion of the linker DNA with particular DNA sequence preferences for alternating dA and dT [15]. Thus, a mild MNase digestion of chromatin gives rise to discrete DNA fragments forming multimers of a constant length of approximately 200 bp resulting in a characteristic "nucleosomal ladder" pattern in most somatic tissues [16, 17]. However, longer chromatin digestion gives rise to shorter DNA length (145–147 bp) corresponding to the size of the DNA wrapped in the nucleosome core particle. Eventually, DNA exposed on the outer surface of the core particle is digested and the particle destroyed [18, 19].

Thus, MNase, in the early steps of digestion, requires a direct access to DNA, and a common assumption is that less condensed "open" chromatin provides more accessible sites to the enzyme than more compact, less accessible, heterochromatin.

Today, ChIP has become an essential technique for the study of protein–DNA interactions in cells. First adapted to map interactions between soluble chromatin-interacting proteins such as transcription factors, ChIP is now frequently used to detect DNA interactions with more insoluble nuclear proteins such as nuclear lamins [13, 20–22]. Sonication-based ChIP protocols are usually appropriate to disrupt the bulk of chromatin fibers tightly associated with the lamin network. This approach generally results in the identification of LADs mostly located in heterochromatic environments at the nuclear periphery with the exclusion of regions with more euchromatic features that might be enriched in the nuclear interior. The MNase-based protocol, relying on genome accessibility, favors enrichment in decondensed chromatin regions whose interactions with lamins are probably weaker, due to their high molecular occupancy by regulatory proteins devoted to transcriptional activity. We argue therefore that the MNase-based ChIP assay is complementary to sonication-based protocols if one wants to detect (in principle) all types of lamin–chromatin interactions throughout the genome. This chapter describes the lamin ChIP protocol with MNase digestion of chromatin (Fig. 1) which we have reported recently [14].

2 Materials

2.1 Laboratory Equipment

1. 1.5 mL centrifuge tubes.
2. 10 mL conical polypropylene tubes.
3. 1 mL Dounce tissue grinder (Wheaton).
4. Refrigerated microcentrifuge.
5. Vortex.
6. Rotator.
7. Thermomixer (e.g., Eppendorf).
8. Agarose gel electrophoresis apparatus.
9. Polyacrylamide gel electrophoresis apparatus.
10. Fujifilm Luminescent Image Analyzer LAS 4000 system.
11. Sonics Vibra-Cell VC 505.
12. Tapered microtip 3 mm for up to 10 mL volume.

2.2 Reagents

1. Milli-Q deionized water.
2. Ice bath.

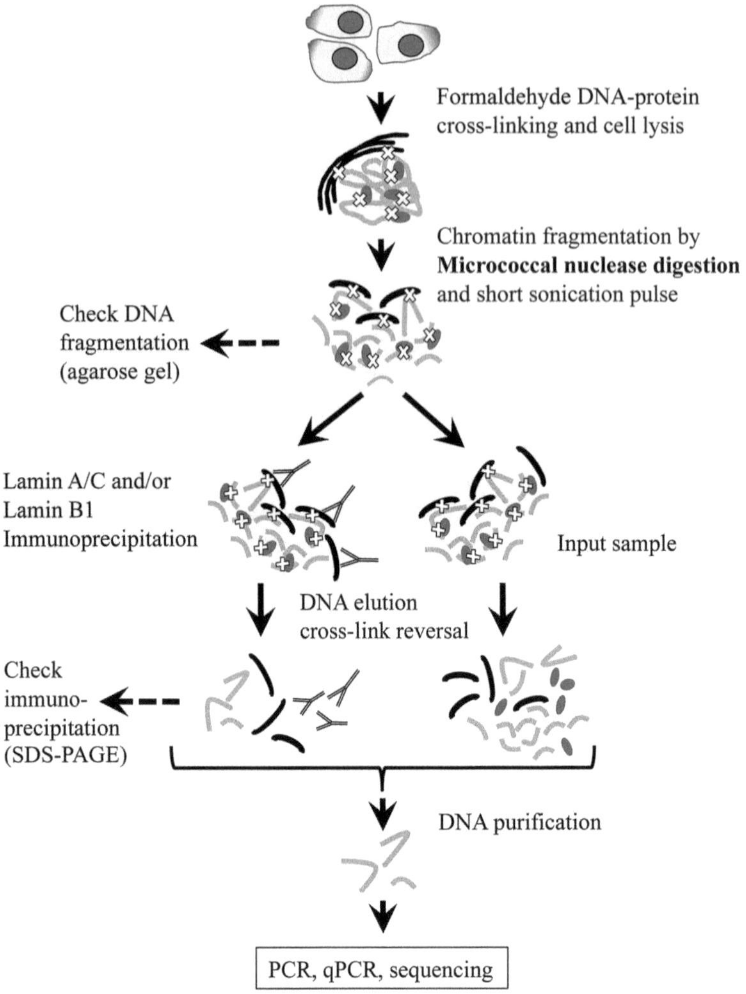

Fig. 1 Main steps of the lamin ChIP protocol using MNase-digested chromatin. Protein–DNA interactions are cross-linked with formaldehyde in cells. Chromatin is digested with MNase, and the nuclear lamin network is disrupted with a short sonication pulse prior to immunoprecipitation with antibodies directed against lamin A/C or lamin B1. After DNA elution and cross-link reversal, enriched-DNA fragments from the lamin-binding sites are purified before analysis

3. 1.25 M glycine.
4. 5 M NaCl.
5. 2.5 M LiCl.
6. 3 M Na acetate, pH 7.0.
7. 1 M Tris–HCl, pH 7.5.
8. 1 M Tris–HCl, pH 8.0.
9. 0.5 M EDTA.
10. 1 M $MgCl_2$.

11. 0.1 M $CaCl_2$.
12. NP-40 (IGEPAL), 10 %.
13. SDS, 20 %.
14. Sodium deoxycholate (DOC), 10 %.
15. Chloroform/isoamylalcohol (24/1).
16. Phenol/chloroform/isoamylalcohol solution (25/24/1).
17. Polyacrylamide carrier.
18. Micrococcal nuclease: stock solution at 600 U/mL in 5 mM Tris–HCl, pH 7.5, 0.025 mM $CaCl_2$.
19. Proteinase K: stock solution at 10 mg/mL in Milli-Q water.
20. RNase A: stock solution at 10 mg/mL in 10 mM Na acetate, pH 5.2.
21. 0.1 M PMSF.
22. 25× stock of complete EDTA-free protease inhibitors.

2.3 Antibodies

1. Rabbit polyclonal anti-lamin A/C [23].
2. Monoclonal anti-lamin A/C (Santa-Cruz, sc-7292X).
3. Rabbit polyclonal anti-lamin B1 (Abcam, ab16048).
4. Rabbit polyclonal anti-histone H3 trimethyl Lys4 (Active Motif, 39159).
5. Protein A/G agarose beads.

2.4 Buffers and Solutions

1. Formaldehyde-culture medium: 0.54 mL of 37 % formaldehyde solution, 20 mL cell culture medium, 1 % nonessential amino acids, 10 % fetal calf serum, and 1 % penicillin/streptomycin.
2. Hypotonic buffer: 10 mM Tris–HCl, pH 7.5, 5 mM NaCl, 1.5 mM $MgCl_2$, 0.5 mM DTT, 0.5 mM PMSF, 1× protease inhibitors.
3. MNase buffer: 20 mM Tris–HCl, pH 7.5, 15 mM NaCl, 1 mM $CaCl_2$.
4. Lysis buffer: 50 mM Tris–HCl, pH 8, 10 mM EDTA, 0.5 % SDS, 0.5 mM PMSF, 1× protease inhibitors.
5. IP buffer: 20 mM Tris–HCl, pH 8, 5 mM EDTA, 150 mM NaCl, 1 % NP-40, 1× protease inhibitors.
6. Wash buffer 1: 20 mM Tris–HCl, pH 8, 5 mM EDTA, 150 mM NaCl, 1 % NP-40, 0.1 % SDS, 1× protease inhibitors.
7. Wash buffer 2: 20 mM Tris–HCl, pH 8, 5 mM EDTA, 500 mM NaCl, 1 % NP-40, 0.1 % SDS, 1× protease inhibitors.
8. Wash buffer 3: 10 mM Tris–HCl, pH 8, 1 mM EDTA, 250 mM LiCl, 1 % NP-40, 1 % DOC, 1× protease inhibitors.

9. Wash buffer 4: 10 mM Tris–HCl, pH 8, 1 mM EDTA.
10. Elution buffer: 20 mM Tris–HCl, pH 7.5, 5 mM EDTA, 50 mM NaCl, 1 % SDS.

3 Methods

This procedure is described for cultured cells grown in 150 mm dishes with 8–10 × 10^6 cells per dish.

3.1 Formaldehyde DNA–Protein Cross-Linking in Cells

1. Remove culture medium from the dishes and quickly wash cells with 10 mL PBS at room temperature (RT) (*see* **Note 1**).
2. Add 20 mL of formaldehyde-culture medium per dish, and incubate for 10 min at RT under gentle shaking. Remove the medium and quickly wash cells with 10 mL of PBS at RT.
3. Add 10 mL of 125 mM glycine per dish, and incubate for 5 min at RT under gentle shaking. Remove the solution and wash cells quickly with 10 mL PBS at RT.
4. Add 2.5 mL of cold PBS containing 0.5 mM PMSF per dish, and scrap cells from the dish with a cell scraper in an ice bath. Collect and split the cell suspension into two 1.5 mL tubes per dish, centrifuge at 600 × *g* for 10 min at 4 °C, and carefully remove the supernatant. Freeze cell pellets in liquid nitrogen and store at −80 °C.

3.2 Cell Lysis

1. Quickly thaw the cell pellets (~10 × 10^6 cells) in an ice bath, resuspend in 250 μL of cold hypotonic buffer (final concentration of 40–45 × 10^6 cells/mL), and incubate for 30 min on ice.
2. Transfer the cell suspension in a Dounce grinder, and apply up to 50 strokes to disrupt the clumps of cross-linked cells.
3. Transfer the cell suspension into a 1.5 mL tube, measure the volume recovered, adjust $MgCl_2$ to a final concentration of 5 mM in the suspension, and add NP-40 to a final concentration of 0.1 % (vol/vol) (*see* **Note 2**).
4. Incubate for 10 min on ice and centrifuge at 2400 × *g* for 10 min at 4 °C. Discard the supernatant.

3.3 Chromatin Fragmentation

1. Resuspend the pellet in a volume of MNase buffer sufficient to obtain a final concentration of 25 × 10^6 cells/mL according to a previously quantified number of cells (*see* **Note 2**), and pre-incubate the cell suspension at 37 °C for 5 min.
2. Add 0.37 U of MNase per 10^6 cells, incubate for 5 min at 37 °C, and stop the digestion by adding EDTA to a final concentration of 10 mM and incubating on ice for 10 min (*see* **Note 3**).

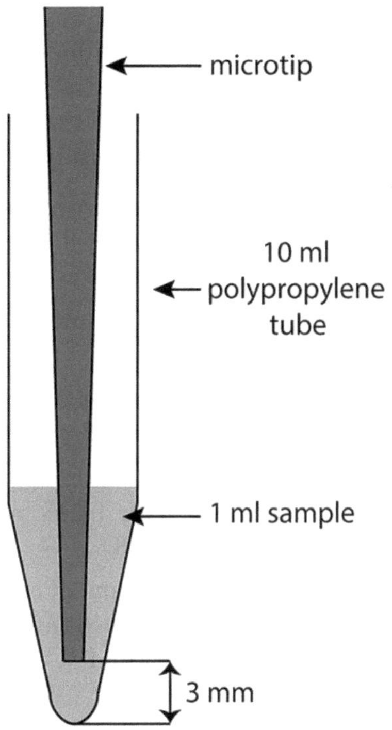

Fig. 2 Sonication installation. Position of the probe inside the propylene tube is shown

3. Dilute the sample with lysis buffer to a final concentration corresponding to 8×10^6 cells/mL, and divide the volume in 1 mL samples in 10 mL conical polypropylene tubes (*see* **Note 4**).
4. Place the sonifier probe at 3 mm from the bottom of the 10 mL propylene tube in melted ice (Fig. 2). Sonicate six times for 10 s each, with pauses of 1 min on melted ice between each pulsing session. Use the following settings on the Sonics Vibra-Cell sonicator: time, 10 s; pulse, 1 s on and 1 s off; and amplitude, 20 %. Repeat for each chromatin sample (if relevant) while leaving the sonicated ones on ice (*see* **Note 5**).
5. Transfer the sonicated samples in 1.5 mL tubes and centrifuge the lysates at $9600 \times g$ for 10 min at 4 °C. Collect and pool the supernatants, and keep them on ice while adding 25 μL of 1× Laemmli sample buffer to each pellet (*see* **Note 6**). Heat the pellets at 95 °C for 5 min and store at −20 °C until use as a control in lamin content analysis. Transfer 1/20 of the volume of the supernatant to another tube to check for DNA fragmentation and lamin solubilization. At this stage, the lysate can be frozen in dry ice and stored at −80 °C in aliquots corresponding to 5×10^6 cells of starting material (*see* **Note 7**).

3.4 Control Analysis of Chromatin Fragment Size and Lamin Solubilization

To assess DNA fragmentation, an aliquot of chromatin lysate (1/20 of the volume taken in **step 5**, Subheading 3.3) is subjected to cross-link reversal while treated with RNase A and to proteinase K digestion before analysis by electrophoresis. The lamin content is directly analyzed by Western blotting after cross-link reversal.

1. Dilute the chromatin aliquot four times in 200 mM NaCl, add RNase A to a final concentration of 50 μg/mL, and incubate for 4 h at 65 °C on a thermomixer at 1300 rpm.
2. Split the chromatin sample in two equal volumes into two 1.5 mL tubes.
3. Add proteinase K to a final concentration of 100 μg/mL to one of the 1.5 mL tube containing half of the lysate and incubate for 1.5 h at 42 °C on a thermomixer at 1300 rpm.
4. Extract the DNA by adding 1 % SDS and 1 M NaCl and mix (*see* **Note 8**).
5. Add 1 volume of chloroform/isoamylalcohol (24/1; vol/vol), vortex at maximum speed for 2 min, centrifuge at 13,000 × *g* for 5 min at 4 °C, and collect the upper phase in a new 1.5 mL tube.
6. Precipitate the DNA by adding 3 μL of polyacrylamide carrier and 2.5 volumes of 100 % ethanol; mix and freeze the DNA sample for 15 min in dry ice. Thaw the tube and centrifuge at 15,000 × *g* for 15 min at RT, then remove the ethanol and wash the precipitate with 1 mL of 80 % ethanol. Recentrifuge at 15,000 × *g* for 5 min at RT, remove the 80 % ethanol, and let the pellet dry at RT for 1 h. Resuspend the pellet in TE buffer (*see* **Note 9**).
7. Load DNA sample onto a 1.8 % agarose gel, and perform electrophoresis at 50 V till the dye front (bromophenol blue in the sample) has reached the bottom of the gel. Detect DNA with SYBR Safe DNA gel stain (Fig. 3a) (*see* **Note 10**).
8. Add 0.5 volume of 3× Laemmli sample buffer to the second 1.5 mL tube containing half of the lysate reserved at **step 2** and heat at 95 °C for 5 min.
9. Analyze the lamin content of the lysate by 12 % SDS-PAGE, and transfer to nitrocellulose and immunoblotting with anti-lamin A/C and anti-lamin B1 antibodies (Fig. 3b).

3.5 Preclearing of Chromatin

All buffers used in Subheadings 3.5–3.7 are kept at 4 °C unless otherwise specified.

1. Determine the amount of protein A/G agarose beads to be used in the preclearing step first by measuring the protein concentration in the lysate using a Bradford or BCA assay: 50 μL of protein A/G agarose beads is commonly used per 500 μg of proteins.

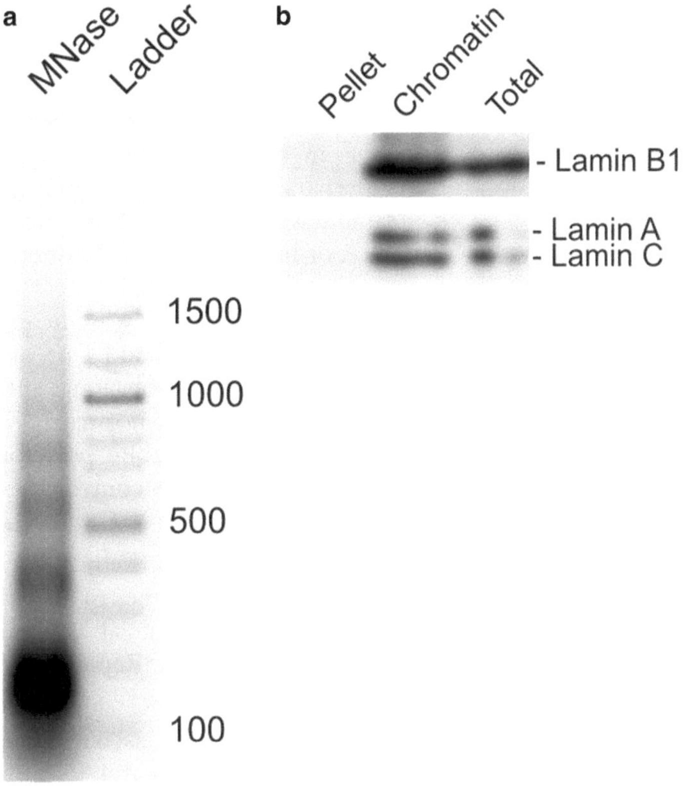

Fig. 3 Qualitative analysis of DNA fragmentation and lamin solubilization. (**a**) Assessment of DNA fragmentation. A chromatin sample digested by MNase and sonicated is analyzed by electrophoresis in 1.8 % agarose gel and SYBR Safe DNA staining. Enrichment in 150–600 bp fragment sizes is observed over a background of longer fragments. (**b**) Lamin content in the chromatin lysate. After cross-link reversal, aliquots of the chromatin lysate and of the pellet (**step 5**, Subheading 3.3) both corresponding to equal amounts of cells were analyzed on 12 % SDS-PAGE, and proteins were immunoprobed with rabbit polyclonal anti-lamin B1 (1/1000 dilution) and anti-lamin A/C (1/5000 dilution) antibodies

2. If one or more aliquots of chromatin lysate (stored at −80 °C; **step 5**, Subheading 3.3) are to be used, prewash the appropriate amount of protein A/G agarose beads: add 10 volumes of IP buffer, invert the tube 10–15 times at RT, centrifuge at 1200 × *g* for 5 min at 4 °C, and discard the supernatant. Repeat this washing step four times.
3. Thaw the desired number of chromatin aliquots and centrifuge at 9600 × *g* for 10 min at 4 °C. Collect and pool the supernatants and dilute five times with the IP buffer to decrease SDS concentration to 0.1 % (*see* **Note 11**). Add protein A/G agarose beads to the lysate and incubate 1 h on a rotator at 10–15 rpm in the cold room. Centrifuge at 1200 × *g* for 6 min at 4 °C,

collect and transfer the supernatant in a new tube, and centrifuge the precleared chromatin lysate once at 1200 ×*g* for 6 min at 4 °C (*see* **Note 12**). Collect and transfer the supernatant containing precleared chromatin in a new tube.

4. Save an aliquot of the precleared chromatin (1/40 of the supernatant volume) in 1.5 mL tube and store it at −20 °C. This sample is the "input DNA" and will be used at later stage for PCR or high-throughput sequencing.

3.6 Chromatin Immunoprecipitation

1. Divide the precleared lysate in approximately 1 mL aliquots in 1.5 mL tubes for IP; add 1 μg of monoclonal anti-lamin A/C antibody or rabbit polyclonal anti-lamin B1 antibody per 10^6 cells, no antibody or normal IgG serum in a negative control tube, and 5 μL of rabbit polyclonal anti-H3K4me3 antibody per IP in a positive control tube; incubate on a rotator at 15 rpm overnight at 4 °C (*see* **Note 13**).
2. Prewash 50 μL of protein A/G agarose beads per mL of IP sample: add 10 volumes of IP buffer, mix by inversion 10–15 times at RT, centrifuge at 1200 ×*g* for 5 min at 4 °C, and discard the supernatant. Repeat this washing step four times.
3. Add the protein A/G agarose beads to the IP sample and rotate at 4 °C for 2 h at 15 rpm on a rotating wheel. Centrifuge at 1200 ×*g* for 6 min at 4 °C and remove the supernatant (*see* **Note 14**).

3.7 Wash of the Immunoprecipitated Complexes

1. Add 1 mL of wash buffer 1 to each bead sample and rotate at 4 °C for 5 min at 15 rpm. Centrifuge at 1200 ×*g* for 6 min at 4 °C and remove the supernatant.
2. Add 1 mL of wash buffer 2 to the beads and rotate at 4 °C for 5 min at 15 rpm. Centrifuge at 1200 ×*g* for 6 min at 4 °C and remove the supernatant. Repeat this step once.
3. Add 1 mL of wash buffer 3 to the beads and rotate at 4 °C for 5 min at 15 rpm. Centrifuge at 1200 ×*g* for 6 min at 4 °C and remove the supernatant. Repeat this step once.
4. Add 1 mL of wash buffer 4 to the beads and rotate at 4 °C for 5 min at 15 rpm. Transfer the beads suspension into a new tube (*see* **Note 15**) and centrifuge at 1200 ×*g* for 6 min at 4 °C and remove the supernatant.

3.8 DNA Elution, RNase and Proteinase K Treatment, Cross-Link Reversal

1. Add 150 μL of elution buffer at RT to each bead sample and vortex briefly. Thaw the reserved input DNA (**step 4**, Subheading 3.5), and bring the volume up to 150 μL with elution buffer. Incubate for 15 min at RT.
2. Transfer 10 % of the immunoprecipitated or input samples volume into a new tube, add one volume of 2× Laemmli sample

buffer, and heat at 95 °C for 5 min. These samples are the protein control and will be used later in SDS-PAGE and immunoblotting analysis (*see* Subheading 3.10).

3. Add RNase A (final concentration 20 U/mL) to the immunoprecipitated or input samples and incubate at 37 °C for 20 min.
4. Add proteinase K (final concentration 50 U/mL) to the immunoprecipitated or input samples, and incubate at 68 °C for 2 h on a thermomixer at 1300 rpm. Centrifuge at 1200 ×*g* for 4 min at RT and transfer supernatant in a new tube. Do not discard the beads.
5. Add another 150 μL of elution buffer to the beads. Incubate at 68 °C for 5 min in the thermomixer as described in **step 4**. Centrifuge at 1200 ×*g* for 4 min and pool the second supernatant with the first one (*see* **Note 16**).

3.9 Purification of ChIP DNA

1. If sample volumes are less than 300 μL, adjust to 300 μL with elution buffer, then add 300 μL of a solution of phenol/chloroform/isoamylalcohol to each sample, and mix thoroughly for 2 min. Centrifuge at RT at 15,000 ×*g* for 5 min, and transfer the aqueous phase into a new tube (*see* **Note 17**). Repeat this step once.
2. Add 300 μL of chloroform/isoamylalcohol to each sample, mix thoroughly for 2 min, and centrifuge at RT at 15,000 ×*g* for 5 min. Transfer the upper phase in a new tube.
3. Make sure that all the samples have the same volume. Complete with H_2O if needed and add 1/10 of the sample volume of 3 M Na acetate pH 7, 3 μL of polyacrylamide carrier, and 2.5 volumes of 100 % ethanol; mix and incubate at −20 °C for a few hours up to overnight.
4. Thaw the tubes and centrifuge at 15,000 ×*g* for 15 min at RT. Discard the ethanol, wash the precipitate with one volume of 80 % ethanol, and recentrifuge at 15,000 ×*g* for 5 min at RT. Discard the ethanol and let the precipitate dry at RT for 1 h; resuspend the precipitate in 10–15 μL PCR grade H_2O per sample. Concentration of each sample may be determined using a NanoDrop spectrophotometer or with any other technique or device. Store DNA samples at −20 °C before qPCR analysis or library preparation for high-throughput sequencing.

3.10 Probing the Chromatin Immunoprecipitation

1. Load the ChIP and input aliquot samples from Subheading 3.8, **step 2**, on a 10 % SDS-PAGE gel for lamin detection and 12 % for histone detection; transfer the proteins to nitrocellulose and immunoprobe with rabbit anti-lamin A/C, anti-lamin B1, and anti-histone H3 trimethyl Lys4, respectively (Fig. 4).

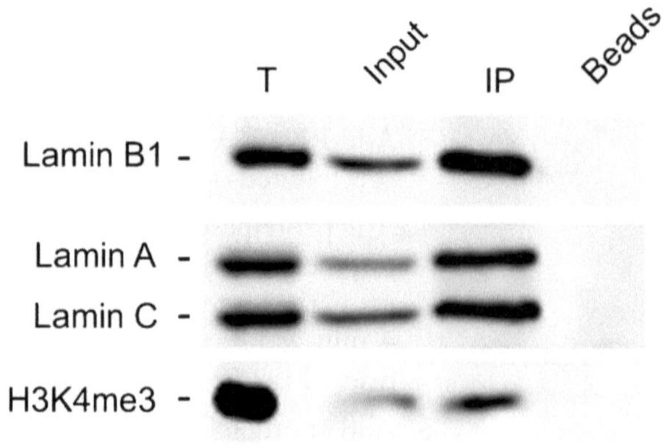

Fig. 4 Control analysis of immunoprecipitated lamins. Proteins present in the precleared chromatin lysate (input; 0.25 % of the starting material), in the immunoprecipitated chromatin lysate (IP; 1.6 % of the starting material), and in the negative immunoprecipitated control (beads; 1.6 % of the starting material) were analyzed by electrophoresis on 10 % or 12 % SDS-PAGE gels, transferred onto nitrocellulose, and revealed with rabbit polyclonal anti-lamin B1 and anti-lamin A/C antibodies as described in Fig. 3 and with an anti-histone H3 trimethyl Lys4 (H3K4me3; 1/1000) antibody. Total cell lysate from 20000 cells (T) was used as a control

4 Notes

1. This cross-linking protocol is scaled for adherent HeLa cells.
2. Usually, the volume recovered at this stage is lower than the starting volume put in the Dounce grinder. This loss of volume should be quantified to estimate the number of cells remaining in the suspension and the volume of $MgCl_2$ and NP-40 needed to adjust their final concentrations to 5 mM and 0.1 % in the sample, respectively.
3. As the digestion conditions may vary upon cell type, optimization of the MNase concentration and incubation time should be established for the desired cell type. It is important to check the DNA fragmentation before proceeding to the immunoprecipitation step (as indicated in Subheading 3.4). Enrichment in fragments of 150–600 bp is recommended.
4. Transfer of 1 mL lysate aliquots in 10–15 mL tubes in this step is intended to prevent possible projections and losses during sonication.
5. This brief sonication step does not modify the fractionation profile of the MNase-digested chromatin but allows the release of lamins into the chromatin supernatant. Note that the sonication protocol reported here is suitable for HeLa cells and the

sonicator model indicated. Optimization of sonication conditions may be required for other cell types and other sonicator models. Samples should not foam as this may damage protein complexes.

6. Centrifugation of the sonicated lysate often leads to the obtention of a small precipitate usually with a gray color due to some metal release of the sonifier probe.
7. A chromatin lysate corresponding to 5×10^6 cells is usually sufficient for 3–4 ChIPs.
8. Addition of SDS and NaCl to the solution may induce the formation of white precipitates that should be dissolved by incubating the sample for a few minutes at 37 °C before adding the chloroform/isoamylalcohol solution.
9. A commercial solution of linear acrylamide polymer (polyacrylamide carrier, *see* Subheading 2.2) was used to precipitate small quantities of DNA. Different inert carriers may also be used, including glycogen or PEG. Nucleic acids such as tRNA must be avoided as they would interfere with measurements of DNA concentration or sequencing analysis.
10. The DNA ladder could also be detected by staining with ethidium bromide.
11. Ionic detergents such as SDS improve lysis by sonication but may be detrimental to immunoprecipitation. A high concentration of SDS is expected to decrease the immunoprecipitation recovery for some antibodies. It is best to decrease fivefold the SDS concentration before proceeding to the immunoprecipitation step.
12. Two centrifugations are performed to ensure that all beads are removed.
13. Different antibodies might have different concentrations and efficiencies. The amount may depend on the antibody used and on recommendations from the manufacturer for commercial antibodies.
14. We recommend taking out 10 μL of each supernatant, and keep them in Laemmli sample buffer at −20 °C for further control analysis.
15. This tube shift step eliminates any nonspecifically bound DNA stuck to the tube wall which may give rise to a background in the analysis.
16. If needed, samples could be frozen at −20 °C at this step.
17. The aqueous phase corresponds to the upper phase after centrifugation. Avoid pipetting the lower phase containing the phenol/chloroform solution.

Acknowledgments

I am grateful to Dr. Jean-Loup Duband and Dr. Ana Ferreiro for helpful comments on the manuscript. This work was supported by the Centre National de la Recherche Scientifique (CNRS), Université Paris Diderot-Paris 7, and the Association Française de lutte contre les Myopathies (AFM).

References

1. Luger K, Mader AW, Richmond RK, Sargent DF, Richmond TJ (1997) Crystal structure of the nucleosome core particle at 2.8 Å resolution. Nature 389:251–260
2. Richmond TJ, Davey CA (2003) The structure of DNA in the nucleosome core. Nature 423:145–150
3. Bednar J, Horowitz RA, Grigoryev SA, Carruthers LM, Hansen JC, Koster AJ, Woodcock CL (1998) Nucleosomes, linker DNA, and linker histone form a unique structural motif that directs the higher-order folding and compaction of chromatin. Proc Nat Acad Sci U S A 95:14173–14178
4. van Steensel B (2011) Chromatin: constructing the big picture. EMBO J 30:1885–1895
5. Lin F, Worman HJ (1995) Structural organization of the human gene (LMNB1) encoding nuclear lamin B1. Genomics 27:230–236
6. Rober RA, Weber K, Osborn M (1989) Differential timing of nuclear lamin A/C expression in the various organs of the mouse embryo and the young animal: a developmental study. Development 105:365–378
7. Burke B, Stewart CL (2013) The nuclear lamins: flexibility in function. Nat Rev Mol Cell Biol 14:13–24
8. Kolb T, Maass K, Hergt M, Aebi U et al (2011) Lamin A and lamin C form homodimers and coexist in higher complex forms both in the nucleoplasmic fraction and in the lamina of cultured human cells. Nucleus 2:425–433
9. Guelen L, Pagie L, Brasset E, Meuleman W et al (2008) Domain organization of human chromosomes revealed by mapping of nuclear lamina interactions. Nature 453:948–951
10. Peric Hupkes D, Meuleman W, Pagie L, Bruggeman SW et al (2010) Molecular maps of the reorganization of genome-nuclear lamina interactions during differentiation. Mol Cell 38:603–613
11. Meuleman W, Peric-Hupkes D, Kind J, Beaudry JB et al (2013) Constitutive nuclear lamina-genome interactions are highly conserved and associated with A/T-rich sequence. Genome Res 23:270–280
12. Kind J, Pagie L, Ortabozkoyun H, Boyle S et al (2013) Single-cell dynamics of genome-nuclear lamina interactions. Cell 153:178–192
13. Lund E, Oldenburg A, Delbarre E, Freberg C et al (2013) Lamin A/C-promoter interactions specify chromatin state-dependent transcription outcomes. Genome Res 23:1580–1589
14. Lund EG, Duband-Goulet I, Oldenburg A, Buendia B, Collas P (2015) Distinct features of lamin A-interacting chromatin domains mapped by ChIP-sequencing from sonicated or micrococcal nuclease-digested chromatin. Nucleus 6:30–39
15. Hörz W, Altenburger W (1981) Sequence specific cleavage of DNA by micrococcal nuclease. Nucleic Acids Res 9:2643–2658
16. Hewish DR, Burgoyne LA (1973) Chromatin sub-structure. The digestion of chromatin DNA at regularly spaced sites by a nuclear deoxyribonuclease. Biochem Biophys Res Commun 52:504–510
17. Noll M, Kornberg RD (1977) Action of micrococcal nuclease on chromatin and the location of histone H1. J Mol Biol 109:393–404
18. McGhee JD, Felsenfeld G (1983) Another potential artifact in the study of nucleosome phasing by chromatin digestion with micrococcal nuclease. Cell 32:1205–1215
19. Cockell M, Rhodes D, Klug A (1983) Location of the primary sites of micrococcal nuclease cleavage on the nucleosome core. J Mol Biol 170:423–446
20. Sadaie M, Salama R, Carroll T, Tomimatsu K et al (2013) Redistribution of the Lamin B1 genomic binding profile affects rearrangement of heterochromatic domains and SAHF formation during senescence. Genes Dev 27:1800–1808
21. Shah PP, Donahue G, Otte GL, Capell BC et al (2013) Lamin B1 depletion in senescent cells triggers large-scale changes in gene expression

and the chromatin landscape. Genes Dev 27: 1787–1799

22. Lund E, Oldenburg AR, Collas P (2014) Enriched domain detector: a program for detection of wide genomic enrichment domains robust against local variations. Nucleic Acids Res 42:e92
23. Chaudhari N, Courvalin JC (1993) Stepwise re-assembly of the nuclear envelope at the end of mitosis. J Cell Biol 122:295–306

Chapter 22

DamID Analysis of Nuclear Organization in *Caenorhabditis elegans*

Georgina Gómez-Saldivar, Peter Meister, and Peter Askjaer

Abstract

The development of genomics and next generation sequencing platforms has dramatically improved our insight into chromatin structure and organization and its fine interplay with gene expression. The nuclear envelope has emerged as a key component in nuclear organization via extensive contacts between the genome and numerous proteins at the nuclear periphery. These contacts may have profound effects on gene expression as well as cell proliferation and differentiation. Indeed, their perturbations are associated with several human pathologies known as laminopathies or nuclear envelopathies. However, due to their dynamic behavior the contacts between nuclear envelope proteins and chromatin are challenging to identify, in particular in intact tissues. Here, we propose the DamID technique as an attractive method to globally characterize chromatin organization in the popular model organism *Caenorhabditis elegans*. DamID is based on the in vivo expression of a chromatin-associated protein of interest fused to the *Escherichia coli* DNA adenine methyltransferase, which produces unique identification tags at binding site in the genome. This marking is simple, highly specific and can be mapped by sensitive enzymatic and next generation sequencing approaches.

Key words *Caenorhabditis elegans*, Chromatin, DamID, Laminopathies, Nuclear envelope, Nuclear lamina, Nuclear organization

1 Introduction

Since the early classification of chromatin into euchromatin and heterochromatin, the concept of a close interplay between chromatin organization and gene expression has been suggested [1]. Nucleosomes are the main structural component of chromatin and the impact of histone posttranslational modifications on gene regulation is well established [2–4]. Moreover, activation or repression of genes relies on interactions between transcription factors and specific sequences within gene bodies or in surrounding regulatory elements [5]. Finally, the nuclear lamina is responsible for anchoring of large chromatin domains (lamina-associated domains; LADs) that typically encompass transcriptionally silent genes [6, 7]. In the last two decades our knowledge on global nuclear

Sue Shackleton et al. (eds.), *The Nuclear Envelope: Methods and Protocols*, Methods in Molecular Biology, vol. 1411, DOI 10.1007/978-1-4939-3530-7_22, © Springer Science+Business Media New York 2016

organization has greatly increased with the development of microarray and next generation sequencing (NGS) technologies. Combined with chromatin immunoprecipitation (ChIP) these advances have provided detailed chromatin profiles across species, developmental stages, and cell types (see for example ENCODE [https://www.encodeproject.org] and modENCODE [http://modencode.org] for references). ChIP is a powerful and popular method but depends on the availability of highly specific and efficient antibodies as well as stringent fixation protocols. As an alternative to ChIP, van Steensel and Henikoff developed DamID for in vivo mapping of chromatin binding sites [8]. The versatility of DamID is reflected by its application to evaluate a wide range of chromatin-associating proteins, such as transcription factors [9–11], components of the RNAi machinery [12], histones [13], and nuclear envelope proteins in *Drosophila* [7, 14], *C. elegans* [15], fission yeast [16], and mammalian cells [6]. DamID can also be used to test association to a single gene of interest using a Southern blot approach [10] or to evaluate the effect of modifications or truncations of chromatin binding proteins (Gómez-Saldivar et al., unpublished data). In addition to not involve fixation or antibodies, DamID has the advantage of being able to identify binding sites within compact chromatin with poor solubility, or when the chromatin-associated proteins are very dynamic or present in low abundance [17].

The principle of DamID is based on the fusion of the enzyme DNA adenine methyltransferase (Dam) from *E. coli* to a protein of interest (POI). Dam methylates adenine at the N6-position within GATC sequences. When the fusion protein is expressed in vivo, it binds directly or indirectly to the native genomic binding sites of the POI and creates specific GmATC methylation tags in the surrounding chromatin (Fig. 1). Through a series of enzymatic reactions, methylated sites are amplified and identified by DNA array or sequencing techniques. DamID is feasible because adenine methylation is very rare in eukaryotic cells and the high frequency of GATC motifs in the genome [18, 19], accessing almost every region of the genome. *C. elegans* has 269,049 GATC sequences per haploid genome [20], corresponding to on average one site for every 374 bp (median 210 bp). To compensate for differences in chromatin compactness and unspecific methylation, the signal from the Dam::POI fusion protein is compared to a diffusible "Dam-only" control, which typically is Dam fused to GFP.

We initially identified chromatin-association profiles for *C. elegans* nuclear envelope proteins by hybridization to tilling-arrays [15, 21] and more recently by NGS [22]. NGS provides definite counts for methylation of each GATC site, from which the chromatin binding profile of the POI can be established. Applying NGS to DamID has required the development of novel scripts for data processing but provides also advantages compared to microarrays:

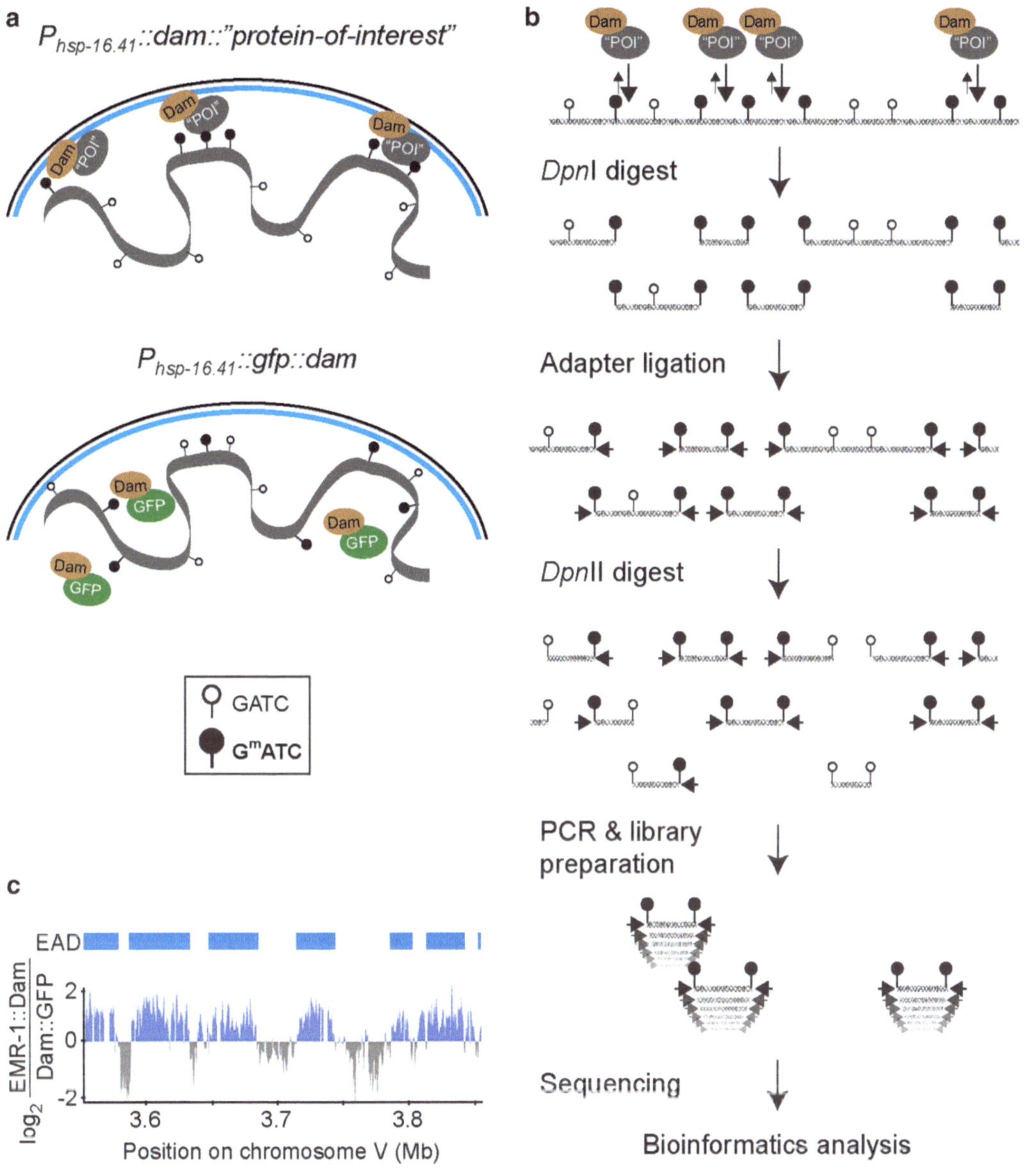

Fig. 1 Principle and workflow of DamID. (**a**) DamID is based on in vivo methylation of genomic GATC sites (*pins*) by Dam fusion proteins expressed at very low concentrations. *Top panels* shows a nuclear envelope "protein of interest" (POI) fused to Dam whereas the *bottom panel* represents a diffusible GFP::Dam control. (**b**) Methylated GATC sites are isolated and amplified trough a series of enzymatic reactions followed by next generation sequencing. (**c**) Comparison of DamID signals from Dam::POI and GFP::Dam (Fig. 2) identifies POI associated domains. The graph illustrates results obtained for emerin/EMR-1 with *blue bars* indicating EMR-1 associated domains (EADs) [15]

sequencing discards hybridization biases and provides absolute measures rather than fluorescence ratios. NGS also allows discarding signals arising from nonspecific DNA breakage during sample preparation because such fragments are most often not flanked by GATC sites.

In this chapter we describe the DamID workflow from sample preparation to bioinformatics analysis. Because sample preparation has been described recently [17, 23] we devote particular attention to data processing (Fig. 2).

2 Materials

2.1 Expression Plasmids for Dam Fusions

1. Plasmid containing the Dam coding sequence and suitable cloning sites, e.g., pBN61 (*Phsp-16.41::dam::myc::MCS::unc-54 3′ UTR*; [15] (*see* **Note 1**).
2. *C. elegans* genomic DNA (gDNA) or vector bearing the gene of interest.
3. Plasmid encoding the Dam-only control, e.g., pBN67 (*Phsp-16.41::gfp::myc::dam*; [15].
4. Standard molecular biology reagents and materials of high analytic grade for cloning (e.g., primers to amplify the gene of interest, high-fidelity polymerase, agarose, restriction enzymes, ligase, competent *E. coli* cells).

2.2 Generation and Validation of DamID Strains

1. Materials required for Mos1-mediated Single-Copy Integration (MosSCI), including microinjection equipment, nematode host strains, co-injection plasmids, etc. [23].
2. Primary antibodies against the Myc epitope (e.g., Sigma-Aldrich C3956 or 9E 10 from Developmental Studies Hybridoma Bank) and nuclear envelope proteins as control (e.g., mAb414 Covance MMS-120R); secondary anti-rabbit and anti-mouse antibodies for immunofluorescence and Western blot analysis [17].

2.3 Nematode Culture

1. According to personal preference nematodes can be grown in liquid medium [17] or on NGM plates [23]; we typically use 3000–4000 animals (~30 mg) per sample, although as little as 20 nematodes can be analyzed [22].
2. *Dam-E. coli* (e.g., strain GM119 or SCS110; *see* **Note 2**).
3. M9 buffer: 22 mM KH_2PO_4, 34 mM Na_2HPO_4, 86 mM NaCl, 1 mM $MgSO_4$.
4. Tween 20.
5. Hypochlorite solution: 1 N NaOH, 30 % household bleach solution.

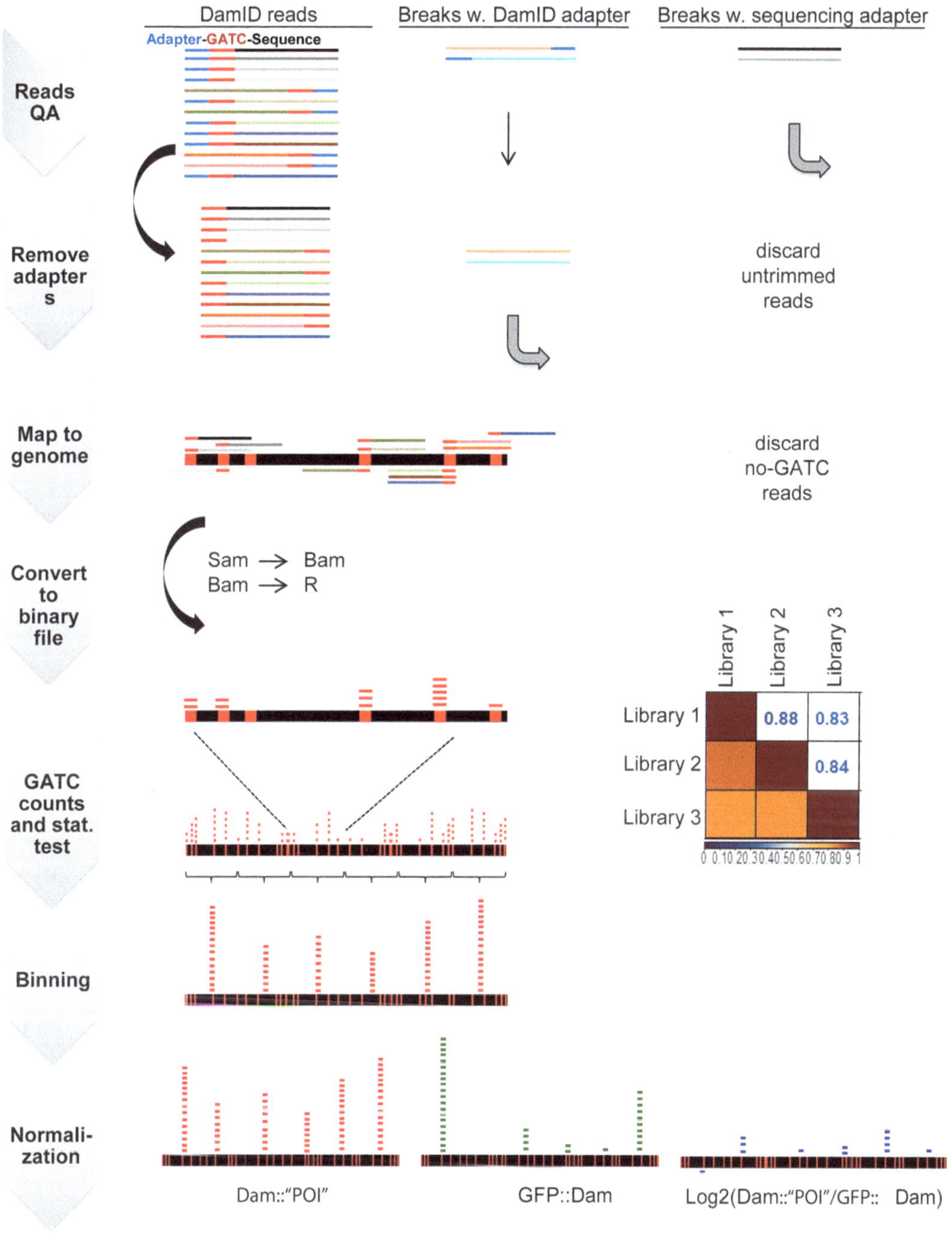

Fig. 2 Work flow to map and count methylated GATC sites. Once the quality of the reads has been evaluated, the adapters are removed from the reads. Next, the clean reads are mapped to the reference genome. The output is converted to a BAM file, which is processed in R. The next step is identification and counting of GATC sites followed by statistical evaluation and evaluation of the correlation between replicates. To increase the correlation, the signals are binned and finally normalized with the total number of genomic GATC sites and with the GFP::Dam control samples

2.4 Purification and Amplification of Dam-Methylated DNA

1. DNeasy Blood and Tissue Kit including RNase A (QIAGEN cat. #69504 and #19101).
2. Ammonium acetate 3 M.
3. 96 % and 70 % ethanol.
4. Thermocycler.
5. *Dpn*I and *Dpn*II restriction enzymes (NEB cat. #R0176S and #R0543S).
6. Primers for the preparation of the AdR double-stranded adapter (50 μM): AdRt (5′-CTAATACGACTCACTATAGG GCAGCGTGGTCGCGGCCGAGGA; 100 μM) and AdRb (5′-TCCTCGGCCGCG; 100 μM).
7. T4 DNA ligase (Roche, cat. # 10799009001, 5 U/μL).
8. Agencourt AMPure XP® (Beckman Coulter, Inc., cat. #A63880).
9. Magnetic particle concentrator.
10. Advantage® cDNA Polymerase Mix (Clontech cat. #639105) and dNTP mix.
11. PCR Primer AdR (5′-NNNNGTGGTCGCGGCCGAGG ATC; 50 μM).
12. QIAquick PCR Purification Kit (QIAGEN cat. #28104).
13. Guanidine hydrochloride 35 %.
14. Standard materials and equipment for DNA agarose gel electrophoresis.

2.5 Library Preparation

1. End-It™ DNA End-Repair Kit (Epicentre cat. #ER0720).
2. QIAquick PCR Purification Kit.
3. Guanidine hydrochloride 35 %.
4. 10× NEB buffer 2.
5. Klenow Fragment (3′–5′ exo-) (NEB cat. #M0212M) and dATP.
6. Primers for the preparation of the Y-shaped Illumina adapters: upper strand (5′-ACACTCTTTCCCTACACGACGCTCTT CCGATCT; 100 μM) and phosphorylated lower strand (P-5′-GATCGGAAGAGCACACGTCT; 100 μM).
7. T4 DNA ligase.
8. MyTaq™ Mix (BIOLINE cat. #BIO-25041).
9. Illumina P7/index primer (index sequence underlined, choose according to NGS facility for easy multiplexing; 5′-CAA GCAGAAGACGGCATACGAGAT<u>NNNNNN</u>GTG ACTGGAGTTCAGACGTGTGCTCTTCCGATCT; 5 μM).
10. Forward P5 primer (5′-AATGATACGGCGACCACCGAGATC TACACTCTTTCCCTACACGACGCTCTTCCGATCT; 5 μM).
11. Agencourt AMPure XP®.
12. QUBIT® fluorometer or NanoDrop.

2.6 DamID Quantification and Bioinformatics Analysis

1. Computer with UNIX environment.
2. Free software tools, e.g., Cutadapt, Bowtie, R/Bioconductor, and Galaxy.

3 Methods

3.1 Construction of DamID Vector

1. Using standard molecular cloning techniques, clone the gene of interest into a Dam expression plasmid of choice, e.g., pBN61 (*see* **Note 3**). The transgene must be under the control of an inducible promoter with low basal activity and a characterized 3′ UTR (*see* **Notes 4** and **5**).
2. Verify the sequence of the insertion.

3.2 Nematode Culture

1. Inject the DamID vector into the appropriate *C. elegans* host strain for Mos-mediated single-copy integration and isolate transgenic lines according to standard MosSCI protocols [24].
2. Verify correct localization and size of the fusion protein by immunofluorescence and Western blot, respectively, using α-Myc antibodies (*see* **Note 6**).
3. Collect embryos from asynchronous cultures by standard hypochlorite treatment. Check the nematodes regularly in a dissection stereoscope. Proceed to the next step when half of the nematodes are broken up.
4. Wash embryos five times in 12 mL M9. Pellet the embryos by centrifugation at 2000 rpm for 3 min. After the last wash, resuspend in 5 mL of M9 with 0.01 % Tween 20.
5. Count number of embryos in 2–10 μL aliquots.
6. Leave embryos to hatch overnight at 16–20 °C with gentle agitation.
7. Assay quantity of hatched L1 larvae in 2–10 μL aliquots.
8. Place 500–1000 L1s per 85 mm plate containing a thick lawn of Dam-*E. coli* bacteria as food source. Use a total of four plates per strain, multiplied by number of replicas (typically three).
9. Incubate nematodes at 20 °C and collect when they are enriched for the life stage to be analyzed (e.g., 53 h for non-gravid young adults; 66 h for accumulation of young embryos). Harvest either embryos [21] or, for better signal/noise ratio, adult nematodes [15].
10. Make aliquots containing ~30 μL embryo or adult material. Remove excess liquid and snap-freeze in liquid nitrogen before −80 °C storage until further processing.

3.3 Purification and Amplification of Dam-Methylated DNA

1. Purify genomic DNA (gDNA) with QIAGEN DNeasy Blood and Tissue Kit. All steps should be as gentle as possible to avoid shearing of the gDNA.
2. Elute the gDNA in 200 μL and determine its concentration. Expected yield is 4–7 μg; concentrate by ethanol precipitation if needed.
3. Incubate 500 ng of adult gDNA with 10 U *Dpn*I in 10 μL for 6 h at 37 °C in a thermocycler to cut methylated $G^{m}ATC$. Include an additional reaction with *Dpn*I for one of the biological samples to be used as ligation control (control A) and also a control reaction without *Dpn*I (control B; *see* **Note 7**). Heat-inactivate *Dpn*I for 20 min at 80 °C.
4. Anneal double-stranded adapters: combine 50 μL of primers AdRt and 50 μL of AdRb. Heat to 95 °C and let cool down slowly to room temperature.
5. Assemble the adapter ligation reaction on ice and incubate at 16 °C overnight. Prepare also a reaction without T4 DNA ligase (control A).

ddH_2O	6.2 μL
*Dpn*I digested gDNA	10 μL
10× Ligation Buffer	2 μL
Double-stranded adapters	0.8 μL
T4 DNA ligase	1 μL

6. Heat-inactivate the T4 DNA ligase for 10 min at 65 °C.
7. Purify the reaction with 36 μL of AMPure XP bead suspension (Agencourt AMPure XP®) using a magnetic particle concentrator and elute in 20 μL.
8. Incubate the ligation reactions with 10 U of *Dpn*II in a volume of 50 μL for 1 h at 37 °C to cut unmethylated GATC sites. Heat-inactivate *Dpn*II for 20 min at 80 °C.
9. Purify the reaction with 90 μL of Agencourt AMPure XP® bead suspension and elute in 25 μL.
10. Amplify methylated DNA using Advantage® cDNA Polymerase Mix. Include a control reaction without DNA template (control C).

ddH_2O	13.75 μL
10× PCR reaction buffer	5 μL
*Dpn*II digested DNA	25 μL
Primer AdR	1.25 μL
dNTP mix (each 2.5 mM)	4 μL
Advantage cDNA Polymerase Mix	1 μL

PCR parameters:

I. 68 °C: 10 min.
II. 94 °C: 1 min.
III. 65 °C: 5 min.
IV. 68 °C: 15 min.
V. 94 °C: 1 min.
VI. 65 °C: 1 min.
VII. 68 °C: 10 min.
VIII. Go to step V 3×.
IX. 94 °C: 1 min.
X. 65 °C: 1 min.
XI. 68 °C: 2 min.
XII. Go to step IX 23× (embryos) or 20× (adults) (*see* **Note 8**).

11. Analyze 5 μL of PCR reaction on an agarose gel. Amplification of Dam-methylated DNA should produce a smear from 300 to 1000 bp.
12. Purify the PCR products using QIAquick PCR Purification Kit. Include an extra wash with 700 μL guanidine hydrochloride 35 % before PE washing (*see* **Note 9**). Elute in 30 μL. Quantify the concentration of the DNA (Fig. 3a).

3.4 Library Preparation

1. Blunt the PCR products with End-It™ DNA End-Repair Kit for 45 min at room temperature.

ddH_2O	to a final volume of 50 μL
PCR product	1 μg
10× End-Repair Buffer	5 μL
dNTP mix	5 μL
ATP	5 μL
End-Repair Enzyme Mix	1 μL

2. Purify the DNA using QIAquick PCR Purification Kit. Include an extra wash with 700 μL guanidine hydrochloride 35 % before PE washing. Elute the DNA in 25 μL.
3. Incubate blunt-ended DNA fragments with Klenow for 30 min at 37 °C to add 3′-A overhangs, then put the tubes on ice.

ddH_2O	18.5 μL
Blunt ended DNA	25 μL
10× NEB buffer 2	5 μL
dATP 10 mM	1 μL
Klenow Fragment	0.5 μL

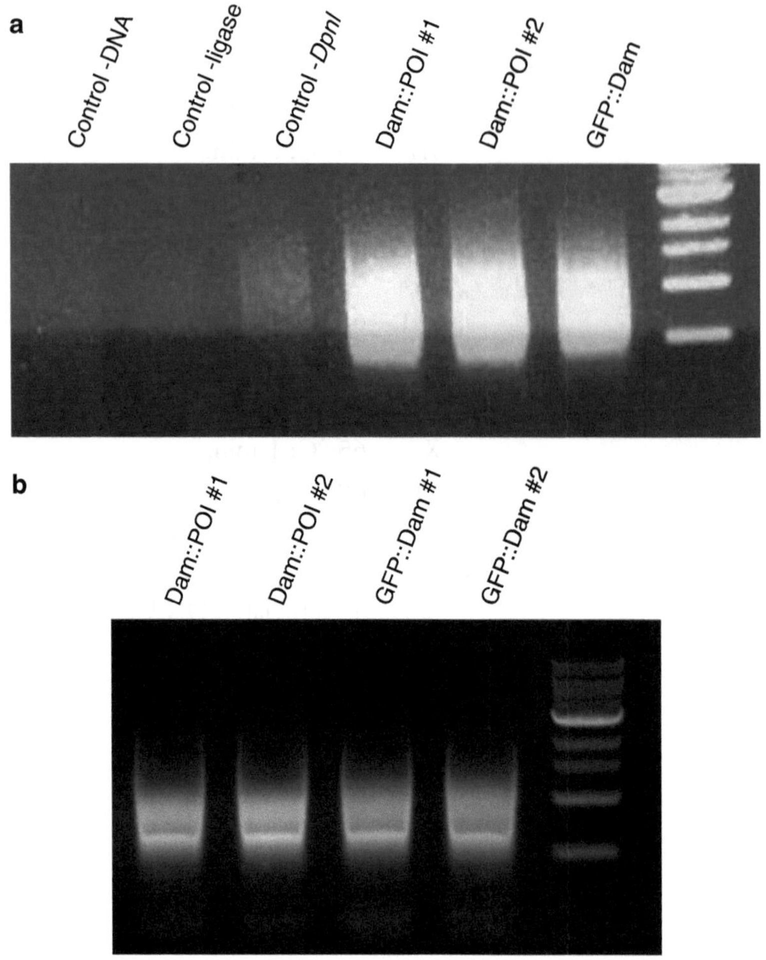

Fig. 3 Visualization of DamID samples on agarose gels. (**a**) Example of PCR-amplified methylated fragments and controls. The DamID samples produce a smear from 300 to 1000 bp whereas the controls "–DNA," "–ligase," and "–*Dpn*I" give no PCR product. (**b**) NGS libraries after amplification by index PCR

4. Purify the DNA using QIAquick PCR Purification Kit. Include an extra wash with 400 μL guanidine hydrochloride 35 % before PE washing. Elute the DNA in 20 μL.
5. Anneal Y-shaped Illumina adapters: combine 1:1 of upper and lower strand primers. Heat to 95 °C and let cool down slowly to room temperature (annealed adapters can be stored frozen).
6. Ligate Y-shaped Illumina adapters. Incubate for 2 h at 25 °C and heat-inactivate T4 ligase for 10 min at 65 °C.

DNA with 3′-A overhang	8 μL (~250 ng)
10× Ligation buffer	1 μL
T4 DNA Ligase	0.5 μL
Illumina Y-Adapter	0.5 μL

7. Purify the DNA using QIAquick PCR Purification Kit. Elute the DNA in 20 μL.
8. Amplify by Illumina Index PCR (*see* **Note 10**) using the forward P5 primer and Illumina TruSeq® Index primers for multiplexing.

DNA with Y-adapter	8 μL (~100 ng)
Illumina Index primer	1 μL
Forward P5 primer	1 μL
2× MyTaq enzyme mix	10 μL

PCR parameters

I. 94 °C: 1 min.

II. 94 °C: 30 s.

III. 58 °C: 30 s.

IV. 72 °C: 30 s.

V. Go to step II 6–10×.

VI. 72 °C: 2 min.

9. Purify the Index PCR material with 14 μL Agencourt AMPure XP® beads and elute in 20 μL.
10. Quantify libraries using preferably QUBIT® fluorometer or NanoDrop and submit for NGS (Fig. 3b).

3.5 DamID Quantification and Bioinformatics Analysis

1. Evaluate the quality of raw reads and discard those reads that do not meet quality standards (*see* **Note 11** and Fig. 2) [25]. This is sometimes performed by the NGS facility, otherwise use programs such as FastQC (http://www.bioinformatics.babraham.ac.uk/projects/fastqc/). Beware that due to the peculiar nature of the DamID-seq libraries (all reads start with the same sequence of the adapter), FastQC reports a low variability in the first 20 nucleotides. To avoid this, FastQC can also be performed on the sequences once the adapter has been removed (in which case only the first four nucleotides show a high bias towards GATC).
2. Remove the DamID adapters using Cutadapt [26]; (https://code.google.com/p/cutadapt/). It reads a FASTA or FASTQ file, and writes the changed sequence to standard output.

Assuming your sequencing data is available as a FASTQ file, use this command line in your UNIX window:

```
$ cutadapt -g CGCGGCCGAG -e 0.15 --discard-untrimmed input.fastq -o output.fastq
```

Parameters:

-g CGCGGCCGAG; the adapter ligated to the 5′ end.

-e 0.15; maximum allowed error rate (default: 0.1). This means that a single error per adapter is allowed.

--discard-untrimmed; discard untrimmed reads, i.e., removes reads resulting from DNA breaks and hence do not carry the DamID adapters.

3. Mapping reads to a reference genome. Currently, the main source for reference genome assembly is from the University of California Santa Cruz (UCSC). To align the reads we use Bowtie [27]; (http://bowtie-bio.sourceforge.net/index.shtml) and the *C. elegans* genome sequence (ce10, corresponding to WormBase release WS220; *see* **Note 12**). The mapping parameters are set to only map unique and best alignments (-m 1, --best):

```
$ /path-to-bowtie-programs/bowtie /path to ce10 bowtie index/genome -m 1 --best -q input_cut.fastq -S output_cut.sam
```

4. Read mapping programs normally use files in FASTQ format as input, and often store output in files with sequence alignment/map (SAM) format; this implies that you need to convert the file to a machine-readable binary file (BAM) before further analysis using the R/Bioconductor program. We use Samtools (http://samtools.sourceforge.net/) to convert the SAM to BAM format:

```
$ samtools view -bS input_cut_mapped.sam > output_cut_mapped.bam
```

5. Mapping GATC sites using RStudio (www.cran.r-project.org and http://www.rstudio.com), the packages "BSgenome.Celegans.UCSC.ce10" and "Biostring", as well as custom-made R scripts (*see* **Notes 13** and **14** and Table 1). The library consists of GATC-flanking DNA sequences that are used to map these to the genome. We first identify all GATC sites in the *C. elegans* genome on either strand using the "DNAstring" function. Next, the intervals corresponding to the reads are checked for the presence of a GATC sequence at the beginning (+ strand) or the end (− strand), providing an additional filtering of true DamID reads versus break-produced sequences. Mapped reads are then assigned to individual GATC in the genome and counted for each genomic GATC (*see* **Note 15**).

6. Comparison of biological replicates and binning. At this step, a correlation coefficient at the GATC level can be calculated, providing a measure of the reproducibility of the observed DamID pattern. This greatly depends on the number of

Table 1
Examples of functions used in custom-made R scripts

Function	Purpose
DNAstring	Identify and store GATC sequences.
vmatchPattern	Find all the hits of GATC pattern in *C. elegans* reference sequence.
which	Give the true indices of a logical object, in this case, strand sense.
Rle	Compute the length (frequency) and values (strand sense) of runs.
cor	Estimate correlations between libraries at GATC level.
corrplot	Generate a graphical display of correlation between libraries at GATC level, with confidence interval.
CreateSlidingRegions	Bins the *C. elegans* genome according to the parameters optStep and optWin.

individual reads, but is normally very high ($R > 0.8$). As methylation of a single GATC carries a certain degree of stochasticity, it is recommended to bin the genome in fragments ranging from 1 to 100 kb, depending on the biological question. The correlation coefficient usually improves when analyzing larger genomic segments, but at the same time resolution is reduced. Thus, running the analysis several times may be advantageous for determining the ideal bin size for the protein of interest.

7. Normalize DamID reads per bin by the total number of DamID reads, taking into account the sequencing depth. To adjust for the accessibility of the individual GATC sites in the genome, normalize the signal for each POI by the Dam-only scores (GFP::Dam).
8. Averaging across replicates. It is good practice to keep replicates separated until the final step (individual Dam::POI and GFP::Dam samples). At this last stage, the average of at least two independent replicates is calculated. The data are then ready to be processed (*see* **Note 16**).
9. Peak Calling. Determination of the chromatin domains to which the POI associates is a central part of DamID analysis. For this, we need to assign regions with significant numbers of mapped reads (peaks). Because peak calling tools have typically been developed for ChIP-seq we must choose or adapt a peak-calling algorithm and normalization method considering: (a) the balance between sensitivity and specificity, (b) the type and dynamics of POI chromatin association (point-source binding [e.g., transcription factors], broadly enriched [e.g., nuclear lamins]). According to the selected parameters we can affect the number and quality of the peaks called. Note that using the

same enrichment metric values, such as p-value or false discovery rate (FDR) threshold does not ensure that the peaks are comparable across libraries. Instead, it is better to use the irreproducible discovery rate (IDR) threshold [28], which also can guide in selecting the best peak calling algorithm and parameter settings. Existing peak callers (e.g., SPP [29], MACS [version 2] [30], SICER [31], RSEG [32], and ZINBA [33]) differ in terms of background modeling and signal smoothing, but most use a windows-based method to define peaks [34]. Finally, using R packages "polyaPeak" and "NarrowPeaks" we can improve the resolution through analysis of the shape of the peaks, including optimizing the bandwidth and peak cutoff parameters, and rerank and refine the final peak-calling list. Once the POI association domains are defined we can continue doing more routine analysis as peak annotation and motif analysis, according to the biological question(-s) to answer.

4 Notes

1. Additional cloning strategies are discussed on the van Steensel laboratory website (http://research.nki.nl/vansteensellab/DamID.htm), which also provides useful tips, FAQs and protocols for DamID in Drosophila and mammalian cells.
2. Using a Dam-*E. coli* as a food source is important to avoid contamination by methylated *E. coli* DNA.
3. To minimize variation in expression levels between different DamID strains, we recommend transgenesis by MosSCI single-copy integration techniques. MosSCI plasmids are available for fusion of the Dam protein to either the N- or the C-terminus of the POI. In both orientations a Myc-tag serves as linker between the POI and Dam. Prior knowledge on the behavior and biological activity of GFP-tagged versions of the POI might be useful (Dam and GFP are of equal size), but verification of correct localization should always be performed.
4. Dam is a highly active enzyme [35] and high methylation levels can be detected at both native binding and non-binding sites when Dam is overexpressed. To avoid false-positive methylation marks as well as potential effects on gene expression or DNA replication DamID experiments are performed with minimal expression of the Dam::POI transgenes. For *C. elegans*, we use an inducible heat shock promoter (*hsp-16.41*) without induction.
5. We use the *unc-54* 3′ UTR in our DamID vectors; other 3′ UTRs could potentially be used to alter the relative signal contribution by different cell types in whole-animal DamID [36].

6. Upon heat shock induction, signals corresponding to the Dam fusion protein should be detectable by immunofluorescence and Western blot. A sample without induction is included as control, which corroborates the low basal transcriptional activity of the *hsp-16.41* promoter during the DamID experiment.
7. Control B is an important indicator of unspecific breaks produced during gDNA purification the technique. Optionally, include also a control with gDNA purified in parallel from wild type nematodes.
8. When using a new batch of Advantage cDNA Polymerase Mix, optimize the total number of PCR cycles required for amplification. Collect 10 μL of the PCR reaction after every two cycles ranging from cycle number 14 to 22 (last cycle).
9. Guanidine hydrochloride is used to ensure the complete removal of PCR primers and primer dimers, thereby avoiding the amplification of these primers during the sequencing reaction.
10. Make two Illumina Index PCR reactions: one to run on a 1 % agarose gel to inspect the quality of the library and one to purify for NGS.
11. Analysis of NGS data is a computational intensive process and requires several software tools, most of which are oriented toward UNIX operating systems. Three main approaches to analyze NGS data, either individually or in combination, exist: (1) Programming to run in a UNIX environment (e.g., Linux, Solaris, Mac OS X, or Windows through Cygwin (https://www.cygwin.com/)), (2) R/Bioconductor packages for the R computing environment. Table 2 lists useful packages available for sequence analysis. (3) Upload data to online servers. Currently, the most popular web server is Galaxy (http://galaxyproject.org/), which hosts several analysis tools.

Table 2
Examples of Bioconductor packages for NGS analysis

Purpose	Packages
Data representation	Biostrings, Bsgenome, GenomicRanges, Iranges, VariantAnnotation
Input/Output	ShortRead (FASTQ), Rsamtools (BAM), rtracklayer (GFF, WIG, BED), VariantAnnotation (VCF)
Annotation	AnnotationHub, biomaRT, ChIPpeakANNo, GenomicFeatures, VarianAnnotation
Alignment	Biostrings, gmapR, Rsubread
Visualization	Gviz, ggbio, corrplot
Peak calling	PolyaPeak, NarrowPeaks, Bayespeak, ChIPpeakAnno

12. The genome database release version is very important when mapping NGS reads. Reads mapped to one version are not directly interchangeable with reads mapped to a different version. The UCSC liftover tool is very useful, but note that repeated remapping of data can yield different results. We recommend using genomes curated at UCSC so that you can easily visualize your data later using the UCSC Genome Browser (http://genome.ucsc.edu).
13. Each replicate should have three to nine million uniquely mapped reads, of which two to eight million reads start with GATC. We are usually able to identify between 65 and 89 % of genomic GATC sites.
14. From this point the data analysis becomes more specific to the technique of DamID-seq (compared to for example ChIP-seq) due to the focus exclusively on GATC sites in the genome. The existing algorithms to analyze ChIP-seq data (e.g., SPP and MACS) are based on the assumption that signals are symmetrically distributed. In DamID-seq experiments, one cannot assume that the adenine methylation signals are symmetrical, because the enzymatic process is affected by many factors, such as chromatin 3D structure [20]. This challenged us to develop new scripts ([22]; available upon request).
15. At this point we generate in R a GenomicRange file (GRanges) containing all the GATC counts as well as a series of genomic features. The GRanges class represents a collection of genomic features that each have a single start and end location on the genome. This includes features or annotations (metadata elements) such as score, transcripts, and exons. These objects can be created by using the GRanges constructor function. Then, we add the data-frame containing the counts of the individual GATC sites as a metadata column in the GRanges file.
16. It is good practice, and a requirement by several journals, to deposit data at the public genomic data repository Gene Expression Omnibus (GEO; http://www.ncbi.nlm.nih.gov/geo/).

Acknowledgement

We gratefully acknowledge funding from the Spanish Ministry of Economy and Competitiveness (BFU2013-42709P) and the European Regional Development Fund to P.A. The Meister laboratory is funded by the Swiss National Foundation (SNF Assistant Professor grant PP00P3_133744), the Swiss Foundation for Muscle Diseases Research, and the University of Bern. G.G.-S. holds a CSIC-JAE Fellowship (JAEPre_2010_00384) and received support from EMBO (ASTF-447-2014) and EC COST Action BM1408 GENiE.

References

1. Jost KL, Bertulat B, Cardoso MC (2012) Heterochromatin and gene positioning: inside, outside, any side? Chromosoma 121:555–563
2. Xu F, Zhang K, Grunstein M (2005) Acetylation in histone H3 globular domain regulates gene expression in yeast. Cell 121:375–385
3. Tessarz P, Kouzarides T (2014) Histone core modifications regulating nucleosome structure and dynamics. Nat Rev Mol Cell Biol 15:703–708
4. Zhou HL, Luo G, Wise JA, Lou H (2014) Regulation of alternative splicing by local histone modifications: potential roles for RNA-guided mechanisms. Nucleic Acids Res 42:701–713
5. Giresi PG, Kim J, McDaniell RM, Iyer VR, Lieb JD (2007) FAIRE (Formaldehyde-Assisted Isolation of Regulatory Elements) isolates active regulatory elements from human chromatin. Genome Res 17:877–885
6. Guelen L, Pagie L, Brasset E, Meuleman W, Faza MB, Talhout W et al (2008) Domain organization of human chromosomes revealed by mapping of nuclear lamina interactions. Nature 453:948–951
7. Pickersgill H, Kalverda B, de Wit E, Talhout W, Fornerod M, van Steensel B (2006) Characterization of the *Drosophila melanogaster* genome at the nuclear lamina. Nat Genet 38:1005–1014
8. van Steensel B, Henikoff S (2000) Identification of in vivo DNA targets of chromatin proteins using tethered dam methyltransferase. Nat Biotechnol 18:424–428
9. Orian A, van Steensel B, Delrow J, Bussemaker HJ, Li L, Sawado T et al (2003) Genomic binding by the *Drosophila* Myc, Max, Mad/Mnt transcription factor network. Genes Dev 17:1101–1114
10. Song S, Cooperman J, Letting DL, Blobel GA, Choi JK (2004) Identification of cyclin D3 as a direct target of E2A using DamID. Mol Cell Biol 24:8790–8802
11. Southall TD, Brand AH (2007) Chromatin profiling in model organisms. Brief Funct Genomic Proteomic 6:133–140
12. Woolcock KJ, Gaidatzis D, Punga T, Bühler M (2011) Dicer associates with chromatin to repress genome activity in *Schizosaccharomyces pombe*. Nat Struct Mol Biol 18(1):94–99
13. Braunschweig U, Hogan GJ, Pagie L, van Steensel B (2009) Histone H1 binding is inhibited by histone variant H3.3. EMBO J 28:3635–3645
14. Kalverda B, Fornerod M (2010) Characterization of genome-nucleoporin interactions in *Drosophila* links chromatin insulators to the nuclear pore complex. Cell Cycle 9:4812–4817
15. Gonzalez-Aguilera C, Ikegami K, Ayuso C, de Luis A, Íñiguez M, Cabello J et al (2014) Genome-wide analysis links emerin to neuromuscular junction activity in *Caenorhabditis elegans*. Genome Biol 15:R21
16. Steglich B, Filion GJ, van Steensel B, Ekwall K (2012) The inner nuclear membrane proteins Man1 and Ima1 link to two different types of chromatin at the nuclear periphery in *S. pombe*. Nucleus 3:77–87
17. Askjaer P, Ercan S, Meister P (2014) Modern techniques for the analysis of chromatin and nuclear organization in *C. elegans*. WormBook 1–35
18. Greil F, Moorman C, van Steensel B (2006) DamID: mapping of in vivo protein-genome interactions using tethered DNA adenine methyltransferase. Methods Enzymol 410: 342–359
19. Greer EL, Blanco MA, Gu L, Sendinc E, Liu J, Aristizábal-Corrales D et al (2015) DNA Methylation on N6-Adenine in *C. elegans*. Cell 161:868–878
20. Sha K, Gu SG, Pantalena-Filho LC, Goh A, Fleenor J, Blanchard D et al (2010) Distributed probing of chromatin structure in vivo reveals pervasive chromatin accessibility for expressed and non-expressed genes during tissue differentiation in C. elegans. BMC Genomics 11:465
21. Towbin BD, Gonzalez-Aguilera C, Sack R, Gaidatzis D, Kalck V, Meister P et al (2012) Step-wise methylation of histone H3K9 positions heterochromatin at the nuclear periphery. Cell 150:934–947
22. Sharma R, Jost D, Kind J, Gómez-Saldivar G, van Steensel B, Askjaer P et al (2014) Differential spatial and structural organization of the X chromosome underlies dosage compensation in *C. elegans*. Genes Dev 28:2591–2596
23. Dobrzynska A, Askjaer P, Gruenbaum Y (2016) Lamin binding proteins in *Caenorhabditis elegans*. Methods Enzymol 569:455–83
24. Frøkjær-Jensen C, Davis MW, Hopkins CE, Newman BJ, Thummel JM, Olesen SP et al (2008) Single-copy insertion of transgenes in *Caenorhabditis elegans*. Nat Genet 40:1375–1383
25. Dai M, Thompson RC, Maher C, Contreras-Galindo R, Kaplan MH, Markovitz DM et al

(2010) NGSQC: cross-platform quality analysis pipeline for deep sequencing data. BMC Genomics 11(Suppl 4):S7

26. Martin M (2011) Cutadapt removes adapter sequences from high-throughput sequencing reads. EMBnetjournal 17:10–12
27. Langmead B, Salzberg SL (2012) Fast gapped-read alignment with Bowtie 2. Nat Methods 9:357–359
28. Ivers KM, Li C, Patel N, Sredar N, Luo X, Queener H et al (2011) Reproducibility of measuring lamina cribrosa pore geometry in human and nonhuman primates with in vivo adaptive optics imaging. Invest Ophthalmol Vis Sci 52:5473–5480
29. Kharchenko PV, Tolstorukov MY, Park PJ (2008) Design and analysis of ChIP-seq experiments for DNA-binding proteins. Nat Biotechnol 26:1351–1359
30. Zhang Y, Liu T, Meyer CA, Eeckhoute J, Johnson DS, Bernstein BE et al (2008) Model-based analysis of ChIP-Seq (MACS). Genome Biol 9:R137
31. Zang C, Schones DE, Zeng C, Cui K, Zhao K, Peng W (2009) A clustering approach for identification of enriched domains from histone modification ChIP-Seq data. Bioinformatics 25:1952–1958
32. Song Q, Smith AD (2011) Identifying dispersed epigenomic domains from ChIP-Seq data. Bioinformatics 27:870–871
33. Rashid NU, Giresi PG, Ibrahim JG, Sun W, Lieb JD (2011) ZINBA integrates local covariates with DNA-seq data to identify broad and narrow regions of enrichment, even within amplified genomic regions. Genome Biol 12:R67
34. Bailey T, Krajewski P, Ladunga I, Lefebvre C, Li Q, Liu T et al (2013) Practical guidelines for the comprehensive analysis of ChIP-seq data. PLoS Comput Biol 9:e1003326
35. Urig S, Gowher H, Hermann A, Beck C, Fatemi M, Humeny A, Jeltsch A (2002) The *Escherichia coli* dam DNA methyltransferase modifies DNA in a highly processive reaction. J Mol Biol 319:1085–1096
36. Mangone M, Manoharan AP, Thierry-Mieg D, Thierry-Mieg J, Han T, Mackowiak SD et al (2010) The landscape of *C. elegans* 3′UTRs. Science 329:432–435

Chapter 23

The Application of DamID to Identify Peripheral Gene Sequences in Differentiated and Primary Cells

Michael I. Robson and Eric C. Schirmer

Abstract

The nuclear envelope interacts extensively with chromatin, though with differences in degree and specificity in different cell types. However, identifying the specific genome sequences associated with individual nuclear envelope associated proteins, particularly nuclear membrane proteins and lamins, has been particularly difficult due to their inherent insolubility and interconnectivity. DamID is a powerful tool developed to bypass many of the inherent difficulties with identifying nuclear envelope protein–chromatin interactions and, as more tissue culture cell types derived from different tissues are examined by DamID, it is increasingly apparent that there are distinct patterns of genome organization in differentiated cell types. However, in applying DamID to both more diverse and/or differentiated cell types a number of technical caveats to the method have been observed which must be circumvented to ensure high quality data is generated. Here we elaborate a detailed methodology to adapt DamID to novel cell types, in particular differentiated cells in culture. Moreover, we highlight heretofore largely ignored variations in the PCR amplified DNA products generated by the DamID procedure and the consequences they have for downstream analysis steps. Thus, the methods described here should serve as a useful resource to researchers new to DamID as well as readily allow its application to an expanded set of cell types and conditions.

Key words Nuclear envelope, DamID, Bacterial dam methylase, Nuclear lamina, Lamin B1, Myogenesis, Myoblasts, Myotubes

Abbreviations

gDNA	Genomic DNA
NE	Nuclear envelope
NET	Nuclear envelope transmembrane protein
PCR	Polymerase chain reaction

1 Introduction

It is now clear that many nuclear envelope (NE)-associated proteins interact with chromatin and can influence various aspects of genome regulation including the spatial organization of the

Sue Shackleton et al. (eds.), *The Nuclear Envelope: Methods and Protocols*, Methods in Molecular Biology, vol. 1411, DOI 10.1007/978-1-4939-3530-7_23, © Springer Science+Business Media New York 2016

genome [1, 2]. However, identifying the genomic sequences present at the site of these interactions using standard approaches such as chromatin immunoprecipitation (ChIP) remains a largely intractable problem since proteins associated with the NE, such as lamins and NE transmembrane proteins (NETs), display a significant degree of interconnectivity and are often inherently insoluble. Although recent advances allowed the application of ChIP to lamin A, this adapted method cannot distinguish the soluble nucleoplasmic pool of lamin A from its larger polymeric peripheral pool [3]. The insolubility of NE components is a significant obstacle for the application of ChIP to NE-associated proteins since an unknown number of interacting sequences at the nuclear periphery are lost. Similarly, the interactivity between NE-associated proteins, with NETs, lamins, and chromatin all sharing interactions with one another [4, 5], further makes interpretation of ChIP results challenging because sequences identified could reflect indirect interactions.

To circumvent these problems, an alternative method was developed by the van Steensel laboratory termed DamID [6, 7]. This method has now been applied to identify DNA in close proximity to a variety of nuclear proteins, including soluble proteins, the polymeric nuclear lamins, and the transmembrane NETs. The DamID method employs the bacterial dam methylase to label proximal GATC motifs. Because this m6A methylation is unique to prokaryotes, the labeled DNA can be specifically isolated from mammalian and other eukaryotic cell types. To ensure the specificity of the labeling, in addition to expressing the dam methylase fused to the protein of interest, the dam methylase is also expressed alone as a soluble nuclear protein to label DNA that is accessible throughout the nucleus. Contrasting the trace of the unfused dam methylase works particularly well when the fusion is to a NE protein because much of the peripheral DNA appears to be less accessible to the free dam methylase. In both cases it is important to express the proteins at a very low level or specificity can be lost; thus, generally expression is not induced but relies upon the leakiness of the *Drosophila melanogaster* minimal heat shock promoter. Following in situ labeling, total cellular DNA is extracted from cells and the uniquely methylated GACT- motifs are specifically cut by the *DpnI* restriction enzyme. DNA regions that are multiply cleaved by *DpnI* (generating fragments smaller than ~2 kb) are then enriched for by specific PCR amplification after ligation of adaptors to the *DpnI*-exposed 5′ and 3′ ends. Because a wide range of fragment sizes would be expected from the genome (unless repetitive elements are involved in binding) the resulting DNA as observed by agarose gel electrophoresis is expected to appear as a smear with no clearly delineated individual bands. The amplified fragments are then either sequenced or identified on whole genome tiling microarrays if available for the subject organism. Plotting the genome sequences enriched in the cells expressing the protein of interest fused to the dam methylase relative to those expressing the dam methylase alone identifies the genomic regions associating with the protein of interest.

The principal advantage of this method over ChIP for the study of NE–chromatin interactions is that DNA is chemically tagged in situ and therefore its identification is not limited by the solubility of the interacting protein. A further advantage is that the method does not depend on the availability of antibodies specific to the protein of interest—a particularly important aspect with regards to the NE where epitope masking in different cell types has often been observed [8]. Finally, DamID avoids the gross cross-linking of ChIP so that interactions may be more likely to be direct, albeit with a resolution limited to the local GATC frequency.

Nonetheless, while highlighting these significant advantages of DamID, it is important to not lose sight of its limitations. While with ChIP the cross-linking may lead to the false-positive identification of distal sequences, if the fusion of the dam methylase tag blocks or alters interactions it can also lead to false identifications. In the case of a NET, loss of specific interactions would likely result in the NET becoming more mobile in the membrane in which case, due to its restriction to the membrane, the NET-dam fusion would simply methylate all DNA at the periphery rather than just specific sequences interacting with the NET. Another limitation is that the method relies upon a PCR amplification step while using ratios of the dam-NET fusion construct to the dam alone construct to bioinformatically determine peripheral sequences; thus, if sequences preferentially methylated by the dam-NET fusion are favored for amplification or repeat sequences are amongst the targets, the PCR could skew the results. Related to this point, the PCR amplification step of DamID would miss any GATC methylated sequences where there is a distance ~3 kb or more between methylated sites; so the method will only work for proteins that have frequent proximal DNA targets within regions of the genome that are not GATC-poor. However, with that said, in *Drosophila melanogaster* GATC sites within 2.5–5 kb of a locus bound by a targeted dam methylase fusion construct have been reported to be methylated. Hence, at least for some proteins, sites of proximity between a protein and DNA will possess multiple contiguous methylated GATC sites [9].

DamID using NE protein fusions is most characterized using lamin B1, which allows the identification of the majority of genomic sequences located at the periphery [6, 10]. The use of lamin B1 is advantageous for this purpose for a number of reasons. Firstly, whereas lamin A has both nucleoplasmic and NE pools, lamin B1 is permanently farnesylated and so is almost exclusively at the NE [11]. Moreover, as lamin B1 is highly abundant in most cell types—lamins are present at roughly 9.5 million copies in an average mammalian cell nucleus [12, 13], it is the best NE protein to yield a comprehensive overview of all chromatin at the nuclear periphery. Most NETs in contrast are thought to have only 1000–10,000 copies per cell. In addition to using dam-lamin B1 fusions, DamID has also been applied with the intent to identify genomic sequences proximal to NETs with the dam methylase fused to emerin, Man1, and lem2 in mammalian cells, *Caenorhabditis elegans* and *Saccharomyces cerevisiae*. However maps of genomic regions associated with these NET-dam fusions were largely similar to those of the lamins, suggesting that due to higher NET-dam than endogenous NET expression or loss of specific tethering they are simply identifying all genomic regions at the periphery.

DamID has previously been used in a number of cellular systems and/or whole organisms ranging from mammalian cells [6, 14, 15] to *Schizosaccharomyces pombe* [16], *C. elegans* [17], *D. melanogaster* [10], and *Arabidopsis thaliana* [18]. However, its application to mammalian differentiation systems and primary cultures of differentiated cells and tissues raises a number of challenges not encountered in these previously engaged tissue culture systems. Thus, the method presented here (Fig. 1) focuses on these challenges and a number of optimizations and controls that must be performed prior to final sequencing to ensure products are of sufficient quality.

Critically, cells must be clear of mycoplasma and other prokaryotic contamination prior to engaging DamID because such contaminants—even at low levels where mycoplasma is undetectable with PCR testing of culture media—will completely saturate the dam methylated signals in the genomic DNA (gDNA). Also, lentiviral transduction conditions must be optimized to allow efficient construct delivery in target cell types without excessive toxicity: for example addition of transduction enhancing agents such as Polybrene can result in toxicity yielding some DamID signal due to ligation of amplification primers to apoptotic fragmented DNA ends unless the Polybrene is carefully titrated before the experiment. Accordingly, following these optimizations a number of controls should be included in the DamID processing steps to confirm amplified DNA is derived from genomic dam methylated DNA isolated from target cells and not from such apoptotic fragmented DNA or contaminating prokaryotic DNA. Finally, once amplified DNA smears are obtained the degree of mitochondrial genomic DNA contamination, an unfortunate by-product of DamID in cell types possessing higher numbers of mitochondria, should be assessed to determine if samples are practical for analysis by sequencing or microarray. Once optimized, DamID can be routinely performed rapidly in multiple simultaneous conditions such as with the knockout or overexpression of NETs.

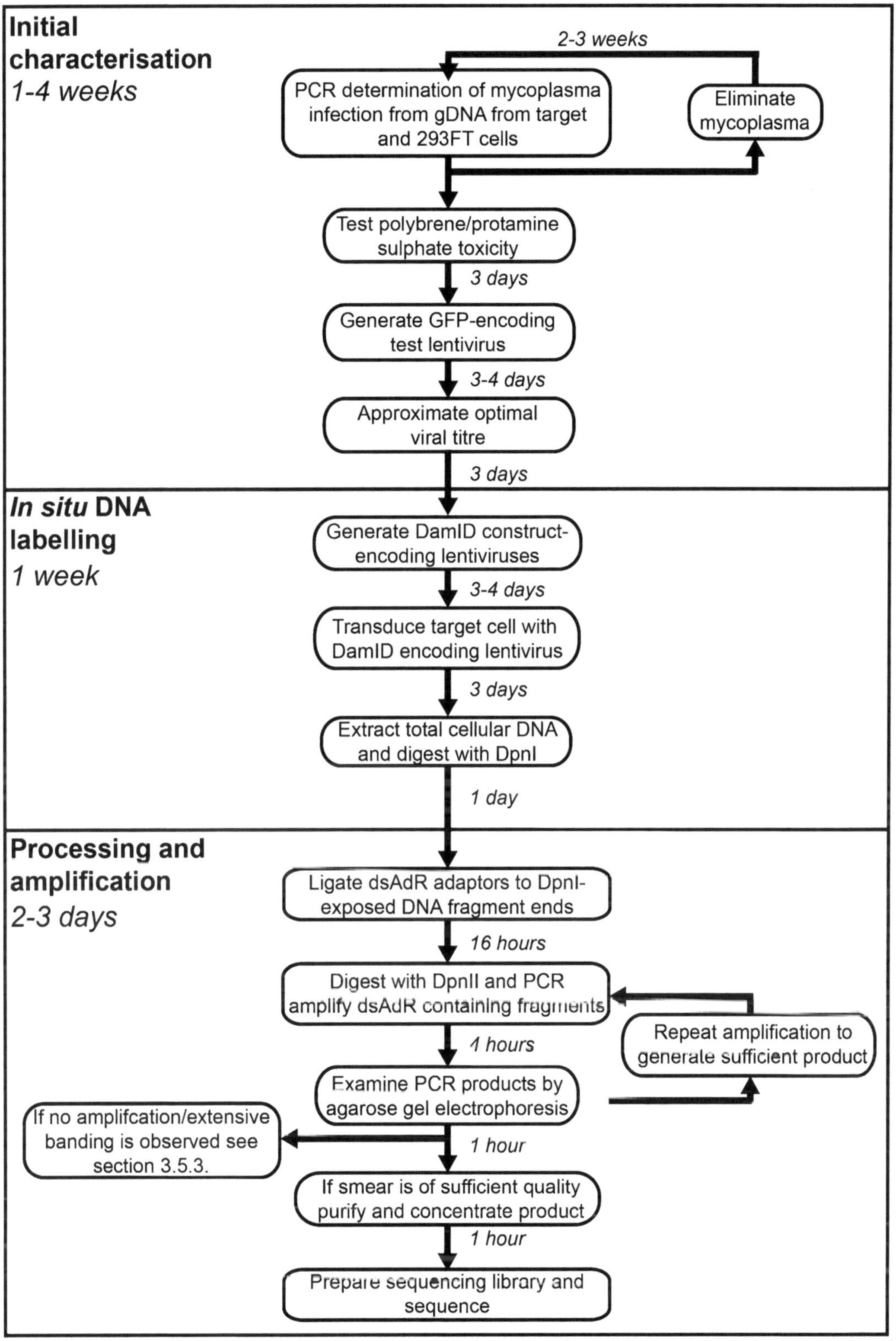

Fig. 1 Flowchart diagram of the DamID methodology to be applied onto a novel cell system. Prior to full-scale experiments quality control and optimization steps are performed to eliminate mycoplasma and ensure cells are transducible. Subsequently, full-scale experiments are performed to label and the amplify target DNA sequences

2 Materials

2.1 Detection of Mycoplasma and Treatment

2.1.1 Hardware

1. Swinging bucket centrifuge capable of at least 1000×*g* centrifugation for pelleting of trypsinized cells and/or larger cell assemblies.
2. Microfuge capable of at least 10,000×*g* centrifugation for purification of gDNA.
3. Thermocycler with heated lid for standard PCR detection of mycoplasma from isolated genomic DNA.
4. Standard labware for agarose gel electrophoresis.
5. Standard labware for tissue culture.

2.1.2 Solutions and Reagents

1. DNeasy Blood and Tissue Lysis kit (*see* **Note 1**).
2. Myco.F (5′-GGGAGCAAACAGGATTAGATACCCT-3′) and Myco.R (5′-TGCACCATCTGTC ACTCTGTTAACCTC-3′) primers for mycoplasma detection. Standard desalted purification is sufficient (*see* **Note 2**).
3. Phusion or Taq polymerase and dNTPs for PCR amplification of mycoplasma detection product.
4. BM cyclins (catalog number 10799050001) (Roche) for treatment of mycoplasma infected cells (*see* **Note 3**).
5. Standard lab reagents for DNA-resolution electrophoresis such as agarose and a sensitive DNA stain such as ethidium bromide.
6. Standard tissue culture media, antibiotics, and other reagents.

2.2 Production of Lentivirus

2.2.1 Hardware

1. Swinging bucket centrifuge capable of at least 4000×*g* centrifugation for clarification of viral supernatants.
2. 0.45 μm^2 low protein binding PES syringe filter, e.g., (catalog number SLHP003RS) (Millipore) and 50 mL syringe for clarification of viral supernatants (*see* **Note 4**).
3. Swinging-bucket or fixed angled ultracentrifuge rotor capable of 55,000×*g*, e.g., Beckman Coulter SW28 rotor with matching Ultra-Clear 25×89 mm centrifuge tubes or Beckman Coulter JA25.50 rotor with 30 mL Oak Ridge round bottomed tubes for virus concentration.
4. Standard tissue culture labware.

2.2.2 Cells and Reagents

1. 293FT cells (Clontech) and culturing media consisting of DMEM supplemented with 10 % fetal bovine serum, 100 units/mL penicillin, 100 μg/mL streptomycin, and 0.5 mg/mL geneticin (*see* **Note 5**).
2. Lipofectamine 2000 and Opti-MEM (Gibco) (*see* **Note 6**).

3. The viral packaging plasmids psPAX2 and pMD2.G and transfer plasmids pLGW-empty, pLGW-Dam, pLGW-Dam-Lamin B1(all made freely available from the van Steensel lab), or pRRLSIN.cPPT.PGK-GFP.WPRE (Addgene). For generation of ecotropic viruses pMD2.G can be replaced with pCAG-Eco (Addgene) added at the same amount by mass.

2.3 Optimization of Transduction Conditions

2.3.1 Hardware Required for Fluorescent Visualization and Optimization of Transduction

1. An epifluorescence microscope capable of at least 400× magnification. Excitation and emission filters suitable for viewing GFP (e.g., 488 nm) and a DNA dye (e.g., 461 nm) fluorescence will be required.
2. Standard tissue culture labware.

2.3.2 Reagents and Solutions for Enhancement of Transduction

1. Polybrene.
2. Protamine sulfate.
3. Hoechst 33342 or 4′,6-diamidino-2-phenylindole (DAPI) for fluorescent labeling of chromatin.
4. Standard microscope slides, coverslips and anti-photobleaching mounting agent.

2.4 Generation and Isolation of Dam-Methylated DNA

2.4.1 Hardware

1. Swinging bucket centrifuge capable of at least 1000 × *g* centrifugation for pelleting of trypsinized cells.
2. Microfuge capable of at least 14,000 × *g* for isolation of gDNA.
3. Standard tissue culture labware.

2.4.2 Solutions and Reagents

1. Cell system of interest with media and tissue culture plastic.
2. DamID constructs packaged into lentiviruses and transduction optimizing reagents, e.g., protamine sulfate or polybrene at predetermined optimal amounts.
3. DNeasy blood and tissue lysis kit (*see* **Note 4**).
4. 3 M potassium acetate pH 5.5, glycogen and 100 % ethanol for precipitation of extracted and column purified genomic DNA.

2.5 Amplification and Purification of Dam-Methylated DNA Fragments

2.5.1 Hardware

1. Thermocycler with heated lid for PCR amplification of dam-methylated DNA fragments.
2. Microfuge capable of at least 14,000 × *g* centrifugation for purification of DamID products.
3. Standard labware for agarose gel electrophoresis and gel documentation.

2.5.2 Solutions and Reagents

1. *Enzymes: DpnI*, T4 DNA ligase, *DpnII*, Antarctic phosphatase.
2. *Primers:*AdRt(5′-CTAATACGACTCACTATAGGGCAGCGT GGTCGCGGCCGAGGA-3′) and AdRb (5′-TCCTCGG CCG-3′) and Adr-PCR primer (5′-GGTCGCGGCCGAGG ATC-3′) purified to standard desalted.
3. cDNA Advantage PCR kit (catalog number 639105) (Clontech). Other reagents tried were not able to amplify DamID products due to variation in size and characteristics of fragments.
4. Reagents for any method for purification of PCR products.
5. Standard reagents for agarose gel electrophoresis and DNA staining.

2.6 Deep Sequencing, Data Analysis, and Presentation

2.6.1 Hardware

1. Deep Sequencing: typically sequencing will be outsourced and many platforms are available. We have used the Illumina HiSeq 2000 platform using 90 bp paired end (90PE) sequencing reactions at five samples per well.
2. Whole genome tiling microarrays: few genome wide tiling arrays of sufficient density are now available for this purpose. For a detailed protocol for labeling of DamID samples for microarray *see* Ref. [7].

2.6.2 Software

1. BWA aligner to assign reads to the reference sequences.
2. Bioconductor Limma package for subsequent data processing.
3. SICER program for identification of peaks in the log2(Lamin B1/Dam) ratio (downloaded freely from http://home.gwu.edu/~wpeng/Software.htm) as used previously in Ref. [14]. Alternatively, the Enriched Domain Detector (EDD) has also been described for the detection of peaks in Lamin A ChIP-seq data sets [19].
4. Integrative genome browser (IGV) for presentation of processed DamID data [20].

3 Methods

3.1 Detection and Treatment of Mycoplasma in Tissue Culture Cells

It is necessary to first ensure cells are clear of prokaryotic intracellular parasites such as mycoplasma. Such infection, in either the lentivirus producing 293FT cells or in the DamID target cells, results in significant contamination of PCR amplified m6A methylated DNA in DamID experiments since such organisms contain an endogenous dam methylase [21]. We have found that even low levels of infection can result in significant contamination of DamID traces resulting in the creation of a distinct underlying banding pattern. In fact, standard PCR mycoplasma detection using culture media was inadequate to detect levels that would saturate DamID experiments and it is necessary thus to isolate DNA from pelleted cells to identify low-level mycoplasma infections. If detected it is necessary to either eradicate the infection or obtain new cells. It is also important to remember that no organism is completely sterile no matter how clean an animal facility and all mammals contain both pathogenic and symbiotic bacteria. Therefore, it is possible that other bacteria will be present in primary culture systems that could interfere with DamID. Thus, when despite the absence of mycoplasma and mitochondrial issues a contaminating banding pattern occurs, it may be possible to remove this by first heavily treating primary cultures with a range of antibiotics.

3.1.1 Detection

1. Isolate total cellular DNA from target cells and 293FT cells using DNeasy blood and tissue kit as described in the manufacturer's instructions (*see* **Note 1**).
2. For each cell line to be tested assemble the following PCR reaction:

   1 μl Total cellular DNA (50-100 ng per reaction)

   4 μL 5× high fidelity (HF) buffer.

   2 μL primer mix (from 5 μM master mix).

   0.2 μL 10 mM dNTPs.

   0.2 μL Phusion polymerase.

   12.6 μL H_2O.
3. In a thermocycler with a heated lid, run reactions using the following program:

Denaturation	98 °C, 30 s
32 cycles of:	
Denaturation	98 °C, 7 s
Annealing	68 °C, 7 s
Extension	72 °C, 5 s
End cycle	
Final extension	72 °C, 10 s

4. Run PCR reactions on a 2 % agarose gel and stain with a DNA dye such as ethidium bromide. If cells are contaminated with mycoplasma a 220 bp amplification product derived from the mycoplasma 16S ribosomal gene can be observed (*see* **Note 7**).

3.1.2 Elimination of Contaminating Mycoplasma

If mycoplasma is detected clearance can be achieved using the BM Cyclin treatment (Roche) as per the manufacturer's instructions. Following treatment, cells should be cultured for 10 days in the absence of BM cyclins and then tested again for contamination to ensure complete clearance (*see* **Notes 8** and **9**).

3.2 Production of Lentivirus

For any DamID experiment the first step is the delivery and expression of exogenous Dam methylase fusion proteins in the target cell type. The favored approach for this is lentiviral transduction because plasmids generated in bacteria will also harbor the m6A methylation. Constructs can be efficiently and stably introduced into a wide range of mammalian cell types, both proliferating and non-dividing, by transduction of lentiviruses pseudotyped with the amphotropic vesicular stomatitis virus glycoprotein (VSV-G). If local health and safety rules make it impractical to generate lentiviruses, then it is essential that extreme care be made in generating plasmid DNA in bacterial strains lacking the dam methylase because when transfected into cells any plasmids harboring G(m6)ATC will be cut by the *Dpn*I restriction enzyme and so will be indistinguishable from gDNA methylated by the ectopically expressed Dam fusion protein. Note that it is not possible to purify completely unmethylated dam methylase-encoding plasmids even from *dam– E. coli* as even the leaky expression of the encoded constructs is sufficient to methylate the plasmid DNA; thus, no matter how carefully prepared, the use of plasmids results in significant contamination of DamID experiments. Such plasmid contaminations will be only a minor problem if whole genome tiling arrays are used, but for deep sequencing, even with carefully prepared plasmid DNA, it can easily quadruple the cost of sequencing to get the necessary depth of coverage. Plasmid constructs can be introduced via liposome and electroporation methods of transfection, but this is also less desirable than lentiviral transduction because, apart from nucleofection, these methods require nuclear disassembly and mitosis to give the plasmid access to the nuclear transcriptional machinery to begin expression and so cannot be applied to post-mitotic differentiated cell types.

Non-replicative, self-inactivating lentiviruses are generated in 293FT cells by joint transfection of a packaging plasmid, an envelope plasmid, and the transfer plasmid, the latter of which encodes the Dam-methylase construct of interest. Virus particles are then pelleted from culture supernatants by high speed centrifugation, resuspended in the desired culture medium and aliquoted in appropriate volumes for storage at –80 °C.

1. Prior to preparation of the transfection culture, grow 293FT cells at logarithmic growth densities in the presence of 500 μg/mL geneticin to obtain seven million cells per virus preparation.
2. Assemble the following DNA-Opti-MEM solution per virus preparation:

   4.6 μg psPAX2.

   2.8 μg PMD2.G.

   7.5 μg transfer vector.

   1.5 mL Opti-MEM.

3. Assemble the following Lipofectamine 2000–Opti-MEM solution per virus preparation in a 15 mL Falcon-type tube:

   36 μL Lipofectamine 2000.

   1.5 mL Opti-MEM.

4. Following a 5-min incubation at room temperature, combine DNA and Lipofectamine solutions and mix gently by inverting the Falcon tube four times. Incubate the DNA–Lipofectamine mixture at room temperature for 20 min.
5. During the 20 min incubation trypsinize 293FT cells and count by haemocytometer. Following pelleting in a swinging bucket centrifuge at room temperature, plate 6 million 293FT cells in 6 mL growth media in a 10 cm diameter tissue culture plate (*see* **Note 10**).
6. Immediately following plating, add the 3 mL DNA–Lipofectamine mixture, mix by rocking and incubate overnight at 37 °C with 5 % CO_2.
7. The next day aspirate media and replace with 15 mL fresh growth media and incubate cells for a further 48 h (*see* **Note 11**).
8. Collect the virus-containing supernatant in a 50 mL Falcon tube. If sufficient numbers of cells have survived, cultures can be incubated an additional 24 h with 10 mL fresh growth medium to generate higher viral titers. During this extra time initial supernatants can be stored at 4 °C.
9. Combine day 2 and day 3 viral supernatants and clarify to remove cellular debris by centrifugation at 1000 × *g* for 12 min in a swinging bucket centrifuge. Transfer supernatant to a fresh 50 mL falcon and discard pellet.
10. Pass clarified supernatants through a 0.45 μm^2 low protein binding PES syringe filter to remove additional cellular debris (*see* **Note 4**).
11. Transfer filter supernatants into a 30 mL Oak Ridge round bottomed tube, topped up with PBS if necessary to adjust the total tube volume to 25 mL and spin at 55,000 × *g* for 75 min at 4 °C in for example a JA 25.5 rotor using a Beckman Avanti JA-25 centrifuge. The lentiviral pellet should appear as a semi-transparent spot roughly 3 mm in diameter on the side of the tube for one 10 cm-dish (*see* **Note 12**).
12. Aspirate the supernatant and cover the lentiviral particle pellet in the centrifuge tube with the desired volume of Opti-MEM, or the culture media required by target cells. To resuspend the pellet agitate the media by shaking on an orbital shaker for 10 min at room temperature (*see* **Note 13**).
13. If not used immediately, aliquots should be frozen at −80 °C (*see* **Note 14**).

3.3 Determination of Transduction Conditions for Target Cell Type

Transduction efficiency varies significantly between different cell types and cell states. As a general rule, increasingly differentiated cell types are more difficult to transduce than proliferative undifferentiated cell types. Similarly, primary cells often require higher titers to achieve satisfactory transduction than standard tissue culture cell lines. Hence, prior to performing DamID on any new cell type it is necessary to perform a transduction titration experiment to determine if the target cell is amenable to DamID by lentiviral transduction and, if so, what fraction of a DamID-viral preparation to use per experiment. Since DamID vectors possess no directly visible means of assessing transduction efficiency, it is necessary to transduce target cells with fluorescence protein-encoding lentiviruses when performing transduction efficiency test experiments. It is also important to note that, as mentioned in the introduction, the most widely used coat protein to achieve entry of lentiviral particles into cells is the VSV-G protein. However, although possessing a wide tropism, if a specific cell type is not transducible by VSV-G-pseudotyped viruses, it is possible to exchange this for other coat proteins from viruses with tropism for those tissues and organisms to achieve efficient infection. For example, use of the murine leukemia virus (MLV) ecotropic envelope glycoprotein encoded in the packaging vector pCAG-Eco (Addgene) allows viruses to efficiently infect specifically mouse cells that express the murine mCat1 protein. It should be possible to determine if the host receptor protein is expressed in your particular cell type at least for human and mouse tissues by accessing expression information from [22] and/or BioGPS [23].

3.3.1 Determination of the Optimal Concentration of Polybrene or Protamine Sulfate

Polybrene and protamine sulfate have been shown to enhance lentiviral transduction by up to five to tenfold [24], presumably by partially negating the negative charge repulsion between the target cell plasma membrane and incoming lentiviral particles to allow more efficient association. However, Polybrene, and to a lesser extent protamine sulfate, are toxic to cells and can in some cases interfere with differentiation. Hence, it is first necessary to determine the maximal amount of either Polybrene and/or protamine sulfate a specific cell type can tolerate without compromising growth and/or normal activity.

1. For testing proliferating cells, plate cells on 13 mm coverslips in a 6-well plate or 6 × 35 mm diameter plates at a density to allow 1–2 days uninterrupted and logarithmic growth 6–8 h before the transduction experiment. For testing differentiated cells, differentiate cells on coverslips in a 6-well plate or 6 × 35 mm diameter plates and perform transductions for the titration experiment 3 days prior to the end of the differentiation time course.

2. To 2 mL of fresh culture medium add either protamine sulfate or Polybrene to a final concentration of 0 μg/mL, 1 μg/mL, 2.5 μg/mL, 5 μg/mL, 7.5 μg/mL, and 10 μg/mL to cells and leave to incubate for 24 h.
3. Replace with fresh media and leave cells to incubate for a further 48 h. During this time observe cells under a light microscope to determine if significant changes to cellular morphology, growth rate, or ability to differentiate have occurred. Ideally, the maximum concentration with no effect on cells should be selected for later steps.

3.3.2 Determination of Optimal Amounts of Lentivirus for Transduction

Having determined the optimal quantity of the charge negating agent, it is next necessary to identify the optimal amount of a viral preparation to achieve sufficient transduction. Once delivered into cells, ectopically introduced Dam methylase-fusion proteins are then expressed at very low levels through the leaky expression of a *D. melanogaster* minimal heat shock promoter (HSP) for 72 h to generate sufficient DNA labeling for subsequent processing.

1. Plate cells as described in Subheading 3.3.1.
2. To 2 mL of fresh culture medium containing the previously selected concentration of protamine sulfate or Polybrene, add 1/40, 1/20, 1/10, 1/4, and 1/2 of a viral preparation to 5 separate wells, leaving 1 well untransduced as a control. Leave cells for 24 h for transduction to occur.
3. Aspirate media and replace with fresh medium appropriate for the cell type. Leave cells for a further 48 h for GFP levels to reach peak levels (*see* **Note 15**).
4. Aspirate media, wash coverslips in PBS, and fix cells for 8 min in 4 % paraformaldehyde, PBS.
5. Counterstain cellular DNA by Hoechst or DAPI and mount coverslips onto slides with a mounting agent such as Vectashield, sealing the edges with nail varnish.
6. View samples on an epifluorescence microscope, using a filter set appropriate for DAPI excitation and emission (this is also suitable for Hoechst 33342) and GFP excitation and emission. GFP expression should be observed in transduced cells (Fig. 2).
7. Count the number of GFP expressing cells in a particular field and the number of DAPI stained nuclei in the same field. Repeat this for five fields. The optimal quantity of a viral preparation for use in DamID experiment can be determined by identifying the minimal quantity of virus sufficient to reach expression of GFP in all cells (*see* **Note 16**).

3.4 Generation and Isolation of Dam-Methylated DNA

Having determined that cells are clear of mycoplasma and identified the optimal quantity of both charge negating agent and virus to use for efficient transduction, a DamID experiment can be

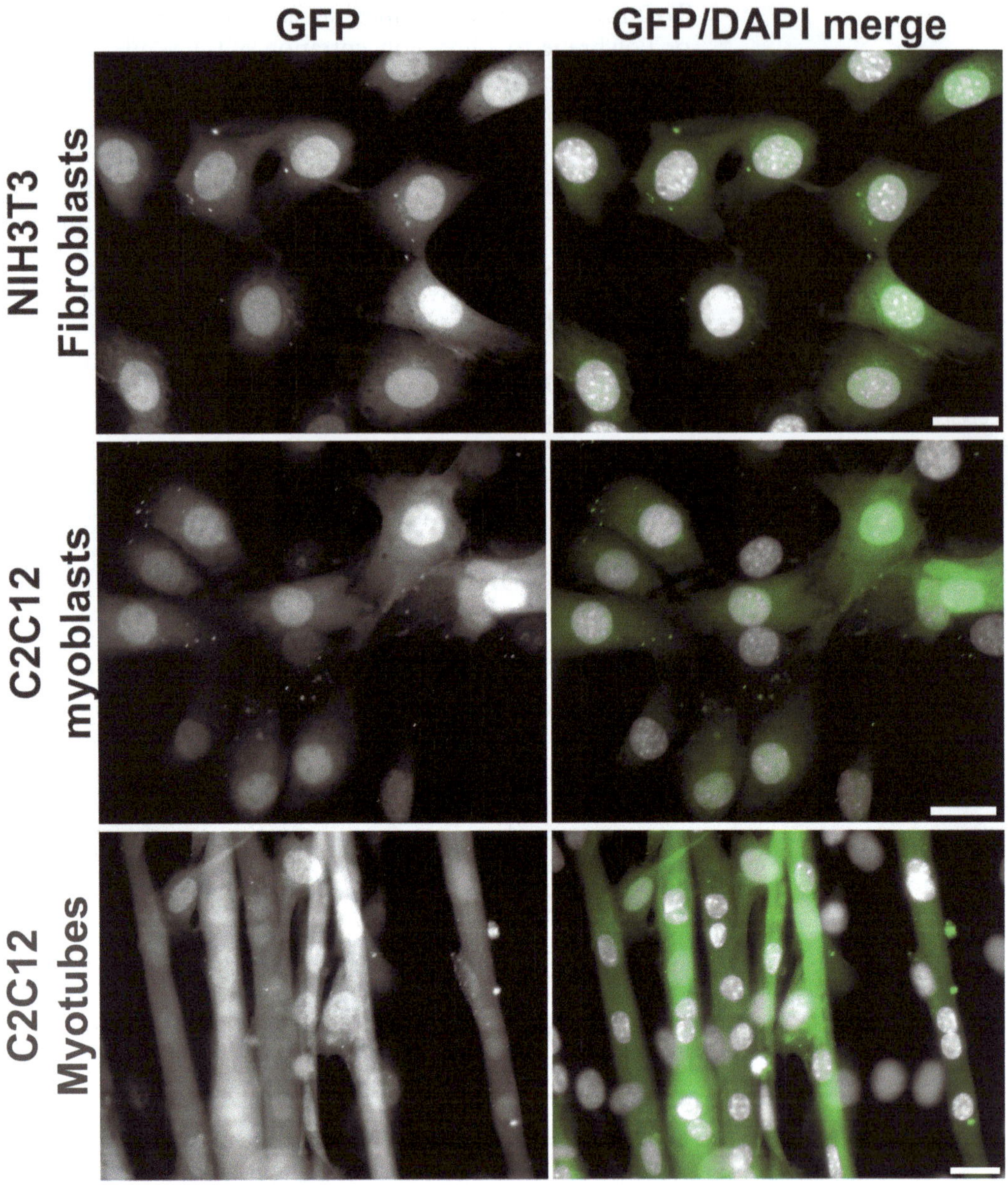

Fig. 2 Following transduction of each cell type/state with GFP encoding lentivirus, efficiency of infection can be assessed under an epifluorescence microscope by comparison of DAPI and GFP fluorescence. Displayed are examples of efficient transduction in NIH 3 T3 fibroblasts, C2C12 myoblasts and C2C12 myotubes. To achieve a similar degree of transduction as NIH 3 T3 fibroblasts, C2C12 myoblasts and myotubes required 5× and 20× the amount of GFP-encoding lentivirus. *Scale bar* is 10 μm

attempted. Cells are plated as previously described in Subheading 3.3.1 and then transduced with dam construct-encoding viruses. 72 h after transduction genomic DNA is extracted. For any DamID experiment, the DNA sequences identified as proximal to the protein of interest are compared to sequences labeled by the soluble dam methylase alone. This experimentally controls for accessibility differences in chromatin in the given cell type and sequence-dependent biases in the dam methylases activity. Additionally, to ensure any subsequently amplified DNA arose only from activity of exogenously introduced dam-encoding constructs, an additional plate of cells is transduced with an empty lentivirus lacking a Dam-methylase open reading frame (ORF). Hence, three transductions are performed for each condition using, an empty lentivirus control, a soluble dam methylase only-encoding lentivirus, and a dam-lamin B-encoding lentivirus.

A sufficient quantity of genomic DNA for each condition can be isolated from two 35 mm diameter plates of sub-confluent and logarithmically growing cells for several attempts at amplification of dam methylated DNA. For differentiated cultures, which often contain significantly more cells because typically confluency is reached before initiating differentiation or they are grown in 3D cultures, a single 35 mm plate generates a comparable quantity of DNA.

1. For proliferating cells: 6–8 h before the transduction plate cells in four 35 mm diameter plates at a density to allow 1–2 days uninterrupted and logarithmic growth. For differentiated cells: differentiate cells in two 35 mm diameter plates and perform the transduction 3 days prior to the stage of differentiation where labeling is desired (*see* **Note 17**).
2. To fresh media containing the predetermined amount of protamine sulfate/Polybrene add the similarly determined quantity of empty, dam alone- and dam-lamin B1-encoding lentiviruses. Return cells to the incubator and leave to incubate for 24 h.
3. Aspirate media and replace with fresh medium appropriate for the cell type. Leave cells for a further 48 h for Dam methylation of target DNA sequences to reach appreciable levels (*see* **Note 18**).
4. Aspirate media, wash cells twice in PBS and trypsinize, being sure to inactivate trypsin by addition of serum-containing media once dissociation is complete. Pellet by centrifugation and resuspend in 1 mL PBS.
5. Repeat centrifugation and resuspend cells in 200 μL PBS.
6. Isolate DNA using the DNeasy blood and tissue lysis kit as per manufacturer's instructions. Be sure to include 100 μg/mL DNase free-RNase and incubate at room temperature for 5 min prior to addition of Buffer AL (*see* **Notes 19** and **20**).
7. Precipitate DNA by addition of 0.1 volumes of 3 M sodium acetate, pH 5.5, and 3 volumes of 100 % ethanol. Place at −20 °C or −80 °C for at least 20 min and then spin at 13,000 × *g* for 30 min at 4 °C in a microfuge (*see* **Note 21**).

8. Aspirate supernatant and wash Eppendorf tubes in 500 μL 70 % ethanol by inversion. Centrifuge sample at 13,000 × *g* for 5 min at 4 °C.
9. Aspirate supernatant ensuring all ethanol has been removed and leave pellet to dry with the Eppendorf lid open for 5 min. If residual ethanol remains re-centrifuge then remove remaining liquid.
10. Resuspend pellet in nuclease free water and heat at 55–65 °C for several hours to allow full re-solubilization of gDNA. This can be aided by occasionally gently flicking the tube.
11. Determine the concentration of DNA by NanoDrop.

3.5 Amplification of Dam-Methylated DNA Fragments

Following extraction, isolated DNA is then processed to specifically amplify DNA fragments containing consecutive methylated GATC sequences. The DNA is first digested by *DpnI* to cut only methylated GATC sequences. Following *DpnI* heat inactivation, adaptors are ligated to these exposed DNA ends to allow their specific amplification. Following ligase heat inactivation, any remaining unmethylated GATC sequences are then digested by addition of *DpnII*. PCR is then employed using primers complementary to the ligated adaptors to specifically amplify *DpnI* fragments. Only fragments that possess adaptors on both ends, i.e., contained two consecutive methylated GATC sequences, are therefore amplified logarithmically. The amplified DNA is then assessed by gel electrophoresis and, ideally, appears as a continuous smear containing DNA fragments ranging from ~100 bp to 2000 bp (Fig. 3).

However, a number of variants to the smear exist which significantly affect the experimental outcome. For example, a smear can also be generated by the presence of partially digested genomic DNA isolated from apoptotic cells in the population. Such degraded DNA presents exposed ends to which the dsAdR can be ligated in a *DpnI*-independent manner. Hence, such DNA fragments will be ligated to adaptors and be amplified, generating amplified DNA that appears to be a smear.

Additionally, in a number of cases a banding pattern can be observed within the smear since the amplified DNA smear is the product of the distribution of fragment sizes produced by cutting the genome at GATC motifs. Hence, specific banding patterns can only be generated by the excessive amplification of either repetitive sequences within the nuclear genome or of smaller, multicopy non-genomic sequences present in the cell. Such sequences can derive from exogenous sources, such as Dam-methylated plasmids transfected into the cell, or contaminating mycoplasma genomes (Fig. 3A). However, surprisingly, the primary source of banding contamination we have experienced comes from the mitochondrial genome (mtDNA), which appears to be methylated by exogenously introduced Dam methylase although it is unclear how the

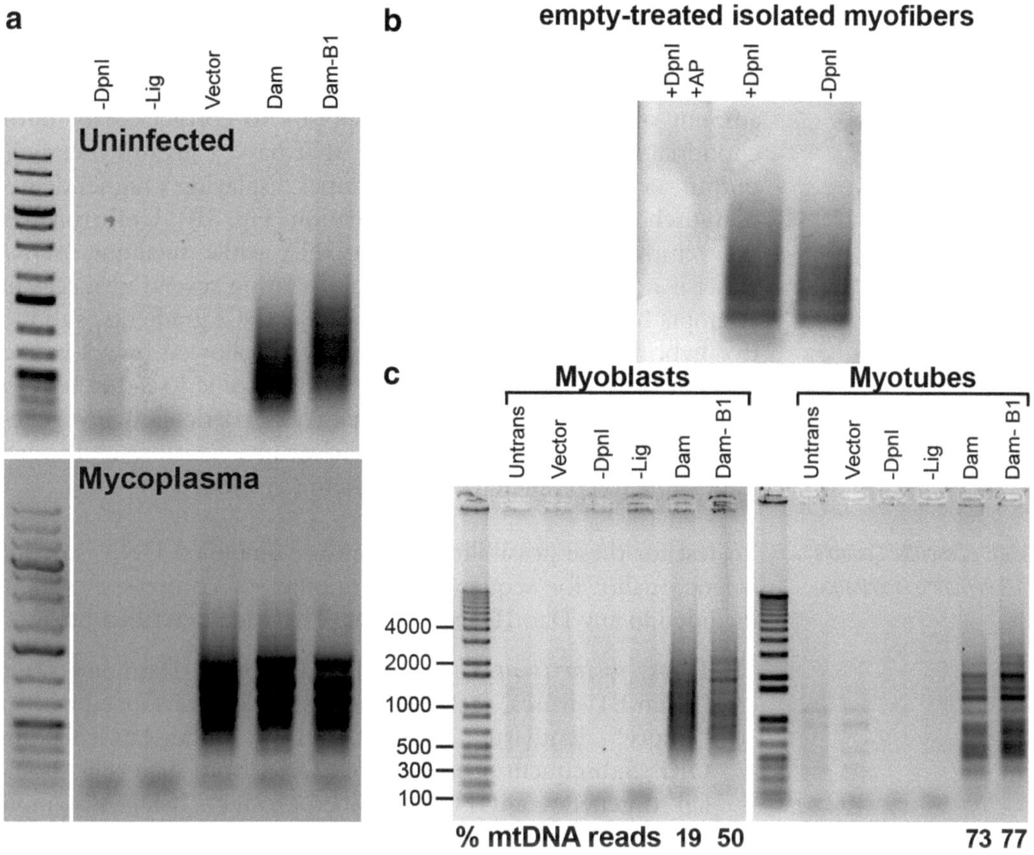

Fig. 3 Examples of DamID amplification product smears with underlying mycoplasma or mtDNA banding patterns. (**a**) In non-mycoplasma infected cells DNA does not amplify using gDNA isolated from an empty vector-treated control because no methylated DNA is present for *DpnI* cleavage. Dam methylated gDNA not treated with DpnI also fails to amplify during PCR as no 5′ or 3′ DNA ends are exposed for adaptors to ligate to. Similarly, non-T4 DNA Ligase treated dam-methylated gDNA lacks the ligated adaptors for amplification to occur also prevented amplification. In mycoplasma infected cells, although no DNA is amplified in *–DpnI* and –Ligase controls, a PCR product is generated in non-Dam methylase-treated cells as the mycoplasma genome, methylated by an endogenous Dam methylase, is present in extracted gDNA material. Since this genome is much smaller, and so has fewer *DpnI* fragments, a significant banding pattern is observed. (**b**) In primary isolated myofibers, which contain a significant number of hypercontracted and consequently apoptotic cells, a significant portion of the gDNA is degraded. Hence, DamID PCR amplification is observed even in cells not treated with the dam methylase in the absence of *DpnI* treatment as free ends are exposed for ligation to the DamID adaptor. This is eliminated upon pre-treatment of DNA with phosphatase as this prevents these exposed DNA ends from ligating to the similarly unphosphorylated DamID adaptor. (**c**) In C2C12 myoblasts treated with the dam methylase alone a normal DamID smear pattern is generated in PCR amplified material. In C2C12 myoblasts treated with Lamin B1 fused to the Dam methylase, and in C2C12 myotubes treated with either form of the methylase, a mtDNA banding pattern is observed. Therefore, the dam methylase labels the mitochondrial genome since amplification is not observed in control samples as with mycoplasma infection. Following sequencing the fraction of total reads derived from the mtDNA was calculated showing significantly higher contamination in the myotube samples

constructs target to mitochondria. The bands observed over the expected smear have been cut out of gels and cloned and confirmed by sequencing to derive from mitochondrial DNA. The amount of this banding pattern appears to correlate with mitochondrial copy number, with cells that have a greater metabolic output such as adipocytes and myotubes displaying a higher degree of mitochondrial banding contamination (Fig. 3B). Unfortunately, the removal of contaminating mtDNA while retaining gDNA intact is non-trivial and at the time of writing several unsuccessful attempts have been made to do so using CsCl gradients, subtractive hybridization and gel electrophoresis followed by gel extraction. In contrast, nuclear isolation by hypotonic lysis can help for many cell types; however, several cell types, particularly myotubes, have mitochondria embedded in nuclear invaginations and so it is impossible to completely remove them (*see* **Note 22**).

3.5.1 Essential Controls to Run During Preparation

To test for these possibilities and ensure amplified DNA is of sufficient quality for sequencing, a number of controls should be included in any DamID experiment. These are described below.

1. *Empty vector treated control.* In addition to Dam and Dam-Lamin B1 viruses, each cell type should be transduced with an "empty" lentivirus lacking a Dam methylase-encoding ORF. Subsequent processing of gDNA from such cells should fail to generate any PCR amplified product. If DNA is amplified then either the 293FT cells used to produce the virus or the target cells themselves are contaminated with a prokaryotic parasite such as mycoplasma (Fig. 3C).
2. *Minus DpnI.* For each cell type an additional aliquot of gDNA from one sample treated with a Dam methylase-encoding virus should be processed without the initial *DpnI* digestion. If an amplification product is observed this would suggest a significant number of exposed DNA ends, such as those generated by apoptosis-induced cleavage of DNA, are available for ligation to the adaptor (*see* **Note 23**) (Fig. 3B).
3. *Minus T4 DNA ligase.* For each cell type an additional aliquot of gDNA from one sample treated with a Dam methylase-encoding virus should be processed without the T4 DNA ligase to ensure amplified DNA is not the result of dsAdR-independent amplification of similar sequences present in the genome (Fig. 3C).

3.5.2 Amplification Steps and Controls

Once DNA has been isolated and quantified it is processed to amplify Dam methylated sequences. However, in addition to amplifying the material for sequencing from each condition it is necessary to include two additional control reactions, using the same processed gDNA for each condition but lacking the *DpnI* digestion and ligation steps, respectively.

1. In a 200 μL PCR tube digest 2.5 μg of isolated genomic DNA with *DpnI* at 37 °C for a minimum of 8 h in the 10 μL reaction described in Table 1:

   One additional reaction lacking *DpnI* should also be performed on gDNA extracted from cells treated with a Dam methylase encoding virus for the dam minus control.

2. Prepare a 50 μM stock of the dsAdR adaptor by combining equal volumes of the 100 μM AdRt and AdRb solutions and denature at 95 °C for 10 min. Leave to cool slowly to room temperature over several hours to allow annealing.

3. Heat inactivate *DpnI* at 80 °C for 25 min and then ligate the dsAdR at 16 °C for 16 h in a thermocycler with the heated lid switched off in the 20 μl reaction described in Table 2:

   One additional reaction lacking T4 DNA ligase should also be performed on gDNA extracted from cells treated with a Dam methylase encoding virus for the minus ligase control.

4. Heat-inactivate T4 DNA ligase at 65 °C for 20 min and digest remaining unmethylated GATC sequences with *DpnII* at 37 °C for 1 h in the 50 μl reaction described in Table 3:

Table 1
Set up of *DpnI* digestion reaction

Reaction component	Total amount (per reaction)	Final concentration	Amount to add (per reaction) (μL)
gDNA (1 μg/μL)	2.5 μg	250 ng/μL	2.5
Buffer 4 (10×)		1×	1.0
DpnI (20 U/μL)	10 U	1 U/μL	0.5
Nuclease-free water			6

Table 2
Setup of adaptor ligation reaction

Reaction component	Total amount (per reaction)	Final concentration	Amount to add (per reaction) (μL)
DpnI digested gDNA	2.5 μg	125 ng/μL	10
Ligation buffer (10×)		1×	2.0
T4 DNA ligase (5 U/μL)	5 U	0.25 U/μL	1.0
dsAdR (50 μM)		2 μM	0.8
Nuclease-free water			6.2

5. On ice assemble the 50 μl reaction described in Table 4 in 200 μl PCR tubes for PCR-amplification of adaptor-ligated *DpnI* fragments with the Adr-PCR primer.
6. Run assembled reactions in a thermocycler using the program described in Table 5:

Table 3
Set up *DpnII* digestion reaction

Reaction component	Total amount (per reaction)	Final concentration	Amount to add (per reaction) (μL)
Ligated DNA	2.5 μg	50 ng/μL	20
DpnII buffer (10×)		1×	5
DpnII (10 U/μL)	10 U	0.2 U/μL	1
Nuclease-free water			24

Table 4
Setup of PCR reaction to amplify Dam methylated DNA fragments

Reaction component	Total amount (per reaction)	Final concentration	Amount to add (per reaction) (μL)
DpnII digested gDNA	250 ng	5 ng/μL	5
cDNA PCR buffer (10×)		1×	5.0
Adr-PCR primer (50 μM)		1.25 μM	1.25
dNTPs (10 mM)		0.2 mM	1.0
Advantage PCR enzyme mix (50×)		1×	1
Nuclease-free water			36.75

Table 5
Program for DamID PCR reaction

Cycle	Denature	Anneal	Extend
Heated lid one			
1			68 °C for 10 min
2	94 °C for 3 min	65 °C for 5 min	68 °C for 15 min
3–6	94 °C for 1 min	65 °C for 1 min	68 °C for 10 min
7–23	94 °C for 1 min	65 °C for 1 min	68 °C for 2 min

7. Take 3 μL of each reaction and load onto a 2 % agarose gel for 40 min at 100 V. Stain the gel with ethidium bromide for 10 min and leave to destain in water for an additional 5 min. Image the gel using local gel documentation apparatus. In Dam-encoding lentivirus treated samples a smear of amplified DNA fragments should be observed ranging from 100 bp to 2000 bp (Fig. 3).
8. Additional PCR reactions are typically required to generate sufficient DNA for sequencing. In general 4–6 accumulated PCR reactions are sufficient to generate ~10 μg DNA for sequencing (*see* **Note 24**). However, before proceeding the pattern of amplification products should be assessed as described in Subheading 3.5.3 to ensure samples are of sufficient quality for further analysis.

3.5.3 Quality Testing of Amplified Dam-Methylated DNA

The results of negative controls and samples as they appear on the agarose gels should be interpreted to ensure samples are of sufficient quality for further analysis. Below are series of scenarios which have been observed together with a means to counter them.

1. If amplification is observed in the empty vector treated control then it is likely that either the 293FT cells or the target cells are contaminated with prokaryotic parasites such as mycoplasma. If so the amplified product will appear as an intense banding pattern matching the *DpnI* digestion pattern of the mycoplasma genome similar to that shown in Fig. 3A. Fresh uninfected cells should be thawed or, if none are available, should be treated with BM cyclins as described (see step 2, Subheading 3.1). If cells were previously shown to not be infected with mycoplasma it is likely cells are infected with another prokaryotic organism. This contamination can be identified by cloning and sequencing the DNA fragments in the DamID PCR product.
2. If amplification is observed in the *DpnI*– control then it is likely that a significant portion of the amplified product is derived from partially degraded DNA extracted from a subpopulation of cells containing apopotically degraded gDNA. We have observed this for example in isolated mouse myofibers where a significant number of myofibers hyper-contract and begin apoptosis (Fig. 3B) (*see* **Note 23**).
3. If amplification is observed in the minus T4 DNA ligase control then it is likely that the amplification product is generated by off-target amplification of genomic sequences by the Adr-PCR primer. If this is the case, then such amplification should also be observed in the empty vector treated control.
4. If amplification is observed only in dam methylase-treated samples but a banding pattern is observed, then a significant frac-

tion of mtDNA contamination is likely present (Fig. 3C). If analyzing DamID PCR products by sequencing, a greater depth of sequencing may be required to generate sufficient quality data. However, if analyzing by microarray such contamination will have a minimal effect as the mitochondrial genome is not represented on genomic tiling arrays. MtDNA contamination may be reduced by isolating nuclei by hypotonic lysis, as described in Ref. [25]. However, in some cell types such as myoblasts and myotubes, many mitochondria remain tightly associated with isolated nuclei and contamination is retained.

3.6 Deep Sequencing, Data Analysis and Presentation

3.6.1 Construction and Deep Sequencing of DamID Libraries

Once the quality of samples is deemed sufficient for sequencing, the PCR product should be purified using for example the QIAquick PCR Purification Kit as per manufacturer's instructions. Samples should then be processed into libraries suitable for the chosen next generation sequencing platform. For example, we use the HiSeq 2000 platform (Illumina) in conjunction with 90 bp paired end (90PE) sequencing reactions with libraries prepared by first fragmenting DamID PCR amplification products, performing end repair and then adding sequencing library adaptors.

The depth of sequencing required will depend on both the amount of mitochondrial contamination and the average reads obtained by your particular sequencing facility and setup. For example, using the HiSeq 2000 platform (Illumina) in conjunction with 90 bp paired end (90PE) sequencing reactions with sequencing performed by Beijing Genomics (BGI), we obtain ~35 million reads per sample with running five samples multiplexed per well. We have found 30–40 million reads is sufficient for robust lamin B1 DamID traces and so the degree of mitochondrial contamination will affect the sequencing depth required. For example, a contamination level of 75 % would require four times the sequencing depth to generate a similar number of functional reads. Also the standard output of a particular sequencing facility must be considered: for example, we have had experiences with other facilities that only get 100 million reads per well whereas BGI typically gets double this number.

3.6.2 Analysis and Presentation of DamID Data

1. Map retrieved reads to the appropriate genome using the BWA aligner.
2. At this stage to determine your depth of coverage it is also important to remove any mitochondrial or other contaminating sequences from the data. If there are less than 30 million reads after removing any contamination then send more sample for sequencing.
3. For analysis, we quantify the "normalized reads" per genomic *DpnI* fragment as the proportion of total reads in that sample, which overlap with the given fragment. This is done for both Dam-lamin B1- and Dam only-treated samples.

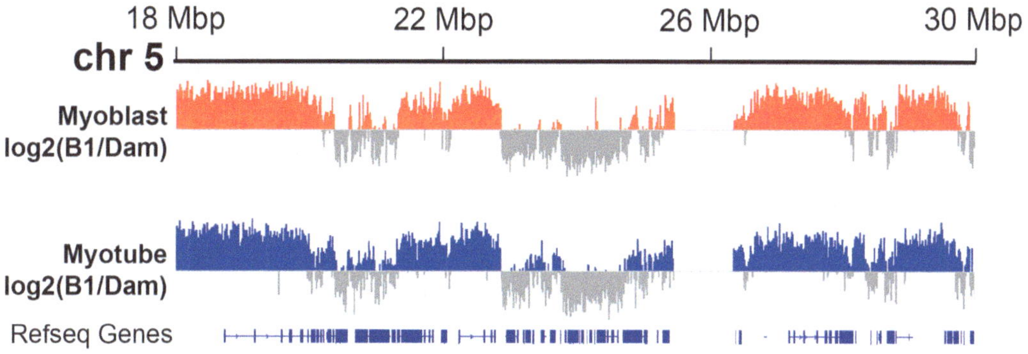

Fig. 4 Example genome browser representation of lamin B1 DamID in C2C12 myoblasts and myotubes. For analysis the genome is represented as individual *DpnI* fragments with signal corresponding to the fragments $\log_2$(Dam-lamin B1/Dam only) value. In most cell types this generates a pattern of lamina-associated domains (LADs) interspersed with non-LAD domains

4. The signal for any given *DpnI* fragment is then represented as the $\log_2$ ratio between the Dam-lamin B1 and Dam only normalized reads, i.e., $\log_2$(Dam-laminB1/Dam only) following quantile normalization using the Bioconductor Limma package.
5. Import the file generated in step 4 into the integrated genome viewer (IGV) [20]. This generates a trace with regions of lamina-associated domains (LADs) interspersed with non-LAD domains (Fig. 4).

4 Notes

1. If the DNeasy kit is inefficient for a specific cell type, phenol–chloroform extractions as described in Ref. [26] can be performed. However, care must be taken to prevent excessive shearing of gDNA.
2. Other prokaryotic contaminations may also result in a banding pattern in DamID PCR product smears and so if cells are mycoplasma negative and still display banding then alternate primers complementing different suspected organisms should be used. Contaminating organisms can also be identified by cloning and sequencing of individual bands from the DamID PCR products. Note the Blunt PCR cloning kit (StrataClone) is less efficient at generating colonies of DamID PCR product DNA fragments.
3. We have only applied the BM cyclins for mycoplasma clearance; however, alternative mycoplasma treatments are available such as Plasmocin (Invivogen) and MycoZap™ (Lonza).

4. Pore sizes of 0.22 μm^2 or filters employing mixed nitrocellulose esters have been noted to reduce viral titers.
5. 293FT cells are a sub-clone of HEK293 human embryonic kidney cells that possess a higher growth rate than standard HEK293 cells allowing more rapid cell culture for viral preparations. They also possess the exogenously delivered and stably expressed large T antigen, which allows for maintenance/amplification of transfected plasmids resulting in higher viral titers. 293FT cells should be maintained in 0.5 mg/mL geneticin to maintain expression of the large T antigen by selection.
6. Calcium phosphate co-precipitation has also been used to transfect 293 T cells for the generation of lentiviruses [7]. However, we have found titers are significantly higher using Lipofectamine.
7. It is important to include a negative control and run samples on a 2 % agarose gel to prevent the primer dimer band being interpreted as a false positive.
8. To achieve complete mycoplasma elimination, be sure to continue the cycle of treatment at least three times as long as suggested by the manufacturer.
9. During treatments of BM cyclins 1 and 2 cells should be passaged as normal. However, in our experience BM cyclin 2 can impair the proliferation and survival of some cell types and so higher than normal initial plating densities may be required when transitioning to the second BM cyclin 2 stage of treatment to ensure that proliferation is not further inhibited by low cell density.
10. During transfections 293FT cells should be plated in media lacking geneticin, penicillin, or streptomycin to promote maximum viability following the addition of Lipofectamine.
11. At this stage cells are highly susceptible to detaching from growth substratum so media should be added gently to prevent a reduction in yield through loss of cells. Adding greater than 10 mL of media provides more nutrients to better sustain cells for a further 48 h.
12. To ensure sterility, Oak Ridge tubes should be washed with bleach for 10 min and then washed repeatedly with ethanol followed by PBS. Note also if centrifugation using a JA 25.5 rotor or ultracentrifugation using a SW28 rotor is not an available option lentivirus particles can also be concentrated using polyethylene glycol (PEG) based concentrators such as LentiX concentration (Clontech). However, in our hands, PEG based concentrators can reduce yields and inhibit processes such as myogenic differentiation.

13. A volume of at least 2 mL is required to ensure pellet is covered in solution in the centrifuge tube for resuspension by rocking. If smaller volumes, and therefore higher concentrations of virus, are required, pellets can be resuspended by gentle pipetting although this does result in a small loss of titer. Since viruses can be resuspended in target cell media the volumes used can be the same as the final volume used during transduction.
14. The ecotropic viral coat protein is less resilient than the more commonly used VSV-G coat protein. Consequently, greater losses in titer occur when freeze thawing ecotropic viruses than VSV-G pseudotyped viruses.
15. The rate and degree to which GFP is expressed following transduction varies significantly between cell types. Testing for the level of GFP expression is done after 3 days because this is the length of time found to be the minimum required for a successful DamID experiment in a number of cell types. However, the length of time required for peak GFP expression may take longer than 3 days in some cell types and so longer times may be considered if expression is deemed insufficient. This longer time post-transduction may be then required for subsequent DamID experiments.
16. It may not be possible to reach 100 % transduction efficiency in some cell types using the entirety of a viral preparation as described in Subheading 3.2. If significantly higher amounts of virus are required, greater quantities can be generated by combining multiple concentrated viral preparations. However, optimal 100 % transduction efficiency is not required for successful DamID. Hence, as long as a significant number of cells are transduced it is often impractical and unnecessary to invest in generating much higher titers, especially as in some cases such high titers can interfere with differentiation or cause toxicity.
17. To generate sufficient methylated gDNA for a DamID experiment it is necessary to leave cells in culture for 72 h following transduction. In our experience, the majority of DNA methylation occurs between 48 and 72 h. However, this does not allow the application of DamID mediated by lentiviral transduction to experiments requiring precise temporal resolution. In such cases ChIP, or an inducible variant of DamID (*see* **Note 18**), may be preferable [3, 27].
18. In our experience the majority of a DamID labeling pattern is established between days 2 and 3 following transduction. This has a major disadvantage in limiting the temporal resolution of the method [7]. To bypass this problem, the van Steensel laboratory has generated a new form of DamID in which Dam

fusion proteins are tagged with the destabilization domain (DD). The addition of the DD targets tagged proteins for rapid proteasomal degradation unless the protein is shielded by the addition of the small cell permeable molecule Shield [27, 28]. This precise control of the expression of constructs allows fusion proteins to be expressed at specific times and at higher levels, generating sufficient enzymatic activity to generate a DamID methylation pattern within a single round of interphase. However, the DD-DamID method requires the selection of individual clones to select lines with optimal levels of protein turnover and so cannot be easily applied to most differentiation systems or primary cells.

19. Greater DNA yields can be recovered if the elution buffer is heated to 55 °C and left on the membrane for 2 min prior to final centrifugation. Finally, use of a second elution using an additional 200 μL of elution buffer allows greater recovery also. Cells can also be lysed directly on the plate without trypsinization by addition of 200 μL PBS and 200 μL Buffer AL.

20. Amplification of methylated DNA fragments for DamID requires the ligation of adaptors to exposed DNA ends made available by *Dpn*I digestions. Although the DNeasy kit shears DNA into large fragments with a number of exposed ends independently of *Dpn*I digestion, this level of fragmentation fails to produce a significant background in DamID experiments. However, excessive gDNA shearing by rough pipetting or an abundance of apoptotic cells in the population may generate sufficient numbers of free DNA ends to allow *Dpn*I-independent addition of adaptors and result in higher backgrounds when sequencing amplified DNA.

21. If restricted by a limited number of cells, enhance the precipitation of low concentrations of DNA through the addition of 2 μg glycogen prior to the addition of ethanol.

22. The presence of mtDNA is only a significant obstacle when using sequencing since all DNA fragments are read. However, if using microarrays, interference from contaminating mtDNA will be limited to genomic sequences of high similarity and will therefore not interfere with the majority of the hybridizations to the genome wide tiling arrays.

23. To eliminate such contamination gDNA can be pretreated in the *Dpn*I Cutsmart buffer prior to the first digestion step with Antarctic phosphatase in the presence of 0.1 mM $ZnCl_2$ for 2 h at 37 °C to eliminate the subsequent *Dpn*I-independent ligation of adaptors to apoptotic DNA cleavage-generated DNA ends. The Antarctic phosphatase must then be heat inactivated for 10 min at 70 °C prior to *Dpn*I digestion.

24. The generation of ssDNA during the PCR, presumably the result of partial amplification of fragments possessing an dsAdR adaptor at only one end or larger fragments than can be amplified in the relatively short OCR amplification timescale, prevents the determination of concentration of double stranded product using a NanoDrop spectrophotometer. The most reliable means of determining an accurate dsDNA concentration is using the Qubit fluorometer with the associated BR dsDNA assay kit.

Acknowledgments

The authors would like to thank Jose de las Heras who contributed significantly to the establishment and analysis of DamID in the Schirmer lab, and Bas van Steensel, Carolyn de Graaf, and Job Kind who provided considerable guidance on the method. MIR was supported by a Wellcome Trust Ph.D. Studentship (093854). Funding for this work was provided by Welcome Trust grants 095209 to ECS and 092076 for the Centre for Cell Biology.

References

1. Zuleger N, Robson MI, Schirmer EC (2011) The nuclear envelope as a chromatin organizer. Nucleus 2:339–349
2. Wong X, Luperchio TR, Reddy KL (2014) NET gains and losses: the role of changing nuclear envelope proteomes in genome regulation. Curr Opin Cell Biol 28:105–120
3. Lund E, Oldenburg AR, Delbarre E, Freberg CT, Duband-Goulet I, Eskeland R, Buendia B, Collas P (2013) Lamin A/C-promoter interactions specify chromatin state-dependent transcription outcomes. Genome Res 23:1580–1589
4. Dechat T, Adam SA, Goldman RD (2009) Nuclear lamins and chromatin: when structure meets function. Adv Enzyme Regul 49:157–166
5. Burke B, Stewart CL (2013) The nuclear lamins: flexibility in function. Nat Rev Mol Cell Biol 14:13–24
6. Guelen L, Pagie L, Brasset E, Meuleman W, Faza MB, Talhout W, Eussen BH, de Klein A, Wessels L, de Laat W, van Steensel B (2008) Domain organization of human chromosomes revealed by mapping of nuclear lamina interactions. Nature 453:948–951
7. Vogel MJ, Peric-Hupkes D, van Steensel B (2007) Detection of in vivo protein-DNA interactions using DamID in mammalian cells. Nat Protoc 2:1467–1478
8. Tunnah D, Sewry CA, Vaux D, Schirmer EC, Morris GE (2005) The apparent absence of lamin B1 and emerin in many tissue nuclei is due to epitope masking. J Mol Histol 36:337–344
9. van Steensel B, Henikoff S (2000) Identification of in vivo DNA targets of chromatin proteins using tethered dam methyltransferase. Nat Biotechnol 18:424–428
10. Pickersgill H, Kalverda B, de Wit E, Talhout W, Fornerod M, van Steensel B (2006) Characterization of the Drosophila melanogaster genome at the nuclear lamina. Nat Genet 38:1005–1014
11. Peter A, Stick R (2012) Evolution of the lamin protein family: what introns can tell. Nucleus 3:44–59
12. Schwanhausser B, Busse D, Li N, Dittmar G, Schuchhardt J, Wolf J, Chen W, Selbach M (2013) Corrigendum: Global quantification of mammalian gene expression control. Nature 495:126–127
13. Schwanhausser B, Busse D, Li N, Dittmar G, Schuchhardt J, Wolf J, Chen W, Selbach M (2011) Global quantification of mammalian gene expression control. Nature 473:337–342

14. Wu F, Yao J (2013) Spatial compartmentalization at the nuclear periphery characterized by genome-wide mapping. BMC Genomics 14:591
15. Peric-Hupkes D, Meuleman W, Pagie L, Bruggeman SW, Solovei I, Brugman W, Graf S, Flicek P, Kerkhoven RM, van Lohuizen M, Reinders M, Wessels L, van Steensel B (2010) Molecular maps of the reorganization of genome-nuclear lamina interactions during differentiation. Mol Cell 38:603–613
16. Steglich B, Filion GJ, van Steensel B, Ekwall K (2012) The inner nuclear membrane proteins Man1 and Ima1 link to two different types of chromatin at the nuclear periphery in S. pombe. Nucleus 3:77–87
17. Gonzalez-Aguilera C, Ikegami K, Ayuso C, de Luis A, Iniguez M, Cabello J, Lieb JD, Askjaer P (2014) Genome-wide analysis links emerin to neuromuscular junction activity in Caenorhabditis elegans. Genome Biol 15:R21
18. Germann S, Juul-Jensen T, Letarnec B, Gaudin V (2006) DamID, a new tool for studying plant chromatin profiling in vivo, and its use to identify putative LHP1 target loci. Plant J 48:153–163
19. Lund E, Oldenburg AR, Collas P (2014) Enriched domain detector: a program for detection of wide genomic enrichment domains robust against local variations. Nucleic Acids Res 42, e92
20. Robinson JT, Thorvaldsdottir H, Winckler W, Guttman M, Lander ES, Getz G, Mesirov JP (2011) Integrative genomics viewer. Nat Biotechnol 29:24–26
21. Lluch-Senar M, Luong K, Llorens-Rico V, Delgado J, Fang G, Spittle K, Clark TA, Schadt E, Turner SW, Korlach J, Serrano L (2013) Comprehensive methylome characterization of Mycoplasma genitalium and Mycoplasma pneumoniae at single-base resolution. Plos Genet 9, e1003191
22. Uhlen M, Fagerberg L, Hallstrom BM, Lindskog C, Oksvold P, Mardinoglu A, Sivertsson A, Kampf C, Sjostedt E, Asplund A, Olsson I, Edlund K, Lundberg E, Navani S, Szigyarto CA, Odeberg J, Djureinovic D, Takanen JO, Hober S, Alm T, Edqvist PH, Berling H, Tegel H, Mulder J, Rockberg J, Nilsson P, Schwenk JM, Hamsten M, von Feilitzen K, Forsberg M, Persson L, Johansson F, Zwahlen M, von Heijne G, Nielsen J, Ponten F (2015) Proteomics. Tissue-based map of the human proteome. Science 347:1260419
23. Wu C, Orozco C, Boyer J, Leglise M, Goodale J, Batalov S, Hodge CL, Haase J, Janes J, Huss JW 3rd, Su AI (2009) BioGPS: an extensible and customizable portal for querying and organizing gene annotation resources. Genome Biol 10:R130
24. Cornetta K, Anderson WF (1989) Protamine sulfate as an effective alternative to polybrene in retroviral-mediated gene-transfer - implications for human-gene therapy. J Virol Methods 23:187–194
25. Korfali N, Fairley EAL, Swanson SK, Florens L, Schirmer EC (2009) Use of sequential chemical extractions to purify nuclear membrane proteins for proteomics identification. Methods Mol Biol 528:201–225
26. Strauss WM (2001) Preparation of genomic DNA from mammalian tissue. *Curr Protoc Immunol* Chapter 10, Unit 10 12
27. Kind J, Pagie L, Ortabozkoyun H, Boyle S, de Vries SS, Janssen H, Amendola M, Nolen LD, Bickmore WA, van Steensel B (2013) Single-cell dynamics of genome-nuclear lamina interactions. Cell 153:178–192
28. Banaszynski LA, Chen LC, Maynard-Smith LA, Ooi AGL, Wandless TJ (2006) A rapid, reversible, and tunable method to regulate protein function in living cells using synthetic small molecules. Cell 126:995–1004

Chapter 24

Visualizing the Spatial Relationship of the Genome with the Nuclear Envelope Using Fluorescence In Situ Hybridization

Craig S. Clements, Ural Bikkul, Mai Hassan Ahmed, Helen A. Foster, Lauren S. Godwin, and Joanna M. Bridger

Abstract

The genome has a special relationship with the nuclear envelope in cells. Much of the genome is anchored at the nuclear periphery, tethered by chromatin binding proteins such nuclear lamins and other integral membrane proteins. Even though there are global assays such as DAM-ID or ChIP to assess what parts of the genome are associated with the nuclear envelope, it is also essential to be able to visualize regions of the genome in order to reveal their individual relationships with nuclear structures in single cells. This is executed by fluorescence in situ hybridization (FISH) in 2-dimensional flattened nuclei (2D-FISH) or 3-dimensionally preserved cells (3D-FISH) in combination with indirect immunofluorescence to reveal structural proteins. This chapter explains the protocols for 2D- and 3D-FISH in combination with indirect immunofluorescence and discusses options for image capture and analysis. Due to the nuclear envelope proteins being part of the non-extractable nucleoskeleton, we also describe how to prepare DNA halos through salt extraction and how they can be used to study genome behavior and association when combined with 2D-FISH.

Key words Fluorescence in situ hybridization, 2D-FISH, 3D-FISH, Genome organization, Chromosome territories, Gene positioning, Nuclear envelope, Nuclear lamins

1 Introduction

Chromosomes and genes are spatially organized within interphase nuclei, interacting with proteinaceous nuclear structures which anchor and tether chromatin [1, 2]. Global methodologies, such as DAM-ID, utilizing the ease of sequencing DNA, can assess the spatial relationship of the whole genome with specific nuclear structures such as nuclear lamins [3–5]. These analyses bring useful information to the field, however as studies are data-rich and expensive, they might be excluded as a method of choice if a large number of samples are required (e.g., steps in a differentiation pathway or large numbers of patient samples). Furthermore, using

Sue Shackleton et al. (eds.), *The Nuclear Envelope: Methods and Protocols*, Methods in Molecular Biology, vol. 1411,
DOI 10.1007/978-1-4939-3530-7_24, © Springer Science+Business Media New York 2016

cells in mixed populations will result in averaged information and the genome, for example, can be positioned differently in proliferating versus non-proliferating cells, e.g., in a primary cell culture [6]. Real-time imaging in live cells [7, 8] can also be an informative approach to visualize genome interaction with nuclear structures; however, this may present challenges [9]. Therefore, despite global and real-time analyses, there is a necessity to analyze individual cells by fluorescence in situ hybridization (FISH) to reveal specific regions of interest in the genome, and analyze their spatial relationship to specific nuclear structures such as the nuclear envelope [10, 11].

FISH is a cytogenetic technique that utilizes labeled DNA or RNA probes to hybridize to specific DNA sequences on chromosomes, or to RNA [12, 13]. It is possible to delineate whole chromosomes, chromosome arms, chromosome bands, gene loci, specific regions of genes or RNA transcripts [14], and analyze their spatial relationship with other FISH signals or nuclear landmarks such as the geometric center of nuclei, nucleoli, or the nuclear envelope. The nuclear envelope can be revealed by indirect immunofluorescence using antibodies [15–17], but the FISH denaturation process can negatively affect nuclear envelope antigens, especially A-type lamins. The use of formamide for DNA denaturation can also destroy the signal from fluorescently tagged proteins but antibodies that recognize the tag itself can be used [18]. Indirect immunofluorescence can be done prior to the FISH, by fixing the antigen–antibody complexes in place with paraformaldehyde; in this scenario, the nuclear envelope can still be revealed with specific antibodies [18–22]. However, the simplest way of revealing the nuclear edge is by using a DNA stain and where it ends is where the edge of the nucleus is; this is however as long as nuclei have maintained their integrity and shape during the procedure [23–26]. Use of phase contrast microscopy or lipophilic dyes can be useful, in cases where the nuclear envelope is compromised, to reveal the nuclear membranes. Abnormal nuclear structures such as blebs can create issues when defining a nuclear edge since they could be defined as an extension of the nuclear envelope or as a separate entity [27].

For 2D-FISH, whereby nuclei are flattened, a standard epifluorescence microscope with a cooled charge coupled device camera will suffice for visualization and imaging of FISH signals. However, for 3D-FISH, microscopy requires optical sections through the z-axis, either using a confocal laser scanning microscope or other microscope systems with a motorized stage or piezo-driven objectives. The 3D datasets then need to be processed through deconvolution packages to remove out of focus fluorescence. There are new opportunities to visualize FISH signals within interphase nuclei using super-resolution microscopy such as 3D-Structural Illumination Microscopy (3D-SIM) [16] or high-throughput flow imaging [28, 29]. The high power lenses that can be added to imaging flow machines, combined with the extended depth of field, enable software to be developed to perform positional analyses of chromosomes and genes in millions of individual interphase nuclei for one sample in 3D. Individual cells can then be sub-categorized into for example proliferating or not, differentiated or not, transgenic or not, by co-staining with specific markers. A further advancement is in high-throughput / high-resolution imaging, an approach called “Deep Imaging”: there, individual cells are imaged at high resolution by automated software and FISH signals positioned by automated analysis software built into the system [30, 32].

Analyzing the position of genes, sub-chromosomal regions or whole chromosomes within interphase nuclei requires the capability to be able to perform measurements and approaches that place FISH signals into specific regions of the nucleus. In 3D, this is usually done using the reconstructed stack of optical sections or a gallery of the 2D optical sections to measure to the nearest point at the nuclear edge in any of the x, y, z axes, or by taking the measurement to the geometric center of the nucleus and expressing this a percentage of the radius. 3D radial analysis packages have been developed [17, 31, 32] to perform measurements between nuclear landmarks and FISH signals, or between FISH signals [33, 34]. Freely available software to position genes and chromosomes includes Smart 3D-FISH [35] and NEMO [36]. However, many laboratories use 2D preparations to perform their spatial positioning analyses since similar nuclear positions for genes and chromosomes are found consistently in 3D, confirming that positional analyses can be performed in 2D [37]. In order to position FISH signals in flattened nuclei different groups have developed their own positional analysis scripts similar to the original created by Paul Perry, MRC Human Genetics Unit, Edinburgh [38], whereby the nucleus is segmented into 5 shells of equal area and the intensity of the FISH signal and DNA signal from the DNA dye DAPI are measured [38–40]. The DNA signal intensity is used to normalize the FISH signal data. Others do not perform any normalization for DNA content across the flattened nuclei but use specific measurements to nuclear landmarks such as the nuclear edge and the nuclear center [41]. With such 2D studies, it is imperative that the data be comparable between samples since they do not give absolute position but an averaged probabilistic position.

By using FISH and delineation of the nuclear envelope or other nuclear structures, it is possible to determine which parts of the genome are colocalized or spatially close to nuclear structures such as the nuclear lamina, nuclear pores and other nuclear components. This chapter outlines methods for 2D- and 3D-FISH in combination with indirect immunofluorescence, and discusses the value of 2D vs 3D analyses. We also include methods for the DNA halo extraction protocol combined with 2D-FISH, making it possible to gather data about which regions of the genome are embedded or attached to insoluble nuclear structures (Fig. 1).

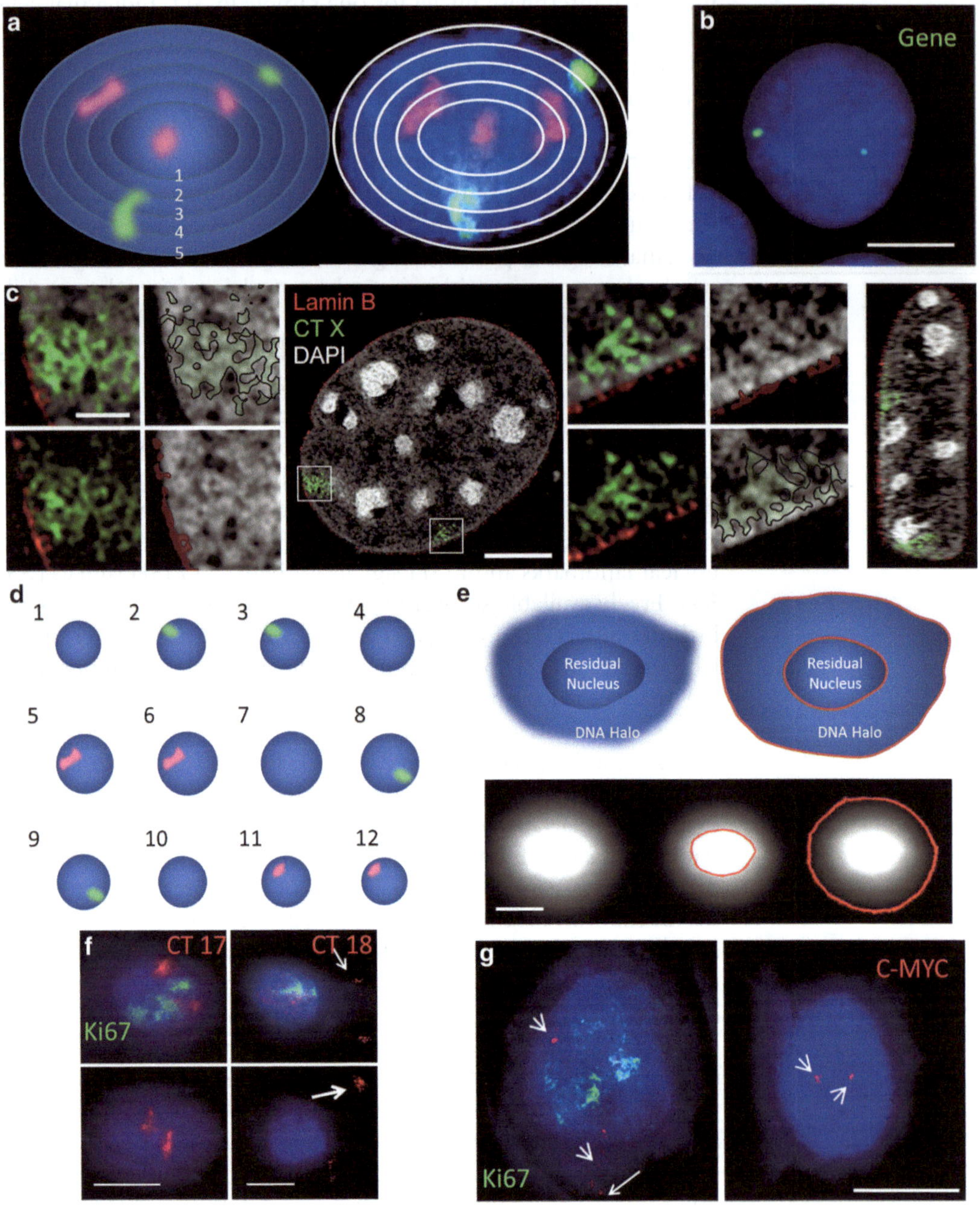

Fig. 1 The composite figure displays images from 2D-FISH protocol to visualize chromosome territories in panel A (*red* and *green*) and gene loci in panel B (*green*). The nucleus is counterstained with DAPI to reveal DNA. This allows the edge of the nucleus to be delineated easily allowing for erosion scripts (*panel A*) to define the edge of the nucleus, as is shown by the cartoon in *panel A. Panel B*(Fig. 1 continued) displays gene loci FISH signals using a BAC probe labeled with biotin and fluorescently tagged streptavidin. One of the gene loci is

2 Materials

Careful selection of probe sequence and type is required when asking questions about which parts of the genome are associated with nuclear structures. Chromosome painting probes are commercially available and ready labeled containing *Cot*1 DNA for suppression of repetitive sequences [42] (*see* **Note 1**). Templates of individual chromosomes, that have been either flow-sorted [43] by size or microdissected are available [44]. Labeling of original template is normally performed by degenerate oligo primer-PCR [45]. Specific centromere and telomere probes are also commercially available as are many gene probes. When performing gene positioning studies we use commercially available bacterial artificial chromosomes (BAC) containing the sequence of interest (*see* **Note 2**). The BAC DNA is released from the bacteria using a commercial kit for DNA extraction, following measurement of the amount of DNA generated. A nick-translation reaction is performed to label the BAC DNA with either biotin or digoxigenin conjugated nucleotides, performed through commercially available kits. Other methods of labeling include random priming and end labeling through PCR methods. The method for labeling probes described here is the DOP-PCR method for whole chromosome painting probes.

2.1 Two-Dimensional Fluorescence In Situ Hybridization

2.1.1 Cell Culture and Fixation

1. Specific cell medium supplemented with 10 % (v/v) fetal bovine serum (FBS) and 2 % (v/v) penicillin/streptomycin.
2. Versene (phosphate buffered saline (137 mM NaCl, 2.7 mM, KCl, 10 mM Na_2HPO_4, 1.8 mM KH_2PO_4; pH 7.4) with 0.2 % (w/v) ethylenediaminetetraacetic acid; EDTA) kept refrigerated at 4 °C, pre-warmed to 37 °C.
3. 0.25 % trypsin produced from a 2.5 % stock solution, which has been diluted (1:10 v/v) in Versene.

Fig. 1 (continued) localized at the nuclear edge, whereas the other maybe at the edge but it is not possible to tell with 2D-FISH and simple measurements. This is why the erosion scripts can be used with a large number of nuclei. Using 3D-FISH the positioning of a FISH signal in the *x*, *y* or *z* coordinates can be revealed by confocal laser scanning microscopy reconstruction or analysis of a gallery of 2D images as shown in the cartoon in *panel D*. Chromosome territories are drawn, that can be seen to touch the nuclear edge in the *x* and *y* planes but also the *z* plane as one of the territories in *red* is at the bottom of the nucleus. *Panel C* was kindly provided by Dr. Marion Cremer, Ludwig-Maximillian University, Munich; it shows the relationship of chromosome *X* territories (*green*) at the nuclear edge with lamin B1 (*red*) in great detail due to the use of super-resolution microscopy (3D-SIM) [16, 51]. *Panels E, F* and *G* display DNA halo images and how to delineate the residual nucleus and DNA halo (*panel E*). *Panel F* displays territories of chromosomes 17 and 18 (*red*) in proliferating primary fibroblasts (Ki67 in *green*) and in senescent fibroblasts (negative for Ki67). Chromosome 18 is seen at some distance from the residual nucleus in the DNA halo in both proliferating and senescent DNA halos. In *panel G* DNA halos are also displayed subjected to 2D-FISH with labeled BAC probes revealing gene loci signals (*red*), again Ki67 is revealed in *green*. Scale bar, 10 μm

4. Fixative: methanol–acetic acid (3:1 v/v), made fresh and made ice-cold.
5. Hypotonic solution: 0.075 M KCl (*see* **Note 3**).

2.1.2 Slide preparation and Denaturation

1. Glass microscope slides that can withstand high and low temperatures.
2. 20× sodium citrate saline (SSC, 300 mM sodium citrate hydrate, 3 M NaCl).
3. Denaturation solution: 70 % (v/v) formamide, 2× SSC, pH 7.0 (*see* **Note 4**).
4. Water baths.
5. Ethanol series comprising of 70, 90, and 100 % (v/v) high quality ethanol at room temperature.
6. Glass Coplin jars that can withstand heat.
7. Oven that can be set at 70 °C.

2.1.3 Degenerate Oligo Primer-PCR

1. PCR machine.
2. 5× GC Buffer containing 2 mM $MgCl_2$ when at 1×.
3. dACGTP mix made up in water at a concentration of 2 mM.
4. dTTP at a concentration of 2 mM.
5. Forward and reverse DOP primers at a concentration of 20 μM (CCGACTCGAGNNNNNNATGGG).
6. Taq DNA polymerase.
7. PCR grade water.
8. Template DNA—chromosome paint.
9. Biotin-16-dUTP or digoxigenin-11-dUTP.
10. Agarose.
11. Agarose mini gel kit.
12. DNA ladder—containing bands at 500 bp and lower.
13. TAE buffer.

2.1.4 Probe Preparation and Hybridization

1. Labeled total human chromosome or BAC DNA probes.
2. *Cot*1 DNA.
3. Herring sperm DNA.
4. 3 M sodium acetate.
5. 4× SSC.
6. High quality ethanol.
7. Hybridization mixture: 50 % formamide, 10 % dextran sulfate, 2× SSC, and 1 % Tween 20.
8. Heating block.

9. 70 % formamide and 2× SSC solution, pH 7.0.
10. Humidified hybridization chamber.
11. Rubber cement.
12. Glass coverslips (22 × 22 mm).
13. Glass Coplin jars (*see* **Notes 5** and **6**).
14. Bench top centrifuge.

2.1.5 Washing 2D FISH and Visualization of Probe

1. Buffer A: 50 % (v/v) formamide, 2× SSC; pH 7.0.
2. Buffer B: 0.1× SSC; pH 7.0.
3. Blocking solution: 4 % bovine serum albumin (BSA).
4. Streptavidin conjugated to a fluorochrome.
5. Antibody reacting with digoxigenin conjugated to a fluorochrome.
6. 4× SSC solution containing 0.05 % Tween 20.
7. Vectashield containing 6-diamidino-2-phenylindole (DAPI).

2.1.6 Ki67 Antigen Marker for Proliferation in Primary Cells

We have found that chromosome positioning changes according to whether cells are proliferating, quiescent or senescent [6, 46–48]. It is possible to identify replicative senescent cells within a proliferating culture grown in high serum since they are anti-Ki67 negative. It is also possible to work with quiescent cells using Ki67 as a marker in a young culture containing fewer senescent cells. After 4 days, the Ki67 is no longer visible. It is also possible to observe three different stages of G1 by Ki67 patterns type Ia, type Ib, and type II [49].

1. 1× PBS.
2. 1× PBS containing 1 % new born calf serum.
3. Anti-Ki67 antigen antibody.
4. Fluorochrome-conjugated secondary antibody raised against the appropriate species.
5. Glass coverslips.

2.1.7 Microscopy and Analysis

1. Epifluorescence microscope equipped with a 100× lens and filter sets to permit the capture of at least fluorochromes emitting photons in the blue, green, and red spectra.
2. Cooled charged coupled device camera.
3. Computer and program to run microscope and camera.
4. Hardware and software to visualize captured images.
5. Image analysis to perform measurements from FISH signal to nearest edge the nucleus or geometric center of the nucleus or erosion analyses as discussed in the Introduction.
6. Programs for data handling, analysis, and presentation.

2.2 Three-Dimensional Fluorescence In Situ Hybridization

2.2.1 Cell Culture and Fixation

1. Cell medium of choice for specific cell type containing 10 % (v/v) fetal bovine serum (FBS) and 2 % (v/v) penicillin/streptomycin.
2. Versene (phosphate buffered saline (137 mM NaCl, 2.7 mM, KCl, 10 mM Na_2HPO_4, 1.8 mM KH_2PO_4; pH 7.4) with 0.2 % (w/v) ethylenediaminetetraacetic acid; EDTA) kept at 4 °C, and pre-warmed to 37 °C.
3. 0.25 % trypsin from a 2.5 % stock solution, which has been diluted (1:10 v/v) in Versene.
4. Chambered tissue culture dishes to hold slides.
5. Glass microscope slides that can withstand high temperatures.
6. 1× phosphate buffered saline.
7. 4 % paraformaldehyde (w/v) (*see* **Note** 7).
8. Permeabilization solution: 0.5 % Triton X-100 (v/v) and 0.5 % saponin (w/v) in 1× PBS.
9. 20 % glycerol in 1× PBS solution.
10. Liquid nitrogen.
11. Glass Coplin jars.

2.2.2 Probe Preparation and Hybridization

See Subheading 2.1.4 for probe preparation for 2D FISH.

2.2.3 Slide Preparation and Denaturation

1. Denaturation buffer A: 70 % formamide, 2× SSC, pH 7.0.
2. Denaturation buffer B: 50 % formamide, 2× SSC, pH 7.0.
3. Glass coverslips (22 × 32 mm).
4. 0.1 N HCl.
5. 2× SSC.
6. Coplin jars.
7. Rubber cement.

2.2.4 Washing and Visualization of Probe in 3D-FISH

See Subheading 2.1.5 for washing and probe visualization for 2D FISH.

2.2.5 Using Ki67 as a Proliferation Marker

See Subheading 2.1.6 for using Ki67 antibody.

2.2.6 Microscopy and Analysis

1. Fluorescence microscope capable of acquiring optical Z-Stacks with high powered lens such as a confocal laser scanning microscope (with at least two lasers), Deltavision, 3D-simulation illumination microscope, 3D-STORM [50].
2. Software for wide-field deconvolution, 3D reconstruction, measurements in 3D.

2.3 FISH in Combination with Indirect Immunofluorescence

The best way to visualize the nuclear envelope is by indirect immunofluorescence using antibodies to nuclear lamins—A-types and B-types, nuclear pores or integral membrane proteins such as emerin, LAP2β, SUN1, SUN 2, MAN1, or other nuclear envelope transmembrane proteins [18] and even nuclear pore proteins. The issue with combining FISH with visualization of proteinaceous structures is that the fixation, and specifically the formamide, can affect the ability of specific antibodies to bind to their antigen post-FISH. We have found this especially difficult for nuclear envelope proteins such as A-type lamins, which for mapping associations with genomes and the nuclear envelope are the presumed first choice. It is not possible to just use an integrated GFP fused protein since GFP is also affected by the FISH process and is no longer visible post-FISH.

There are two ways to get around this problem: either use the primary and fluorescently tagged secondary antibodies and fix them prior to the FISH protocol, or use the primary antibody prior to the FISH protocol, fix them in place with paraformaldehyde and then carry out the FISH protocol, following the hapten visualization step, or concomitant with it, use the fluorescently labeled secondary antibody to reveal the primary antibody bound to the antigen. As mentioned above, GFP fluorescence does not survive the FISH denaturation process but it possible to use an antibody to GFP to reveal the exogenous protein.

The 2D-FISH protocol could, given that it uses acetic acid, alter structure of proteins. Therefore, it would be necessary to perform a pilot experiment to determine if the antigen of choice at the nuclear envelope survives fixation and the FISH denaturation. With respect to analyzing genome interaction with the nuclear envelope, it is more appropriate to use 3D preserved cells or nuclei.

1. Reagents for 2D or 3D FISH (*see* Subheadings 2.1 and 2.2).
2. Primary antibodies to specific nuclear envelope antigen, Ki67 or GFP.
3. Secondary antibodies reacting with species primary antibodies are made in 4 % paraformaldehyde.
4. 1× phosphate buffered saline.
5. Glass Coplin jars.
6. New born calf serum.
7. Mounting medium containing a DNA counterstain.
8. Humid incubation chamber.
9. Glass coverslips.

2.4 DNA Halo Preparations

Although this chapter concerns visualizing the interaction of the genome with the nuclear envelope we have decided to include a protocol to look at DNA halos which is a technique we have been using in our laboratory to give a further insight into the behavior and binding of the genome to the nucleoskeleton (also known as the nuclear matrix). Many of the nuclear envelope proteins and complexes at the nuclear edge are part of the nuclear matrix and indeed, a number of nuclear envelope proteins are also found deep within the nucleoplasm as part of the unextractable nucleoskeleton, such as the nuclear lamins.

1. Temperature resistant glass microscope slides or poly-l-lysine coated slides.
2. Chambered tissue culture dishes.
3. Glass Coplin jars.
4. CSK buffer: 10 mM Pipes pH 7.8; 100 mM NaCl, 0.3 M sucrose, 3 mM $MgCl_2$, 0.5 % Triton X-100.
5. Phosphate buffered saline 10×, 5×, 2×, and 1×.
6. Extraction buffer: 2 M NaCl, 10 mM Pipes pH 6.8, 10 mM EDTA, 0.1 % digitonin, (*see* **Note 8**) 0.05 mM spermine, 0.125 mM spermidine.
7. Ethanol series 10, 30, 70, and 95 % made up in water
8. *See* Subheadings 2.1.4 and 2.1.5 for reagents required for 2D-FISH
9. Epifluorescence microscope capable of taking 2D images.
10. CCD camera with hardware and software to control microscope and camera.
11. Image analysis software.

3 Methods

3.1 Two-Dimensional Fluorescence In Situ Hybridization

3.1.1 Cell Culture and Fixation

1. Culture adherent cells for at least 1 week after recovering from liquid nitrogen storage (*see* **Note 9** for suspension cells).
2. Remove medium and wash the cells once with Versene.
3. Harvest cells by incubation 0.25 % trypsin in Versene; neutralize the trypsin by adding at least an equal volume of medium.
4. Centrifuge at 400×*g* for 5 min; remove the supernatant. Resuspend cells in a small amount of remaining medium.
5. Treat harvested cells with a hypotonic solution (0.075 M KCl) for 15 min at room temperature to swell the cells (*see* **Note 10**).
6. Centrifuge the cell suspension at 300×*g* for 5 min using a bench top centrifuge.
7. To fix the cells, remove the supernatant and resuspend the cell pellet in a small amount of buffer left to avoid cell clumping.

8. Add ice-cold methanol: acetic acid (3:1 v/v) drop-wise to the cells with gentle shaking.
9. Incubate cells at 4 °C or on ice for at least 1 h, up to 18 h (*see* **Note 11**).
10. Centrifuge cells at 300×*g* at 4 °C for 5 min and repeat the fixing steps 8, 9 and 10 four to five times, but only leaving the cells in the fix solution on ice for 5–15 min until more than 90 % of the cells have lost their cytoplasm. Use phase contrast with X40 lens to view the slides prior to use. If there are still cells with cytoplasm, keep repeating the fixation steps until there are fewer than 10 %.
11. Store fixed cell suspensions at −20 °C until required.

3.1.2 Slide Preparation and Denaturation

1. Drop the cells onto humid or damp slides from a height, air dry the slides and then age the cells at 70 °C for 1 h or for 2 days at room temperature.
2. Pass the aged slides through an ethanol row of 70, 90 and 100 % ethanol for 5 min in each solution, followed by air drying.
3. Pre-warm the slides at 70 °C for 5 min.
4. Incubate in denaturating solution (70 % (v/v) formamide, 2× SSC, pH 7.0 at 70 °C for 2 min.
5. Immediately plunge the slides into ice-cold 70 % ethanol for 5 min and then again pass through the ethanol row (90 and 100 % ethanol).
6. Air-dry the slides and keep them warm on a warm plate until hybridization with the probe.

3.1.3 DOP-PCR Protocol

1. To make 50 μl of secondary template from the primary template, add together in a small Eppendorf tube that fits the PCR machine to be used: 10 μl DOP-PCR buffer, 5 μl 2 mM dACGTP, 5 μl 2 mM dTTP, 5 μl 20 μM DOP primers, 23 μl PCR grade water, 1 μl of chromosome paint template and 1 U of *Taq* polymerase (to be added last).
2. Place into PCR machine.
3. For amplification of primary product into secondary template use the following temperature and times: 1 cycle of 95 °C for 3 min followed by 98 °C for 20 s; 31 cycles of 62 °C for 1 min followed by 72 °C for 30 s; 1 cycle of 72 °C for 5 min; 4 °C hold.
4. To check the amplification and length of the secondary template, run 3 μl of the PCR product in a 1 % agarose gel with markers that give you the ability to assess smears of PCR product between 200 and 500 bp.

5. Use the unlabeled secondary template DOP-PCR product as the next template DNA and add 5 μl to a plastic PCR tube the correct size for your machine chambers.
6. To the secondary template, add 10 ml of DOP-PCR buffer, 5 ml dACGTP (2 mM), 2 μl dTTP (2 mM) and 10 μl of biotin-16-dUTP or digoxigenin-11-dUTP, 5 μl of DOP-PCR primers, and 1 unit of *Taq* polymerase.
7. For amplification of primary product into secondary template use the following temperature and times: 1 cycle of 95 °C for 3 min followed by 98 °C for 20 s; 34 cycles of 62 °C for 1 min followed by 72 °C for 30 s; 1 cycle of 72 °C for 5 min; 4 °C hold.
8. Using a small aliquot check that there are PCR bands between 200 and 500 bp in an agarose gel.

3.1.4 Probe Denaturation and Hybridization

1. Prepare the probe mixture by ethanol precipitation of the labeled chromosome paint (8 μl per slide) with the addition of *Cot*-1 DNA (7 μl per slide), herring sperm (3 μl per slide), 1/20th volume of 3 M sodium acetate, and 2× volume of high quality ethanol (ice-cold).
2. Incubate the mixture at −80 °C for at least 30 min; centrifuge at full speed in a standard microfuge for 30 min at 4 °C.
3. Wash the DNA pellet with ice-cold 70 % ethanol and centrifuge at full speed in a standard microfuge for 15 min at 4 °C.
4. Dry the pellets at 50 °C on a hot block (*see* **Note 11**).
5. Dissolve the probes in 12 μl (per slide) of hybridization mix overnight at room temperature.
6. Denature the probes at 75 °C for 5 min; allow the *Cot*-1 DNA to anneal to the repetitive sequences at 37 °C for at least 10 min but no longer than 2 h.
7. Incubate slides at 70 °C in 70 % formamide, 2× SSC, pH 7.0 for 2 min.
8. Slides are plunged into ice-cold 70 % ethanol and then placed in 90 and 100 % ethanol for 5 min each at room temperature.
9. Apply 12 μl of the probe to the slide and cover with pre-warmed 22 × 22 mm coverslip and seal with rubber cement.
10. Allow the slides to hybridize with the probe in an humidified hybridization chamber at 37 °C for 2 days.

3.1.5 Washing 2D-FISH Slides

1. Post hybridization, gently remove the rubber cement and wash the slides three times for 5 min each in buffer A (50 % (v/v) formamide, 2× SSC, pH 7.0) preheated to 45 °C.
2. Wash the slides in buffer B (0.1× SSC, pH 7.0) pre-warmed at 60 °C but used in the 45 °C water bath, three times for 5 min each before transferring to 4× SSC at room temperature.

3. Incubate the slides with a 100 μl blocking solution (4 % bovine serum albumin; BSA, w/v) for 10 min at room temperature.
4. Incubate the slides in 100 μl of fluorescently tagged streptavidin or anti-digoxigenin conjugated to a fluorochrome diluted in 1 % BSA for 1 h at room temperature.
5. Wash the slides in a 4× SSC solution containing 0.5 % Tween 20 in the dark at 42 °C with three changes of the solution every 5 min.
6. Mount the slides in mounting medium containing DAPI.

3.1.6 Ki67 Antigen Marker for Proliferation in Primary Cells

1. Once cells have been washed, hapten visualized and slides checked positive for the FISH signals, soak the coverslips in 1× PBS for 30 min at room temperature.
2. Dilute primary Ki67 antibody in 1× PBS containing 1 % new born calf serum and add 100 μl to the slide and cover with a 22×40 mm coverslip.
3. Incubate for 1 h at room temperature or 30 min at 37 °C or overnight at 4 °C.
4. Wash in 1× PBS for 15 min with 3 changes.
5. Place 100 μl of appropriate secondary antibody on the slides, cover with a 22×40 mm coverslip and incubate 1 h at room temperature, or 30 min at 37 °C, or at 4 °C for 4 h.
6. Mount in suitable mounting medium containing DAPI (*see* **Note 12**).

3.1.7 Microscopy and Analysis

1. Examine the slides using 100× oil immersion lens.
2. Cells can be imaged according to their Ki67 (proliferative status).
3. Select nuclei randomly by following a rectangular scan pattern, and capture gray-scale images of these nuclei.
4. Capture at least 50–60 images per slide and convert into TIFF or PICT format.
5. Pass the images through an erosion analysis script (see Introduction). These scripts are devised to divide each captured nuclei in five concentric shells of equal area, the first shell starting from the periphery of the nucleus going to the interior of the nucleus (fifth shell). The script measures the pixel intensity of DAPI and the chromosome probe in these five shells and puts the data obtained into a table.
6. Normalize the probe signal by dividing the percentage of the probe by the percentage of DAPI signal in each shell. Thus, the normalized proportion of probe is calculated in all five shells for at least 50 nuclei.

7. Plot histograms using these data, with standard error bars (+/– SEM).
8. Alternatively, make measurements to the nearest edge or geometric center of the nucleus to position genes/chromosomes. These types of analyses can be meaningful in 2D especially when studying peripherally located FISH signals.

3.2 Three-Dimensional Fluorescence In Situ Hybridization (3D-FISH)

3.2.1 Cell Culture and Fixation

1. Culture adherent cells on sterile glass slides placed in a chambered tissue culture dish for at least 2 days at 37 °C, 5 % CO_2 at a starting density of 1×10^5 cells per slide.
2. Wash slides three times in 1× PBS.
3. Fix in 4 % paraformaldehyde (w/v) for 10 min at room temperature.
4. Wash slides three times in 1× PBS.
5. Permeabilize cells with 0.5 % Triton X-100 (v/v) and 0.5 % saponin (w/v) in 1× PBS solution for 20 min at room temperature.
6. Wash slides three times in 1× PBS.
7. Incubate slides in a solution of 20 % glycerol in 1× PBS solution for at least 30 min at 4 °C
8. Snap-freeze slides in liquid nitrogen for 15–30 s.
9. Repeat step 8–9 another 4–5 times without the 30 min incubation in glycerol.
10. Place slides in boxes and store at –80 °C until required.

3.2.2 Probe Preparation and Hybridization

See Subheading 3.1.4

3.2.3 Slide Preparation and Denaturation

1. Thaw slides taken out of –80 °C storage at room temperature.
2. Repeat the freeze–thaw process in liquid nitrogen for another 4–5 times, soaking the slides in 20 % glycerol between each freeze–thaw if this has not been done prior to freezing.
3. Wash excess glycerol from the slides using three changes of 1× PBS for 10 min each, followed by depurination in 0.1 N HCl for 5 min at room temperature with shaking (*see* **Note 13**).
4. Wash off excess acid with 2× SSC for 15 min with three changes of the buffer and then incubate slides in 50 % formamide, 2× SSC, pH 7.0 solution overnight.
5. Denature the slides by incubation in denaturation buffer A (70 % formamide, 2× SSC, pH 7.0) pre-warmed at 73 °C for precisely 3 min.
6. Rapidly transfer the slides to denaturation buffer B (50 % formamide, 2× SSC, pH 7.0) pre-warmed at 73 °C for 1 min. During this time, add 10 μl of previously denatured probe that

has been allowed to partially re-anneal at 37 °C to a pre-warmed 22 × 22 mm coverslip.

7. Once the slide is ready after denaturation, remove excess denaturing solution with a tissue without disturbing the cells.
8. Present the slide immediately to the probe on the coverslip.
9. Seal the coverslip on the slide using rubber cement and leave to hybridize on the slide in a pre-warmed humidified chamber at 37 °C for 2 days.

3.2.4 Washing 3-D FISH

Same as for 2D-FISH; *see* Subheading 3.1.5.

3.2.5 Using Ki67 as a Marker for Proliferation

Same as for 2D-FISH; *see* Subheading 3.1.6.

3.2.6 Microscopy and Analysis

1. Capture the images of nuclei using a microscope capable of capturing in three dimensions.
2. Collect stacks of optical sections with an axial distance of 0.2 μm for high resolution from random nuclei (*see* **Note 14**)
3. For confocal images you could obtain stacks of 8-bit gray-scale 2D images with eight averages from each optical image.
4. Assess the positioning of chromosomes in relation to the nuclear edge by performing measurements using image analysis packages as described in the introduction, whereby the distance between the geometric center of each chromosome territory/gene signal and the nearest nuclear edge can be measured.
5. Perform measurements for at least 20 nuclei for each sample.
6. Plot frequency distribution curves with the distance between the center of chromosome territory and the nearest nuclear periphery on the x-axis and the frequency on the y-axis.

3.3 3D-FISH Combined with Indirect Immunofluorescence

1. Place slides with fixed cells into a Coplin jar full of 1× PBS.
2. Place slides in an humidified chamber
3. Pipette 20–100 μl of diluted primary antibody (in PBS 1 % new born calf serum or similar) and cover with a glass coverslip.
4. Incubate at room temperature for 1 h, at 37 °C for 30 min or overnight at 4 °C.
5. Remove coverslips gently and place slides in glass Coplin jar in 1× PBS. Place on a rocker at a gentle speed changing buffer three times for a total wash time of 15 min.
6. Carefully wipe away PBS from the back of slides and replace in humidified chamber and repeat steps 3–5 but with the appropriate fluorochrome-conjugated secondary antibody. Incubate at room temperature for 1 h, 37 °C for 30 min or 4 °C for no longer than 4 h.

7. After the final wash, incubate slides in 4 % paraformaldehyde made up in 1× PBS for 10 min at room temperature.
8. Place slides in 2× SSC to prevent drying out before proceeding to the 3D FISH protocol (*see* **Note 15**) for description of doing primary antibody staining before FISH and secondary antibody after FISH.

3.4 DNA Halo Preparations

1. Either grow adherent cells directly onto glass microscope slides in culture or allow suspension cells to settle and attach to poly-l-lysine coated slides.
2. Rinse slides twice in 1× PBS.
3. Add CSK buffer for 15 min.
4. Replace CSK buffer with extraction buffer and leave for 4 min.
5. Rinse slides consecutively in 10×, 5×, 2×, and 1× PBS for 1 min each.
6. Take slides through an ethanol series comprising 10, 30, 70, and 95 % ethanol.
7. Air-dry slides and bake for 2 h at 70 °C. The slides are now ready to be subjected to 2D-FISH.

3.4.1 Measuring the Ratio of the Residual Nucleus to the Maximum Extent of the DNA Halo

1. At 100× magnification, select at least 50 images of extracted cell nuclei per sample at random.
2. Capture of 8-bit gray-scale 2D images of each DNA halo in TIFF format using a high resolution digital camera.
3. Open the image in NIH ImageJ software or similar.
4. Configure ImageJ to analyze area (Analyse > Set Measurements, tick the area checkbox).
5. Using the adjust threshold tool (Image > Adjust > Threshold), move the top slider to cover only the residual nucleus.
6. Measure the area using analyze particles (Analyze > Analyse Particles) and record the measurement.
7. Using the adjust threshold tool again, move the top slider to cover the entire area of the DNA halo, noting that the DNA halo will likely be very pale at its greatest extent.
8. Repeat step 6 and open the next image.
9. The ratio of the residual nucleus to the maximum extent of the DNA halo is calculated by dividing the total area of the DNA halo by the area of the residual nucleus.
10. Plot histograms using these data and standard error bars representing +/− SEM.

3.4.2 Analyzing Positioning of Whole Chromosomes Within a DNA Halo

1. Using an image analysis program the images are cropped and split into individual color channels, i.e., red, blue, and green.
2. Save each channel separately.

3. Identify and select the residual nucleus in an imaging software program to perform density threshold and binary functions to reveal not only the edge of the nucleus but also the center.
4. The distance from the nuclear center to each furthest chromosome territory edge (ctE) can then be determined.
5. These distances are divided by the distance from the nuclear center to each respective nuclear edge (NE).

4 Notes

1. Sometimes commercial paints scrimp on the Cot-1 DNA and it may be necessary to add more Cot1 to avoid repetitive elements in the genome from being hybridized to.
2. The BAC gives you your gene of interest but be aware that BACs may contain other sequence adjacent to the gene of interest.
3. This concentration of KCl in the hypotonic solution maybe too high for some cells; in some instances, this may need to be reduced to 0.05 M KCl (e.g., with snail cells).
4. Formamide is a teratogen and its use should be forbidden to pregnant women. Use in a fumehood otherwise.
5. Use glass Coplin jars that can withstand high heat, i.e., >75 °C: otherwise they crack.
6. To avoid large amounts of formamide in the denaturing step, it is possible to use a heating block and place the probe and sample together and co-denature them. Alternatively it is possible to purchase a programmable temperature controlled slide processing system that will perform the denaturing steps and the hybridization step as well.
7. The Cremer laboratory recommends 2 % paraformaldehyde as it preserved 3D structure better when using super-resolution microscopy [51].
8. Digitonin is a dangerous reagent and requires a careful and comprehensive risk assessment.
9. Suspension cells can be used for any FISH methodologies. They need to be placed on a glass microscope slide. This can be done by cytospinning but this disturbs three-dimensionality. Cells can also be placed on slides coated with poly-l-lysine and allowed to stick. These cells can then be subjected to 3D protocols.
10. It is very important that cells are fully suspended and triturated so that they are separated and not clumped together.
11. The DNA must not be over-dried as it will not be possible to resuspend it in the hybridization mix. It is best to keep an eye on the probe as it dries and take it off the heating block when the liquid has evaporated but the DNA looks "glassy."

12. At this stage it is also possible to improve the FISH process by using an RNAseA treatment for 30 min. Only do this if you are not detecting RNA obviously but also bear in mind that RNA is part of the nuclear structure and a component of chromosomes themselves.
13. For protocols specifically for super-resolution microscopy [18].
14. The Ki67 antibody can disassociate from its antigen after a few hours. 4 % paraformaldehyde (5 min at room temperature) can be used to fix the antibody in place after washing off excess antibody.
15. It is also possible to perform the primary antibody step first prior to the FISH procedure and fix it into place with 4 % paraformaldehyde for 10 min at room temperature. The diluted secondary antibody should be mixed with the streptavidin or anti-digoxigenin antibody after the FISH washing step, and as part of the visualization step.

Acknowledgements

We would like to thank Dr Marion Cremer for allowing us to include some of their 3D-SIM super-resolution images of chromosomes and nuclear envelope, Dr Karen Meaburn for helpful discussions and SPARKs children's charity for funding CSC, Brunel University London Progeria Research Fund for partial funding of UB, The Gordon Memorial Trust for supporting MHA and the EURO-laminopathies consortium FP6 for supporting LSG.

References

1. Bourne G, Moir C, Bikkul U et al (2013) Interphase chromosome behavior in normal and diseased cells. In: Yurov Y (ed) Human interphase chromosomes: the biomedical aspects. Springer, New York, pp 9–33
2. Foster HA, Bridger JM (2005) The genome and the nucleus: a marriage made by evolution. Chromosoma 114:212–229
3. Németh A, Conesa A, Santoyo-Lopez J et al (2010) Initial genomics of the human nucleolus. PLoS Genet 6:1–11
4. Ikegami K, Egelhofer TA, Strome S, Lieb JD (2010) Caenorhabditis elegans chromosome arms are anchored to the nuclear membrane via discontinuous association with LEM-2. Genome Biol 11:1–20
5. Guelen L, Pagie L, Brasset E et al (2008) Domain organization of human chromosomes revealed by mapping of nuclear lamina interactions. Nature 453:948–951
6. Bridger JM, Arican-Goktas HD, Foster HA et al (2014) The non-random repositioning of whole chromosomes and individual gene loci in interphase nuclei and its relevance in disease, infection, aging, and cancer. Adv Exp Med Biol 773:263–279
7. Chubb JR, Boyle S, Perry P, Bickmore WA (2002) Chromatin motion is constrained by association with nuclear compartments in human cells. Curr Biol 12:439–445
8. Zink D, Cremer T, Saffrich R et al (1998) Structure and dynamics of human interphase chromosome territories in vivo. Hum Genet 102:241–251
9. Wallace LM, Moreo A, Clark KR, Harper SQ (2013) Dose-dependent toxicity of humanized

renilla reniformis GFP (hrGFP) limits its utility as a reporter gene in mouse muscle. molecular therapy—nucleic acids 2:1–6.

10. Elcock LS, Bridger JM (2010) Exploring the relationship between interphase gene positioning, transcriptional regulation and the nuclear matrix. Biochem Soc Trans 38:263–267
11. Elcock LS, Bridger JM (2008) Exploring the effects of a dysfunctional nuclear matrix. Biochem Soc Trans 36:1378–1383
12. Bridger JM, Kalla C, Wodrich H et al (2005) Nuclear RNAs confined to a reticular compartment between chromosome territories. Exp Cell Res 302:180–193
13. Lampel S, Bridger JM, Zirbel R, Mathieu U, Lichter P (1997) Nuclear RNA accumulations contain released transcripts and exhibit specific distributions with respect to Sm antigen foci. DNA Cell Biol 16:1133–1142
14. Volpi EV, Bridger JM (2008) FISH glossary: an overview of the fluorescence in situ hybridization technique. Biotechniques 45:385–386
15. Marshall WF, Dernburg AF, Harmon B, Agard DA, Sedat JW (1996) Specific interactions of chromatin with the nuclear envelope: positional determination within the nucleus in Drosophila melanogaster. Mol Biol Cell 7:825–842
16. Markaki Y, Smeets D, Fiedler S et al (2012) The potential of 3D-FISH and super-resolution structured illumination microscopy for studies of 3D nuclear architecture. Bioessays 34:412–426
17. Luo L, Gassman KL, Petell LM et al (2009) The nuclear periphery of embryonic stem cells is a transcriptional permissive and repressive compartment. J Cell Sci 122:3729–3737
18. Zuleger N, Boyle S, Kelly DA et al (2013) Specific nuclear envelope transmembrane proteins can promote the location of chromosomes to and from the nuclear periphery. Genome Biol 14:1–48
19. Lund E, Oldenburg AR, Delbarre E et al (2013) Lamin A/C-promoter interactions specify chromatin state-dependent transcription outcomes. Genome Res 23:1580–1589
20. Bridger JM, Herrmann H, Muenkel C, Lichter P (1998) Identification of an interchromosomal compartment by polymerization of nuclear-targeted vimentin. J Cell Sci 111:1241–1253
21. Bridger JM, Lichter P (1999) Analysis of mammalian interphase chromosomes by FISH and immunofluorescence. In: Bickmore W (ed) Chromosome structural analysis: a practical approach. Oxford University Press, Oxford, UK, pp 103–123
22. Solovei I, Cremer M (2010) 3D-FISH on cultured cells combined with immunostaining. Methods Mol Biol 659:117–126
23. Hatch E, Hetzer M (2014) Breaching the nuclear envelope in development and disease. J Cell Biol 205:133–141
24. Bridger JM, Kill IR (2004) Aging of Hutchinson–Gilford progeria syndrome fibroblasts is characterised by hyperproliferation and increased apoptosis. Exp Gerontol 39:717–724
25. Goldman RD, Shumaker DK, Erdos MR et al (2004) Accumulation of mutant lamin A causes progressive changes in nuclear architecture in Hutchinson–Gilford progeria syndrome. Proc Natl Acad Sci U S A 101:8963–8968
26. Shimi T, Pfleghaar K, Kojima S et al (2008) The A- and B-type nuclear lamin networks: microdomains involved in chromatin organization and transcription. Genes Dev 22:3409–3421
27. Bercht Pfleghaar K, Taimen P, Butin-Israeli V et al (2015) Gene-rich chromosomal regions are preferentially localized in the lamin B deficient nuclear blebs of atypical progeria cells. Nucleus 6:66–76
28. Lalmansingh A, Arora K, DeMarco R et al (2013) High-throughput RNA FISH analysis by imaging flow cytometry reveals that pioneer factor Foxa1 reduces transcriptional stochasticity. PLoS One 8:1–12
29. Basiji DA, Ortyn WE, Liang L, Venkatachalam V, Morrissey P (2007) Cellular image analysis and imaging by flow cytometry. Clin Lab Med 27:653–670
30. Roukos V, Pegoraro G, Voss TC, Misteli T (2015) Cell cycle staging of individual cells by fluorescence microscopy. Nat Protoc 10:334–348
31. Shiels C, Adams NM, Islam SA, Stephens DA, Freemont PS (2007) Quantitative analysis of cell nucleus organisation. PLoS Comput Biol 3(7), e138
32. Bewersdorf J, Bennett BT, Knight KL (2006) H2AX chromatin structures and their response to DNA damage revealed by 4Pi microscopy. Proceedings of the National Academy of Sciences of the United States of America 103:18137–18142
33. Jost K, Haase S, Smeets D et al (2011) 3D-Image analysis platform monitoring relocation of pluripotency genes during reprogramming. Nucleic Acids Res 39:1–8
34. Bolzer A, Kreth G, Solovei I et al (2005) Three-dimensional maps of all chromosomes in human male fibroblast nuclei and prometaphase rosettes. PLoS Biol 3:826–842

35. Gué M, Messaoudi C, Sun JS, Boudier T (2005) Smart 3D-FISH: automation of distance analysis in nuclei of interphase cells by image processing. Cytometry A 67:18–26
36. Iannuccelli E, Mompart F, Gellin J, Lahbib-Mansais Y, Yerle M, Boudier T (2010) NEMO: a tool for analyzing gene and chromosome territory distributions from 3D-FISH experiments. Bioinformatics 26:696–697
37. Foster HA, Griffin DK, Bridger JM (2012) Interphase chromosome positioning in in vitro porcine cells and ex vivo porcine tissues. BMC Cell Biol 15:13–30
38. Croft JA, Bridger JM, Boyle S et al (1999) Differences in the localization and morphology of chromosomes in the human nucleus. J Cell Biol 145:1119–1131
39. Mehta IS, Kulashreshtha M, Chakraborty S, Kolthur-Seetharam U, Rao BJ (2013) Chromosome territories reposition during DNA damage-repair response. Genome Biol 14:1–15
40. Skinner BM, Robertson LB, Tempest HG et al (2009) Comparative genomics in chicken and Pekin duck using FISH mapping and microarray analysis. BMC Genomics 10:1–11
41. Federico C, Cantarella CD, Di Mare P, Tosi S, Saccone S (2008) The radial arrangement of the human chromosome 7 in the lymphocyte cell nucleus is associated with chromosomal band gene density. Chromosoma 117:399–410
42. Lichter P, Cremer T, Borden J, Manuelidis L, Ward DC (1988) Delineation of individual human chromosomes in metaphase and interphase cells by in situ suppression hybridization using recombinant DNA libraries. Hum Genet 80:224–234
43. Harris P, Boyd E, Ferguson-Smith MA (1985) Optimising human chromosome separation for the production of chromosome-specific DNA libraries by flow sorting. Hum Genet 70:59–65
44. Meltzer P, Bittner M (2001) Chromosome microdissection. Curr Protoc Hum Genet. Chapter 4:Unit4.8
45. Telenius H, Carter NP, Bebb CE et al (1992) Degenerate oligonucleotide-primed PCR: general amplification of target DNA by a single degenerate primer. Genomics 13:718–725
46. Bridger JM, Boyle S, Kill IR, Bickmore WA (2000) Re-modelling of nuclear architecture in quiescent and senescent human fibroblasts. Curr Biol 10:149–152
47. Mehta IS, Figgitt M, Clements CS, Kill IR, Bridger JM (2007) Alterations to nuclear architecture and genome behavior in senescent cells. Ann N Y Acad Sci 1100:250–263
48. Mehta IS, Amira M, Harvey AJ, Bridger JM (2010) Rapid chromosome territory relocation by nuclear motor activity in response to serum removal in primary human fibroblasts. Genome Biol 11:1–23
49. Bridger JM, Kill IR, Lichter P (1998) Association of pKi-67 with satellite DNA of the human genome in early G1 cells. Chromosome Res 6:13–24
50. Schermelleh L, Heintzmann R, Leonhardt H (2010) A guide to super-resolution fluorescence microscopy. J Cell Biol 190:165–175
51. Markaki Y, Smeets D, Cremer M, Schermelleh L (2013) Fluorescence in situ hybridization applications for super-resolution 3D structured illumination microscopy. Methods Mol Biol 950:43–64

Chapter 25

Visualization of Genomic Loci in Living Cells with a Fluorescent CRISPR/Cas9 System

Tobias Anton, Heinrich Leonhardt, and Yolanda Markaki

Abstract

The discovery that the RNA guided bacterial endonuclease Cas9 can be harnessed to target and manipulate user-defined genomic sequences has greatly influenced the field of genome engineering. Interestingly, a catalytically dead Cas9 (dCas9) can be employed as a targeted DNA-binding platform to alter gene expression. By fusing this dCas9 to eGFP, we and others could show that the CRISPR/Cas9 system can be further expanded to label and trace genomic loci in living cells. We demonstrated that by exchanging the sgRNA, dCas9-eGFP could be specifically directed to various heterochromatic sequences within the nucleus. Here, we provide a basic protocol for this versatile tool and describe how to verify new dCas9-eGFP targets.

Key words CRISPR/Cas9, sgRNA, In vivo labeling, Repetitive sequences

1 Introduction

To study nuclear architecture and the spatiotemporal organization of chromatin, it is essential to have tools that allow visualization of proteins as well as specific genomic loci. Whereas nuclear proteins can be readily imaged in living cells by fusing the target protein to a fluorescent tag, in vivo visualization of DNA sequences has been challenging. Fluorescent in situ hybridization (FISH) represents a very reliable and prominent tool to label specific loci. However, this method relies on fixed samples and thereby only represents a "snapshot" in cellular development. Moreover, the relatively harsh fixation and denaturation of DNA may potentially alter the native context of the targeted locus [1]. To overcome these limitations, fluorescently tagged proteins can be employed, which recognize DNA in a sequence specific manner. For example, it has been shown that individual Cys_2His_2 zinc finger modules, each recognizing 3 bp, can be tandemly arranged to form a polydactyl zinc finer protein (PZF), which in turn recognizes a user-defined DNA sequence [2–4]. Yet testing PZFs for their target specificity

Sue Shackleton et al. (eds.), *The Nuclear Envelope: Methods and Protocols*, Methods in Molecular Biology, vol. 1411,
DOI 10.1007/978-1-4939-3530-7_25, © Springer Science+Business Media New York 2016

renders this method very time-consuming [5, 6]. Similar to PZFs, designer TAL (transcription activator-like) effectors (dTALEs) have been used to label distinct sequences in live cells [7–9]. Here, repeat variable diresidues (RVDs), located in the central repeat domain of the protein, confer target specificity [10]. However, due to the tandemly arrangement of these repeats, designing new dTALEs can be tedious and often requires multistep cloning approaches [11, 12].

In this chapter, we describe a novel method to label and trace endogenous repetitive sequences (Fig. 1) [13–15], which is based on the bacterial CRISPR/Cas9 adaptive immune system [16]. Whereas in dTALE or PZF based approaches the protein itself confers target specificity, in this case an inactive version of the endonuclease Cas9 (dCas9) is directed to its target by homologous base pairing between a single guide RNA (sgRNA) and the respective genomic sequence [17, 18]. The fact that the sgRNA, and with it the sequence to be targeted, is easily interchangeable, greatly enhances the practicality of this system by minimizing cloning efforts.

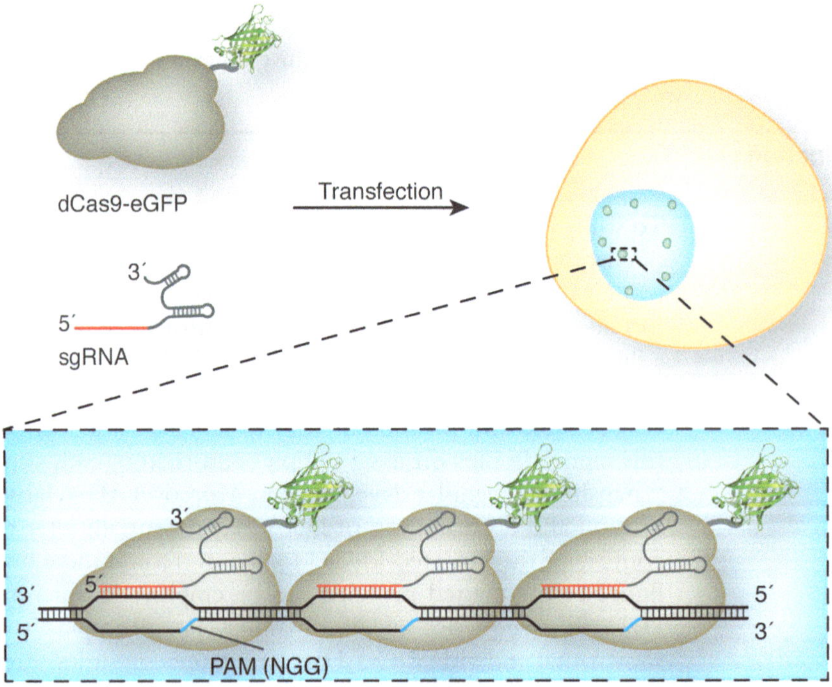

Fig. 1 Schematic overview of the CRISPR labeling approach. dCas9-eGFP is guided to the desired locus by a co-transfected sgRNA. Note that target specificity is mediated by homologous base pairing between DNA (*black*) and RNA (*red*). Due to the repetitive organization of this locus, one specific sgRNA is sufficient to ensure bright signals

2 Materials

2.1 Cell Culture

1. Dulbecco's modified Eagle's medium (DMEM).
2. Fetal bovine serum (FBS).
3. 200 mM L-glutamine.
4. 100× penicillin–streptomycin (Pen–Strep).
5. Trypsin (0.25 % trypsin–EDTA solution).
6. Phosphate-buffered saline (PBS).
7. Blasticidin S (10 mg/mL) (optional).

2.2 Medium

C2C12 medium: 390 mL DMEM supplemented with 100 mL FBS (final concentration: 20 %), 5 mL L-glutamine (final concentration: 2 mM), and 5 mL Pen–Strep (final concentration: 1×).

2.3 Transfection (See Note 1)

1. Lipofectamine® 3000 reagent.
2. Opti-MEM serum-free medium.
3. CAG-dCas9-eGFP plasmid [14].
4. U6-MaSgRNA plasmid [14] (optional).
5. U6-MiSgRNA plasmid [14] (optional).
6. U6-TelgRNA plasmid [14] (optional).

2.4 Buffers and Solutions for Immuno-FISH

1. 20× PBS: 2.74 M NaCl, 53.7 mM KCl, 130 mM Na_2HPO_4, 35 mM KH_2PO_4. For working dilution (1×): dilute 1:20 in ddH_2O.
2. Fixation solution: 4 % formaldehyde in 1× PBS (4 % FA).
3. Permeabilization solution: 0.5 % Triton X-100 in 1× PBS.
4. PBST: 0.02 % Tween 20 in 1× PBS.
5. Immunofluorescence blocking solution: 2 % bovine serum albumin (BSA), 0.5 % fish skin gelatin (FSG) in PBST.
6. 20 % glycerol solution in 1× PBS.
7. 0.1 N HCl.
8. 20× saline-sodium citrate (20× SSC) pH 7: 3 M NaCl, 0.3 M sodium citrate, pH 7.
9. 4× SSCT: dilute 20× SSC 1:5 in ddH_2O and supplement with 0.02 % Tween 20.
10. 2× SSC: dilute 20× SSC 1:10 with ddH_2O.
11. 0.1× SSC: dilute 20× SSC 1:200 with ddH_2O.
12. 50 % formamide in 2× SSC: 50 mL 20× SSC, 250 mL formamide, 200 mL ddH_2O. Adjust pH to 7 and store at −20 °C.
13. DAPI counterstaining solution: 2 μg/mL DAPI in 1× PBST.

2.5 Microscopy

While many imaging setups would work, a standard fluorescence microscope should suffice. Ideally, the microscope should be equipped with a CCD camera, a 63×/1.4 NA oil immersion objective and filters for eGFP excitation/emission. When CRISPR imaging is combined with Immuno-FISH or any other immunofluorescence detection method, appropriate filter sets should be installed. For live cell experiments the microscope should additionally be equipped with a live cell chamber, which guarantees stable conditions (37 °C, 5 % CO_2) (*see* **Note 2**).

1. Image analysis software.

2.6 Cloning

1. U6-gRNA plasmid [14].
2. Restriction enzymes: DpnI, SacI, and NsiI.
3. dNTPs (2,5 mM each).
4. 100 % dimethyl sulfoxide (DMSO).
5. Phusion High Fidelity polymerase (*see* **Note 3**).
6. DNA purification kit.

2.7 General Laboratory Equipment

1. p100 cell culture plates.
2. Six-well cell culture plates.
3. 12-well cell culture plates.
4. FACS machine (optional).
5. PCR cycler.
6. Fine-tip forceps.
7. Borosilicate coverslips (# 1.5).
8. Glass microscope slides.
9. Dewar container for liquid N_2 transport.
10. Heating block with flat surface for microscope slides.
11. Water bath.
12. Humidified chamber.
13. Heated incubation chamber.
14. Shaking platform.
15. Floating tin box.
16. Soft tissue paper.
17. Rubber cement.
18. Non-hardening antifade mounting medium (e.g., Vectashield).
19. Transparent nail polish.
20. Parafilm.

3 Methods

3.1 Cloning

3.1.1 Primer Design

1. Choose appropriate dCas9-eGFP target sequence ($N_{17\text{-}28}$NGG) (*see* **Notes 4** and **5**; Fig. 2).
2. Order HPSF purified DNA oligonucleotides (for comparison, *see* **Note 6**; Fig. 2):

   Forward: 5′-$N_{17\text{-}28}$ GTTTTAGAGCTAGAAATAGCAAG-3′

   Reverse: 5′-$N_{17\text{-}28\,(\text{rev. comp.})}$ CGGTGTTTCGTCCTTTCCAC-3′.

3.1.2 PCR and Transformation

1. Set up PCR mix for sgRNA plasmid with new target sequence, as shown in Table 1:
2. Run rolling cycle PCR with the following program: 1 cycle of 98 °C for 2 min; 30 cycles of: 98 °C for 15 s, 62 °C for 15 s and 72° for 1.5 min; 1 cycle of 72 °C for 2 min; 4 °C hold.
3. Purify PCR product to remove DMSO.
4. Digest up to 2 μg of purified PCR product with *Dpn*I for at least 1 h at 37 °C. This step ensures the removal of any residual PCR template, which would lead to a smaller number of positive clones after transformation.

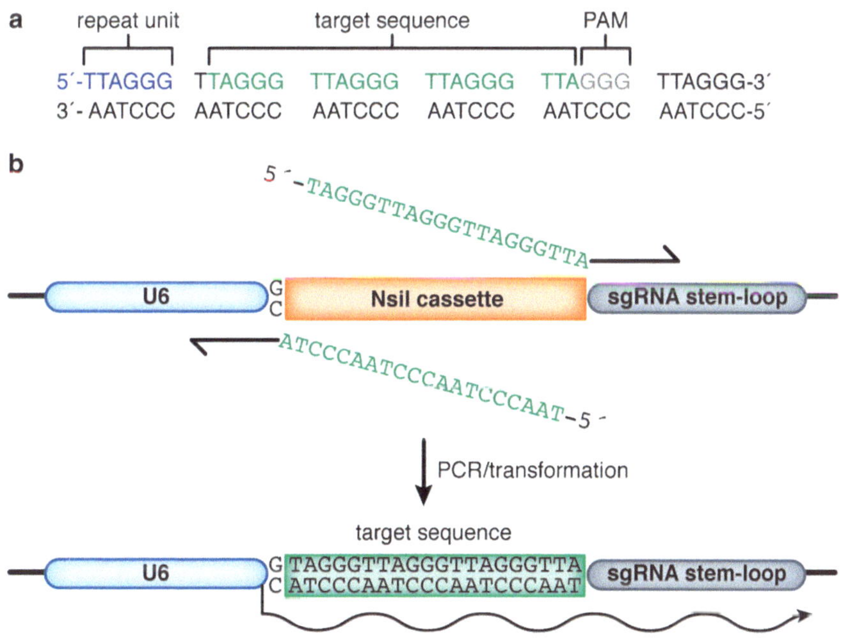

Fig. 2 Overview of gRNA design. (**a**) Example of a repetitive genomic locus (telomeres). Indicated are the basic repeat unit (*blue*), the target sequence (*green*), and the PAM (*gray*, NGG). (**b**) Schematic representation of the U6-gRNA plasmid. Primers are designed in a way that the target sequence is included in non-annealing regions. After PCR and transformation, the *NsiI* cassette is exchanged by the target sequence

Table 1
Set up of PCR mix

Component	Volume in µL (final conc.)
H_2O	32.5
5× buffer	10 (1×)
DMSO	2.5 (5 %)
Forward primer	1 (0.2 µM)
Reverse primer	1 (0.2 µM)
U6-gRNA plasmid	1 (1 ng)
dNTPs	1 (50 µM)
Polymerase	1 (1 µ)

5. Heat-inactivate restriction enzyme according to manufacturer´s instructions.
6. Use up to 5 µL of heat-inactivated reaction to transform 50 µL chemically competent *E. coli* culture. Refer to standard transformation protocols and plate bacteria on LB agar containing ampicillin (100 µg/mL). Incubate the bacteria over night at 37 °C.
7. Pick 3–12 colonies and inoculate 2 mL LB-medium supplemented with 100 µg/mL ampicillin. Incubate cultures over night at 37 °C while shaking at 200 rpm.
8. Perform plasmid preparation according to standard protocols and digest plasmid DNA with *Sac*I and *Nsi*I. Run digested DNA on a 1 % agarose gel (*see* **Note** 7).

3.2 Cell Culture

1. Culture C2C12 cells at 37 °C and 5 % CO_2 on p100 plates in 10 mL C2C12 medium.
2. Split cells every 2–3 days, typically in ratio of 1:6 to 1:8. For this, aspirate medium and wash cells with 10 mL PBS. Add 1 mL of trypsin solution and incubate for 5 min at 37 °C. After trypsinization, resuspend cells in fresh medium and transfer an appropriate amount to a new culture dish. Add medium to a final volume of 10 mL.

3.3 Transfection

In general, any transfection reagent may be used. However, we recommend testing the transfection efficiency beforehand. According to our experience, cells grown to a confluency of ~60–80 % are most suitable for transfection and result in > 20 % transfected cells.

3.3.1 Transient Transfection

1. Seed cells 1 day prior to transfection in a six-well plate. For fixed-cell experiments, seed the cells on coverslips.
2. Co-transfect cells according to manufacturer's instructions with the CAG-dCas9-eGFP plasmid and a U6-sgRNA plasmid, which encodes the desired target sequence.
3. Collect images 24–48 h post-transfection.

3.3.2 Stably dCas9-eGFP Expressing Cells (Optional)

As the dCas9-eGFP target is easily exchanged by switching the sgRNA, it might be preferable to establish a cell line, which stably expresses dCas9-eGFP (Fig. 3).

1. Seed cells 1 day prior to transfection in a p100 plate.
2. Transfect cells according to manufacturer's instructions with the CAG-dCas9-eGFP plasmid.
3. After 24 h, supplement medium with 10 μg/mL blasticidin S.
4. FACS sort GFP-positive cells 2–3 weeks after transfection. Time of sorting depends on recovery of blasticidin resistant cells, after an initial wave of cell death.
5. Culture sorted cells in C2C12 medium supplemented with 5 μg/mL blasticidin S.

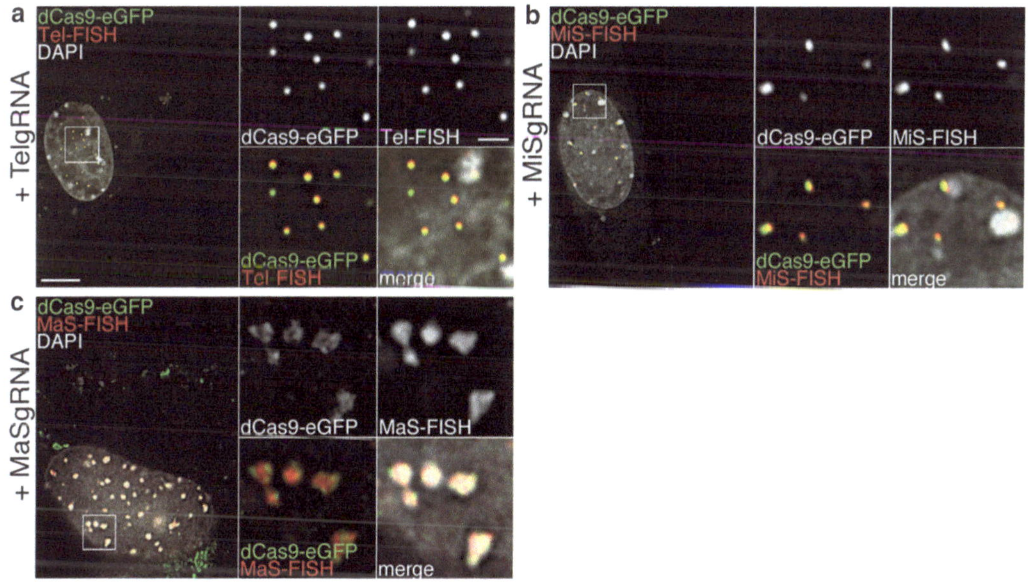

Fig. 3 Example data set for verification of CRISPR labeling via Immuno-FISH. C2C12 cells, stably expressing dCas9-eGFP (*green*) were transfected with (**a**) TelgRNA, (**b**) MiSgRNA or (**c**) MaSgRNA. Cells were FISH-treated with the corresponding probes (*red*) and counterstained with DAPI (*gray*). Galleries show 4× magnifications of boxed areas. Scale bar, 5 μm; scale bar in magnifications, 1 μm

3.4 Immuno-FISH

When dCas9-eGFP targeting has to be confirmed by FISH, we recommend seeding and transfect the cells in a 12-well cell culture plate on coverslips (Fig. 3). The cells should be fixed 24–48 h after transfection. For FISH-probe generation, *see* refs. [19, 20]. A modified Immuno-FISH protocol is presented below [1]. Unless otherwise noted, all washing steps are performed with 1 mL of buffer/solution.

1. Wash cells three times with 1× PBS.
2. Fix cells with 4 % FA for 10 min at room temperature (RT).
3. Stepwise exchange fixative with PBST: add an equal amount of PBST directly to the fixative and gradually remove the solution from the well. Do not remove all the solution as fixed samples are prone to drying out, which may lead to a deformed cell morphology. Repeat this step three times, until the fixative has been completely exchanged with PBST.
4. Permeabilize cells with permeabilization solution for 15 min at RT.
5. Incubate coverslip for at least 1 h in 20 % glycerol solution at RT.
6. Snap-freeze coverslip in liquid nitrogen. Dip the coverslip into a Styrofoam-box filled with liquid N_2 using the fine-tip forceps. Place the coverslip (cell-side up) for a few seconds on a soft tissue paper to defreeze. Return the coverslip to the well containing the 20 % glycerol solution. Repeat this step three times.
7. Wash three times with PBST.
8. Incubate coverslips in IF blocking buffer for 1 h at RT to block unspecific binding sites.
9. Incubate the coverslip in a dark humidified chamber with primary antibody (anti-GFP) for 1 h at RT. For this, firmly attach Parafilm on a smooth surface (e.g., lid of a six-well cell culture plate), place a drop (~50 μL) of diluted antibody onto the Parafilm and carefully place the coverslip (cell-side facing downwards) onto the drop.
10. Wash four times with PBST.
11. Incubate the coverslip in humidified chamber with appropriate secondary antibody (e.g., coupled to Alexa 488) as described in **step 9**.
12. Wash four times with PBST.
13. Post-fix the sample with fixation solution for 10 min at RT.
14. Wash twice with 1× PBS.
15. Denature samples with 0.1 N HCl for 5 min at RT.
16. Wash twice with 1× PBS.

17. Wash three times with 2× SSC for 3 min.
18. Incubate coverslips for 1 h at 37 °C with pre-warmed 50 % formamide in 2× SSC.
19. Denature FISH-probe at 94 °C for 3 min and place probe on ice.
20. Place a drop of FISH-probe on a clean microscope slide and mount coverslip.
21. Seal coverslip with rubber cement.
22. Denature the slide for 2 min at 76 °C.
23. Transfer the slide to a floating tin box and hybridize the probe over night at 37 °C in a water bath.
24. Carefully unmount coverslip by rehydrating with 2× SSC dropped around the edges of the coverslip.
25. Wash three times with 2× SSC for 15 min at RT.
26. Wash with preheated 0.1× SSC (60 °C). Repeat this step three times.
27. Wash three times with 4× SSCT.
28. Wash twice with PBST.
29. Counterstain nuclei with 2 μg/mL DAPI for 5–8 min at RT.
30. Wash two times with PBST
31. Mount coverslips on a drop of Vectashield (~10 μL) placed on a clean microscope slide.
32. Seal with see-through nail polish.

4 Notes

1. sgRNAs targeting MaS, MiS, and Tel may be used as positive controls to test labeling and transfection efficiency. For designing new sgRNAs, *see* Subheading 3.1.
2. Turn live cell chamber on at least 4 h prior to imaging session. This is done to ensure that CO_2 flux and temperature are stable.
3. We recommend using a High Fidelity polymerase to minimize the risk of incorrect incorporated nucleotides.
4. Preferentially, the genomic target sequence is highly repetitive (e.g., telomeric repeats or LINEs) to ensure bright signals. However, it has been shown that ~30 consecutive repeats are sufficient to visualize a genomic locus [13].
5. Although it is necessary that a transcript driven by the U6 promoter starts with a G, this base does not have to be part of the target sequence, as it is already included in the reverse primer sequence.

6. Forward and reverse primer for TelgRNA. The telomere target sequence is underlined. The part of the primer which anneals to the U6-gRNA PCR template is written in *italic* letters (Fig. 2).

   Forward: 5′- <u>TAGGGTTAGGGTTAGGGTTA</u> *GTTTTAGAG CTAGAAATAGCAAG* -3′

   Reverse: 5′- <u>TAACCCTAACCCTAACCCTA</u> *CGGTGTTTCG TCCTTTCCAC* -3′

7. Since the *Nsi*I cassette of the U6-gRNA plasmid is exchanged by the target sequence, positive clones show a linearized band at ~2.9 kb. In contrast, negative clones show two bands at ~2.6 kb and ~0.35 kb.

Acknowledgements

This work was supported by the Deutsche Forschungsgemeinschaft (DFG, SFB 1064 and Nanosystems Initiative Munich, NIM), and T.A. thankfully acknowledges the Graduiertenkolleg GRK1721.

References

1. Markaki Y, Smeets D, Cremer M, Schermelleh L (2013) Fluorescence in situ hybridization applications for super-resolution 3D structured illumination microscopy. Methods Mol Biol 950:43–64
2. Klug A (2010) The discovery of zinc fingers and their development for practical applications in gene regulation and genome manipulation. Q Rev Biophys 43:1–21
3. Pabo CO, Peisach E, Grant RA (2001) Design and selection of novel Cys2His2 zinc finger proteins. Annu Rev Biochem 70:313–340
4. Segal DJ, Barbas CF 3rd (2000) Design of novel sequence-specific DNA-binding proteins. Curr Opin Chem Biol 4:34–39
5. DeFrancesco L (2011) Move over ZFNs. Nat Biotech 29:681–684
6. Segal DJ, Dreier B, Beerli RR, Barbas CF 3rd (1999) Toward controlling gene expression at will: selection and design of zinc finger domains recognizing each of the 5′-GNN-3′ DNA target sequences. Proc Natl Acad Sci U S A 96:2758–2763
7. Miyanari Y, Ziegler-Birling C, Torres-Padilla ME (2013) Live visualization of chromatin dynamics with fluorescent TALEs. Nat Struct Mol Biol 20:1321–1324
8. Ma H, Reyes-Gutierrez P, Pederson T (2013) Visualization of repetitive DNA sequences in human chromosomes with transcription activator-like effectors. Proc Natl Acad Sci U S A 110:21048–21053
9. Thanisch K, Schneider K, Morbitzer R, Solovei I, Lahaye T, Bultmann S, Leonhardt H (2014) Targeting and tracing of specific DNA sequences with dTALEs in living cells. Nucleic Acids Res 42, e38
10. Boch J, Scholze H, Schornack S, Landgraf A, Hahn S, Kay S, Lahaye T, Nickstadt A, Bonas U (2009) Breaking the code of DNA binding specificity of TAL-type III effectors. Science 326:1509–1512
11. Cermak T, Doyle EL, Christian M, Wang L, Zhang Y, Schmidt C, Baller JA, Somia NV, Bogdanove AJ, Voytas DF (2011) Efficient design and assembly of custom TALEN and other TAL effector-based constructs for DNA targeting. Nucleic Acids Res 39, e82
12. Morbitzer R, Elsaesser J, Hausner J, Lahaye T (2011) Assembly of custom TALE-type DNA binding domains by modular cloning. Nucleic Acids Res 39:5790–5799
13. Chen B, Gilbert LA, Cimini BA, Schnitzbauer J, Zhang W, Li GW, Park J, Blackburn EH, Weissman JS, Qi LS et al (2013) Dynamic imaging of genomic loci in living human cells by an optimized CRISPR/Cas system. Cell 155:1479–1491
14. Anton T, Bultmann S, Leonhardt H, Markaki Y (2014) Visualization of specific DNA

sequences in living mouse embryonic stem cells with a programmable fluorescent CRISPR/Cas system. Nucleus 5:163–172

15. Ma H, Naseri A, Reyes-Gutierrez P, Wolfe SA, Zhang S, Pederson T (2015) Multicolor CRISPR labeling of chromosomal loci in human cells. Proc Natl Acad Sci U S A 112:3002–3007
16. Westra ER, Swarts DC, Staals RH, Jore MM, Brouns SJ, van der Oost J (2012) The CRISPRs, they are a-changin': how prokaryotes generate adaptive immunity. Annu Rev Genet 46:311–339
17. Cong L, Ran FA, Cox D, Lin S, Barretto R, Habib N, Hsu PD, Wu X, Jiang W, Marraffini LA et al (2013) Multiplex genome engineering using CRISPR/Cas systems. Science 339:819–823
18. Mali P, Yang L, Esvelt KM, Aach J, Guell M, DiCarlo JE, Norville JE, Church GM (2013) RNA-guided human genome engineering via Cas9. Science 339:823–826
19. Cremer M, Grasser F, Lanctot C, Muller S, Neusser M, Zinner R, Solovei I, Cremer T (2008) Multicolor 3D fluorescence in situ hybridization for imaging interphase chromosomes. Methods Mol Biol 463:205–239
20. Byron M, Hall LL, Lawrence JB (2013) A multifaceted FISH approach to study endogenous RNAs and DNAs in native nuclear and cell structures. Curr Protoc Hum Genet Unit 4:15

Chapter 26

Methods to Monitor DNA Repair Defects and Genomic Instability in the Context of a Disrupted Nuclear Lamina

Susana Gonzalo and Ray Kreienkamp

Abstract

The organization of the genome within the nuclear space is viewed as an additional level of regulation of genome function, as well as a means to ensure genome integrity. Structural proteins associated with the nuclear envelope, in particular lamins (A- and B-type) and lamin-associated proteins, play an important role in genome organization. Interestingly, there is a whole body of evidence that links disruptions of the nuclear lamina with DNA repair defects and genomic instability. Here, we describe a few standard techniques that have been successfully utilized to identify mechanisms behind DNA repair defects and genomic instability in cells with an altered nuclear lamina. In particular, we describe protocols to monitor changes in the expression of DNA repair factors (Western blot) and their recruitment to sites of DNA damage (immunofluorescence); kinetics of DNA double-strand break repair after ionizing radiation (neutral comet assays); frequency of chromosomal aberrations (FISH, fluorescence in situ hybridization); and alterations in telomere homeostasis (Quantitative-FISH). These techniques have allowed us to shed some light onto molecular mechanisms by which alterations in A-type lamins induce genomic instability, which could contribute to the pathophysiology of aging and aging-related diseases.

Key words Genomic instability, DNA repair, DNA damage response, Western blot, Immunofluorescence, Neutral comet assay, Q-FISH, Nuclear lamina

1 Introduction

Efficient DNA repair is intimately linked to the integrity of the nuclear lamina, a meshwork of intermediate filaments under the inner nuclear membrane that also extends throughout the nucleoplasm. The nuclear lamina is formed by A-type lamins (lamin A/C), B-type lamins (lamin B1/B2), and lamin-associated proteins. A whole body of evidence indicates that cells with a disrupted nuclear lamina due to mutation or loss of lamins exhibit genomic instability [1–4].

Many studies have focused on the characterization of how expression of unprocessed prelamin A or mutant lamin A proteins affect DNA repair. A landmark study demonstrated delayed recruitment of DNA repair factors 53BP1 and RAD51 to sites of DNA

Sue Shackleton et al. (eds.), *The Nuclear Envelope: Methods and Protocols*, Methods in Molecular Biology, vol. 1411,
DOI 10.1007/978-1-4939-3530-7_26, © Springer Science+Business Media New York 2016

damage (γH2AX foci) induced by radiation in cells with a disrupted nuclear lamina [5]. Other studies showed aberrant accumulation of the repair protein XPA at DNA lesions, which activates ATM- and ATR-dependent signaling cascades contributing to proliferation arrest in cells deficient in A-type lamins [6]. Additional suggested mechanisms include reduction of DNA-PK holoenzyme, a key factor in non-homologous end-joining (NHEJ) repair [7], delayed recruitment to DNA double-strand breaks (DSBs) of factors of the MRN complex (NBS1 and MRE11), which are necessary for homologous recombination (HR) [8], and defects in chromatin-modifying activities such as the NuRD complex and the histone acetyltransferase Mof [9, 10]. Recently, increased H3K9me3 levels due to higher activity of histone methyltransferase Suv39h1 has been linked to genomic instability and premature senescence [11].

Our studies have also revealed mechanisms behind DNA repair defects in lamin A/C-deficient cells. We found that loss of lamin A/C leads to activation of cathepsin L-mediated degradation of 53BP1 and reduced expression of BRCA1 and RAD51 [12–17]. We also showed that loss of lamin A/C leads to altered distribution of telomeres in the 3D nuclear space, which is accompanied by telomere shortening. Moreover, an increase in aneuploidy and in the frequency of chromosome and chromatid breaks was observed in lamin A/C-deficient cells. Altogether, these studies identified a plethora of factors whose levels and recruitment to sites of DNA damage are altered in cells with a disrupted nuclear lamina, leading to defects in the two main mechanisms of DNA DSB repair: HR (homologous recombination) and NHEJ (non-homologous end-joining). As a consequence, cells exhibit a permanent checkpoint activation that induces proliferation arrest.

Here, we describe methods utilized in lamin A/C-deficient cells to monitor expression of DNA repair factors (Western blot), and their recruitment to sites of DNA damage (immunofluorescence), as well as the kinetics of DNA DSB repair (neutral comet assays after ionizing radiation). Alterations in the lamina can also induce shortened and dysfunctional telomeres, which can be assessed by quantitative-fluorescence in situ hybridization (Q-FISH). FISH also allows the monitoring of genomic instability (aneuploidy, chromosome end-to-end fusions, chromosome and chromatid breaks, and other chromosomal aberrations) in metaphase spreads.

2 Materials

All techniques, with the exception of Western blotting, require a fluorescence microscope with camera.

2.1 Western Blotting Components

1. RIPA buffer: 50 mM Tris–HCl (pH 8.0), 150 mM NaCl, 1 % NP-40, 0.5 % sodium deoxycholate, 0.5 % SDS. Prepare 50 mL by mixing 2.5 mL 1M TRIS–HCl (pH 8.0), 1.5 mL 5 M

NaCl, 500 μL NP-40, 2.5 mL DOC 10 %, 2.5 mL 10 % SDS, and 40.5 mL deionized water. Store at 4 °C.

2. Fresh RIPA Solution: 2 mL RIPA buffer, 20 μL PMSF, 4 μL DTT, 20 μL 100× protease and phosphatase inhibitor cocktail. Prepare fresh for each Western blot (*see* **Note 1**).
3. 1 mL syringes.
4. Needles: 26 G × 1/2 in; 30 G × 1/2 in.
5. Bovine Gamma Globulin Standard (BGG).
6. Protein Assay Dye Reagent: Dilute Protein Assay Dye Reagent Concentrate 1:4 with deionized water.
7. 4× Laemmli buffer.
8. Tris-glycine running buffer (10×): Add 29 g TRIS base, 144 g glycine, 10 g SDS and bring final volume to 1 L with deionized water.
9. Precision Plus Protein™ Kaleidoscope™ Standards.
10. Gel 4–15 % Criterion™ TGX™ (18 wells of 30 μL).
11. Trans-Blot® Turbo™ Transfer Starter System.
12. Trans-Blot® Turbo™ Midi Nitrocellulose Transfer Kit.
13. Transfer buffer: Mix 40 mL 5× transfer buffer and 40 mL ethanol, and bring to 200 mL with deionized water.
14. Ponceau S solution.
15. PBS-T: Add 800 μL Tween™ 20 to 1 L PBS.
16. Blotting buffer: Add 3 g of blotting grade buffer to 100 mL of PBS-T.
17. Antibodies: 53BP1, BRCA1, vinculin, RAD51, γH2AX, donkey anti-rabbit IgG-HRP, bovine anti-goat IgG-HRP, goat anti-mouse IgG-HRP.
18. Boxes or trays suitable for incubation of blotting membranes.
19. Clarity Western ECL Substrate (*see* **Note 2**).
20. Membrane development: multi-application gel imaging system or classical autoradiography film and cassette (*see* **Note 3**).

2.2 Components for Detection of DNA Repair Foci

1. Unifrost microscope slides 75 × 25 × 1 mm.
2. Glass coverslips 9 × 9 mm. Coat coverslips with HistoGrip following the manufacturers' instructions. Histogripped coverslips can be stored after preparation in a sterile 50 mL conical tube (*see* **Note 4**).
3. 60 mm (p60) tissue cell culture plate.
4. Fine point micro tweezers for handling coverslips.
5. 10 % Triton X-100 solution: Dilute Triton X-100 1:10 in PBS.
6. 10 % BSA solution: Dissolve 10 g of bovine serum albumin in PBS to a final volume of 100 mL. Mix solution until BSA is

entirely dissolved in solution. Make aliquots and store them at −20 °C.

7. Fixing buffer: 3.7 % formaldehyde, 0.2 % Triton X-100 in PBS. Use 4 mL for each p60. For 20 mL, mix 2 mL 36.5–38 % formaldehyde solution, 400 μL Triton X-100 10 %, and 17.6 mL PBS.
8. Blocking solution: 2 % BSA, 0.1 % Triton X-100 in PBS. For 20 mL, mix 4 mL BSA 10 %, 200 μL Triton X-100 10 %, and 15.8 mL PBS.
9. Antibody diluting solution: 1 % BSA, 0.1 % Triton X-100 in PBS. For 20 mL, mix 2 mL BSA 10 %, 200 μL Triton X-100 10 %, and 17.8 mL PBS.
10. Antibodies: γH2AX, 53BP1, BRCA1, RAD51, goat anti-rabbit secondary antibody Alexa Fluor® 488 conjugate, goat anti-mouse IgG secondary antibody Alexa Fluor® 594 conjugate.
11. DAPI Mounting Medium: Mix two drops of VECTASHIELD Mounting Medium with one drop of VECTASHIELD Mounting Medium containing DAPI.

2.3 Components for Neutral Comet Assay

1. PBS and Lysis Solution. Bring to 4 °C before use.
2. Low-Melting Agarose. Prepare 1 mL aliquots. One hour before use, melt aliquots at 95 °C and then immediately move to thermal block at 37 °C (*see* **Note 5**).
3. TBE electrophoresis buffer: Prepare 1× Tris–Borate–EDTA buffer by diluting 5× concentrate in deionized water. Bring to 4 °C before use.
4. Staining dish and slide staining rack.
5. SYBR® Gold Nucleic Acid Gel Stain (10,000× Concentrate in DMSO). For use, dilute concentrate in TE buffer (10 mM Tris–HCl pH 7.5, 1 mM EDTA). Diluted stock is stable for several weeks if stored at 4 °C in dark.
6. Comet assay analysis by CometScore program.

2.4 Components for Q-FISH

1. Colcemid—10 μg/mL.
2. Potassium chloride. Prepare 0.56 % KCl by adding 0.56 g to 100 mL of deionized water.
3. Methanol–acetic acid fixing solution: Mix methanol with acetic acid at 3:1 and keep at −20 °C. Make fresh for each Q-FISH.
4. Acidified pepsin: Mix 200 mg pepsin, 200 mL deionized water, and 168 μL of concentrated HCl.
5. 4 % formaldehyde fixing solution: For 800 mL, mix 94.4 mL of 35 % formaldehyde with 705.6 mL PBS.
6. Telomere probe mix (Table 1):

7. Maleic acid buffer: 100 mM maleic acid, 150 mM NaCl (pH 7.5). Adjust with NaOH. For 200 mL, mix 6 mL 5 M NaCl, 2 g maleic acid and bring to 200 mL with water. Aliquot and store at −20 °C.
8. BM 10 %: 10 g blocking reagent in 100 mL maleic acid buffer pH 7.5. Dissolve on a heating block or microwave. Aliquot and store at −20 °C.
9. $MgCl_2$ buffer: 25 mM $MgCl_2$, 9 mM citric acid, 82 mM Na_2HPO_4, pH 7.0. For 100 mL, mix 2.5 mL of 1 M $MgCl_2$, 9.0 mL 0.1 M citric acid, 8.2 mL of 1 M Na_2HPO_4, and 80.3 mL deionized water.
10. Formamide–BSA wash solution (Table 2):
11. TBS–Tween 20 wash solution: Make 10× TBS stock, with final concentration 1.5 M NaCl and 1 M TRIS pH 7.0–7.5. Dilute stock 10× TBS 1:10 in deionized water and add Tween-20 (0.08 %).
12. Staining dish and slide staining rack.
13. DAPI Mounting Medium: Mix two drops of VECTASHIELD Mounting Medium with one drop of VECTASHIELD Mounting Medium with DAPI.

Table 1
Composition of the telomere probe mix

Stock	10 slides (μL)	20 slides (μL)
Tris 1M pH 7.4	2.5	5.0
$MgCl_2$ buffer	21.4	42.8
Deionized formamide	175.0	350.0
Probe 25 μg/mL (telomeric)	5.0	10.0
BM 10 % (Blocking reagent)	12.5	25.0
Deionized water	33.6	67.2

Table 2
Composition of the formamide–BSA solution

Stock	400 mL	800 mL
Formamide	280 mL	560 mL
TRIS 1M, pH 7.2	4 mL	8 mL
BSA 10 % (in water)	4 mL	8 mL
Deionized Water	112 mL	224 mL

14. Nail polish.
15. TFL-Telo computer program. This is an application program developed by Peter Lansdorp and used to estimate the length of telomeres from captured images of metaphases that have been stained with a telomere probe for quantitative in situ hybridization analysis [18]. With the program, the integrated fluorescence intensity value for each telomere, which is proportional to the number of hybridized probes, is calculated.

3 Methods

3.1 Western Blotting for DNA Repair Factors

Day 1

1. Collect at least 600,000 cells in conical tube. Centrifuge at 300 × *g* for 5 min. Discard supernatant and resuspend pellet in PBS. Centrifuge at 300 × *g* for 5 min. Discard supernatant and collect the cell pellet.
2. Add fresh RIPA solution, according to size, to each cell pellet. The amount of RIPA added ranges between 60 and 150 μL, although this depends upon the number of pelleted cells. Generally, 10^6 cells will be suspended in ~100 μL to obtain an adequate protein concentration. Mix cell lysates well by pipetting up and down.
3. Transfer lysates to Eppendorf tubes. Sonicate with "High" setting for 7.5 min at 4 °C.
4. Shear lysates with ten passes through a 26-gauge needle. Then, shear samples with ten passes through a 30-gauge needle (*see* **Note 6**).
5. Determine protein concentration by Bradford assay, as per manufacturer's instructions. Read absorbance at 595 nm.
6. Prepare samples for gel loading. Load ~50 μg in 30 μL per well in 4–15 % Criterion™ TGX™ gel with 18 wells per gel (*see* **Note 7**). Mix 50 μg of protein, 7.5 μL 4× Laemmli buffer, and take the volume to 30 μL with RIPA buffer.
7. Heat tubes to 90 °C for 5 min.
8. Vortex samples well and centrifuge.
9. Set up electrophoresis chamber: place gel in chamber and add 1× Tris-Glycine running buffer to compartment; remove comb and load Kaleidoscope™ Standards ladder and samples.
10. Run gel with voltage set anywhere from 90 V to 180 V. Run gel as far as needed to allow adequate separation of proteins of interest for investigation.

11. After electrophoresis is complete, remove cartridge and open on sides by using spatula.
12. Assemble transfer cassette for Trans-Blot® Turbo™ Transfer Starter System. Place membrane stack in transfer buffer and add first to cassette. Place nitrocellulose membrane in transfer buffer and add to stack. Place gel on membrane (*see* **Note 8**). Place another membrane stack in transfer buffer and add to cassette. Close cassette and insert into Trans-Blot® Turbo™ Transfer Starter System.
13. Run Transfer System at 1.3 A and 25 V for 25 min.
14. Remove membrane from cassette and place in tray. Add Ponceau-S for ~30 s (*see* **Note 9**).
15. Wash off Ponceau-S with deionized water and cut membrane. Place various sections of membrane in Western blot boxes.
16. Wash membranes in blotting buffer for 15 min.
17. Incubate samples overnight, shaking at 4 °C, in primary antibody dissolved in blotting buffer.

Day 2

18. Wash membranes 3× with PBS-T, 10 min each time.
19. Incubate samples for 1 h with secondary antibodies dissolved in blotting buffer.
20. Wash membranes 3× with PBS-T, 10 min each time.
21. Develop membranes by incubating membrane with Clarity Western ECL Substrate or ECL of choice (*see* **Note 2**).
22. Visualize blot using either gel imaging system or autoradiography film in dark room. An example of results obtained by Western blotting is shown in Fig. 1.

3.2 Formation and Resolution of DNA Repair Foci

Day 1

1. HistoGrip coverslips following manufacturer's instructions.
2. Drop 8 coverslips coated with HistoGrip in a p60 for each set of conditions to be investigated, as shown in Fig. 2 (*see* **Note 10**).
3. Plate cells at 70–80 % confluency in p60. Ensure that after cells are plated, all coverslips are attached to bottom of plate and are not floating. Leave overnight for cells to attach.

Day 2

4. Add treatment/irradiate all p60s together with desired level of treatment/dosage (*see* **Note 11**).
5. After treatment/irradiation, place cells back in incubator and allow given amount of time to repair DNA damage. If there is a 0 min time point for irradiation, irradiate those cells on ice or place them on ice immediately after irradiation.

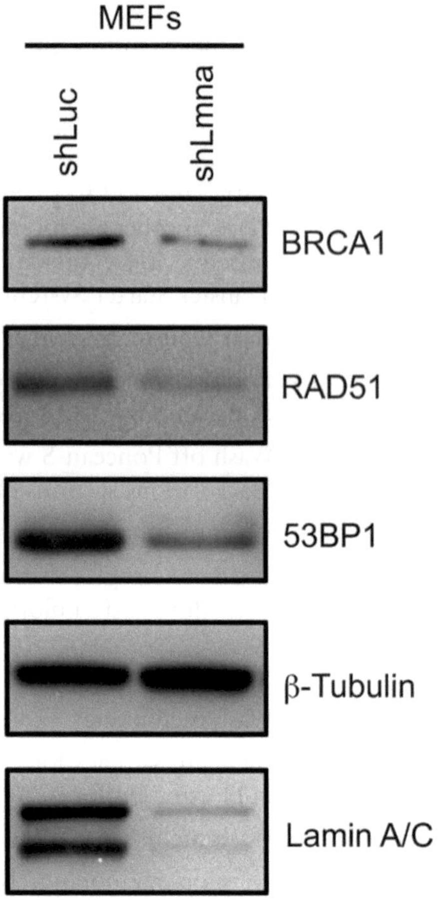

Fig. 1 Detection of DNA repair factors by immunoblotting. Immunoblots show changes in several DNA repair factors upon depletion of A-type lamins (lamin A/C) via lentiviral transduction with specific shRNAs in mouse embryonic fibroblasts (MEFs). Decreased levels of BRCA1 and RAD51 are associated with defects in homologous recombination (HR), and decreased 53BP1 levels with defects in non-homologous end joining (NHEJ)

6. When the desired amount of repair time has elapsed, aspirate media and wash cells with PBS. Then, immediately fix cells with fixing buffer for 10 min at room temperature (*see* **Note 11**).
7. Wash coverslips 3× with PBS, 5 min each time. After washing, cells can be left at 4 °C in PBS until all other time points have been fixed.
8. Incubate coverslips for 1 h at 37 °C in blocking solution.
9. Wash coverslips 3× with PBS, 5 min each time. At this point, cells can be stored in PBS at 4 °C for up to a week before proceeding. For longer storage, keep coverslips in PBS with 0.02 % sodium azide.

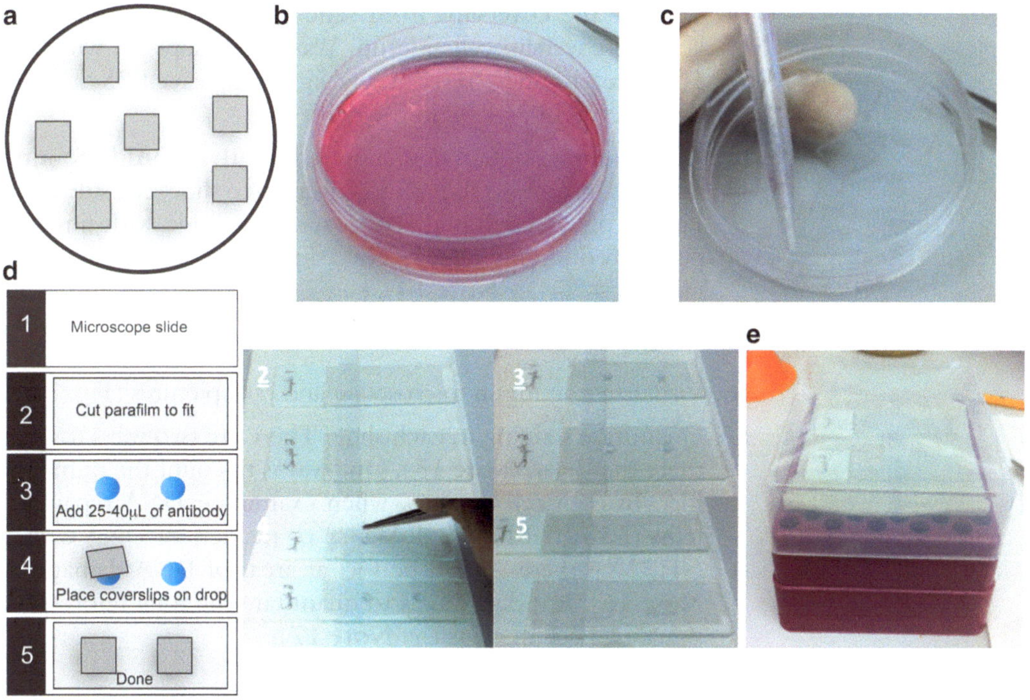

Fig. 2 Immunofluorescence of cells growing on coverslips. (**a**) Place eight coverslips in each 60 mm plate (p60) for immunofluorescence (IF). (**b**) Plate cells on top of coverslips. (**c**) When ready to fix cells, aspirate media carefully. (**d**) Perform incubation with antibodies of interest as in figure: (1) obtain microscope slides; (2) cover slide with Parafilm; (3) add 25–40 μL drops of diluted antibodies; (4) place coverslips up-side down on antibody drop; (5) repeat for all coverslips. (**e**) Place slides in a humidity chamber adding a wet paper towel, cover, and incubate at 37 °C

10. Prepare microscope slides for immunofluorescence (Fig. 2): put Parafilm over a microscope slide to cover it. Pipette 25–40 μL of the diluted antibody in two separate spots on Parafilm. Our typical antibody dilutions, diluted in antibody diluting solution, are: 1:50 (γH2AX), 1:100 (BRCA1), 1:200 (RAD51), 1:1000 (53BP1). Use tweezers to lower the coverslip (cell side down) onto the droplets (*see* **Notes 12** and **13**).
11. Incubate for 1 h in primary antibody at 37 °C in humidity chamber made by placing a wet paper towel into a sealed plastic box, together with the slide(s) (Fig. 2).
12. Transfer coverslips from slide to a new p60 containing PBS, keeping coverslip cell side up. Wash coverslips 3× with PBS, 10 min each time. Meanwhile, prepare slide with new piece of Parafilm and two drops of secondary antibody. We use 1:1000 secondary antibody dilutions.
13. Incubate coverslips in secondary antibody for 1 h in humidity chamber.

14. Transfer the coverslips from slide to a p60 containing PBS, keeping coverslip cell side up. Wash coverslips 3× with PBS, 10 min each time. To reduce background, leave PBS on coverslips overnight at 4 °C in the dark.
15. After washing, obtain new slide. Add 7 μL of DAPI mounting medium directly to slide. Add coverslip to slide, cell side down. Pipette off excess liquid. Fix coverslip to slide with fingernail polish by outlining edge of coverslip. Leave slides in dark overnight at 4 °C (*see* **Note 14**).

Day 3

16. Visualize staining in microscope and take pictures (Fig. 3).
17. Quantitate staining in each slide. There are two ways to quantitate the IF (*see* **Note 15**). One way is to count the number of foci. In each cell nucleus, when examining DNA repair proteins (53BP1, BRCA1, RAD51) or markers of DNA damage (γH2AX), there should be foci at areas of DNA damage (*see* **Note 16**). Another way is to quantitate the total fluorescence signal in a cell nucleus (*see* **Note 17**).

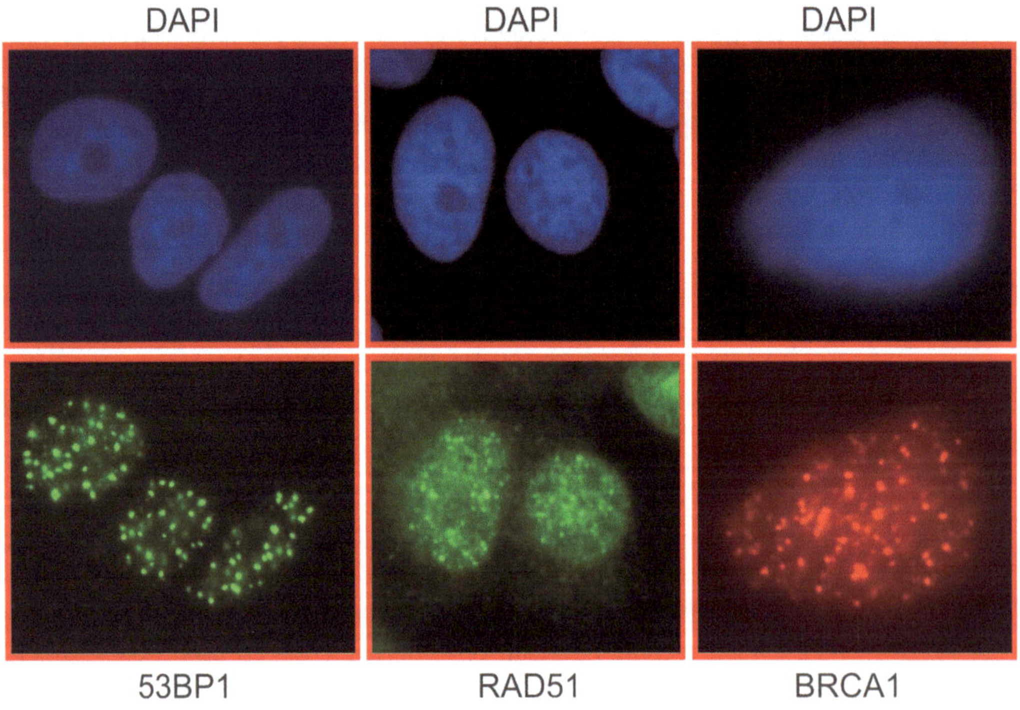

Fig. 3 Detection of DNA repair foci by immunofluorescence. Cells were irradiated with 3 Gy and processed for immunofluorescence with antibodies recognizing the DNA repair factors 53BP1, RAD51, and BRCA1. All proteins form foci at sites of radiation-induced DNA damage. We score cells as positive for DNA repair foci when they show at least 5 foci

3.3 Neutral Comet Assay

Day 1

1. Plate ~600,000 cells in a p60 and add treatment interested in investigating. Prepare 4–6 plates per sample (*see* **Note 18**).
2. Irradiate sample with 8 Gy. Write down the exact time the plates came out of irradiator and the subsequent time they are to be harvested, depending upon the amount of time to be given to repair (*see* **Note 19**).
3. Trypsinize cells to a single cell suspension. Dilute approximately 1:1 in PBS. Do not use media! Immediately place cell suspension on ice (*see* **Note 20**).
4. Place slides on thermal block at 37 °C.
5. Mix 10 μL cell suspension and 100 μL melted agarose in Eppendorf tube. Mix well and drop 50 μL in each window on the comet assay slide (*see* **Note 21**).
6. Place the slides immediately at 4 °C. Let them cool down for 10–30 min (*see* **Note 22**).
7. Run electrophoresis at 35 V for 30 min using TBE buffer.
8. Wash slides 2× with deionized water, 10 min each time.
9. Wash slides 1× with ethanol for 5 min. Keep slides in a dry box until the next day.

Day 2

10. Stain slides with SYBR Gold.
11. Wash membranes 2× with deionized water, 10 min each time.
12. Take pictures of at least 20–25 comets for each experimental condition using fluorescence microscope (*see* **Note 23**).
13. Analyze images with CometScore™, a free software that measures a number of parameters such as Olive tail moment, tail percentage intensity, and tail length, which overall serve to determine the extent of unrepaired DNA damage (double-strand breaks if the comet assay is performed under neutral conditions). Olive tail moment is used most frequently to monitor DNA repair kinetics, and to identify cells that are deficient in DNA repair (Fig. 4).

3.4 Q-FISH

Day 1: Preparation of Metaphases

1. Add colcemid (100 ng/mL) to cells in culture. Incubate for 2–4 h at 37 °C (*see* **Note 24**).
2. Collect 10 mL of media from cells and transfer to a conical vial. Wash cells with 5 mL PBS and combine with 10 mL media (*see* **Note 25**).
3. Trypsinize cells. Inactivate trypsin with the medium in the conical tube and collect cells by centrifugation for 8 min.

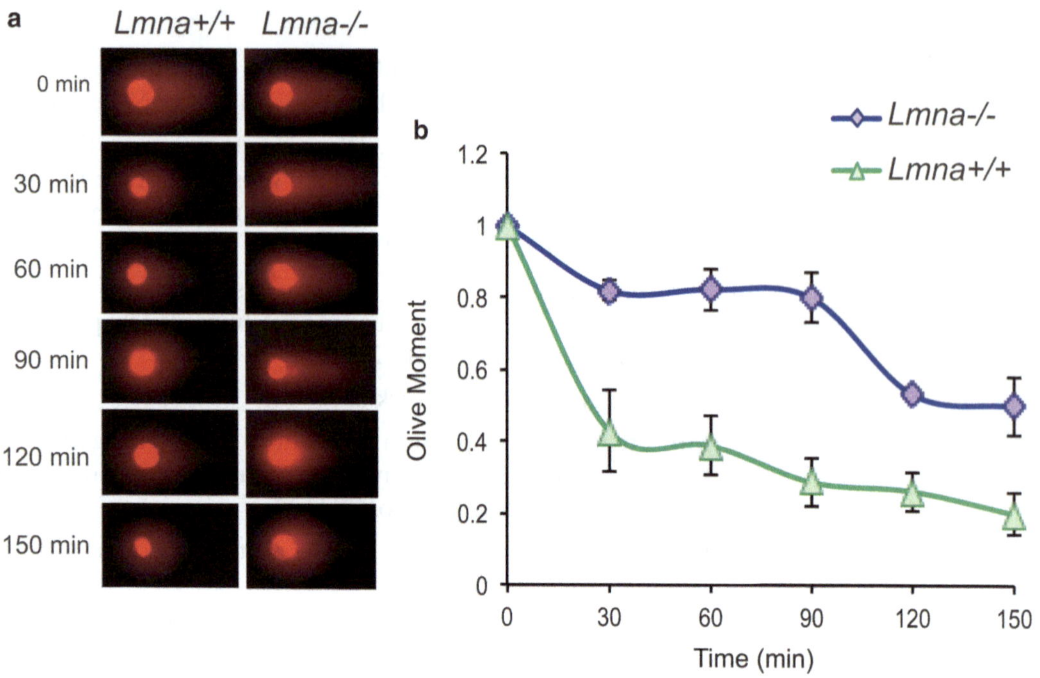

Fig. 4 DNA repair deficiencies in lamin A/C-deficient cells. (**a**) Representative images of individual DNA comets from *Lmna*$^{+/+}$ and *Lmna*–/– MEFs, indicating deficiencies in DNA double-strand break (DSB) repair in *Lmna*–/– cells. (**b**) Graph shows average olive moments over a period of time in *Lmna*$^{+/+}$ and *Lmna*–/– MEFs. *Lmna*$^{+/+}$ cells show the characteristic biphasic mode of DNA DSB repair. Note how *Lmna*–/– cells exhibit lower kinetics of repair

4. Aspirate supernatant, leaving 1 mL in the tube; resuspend cells by tapping the tube.
5. Add dropwise, while gently vortexing, 9 mL of 0.56 % KCl preheated to 37 °C.
6. Keep in water bath at 37 °C for 10–12 min.
7. Add 3 drops of fresh methanol–acetic acid fixing solution at 4 °C (*see* **Note 26**).
8. Sediment cells by centrifugation for 8 min.
9. Aspirate supernatant until only 1 mL remains.
10. Add 2 mL fresh methanol–acetic acid fixing solution while gently vortexing. Add another 9 mL (*see* **Note 27**).
11. Repeat **steps 8–10**. Samples can be kept at −20 °C until preparation of metaphases.
12. Sediment cells by centrifugation for 8 min.
13. Aspirate supernatant until only 1 mL remains.
14. Resuspend cells and add 10 mL fresh methanol–acetic acid fixing solution while vortexing.
15. Sediment cells.

16. Aspirate supernatant until only 1 mL or less remains, depending on size of cell pellet (*see* **Note 28**).
17. Aspirate cells with a Pasteur pipette where a capillary end has been created.
18. Wet a glass slide in 45 % Acetic acid and drain. Let some drops of the cell solution fall on the slide from the maximum height possible (*see* **Note 29**).
19. Let slides dry overnight. Check for metaphases in the microscope (*see* **Note 30**).

Day 2: Metaphase Hybridization

20. Prepare acidified pepsin and incubate for 15 min at 37 °C.
21. Wash slides in PBS for 15 min in shaker (*see* **Note 31**).
22. Fix cells in 4 % formaldehyde fixing solution for 2 min.
23. Wash slides 3× with PBS, in shaker, 5 min each time.
24. Digest with preheated pepsin 10 min at 37 °C in water bath.
25. Wash slides 2× with PBS, in shaker, 5 min each time.
26. Fix cells in 4 % formaldehyde fixing solution for 2min.
27. Wash slides 3× with PBS, in shaker, 5 min each time.
28. Dehydrate slides: wash 5 min each in 70 %, 90 % and 100 % ethanol.
29. Air-dry for 5–20 min.
30. Prepare telomere probe mix.
31. Add two drops (10–15 μL) of probe mix to a long cover slide. Turn slide upside down onto the cover, so that the probe extends by diffusion (*see* **Note 32**).
32. Denature at 80 °C for 3 min (*see* **Note 33**).
33. Make a wet chamber by covering the walls of a big cylinder with wet paper towels. Put slides into the chamber with covers facing down. Seal cylinder with Saran Wrap. Incubate in dark for 2 h at room temperature.
34. Wash 2× 15 min each with formamide–BSA wash solution, while vortexing. If covers do not separate from slides after 5 min, use tweezers to remove them.
35. Wash slides 3× 5 min with PBS in shaker.
36. Dehydrate slides: wash 5 min each in 70 %, 90 % and 100 % ethanol.
37. Air-dry slides.
38. Add two drops (10–15 μL each) of DAPI mounting medium to a long coverslip; place slides over the coverslip; let dry for 5 min.
39. Seal cover to slides with nail polish.
40. Keep samples at 4 °C in the dark.

Day 3: Microscopy and Analysis

41. Take pictures of metaphases (Fig. 5a).
42. Analyze pictures: Analyze telomere length TFL-Telo following software instructions. Analyze metaphases for genomic insta-

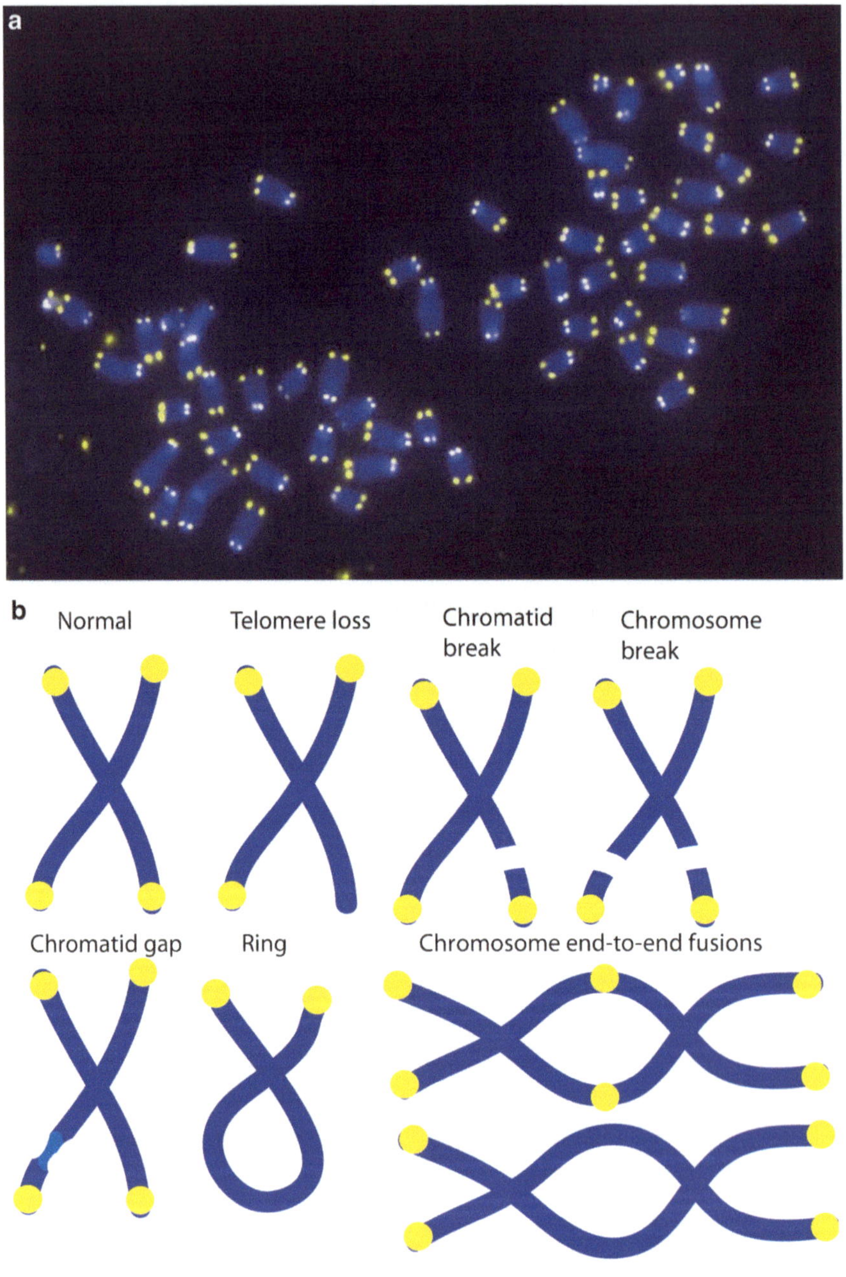

Fig. 5 Monitoring chromosomal instability. (**a**) Metaphase spread processed for FISH with a telomere probe (*yellow*). (**b**) Examples of chromosomal aberrations that can be monitored in metaphase spreads to determine the extent of genomic instability in different cells

bility by looking for telomere loss, chromatid breaks, chromosome breaks, gaps, fusions, and other markers of genomic instability (Fig. 5b).

4 Notes

1. While the RIPA buffer can be prepared and stored for months at 4 °C, make the RIPA working solution, with PMSF, DTT, and protease & phosphatase inhibitors fresh before each blot.
2. ECL used for ideal signal can vary depending upon protein of interest and the amount of protein loaded. Clarity Western ECL is a good starting point. If the signal is weak, Immobilon Western Chemiluminescent HRP Substrate is stronger and works well for less abundant proteins. If the Clarity Western ECL signal is too strong, Pierce™ ECL Western Blotting Substrate produces a weaker signal and works well for more abundant proteins, like vinculin.
3. We develop our Western blots with the Syngene PXi. We like this method because you can view the blot as it is developing. However, a cheaper alternative is to use autoradiography films which also work well. The developing method might also impact ECL choice.
4. We previously coated coverslips with poly-L-lysine. However, we found that HistoGrip works much better than poly-L-lysine, especially for cells that do not normally attach tightly to coated surfaces.
5. Be cautious not to keep agarose at 95 °C for too long. Agarose that is too hot can impact comet tail length.
6. After sonication and processing with needles, the cell lysate should have a similar viscosity to water, whereby it can be easily pipetted dropwise. If the sample is too viscous, more RIPA buffer should be added.
7. The ideal amount of protein to load depends upon the protein being investigated. It is imperative that the amount of protein loaded falls within the linear and quantitative dynamic range for the protein of interest given the experimental conditions. This linear range must be determined by performing a dilution series of protein and determining if the signal generated from the blot corresponds to the dilution.
8. As each successive component is added to the cassette, it is crucial to ensure that everything is flat by using a roller. When adding the gel to the cassette, make sure there is ample buffer on the membrane to allow easy repositioning of the gel if necessary. After adding the gel to the membrane, ensure that there are no air bubbles between the gel and the membrane, as this

could prevent transfer in those areas. Before closing the cassette, dump excess liquid from the cassette. Too much extra liquid could reduce efficiency of transfer.

9. Ponceau-S staining reveals success for protein transfer. If transfer occurred correctly, protein bands should be evident, including high molecular weight bands. When staining, take precaution that the membrane does not become dry for an extended period of time.
10. Prepare one p60 for each set of conditions being investigated. For example, to investigate a particular DNA repair protein in wild-type and knock-out cells at 0 min, 30 min, 1 h, 6 h, 12 h, and 24 h after irradiation, plate 12 plates total, six for the wild-type and six for the knock-out. Pick time points based upon cell type and protein of interest, as has previously been done [17, 19].
11. Since the goal is to monitor kinetics of DNA repair, the amount of damage should be enough to induce damage but not kill the cell. When irradiating, we normally irradiate between 0.5 Gy and 3 Gy for human normal fibroblasts. However, the optimal dose of irradiation will need to be determined for each cell type.
12. When washing cells, be careful to minimize detachment of cells. If any solution is added to p60 too quickly, cells can detach from coverslips, hindering your ability to perform immunofluorescence. The Parafilm prevents the antibody from spreading out on the slide and allows the coverslip to float on the antibody solution.
13. We usually have at least two coverslips per condition. That way, if one coverslip breaks, there is another coverslip to use. Also, there are more cells to analyze this way.
14. There is usually a moderate amount of background present if you look at the cells immediately after fixing the coverslips to the slide. Letting the coverslips sit overnight before viewing usually increases the quality of the immunofluorescence.
15. Generally, count at least 200 cells/condition when quantitating.
16. One method for quantitating immunofluorescence for DNA repair proteins involves quantitating the number of foci present in the cell nucleus. This is usually done in two ways: count the number of foci per cell and determine average number of foci per cell; alternatively, set a cut-off point and determine whether the cells have more foci (termed positive) or less foci (termed negative) than the cutoff point. Then, determine the percentage of cells that are positive or negative. Generally, we use five as our cut-off point when determining whether cells are positive or negative for a given DNA repair protein, although this cut-off can be changed depending on the given situation.

17. The intensity of the immunofluorescence signal can be quantitated in each situation. Programs like ImageJ can measure the intensity of signal for a given area, and DAPI staining can allow for determination of the specific area of the nucleus.
18. The neutral comet assay will monitor kinetics of repair for DNA double strand breaks. Consequently, for each treated and untreated sample, we usually have plates for repair times of 0, 30, 60, 90, 120, and 150 min.
19. Induce a large amount of damage for neutral comet assays. However, the amount irradiated will vary based on cell type. If there is no repair by 120 min, reduce the dose of irradiation. Also, we usually irradiate our samples so that they can all be embedded on the slide at the same time. Thus, we irradiate our 150 min repair sample first. 30 min later, we irradiate our 120 min sample, etc. Cells for the 0 time point are trypsinized before irradiation and irradiated on ice.
20. Choose a dilution so that you will have cells on your slide, but not so many that it will be hard to analyze. Confluency will need to be optimized to determine the correct dilution of trypsin:PBS to determine adequate cell plating without overplating.
21. Once the suspension is added to the slide, quickly distribute suspension with pipette tip throughout slide. Be sure that agarose is secured to edge of slide windows to ensure that electrophoresis will work properly. Add suspension right against the edge and work it up onto red border to ensure that the edges are sealed as suspension is being spread across the slide.
22. It is important that the slides are thoroughly cooled. If electrophoresis is begun with warm TBE or slides where the agarose is not cool, the agarose may melt or dissociate from the slide.
23. The best pictures are taken from the center of the slide windows. Pictures should reveal a comet-shaped DNA pattern as a result of electrophoresis (Fig. 4). The comet head will contain high molecular weight and intact DNA, while the tail will contain the leading ends of migrating fragments.
24. Colcemid arrests cells in metaphase. Since telomeres are best analyzed during metaphase, it is best to arrest many cells in metaphase. The amount of cells arresting with a given colcemid treatment is a function of how quickly cells are cycling. Thus, slower cycling cells will need to be kept in colcemid for a greater amount of time. As a starting point, we keep lymphocytes and immortalized cells in colcemid for 2 h. Fibroblasts and primary cells are kept in colcemid for 4 h. This time may need to be lengthened if very few cells arrest in metaphase.
25. Cells in mitosis do not attach well to the plate. It is important to keep the media and the PBS wash, since it might contain mitotic cells.

26. It is important that fixative is made fresh every time. Also, cover tube label with transparent tape to ensure that methanol–acetic acid does not drip on outside and remove label while vortexing.
27. If there are too many tubes, add 2 mL to all of them first, and then add the rest of the solution. If there are only a few tubes, add 11 mL to each tube.
28. As you become familiar with Q-FISH, you will figure out the best cell density for your experiments. However, start by leaving 1 mL, and if this is too dilute, use smaller volumes in future trials.
29. Drop metaphases from the maximum height possible. Extend one arm as far down as possible, and place the pipette bulb near your eye so that you can look down on your slide as you drop metaphases. There will be a greater separation of metaphases if dropped from a greater height. Place two drops per slide.
30. If there are not enough metaphases, re-drop metaphases from a smaller starting volume in **step 16**.
31. Perform all washes and other treatments of slides in staining dish.
32. Take care not to make bubbles. If bubbles are present, use a pipette tip to tease them out. However, take care not to move the coverslip.
33. The exact time here is important. Use a timer to ensure slides are on the thermal block for exactly 3 min. When we do this with many slides, we add slides to the thermal block every 5 s to ensure we are able to collect them all after they have been at 80 °C for exactly 3 min.

Acknowledgment

This work was supported by NIGMS Grant RO1 GM094513-01, DOD BCRP Idea Award BC110089, and Presidential Research Award from St Louis University. R.K. is recipient of the William S. Sly Fellowship in Biomedical Sciences. The authors declare no conflict of interest.

References

1. Cau P, Navarro C, Harhouri K, Roll P, Sigaudy S, Kaspi E, Perrin S, De Sandre-Giovannoli A, Levy N (2014) Nuclear matrix, nuclear envelope and premature aging syndromes in a translational research perspective. Semin Cell Dev Biol doi: 10.1016/j.semcdb.2014.03.022. [Epub ahead of print]
2. Oberdoerffer P, Sinclair DA (2007) The role of nuclear architecture in genomic instability and ageing. Nat Rev Mol Cell Biol 8:692–702
3. Mekhail K, Moazed D (2010) The nuclear envelope in genome organization, expression and stability. Nat Rev Mol Cell Biol 11: 317–328

4. Capell BC, Collins FS (2006) Human laminopathies: nuclei gone genetically awry. Nat Rev Genet 7:940–952
5. Liu B, Wang J, Chan KM, Tjia WM, Deng W, Guan X, Huang JD, Li KM, Chau PY, Chen DJ, Pei D, Pendas AM, Cadinanos J, Lopez-Otin C, Tse HF, Hutchison C, Chen J, Cao Y, Cheah KS, Tryggvason K, Zhou Z (2005) Genomic instability in laminopathy-based premature aging. Nat Med 11:780–785
6. Liu Y, Wang Y, Rusinol AE, Sinensky MS, Liu J, Shell SM, Zou Y (2008) Involvement of xeroderma pigmentosum group A (XPA) in progeria arising from defective maturation of prelamin A. FASEB J 22:603–611
7. Liu GH, Barkho BZ, Ruiz S, Diep D, Qu J, Yang SL, Panopoulos AD, Suzuki K, Kurian L, Walsh C, Thompson J, Boue S, Fung HL, Sancho-Martinez I, Zhang K, Yates J 3rd, Izpisua Belmonte JC (2011) Recapitulation of premature ageing with iPSCs from Hutchinson-Gilford progeria syndrome. Nature 472:221–225
8. Constantinescu D, Csoka AB, Navara CS, Schatten GP (2010) Defective DSB repair correlates with abnormal nuclear morphology and is improved with FTI treatment in Hutchinson-Gilford progeria syndrome fibroblasts. Exp Cell Res 316:2747–2759
9. Pegoraro G, Kubben N, Wickert U, Gohler H, Hoffmann K, Misteli T (2009) Ageing-related chromatin defects through loss of the NURD complex. Nat Cell Biol 11:1261–1267
10. Krishnan V, Chow MZ, Wang Z, Zhang L, Liu B, Liu X, Zhou Z (2011) Histone H4 lysine 16 hypoacetylation is associated with defective DNA repair and premature senescence in Zmpste24-deficient mice. Proc Natl Acad Sci U S A 108:12325–12330
11. Liu B, Wang Z, Zhang L, Ghosh S, Zheng H, Zhou Z (2013) Depleting the methyltransferase Suv39h1 improves DNA repair and extends lifespan in a progeria mouse model. Nat Commun 4:1868
12. Gonzalez-Suarez I, Gonzalo S (2010) Nurturing the genome: A-type lamins preserve genomic stability. Nucleus 1:129–135
13. Gonzalez-Suarez I, Redwood AB, Gonzalo S (2009) Loss of A-type lamins and genomic instability. Cell Cycle 8:3860–3865
14. Gonzalez-Suarez I, Redwood AB, Grotsky DA, Neumann MA, Cheng EH, Stewart CL, Dusso A, Gonzalo S (2011) A new pathway that regulates 53BP1 stability implicates cathepsin L and vitamin D in DNA repair. EMBO J 30:3383–3396
15. Gonzalez-Suarez I, Redwood AB, Perkins SM, Vermolen B, Lichtensztejin D, Grotsky DA, Morgado-Palacin L, Gapud EJ, Sleckman BP, Sullivan T, Sage J, Stewart CL, Mai S, Gonzalo S (2009) Novel roles for A-type lamins in telomere biology and the DNA damage response pathway. EMBO J 28:2414–2427
16. Redwood AB, Gonzalez-Suarez I, Gonzalo S (2011) Regulating the levels of key factors in cell cycle and DNA repair: new pathways revealed by lamins. Cell Cycle 10: 3652–3657
17. Redwood AB, Perkins SM, Vanderwaal RP, Feng Z, Biehl KJ, Gonzalez-Suarez I, Morgado-Palacin L, Shi W, Sage J, Roti-Roti JL, Stewart CL, Zhang J, Gonzalo S (2011) A dual role for A-type lamins in DNA double-strand break repair. Cell Cycle 10: 2549–2560
18. Poon SS, Martens UM, Ward RK, Lansdorp PM (1999) Telomere length measurements using digital fluorescence microscopy. Cytometry 36:267–278
19. Grotsky DA, Gonzalez-Suarez I, Novell A, Neumann MA, Yaddanapudi SC, Croke M, Martinez-Alonso M, Redwood AB, Ortega-Martinez S, Feng Z, Lerma E, Ramon Y Cajal T, Zhang J, Matias-Guiu X, Dusso A, Gonzalo S (2013) BRCA1 loss activates cathepsin L-mediated degradation of 53BP1 in breast cancer cells. J Cell Biol 200:187–202

Part V

Nucleo-Cytoplasmic Transport

Chapter 27

High-Resolution Scanning Electron Microscopy and Immuno-Gold Labeling of the Nuclear Lamina and Nuclear Pore Complex

Martin W. Goldberg

Abstract

Scanning electron microscopy (SEM) is a technique used to image surfaces. Field emission SEMs (feSEMs) can resolve structures that are ~0.5–1.5 nm apart. FeSEM, therefore is a useful technique for imaging molecular structures that exist at surfaces such as membranes. The nuclear envelope consists of four membrane surfaces, all of which may be accessible for imaging. Imaging of the cytoplasmic face of the outer membrane gives information about ribosomes and cytoskeletal attachments, as well as details of the cytoplasmic peripheral components of the nuclear pore complex, and is the most easily accessed surface. The nucleoplasmic face of the inner membrane is easily accessible in some cells, such as amphibian oocytes, giving valuable details about the organization of the nuclear lamina and how it interacts with the nuclear pore complexes. The luminal faces of both membranes are difficult to access, but may be exposed by various fracturing techniques. Protocols are presented here for the preparation, labeling, and feSEM imaging of *Xenopus laevis* oocyte nuclear envelopes.

Key words Nuclear envelope, Nuclear pore complex, Scanning electron microscopy, Nuclear lamina

1 Introduction

Many important biological structures and processes occur at surfaces or interfaces. One of the best methods for obtaining medium resolution (in the order of 1–5 nm) images of large and complex membrane associated structures is field emission scanning electron microscopy (feSEM).

The nuclear envelope (NE) is a particularly complex membrane system. It consists of two membranes, the inner nuclear membrane (INM) and outer nuclear membrane (ONM). The ONM is linked to the rough endoplasmic reticulum (rER). The INM is connected to the ONM via the pore membrane which defines small annuli connecting the cytoplasm to the nucleoplasm. The pore membrane facilitates the movement of integral membrane proteins between the ONM and INM [1]. Embedded within

Sue Shackleton et al. (eds.), *The Nuclear Envelope: Methods and Protocols*, Methods in Molecular Biology, vol. 1411,
DOI 10.1007/978-1-4939-3530-7_27, © Springer Science+Business Media New York 2016

the pore membrane is a large complex protein structure, the nuclear pore complex (NPC). Attached to the NPC and lining the INM is the nuclear lamina. B-type lamins (which are present in most cells) form a two-dimensional network on the nucleoplasmic surface of the INM [2, 3], whereas A-type lamins, generally found in more differentiated cells [4], may assemble on this network with a less certain organization. Lamins and NPCs interact with SUN domain proteins, thought to exist primarily in the INM, which in turn bind KASH domain proteins thought to reside primarily in the ONM. This forms a nucleocytoplasmic link that then attaches to the cytoskeleton [5].

All these structures and connections occur at membrane surfaces; therefore if the surface of interest can be exposed, surface imaging methods, such as feSEM, can provide valuable information. Although cryo electron tomography [6] is providing unique insights into the structure of the NPC scaffold, the structure of peripheral components such as the basket, cytoplasmic filaments, and central channel are more elusive, presumably due to their disordered and variable nature. Such structures, on the other hand, can be routinely imaged directly and simply by feSEM [7]. Similarly, although some of the fine details of nuclear lamina structure and organization can be studied by electron tomography [8] and X-ray crystallography [9], the organization of filaments on the NE membrane can be best determined by feSEM. Importantly, once the preparation procedures have been worked out, it is generally relatively simple to prepare and image large numbers of samples, allowing more complex experiments to be performed.

1.1 Field Emission Scanning Electron Microscopy

In feSEM, electrons are accelerated and focussed into a fine beam, which is scanned across the surface of the sample. When these high-energy electrons hit the sample they either transfer energy to electrons in the sample, which can result in release of low energy electrons from the surface (secondary electrons) that can be detected, or they can be "reflected" without losing their energy (backscattered electrons) and detected by a different detector. Secondary electrons generally provide high-resolution information, whereas backscattered electrons can be used to determine the position of dense, high atomic number objects such as gold particles that can be attached to antibodies.

The primary electron beam scans the sample surface causing emission of negatively charged secondary electrons. The secondary electrons are accelerated to the detector by a high positive voltage. Here they strike a phosphorescent scintillator, generating photons, which are converted to an amplified electron signal within a photomultiplier. The output signal, which is proportional to the secondary electron emission from each part of the scanned sample, is used to build up an image on a screen. Resolution depends on how fine the beam can be focussed as well as the level of noise. Noise depends on many factors, but using the very bright electron source of a field emission gun is crucial for imaging fine details of the nuclear lamina and NPCs, where you need to distinguish structural components that are only a few nanometers, apart.

FeSEMs can be operated using a wide range of accelerating voltages, which determines the speed of the electrons in the primary beam. There are a number of factors that must be considered when deciding on the accelerating voltage and sometimes it is wise to try different voltages to see what works best for a particular sample. High accelerating voltages provide high instrument resolution. However, particularly with biological material, these high energy electrons can penetrate deep into the sample generating signal that appears to come from below the surface as well as noise. If the sample is electrically non-conductive, it can also lead to a build up of charge within the sample, creating artifacts and instability. This is particularly a problem with thicker samples. Such problems may be alleviated by coating the sample with a thin film of metal, but, in order to retain high resolution detail, it is necessary to keep such coats very thin (a few nanometers at most). Such thin coats are not very effective at preventing electron penetration. Metal coats also greatly enhance the generation of secondary electrons and so are routinely used for most samples. Traditional metal coats such as gold are no longer appropriate because they form aggregates on the sample which are not only discontinuous when thin, but also mask fine detail. The grain size from platinum is much smaller than gold, and platinum, or platinum alloys, can be useful, particularly for bulky samples that are likely to suffer from charge build up. Platinum coat thickness of about 2 nm may be used to see reasonably fine detail of the specimen, but it may be necessary to compromise on resolution by using 5–10 nm coats in order to prevent charge build up in bulky and/or non-conductive samples. Platinum will obscure finer details of the NPC, and in particular, in situ lamin filaments are difficult to distinguish. Chromium, and other refractory metals provide the finest continuous coats at 1–2 nm thickness [10] and it is essential to use refractory metals (chromium, tantalum, or tungsten) to be able to image details of the NPC or nuclear lamina.

Low primary electron beam accelerating voltages may, in some samples, alleviate the need for coating, but at the cost of lower instrument resolution. Buildup of charge in the specimen can, in certain samples, be more of a problem when using low voltages, presumably because electrons become trapped at the sample surface. For these reasons we routinely coat samples and start imaging using high voltages to maximize the resolution, then try low and intermediate voltages if problems with noise and charging are encountered.

The substrate on which samples are deposited is important and must be conductive to prevent charge accumulation, but it also must not generate significant secondary or backscatter electrons that will contribute to background noise. For the former reason glass coverslips are not suitable. Our preferred substrate is the silicon chip which is electrically conductive and on which cells can usually be grown in exactly the same way as on glass, but with the limitation that they are not transparent.

For immuno-gold labeling experiments it is usually necessary to use high or intermediate voltages because most backscatter detectors, which are used to detect the position of the gold particles, do not work at low voltages. Such detectors are designed to distinguish high energy backscattered electrons from low energy secondary electrons and therefore filter out electrons below certain energies. For this reason it is also essential that there is backscatter contrast between the immuno-gold particles and the surrounding material. The need for this contrast makes gold and platinum completely inappropriate metal coats for high-resolution immuno-gold SEM. Even refractory metals like tantalum have an atomic number too close to gold so that the two metals cannot be distinguished. Therefore gold particles are difficult, or impossible, to identify when samples are coated with platinum, gold, tantalum, or tungsten. Chromium thus is the best metal for coating when immuno-gold labeling is being undertaken. It is grainless, even when very thin (less than 1 nm), and most importantly it has a relatively low atomic number. Therefore, gold nanoparticles (5–15 nm diameter) can be unequivocally identified with a backscatter detector when coated with a thin film (1–2 nm) of chromium. Chromium nonetheless has a few drawbacks. Because of its low atomic number, the secondary electron signal is relatively low. Most significantly, chromium oxidizes readily. This means that the samples have to be imaged immediately after coating and may only be good for a few hours to a few days. Samples may be usable after a few weeks, but we have not found a convenient way to store them long term. Another consequence of the oxidation is that the vacuum conditions during sputter coating have to be very stringent. A vacuum of at least 5×10^{-7} mBar, preferably using some sort of cryo-pumping system, is required prior to introducing ultrapure argon gas for sputter coating.

1.2 Nuclear Envelopes for feSEM

We have imaged NEs from a wide variety of organisms and tissues. Amphibian, as well as fish and bird, oocytes and other cells with large nuclei are particularly suitable for studying the fine structural organization of the NPC and lamina [11–14]. Amphibian oocytes are large single cells containing a large nucleus (germinal vesicle) which are produced in the thousands. Because of its large size, the nucleus can be isolated by manual manipulation and the NE spread on an electron microscopy mount within a minute or two. There is also very little chromatin attached to the NE in amphibian oocytes, unlike most other cells. Although this makes this an unusual NE, it does mean the INM is uniquely accessible to surface imaging, providing particular insight into the organization of the nuclear lamina [2, 3]. The germ cell lamina is also unusual in consisting of only B-type lamins [4]. Although this gives clear access to the B-type lamina, thus improving its structural resolution, it tells us nothing about the organization of A-type lamins. We can however use this system to study A-type lamin organization by ectopic expression after microinjection of A-type lamin genes [3, 15]. Although this system can provide some information on the interaction of the NE with intra-nuclear structures [16], is unable to provide information on the interaction of the lamina and NPCs with chromatin because of its lack of attachment in these cells to the NE. To investigate lamin/NPC-chromatin interactions we generally have to go to somatic cells, using dry or wet fracturing techniques.

2 Materials

2.1 Obtaining Oocytes

1. Mature female *Xenopus laevis* (other amphibia, birds, and fish can be used).
2. 0.2 % MS222 (3-aminobenzoic acid ethyl ester methane sulfonate) in water containing 10 mM Hepes, check and adjust pH to about neutral (~7.5). MS222 without buffer is acidic and will cause the animal distress.
3. OR-2 Medium (*see* ref. 17): 82.5 mM NaCl, 2.5 mM KCl, 1 mM $MgCl_2$, 1 mM Na_2HPO_4, 5 mM Hepes, 3.8 mM NaOH, pH 7.2. Prepare from stocks: Solution A (825 mM NaCl, 25 mM KCl, 10 mM $MgCl_2$, 50 mM Hepes, 38 mM NaOH) and solution B (10 mM Na_2HPO_4). Add 1 volume solution A to 8 volumes distilled water and then add 1 volume solution B.

2.2 Isolating Nuclei and Spreading NEs

1. NE isolation buffer: 83 mM NaCl, 17 mM KCl, 10 mM Hepes–KOH pH 7.2.
2. Low Salt Buffer: 1 mM EDTA, 10 mM Hepes–KOH, pH 7.4.
3. Silicon chips/mounts or a silicon wafer cut by scoring with a diamond scribe and breaking into the appropriate size for mounting in the SEM.
4. Tweezers Dumoxel 5/45.
5. Tweezers Dumont GG.
6. Diamond scribing pen.
7. Glass petri dish.
8. Acetone.
9. Filter paper (Grade 1).
10. 35 mm petri dishes.

11. Stereomicroscope with oblique cool illumination and black stage.
12. Dissecting scissors.
13. Dissecting needles that have been sharpened with a wet stone or other sharpening device to a point of ~0.1 mm diameter, used to pierce a hole in the oocyte that is about the same size as the nuclear diameter.
14. Pasteur pipettes.
15. Bunsen burner.
16. Cocktail sticks.
17. Sticky tape.
18. NE fix: 2 % EM-grade glutaraldehyde, 0.2 % EM-grade tannic acid, 0.1 M Hepes pH 7.2.

2.3 Immuno-Gold Labeling NEs

1. PBS: 137 mM NaCl, 2.7 mM KCl, 10 mM Na_2PO_4, 2 mM KH_2PO_4 pH 7.4.
2. Immuno-Fix: 4 % EM-grade paraformaldehyde in PBS.
3. 100 mM glycine in PBS.
4. 1 % BSA or fish skin gelatine in PBS.
5. Filter paper (Grade 1).
6. 35 mm petri dishes.
7. Parafilm.
8. Primary antibody.
9. 5 nm or 10 nm colloidal gold conjugated to secondary antibody (*see* **Note 1**).

2.4 Processing for feSEM

1. 0.1 M sodium cacodylate pH 7.2
2. 0.1 % OsO_4 (EM grade) in 0.1 M sodium cacodylate pH 7.2.
3. 50, 70, 95, 100 % ethanol.
4. 100 % dried ethanol: pour 100 % ethanol into a 1 L dry bottle containing ~30 g of dried 3 Å molecular sieve, leave overnight to absorb water and to allow small particles of molecular sieve to settle, then store ethanol containing molecular sieve at room temperature.
5. Critical point dryer supplied with a high purity liquid CO_2 cylinder (with an internal dip tube) with less than 5 p.p.m. water, and with a gas flow meter attached to the gas outlet.
6. High-resolution coating unit, capable of pumping to a vacuum of at least 5×10^{-7} mBar, preferably using some sort of cryo-adsorption pump (e.g., a liquid nitrogen cooled plate). It should have a magnetron sputter head containing a chromium target. There should be a moveable shutter between the chromium

target and the specimen, a tilting turntable, and it should be supplied with 100 % pure research grade argon. An airlock for sample introduction is useful to prevent exposure of the chromium target to air, which causes oxidation of the target surface.

3 Methods

3.1 Obtaining Oocytes

1. Place 1 mature female *Xenopus laevis* in a small tank containing MS222.
2. After 20–30 min the animal should be fully anesthetized (*see* **Note 2**).
3. Place animal on its back on damp paper towels and carefully wipe abdomen with ethanol (*see* **Note 3**).
4. Using a sterile scalpel, make a 10–15 mm incision in the outer skin layer to one or other side of the midline (about 1 cm on an average sized adult frog), and then cut through the muscle with a slightly smaller incision.
5. The ovary should be visible just under the muscle and appears as a mass of oocytes tightly held together by connective tissue. Individual oocytes are ~1 mm in diameter and have sharply demarcated dark colored and light colored hemispheres. Using sterile tweezers, gently ease out enough ovary for the experiments and cut off with sterile scissors.
6. Place the piece of ovary in OR2 buffer and wash two to three times.
7. If regulations allow and if desired, it is possible, after partial ovarectomy, to suture the muscle layer then the skin separately, then return the animal to clean water for recovery, with several changes of water to wash away residual MS222. Alternatively the animal can be euthanized while still under anesthetic, in accordance with local regulations.

3.2 Isolating, Placing and Spreading the Nucleus on a Silicon Chip

3.2.1 Isolation of the Nucleus and Placing onto a Silicon Chip

1. Scratch numbers on silicon chips using the diamond pen.
2. Using Dumont GG tweezers remove silicon chips from plastic backing and place in glass petri dish containing acetone.
3. Pick up silicon chips with tweezers and wipe dry to clean thoroughly (*see* **Note 4**).
4. Place chips on filter paper until ready to use.
5. Next, it is necessary to prepare a Pasteur pipette with a reduced diameter as follows: hold a standard glass Pasteur pipette in the hottest part of a Bunsen burner flame (the region just above the blue cone at the bottom of the flame) and turn it slowly so that the heat is distributed evenly. Hold one end with one hand

and the other with the other hand. When you feel/see the pipette is getting soft and malleable, but before it melts, pull the two ends quickly and smoothly in opposite directions to about arms' length (1–1.5 m length). Next, after cooling, using a diamond pen, score all around the glass about 6–10 cm from the tapered part, so that it can be broken cleanly to give a bore diameter of a little less than 1 mm. You may have to practice a few times to get the ideal pipette (*see* **Note 5**).

6. Another important tool that can be easily constructed in the lab are the fine glass needles used to spread the isolated NEs. We make these from glass Pasteur pipettes in a similar way to the drawn-out pipettes: hold each end of one Pasteur pipette and heat the middle wide part in the hot region of a Bunsen flame, while turning, until it melts then pull each end to arms length with a fast jerking movement. This will give you a fine glass thread that is thicker at the ends and very thin in the middle. After cooling, break off ~5 cm pieces that are ~0.1 mm in diameter. This does not need to be accurately measured, but must be thin enough to carry out fine manipulations, and not so thin that it bends too easily under the surface tension of liquid. Then attach the 5 cm glass needle to a wooden cocktail stick with sticky tape, so that 2–3 cm protrudes (*see* **Note 6**).
7. Fill four small petri dishes with NE isolation buffer to a level just below the top (*see* **Note 7**).
8. Place two dishes on the bench and 2 on the microscope stage (*see* **Note 8**).
9. Place a chip in one of the dishes on the microscope stage and put to one side.
10. Cut off a piece of ovary 0.5–1.0 cm across (containing about 20–50 oocytes) and wash two times briefly in NE isolation buffer in the dishes on the bench.
11. Place in the dish on the microscope stage under the field of view.
12. Examine ovary piece at low zoom. It will appear like a bunch of grapes. The majority of oocytes will be mature stage VI oocytes of about 1.2 mm in diameter and have clear dark and light hemispheres. The oocytes, of all sizes, will be attached to each other by connective tissue.
13. You should use the largest mature stage VI oocytes (1–1.2 mm diameter).
14. Using fine angled tweezers (Dumoxel 5/45), grip a piece of connective tissue within the ovary close to the chosen oocyte to hold the oocyte in place while puncturing it. The nucleus is in the "animal pole" which is the dark pigmented hemisphere, so pick an oocyte where the dark half is facing up.

15. Pierce the dark hemisphere with a sharpened dissecting needle (*see* **Note 9**).
16. Ideally the nucleus should simply emerge from the hole. It can be recognized as a bluey flexible sphere that may appear knobbly at high magnification. Sometimes yellow/white yolk will come out first (*see* **Note 10**).
17. Fill the drawn out Pasteur pipette with NE isolation buffer, ensuring no air bubbles are present in the pipette.
18. Place the pipette under that surface of the liquid and expel a small amount of buffer.
19. Pick up the nucleus by sucking it into the pipette making sure it remains near the open end of the pipette (*see* **Notes 11–13**).
20. Raise the pipette out of the dish and move it aside, then move the other dish containing the chip into the field of view.
21. Touch the end of the pipette into the liquid in the dish over the chip, letting the nucleus fall out onto the chip. Do not expel it forcibly or you may lose sight of the nucleus.
22. If the nucleus comes into contact with a liquid/air interface it will lyse and be lost; therefore, it is important not to have any air bubbles in the pipette.
23. The nucleus should stick naturally to the chip. Adhesion can be checked by blowing buffer over the chip using the pipette. However, be sure not to expel an air bubble over it which will destroy it.

3.2.2 Spreading the Nuclear Envelope

1. To spread the NE, use a fine glass needle to pierce the nucleus near its contact point with the silicon chip, then carefully tear it open, releasing the nuclear contents and exposing the inner face of the NE.
2. Transfer the chip into a fresh petri dish containing NE isolation buffer.
3. Do not attempt to put more than one NE on each chip (*see* **Note 14**).
4. Specifically for imaging the nuclear lamina, chips should be transferred into Low Salt Buffer for 2 min (Fig. 1a) (*see* **Note 15**).
5. For imaging NPCs, the Low Salt Buffer step is not necessary and should be avoided if not required as prolonged incubation in Low Salt buffer may cause disruption to NPC components such as the basket.
6. Alternatively the visibility of the lamina can be enhanced by removing the membranes with a 5 min incubation in NE isolation buffer containing 0.5 % Triton X-100 (Fig. 1b).
7. For samples that will be immuno-gold labeled, transfer to Immuno-Fix for 10 min, then proceed to Subheading 3.3 (*see* **Note 16**).

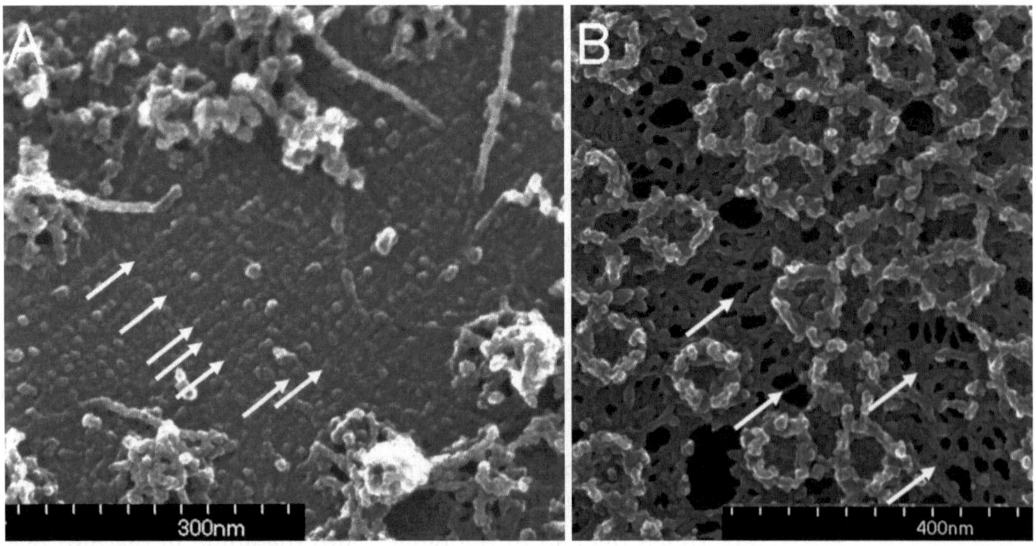

Fig. 1 FeSEM of isolated Xenopus oocyte NEs showing lamin LIII filaments on the surface of the inner nuclear membrane (**a**), and from the cytoplasmic face after Triton X100 extraction (**b**). Lamin filaments are indicated by the *white arrows*

8. For samples that will not be labeled, place in NE Fix for at least 10 min and proceed to Subheading 3.4 (*see* **Note 17**).

3.3 Immuno-Gold Labeling

1. First, construct a simple moist chamber by placing 2–3 sheets of filter paper in a 9 cm petri dish lid. Pour in distilled water and then discard to leave damp filter paper. Place a dry glass slide or some Parafilm to partially cover the damp filter paper. Use the bottom of the petri dish as a lid.
2. Place sample chip in 100 mM Glycine in PBS in a small petri dish for 5 min at room temperature.
3. Transfer to 1 % BSA, or 1 % fish skin gelatine in PBS for 20 min at room temperature.
4. Briefly place chip face up onto dry filter paper and move once to dry back of chip (*see* **Note 18**).
5. Place chip onto slide/Parafilm in moist chamber.
6. Pipette 10 μL primary antibody onto chip, making sure it remains as a drop on the chip and does not run over the sides, and incubate, with the moist chamber closed, for at least 60 min (*see* **Note 19**).
7. Wash 3× 5 min and once for 15 min in PBS by transferring chips between four dishes filled with PBS.
8. Briefly place chip on filter paper and move once to dry back of chip.
9. Place in moist chamber on dry slide/Parafilm.

10. Pipette on 10 μL secondary-gold antibody (*see* **Note 20**).
11. Wash 3× 5 min then 1× 15 min in PBS.
12. Place in NE fix for at least 10 min and proceed to Subheading 3.4.
13. Consideration should be given to controls. We always included a sample that is incubated with the secondary antibody without the primary antibody to ensure there is no nonspecific binding of the secondary with the sample. Controls for the primary antibody are more difficult, but only antibodies that have been well characterized by Western blot and/or immunofluorescence should be used. If possible, pre-incubation of the antibody with the antigen (i.e., the peptide used to raise the antibody) is a good negative control. Knowledge about where you expect to find the labeling may also be useful.

3.4 Processing for SEM

3.4.1 Osmium Tetroxide Staining and Dehydration

1. In the fume hood, fill two petri dishes with sodium cacodylate buffer and one with 0.1 % OsO_4 (SAFETY WARNING: OsO_4 is highly toxic and volatile and should be kept in the fume hood and disposed of appropriately in accordance with local regulations. Gloves must be worn).
2. Transfer chips into first dish for 1 min, then the second for 1 min, then into the OsO_4 for 10 min.
3. Set out six petri dishes. Fill 1 with water and remainder with 50, 70, 95, 100, 100 % ethanol respectively.
4. Transfer chips to each of these for 2 min each.

3.4.2 Critical Point Drying (CPD)

It is necessary to dry biological samples in a way that preserves structure. Air drying results in damage due to crushing by surface tension. Critical point drying avoids this. Samples are placed in ethanol in a high pressure chamber. Ethanol is exchanged for liquid CO_2 under pressure, then the chamber is warmed to 40 °C, where the CO_2 passes through the critical point and becomes a gas. The gas is then released leaving a dry sample.

1. Place CPD carrier in a glass petri dish and fill with ethanol that has been stored with molecular Sieve (to remove water).
2. Transfer silicon chips with samples to carrier, making sure to not let them air dry.
3. Fill CPD chamber with the molecular sieve dried 100 % ethanol.
4. Transfer carrier into CPD chamber.
5. Cool chamber to about 10 °C.
6. Open liquid CO_2 inlet valve and fill chamber with liquid CO_2 which should form a layer on top of the ethanol.
7. Open liquid outlet valve and watch level of the liquid, making sure it does not drop below the level of sample carrier.

8. Close outlet valve.
9. Open liquid CO_2 inlet valve again and fill chamber with liquid CO_2.
10. Repeat **steps 7–9** for six to ten times in order to replace all ethanol with CO_2.
11. Leave for 30–60 min (*see* **Note 21**).
12. Repeat **steps 7–9** for six to ten times.
13. Allow liquid to rise to just below the level of the sight glass and close all valves.
14. Stop cooling, start heating and close valves to CO_2 cylinder.
15. When temperature reaches 40 °C, open the gas outlet valve, but with the needle valve fully closed.
16. Slowly open the needle outlet valve to give a flow rate of about 150 L/h.
17. Maintain the gas flow by slowly opening the valve further as the pressure in the chamber drops.
18. When the pressure reaches zero open the lid and remove dry specimens.

3.4.3 Chromium Coating

For high-resolution feSEM it is necessary to coat samples with a thin film (1–2 nm) of chromium. We us the Cressington 328UHR chromium magnetron sputter coater, which has a second sputter head for platinum (or any other metal). Other coaters can be used, as long as they can pull a vacuum of at least 5×10^{-7} mBar and preferably have some sort of cryo-pumping system (such as a liquid nitrogen or helium cool plate). This is important to remove any trace of O_2 and water from the system.

1. Obtain a vacuum of at least 5×10^{-7} mBar.
2. Insert specimens.
3. Tilt stage to 45° and rotate.
4. Move shutter over the target.
5. Introduce argon to a pressure of about 8×10^{-3} mBar.
6. Using a current of 80 mA start sputtering onto the shutter for about 15 s.
7. Open shutter and deposit 1–2 nm of chromium, using the film thickness monitor to indicate thickness.
8. Close shutter and turn off current to sputter head.
9. Turn off argon and remove specimens (*see* **Note 22**).

3.5 Scanning Electron Microscopy

There are a number of suitable instruments available for imaging details of the NE and NPCs. Generally these should have a field emission electron source and an instrument resolution in the range of 0.5–1.5 nm. For immuno-gold labeling the instrument must also have a high resolution backscatter detector. We use the Hitachi S5200 with an "Autrata" yttrium aluminum garnet (YAG) backscatter detector and an "in-lens" stage which provides very low noise, stable images. The in-lens stage is limited for sample size to mounts that are 5 × 5 mm, but this is not a problem for small samples like NEs.

3.5.1 Imaging Structure

1. Insert sample into SEM.
2. For straight imaging of the structure of isolated NEs we use the maximum accelerating voltage (30 kV), maximum beam current (20 μA) and minimum probe diameter. This gives the maximum resolution and, for these samples which are thin, the best signal–noise ratio (*see* **Notes 23** and **24**).
3. For more bulky samples, such as isolated whole nuclei it may be useful to try lower accelerating voltages to avoid problems with charge accumulation and excessive noise.

3.5.2 Imaging Immuno-Gold Particles

1. For immuno-gold labeled samples we find, in our instrument, an intermediate accelerating voltage (10 kV) gives the best backscatter signal, possibly because if higher accelerating voltages are use, the electrons have too much energy and pass through the gold nanoparticle instead of being backscattered. However, the YAG detector is no longer efficient much below this voltage. For efficient detection of small gold particles (5–10 nm) it is sometimes necessary to compromise on resolution by using a large probe diameter. It is also essential for the SEM to be perfectly aligned and astigmatism corrected. Especially at lower magnifications (30–50 kX), the gold particles may only be visible using slow scan speeds.
2. For precise localization of gold particles within the structure, it is useful to superimpose the backscatter image on the secondary image. However, the only important signal in the backscatter image is that coming from the gold particles. The rest is noise or unwanted signal which will degrade the combined image. Image J (now known as Fiji, download at http://fiji.sc/Fiji) can be used to remove this unwanted information/noise and to superimpose the two images. The procedure is outlined in Fig. 2.

3.5.3 3D SEM

1. Most SEM specimen stages can be tilted to 35° or even up to 90°. Tilting the specimen gives a 3D perspective and can yield images where height information can be obtained, for instance the height of the NPC basket (Fig. 3a) or cytoplasmic filaments (Fig. 3b).
2. Real 3D images can also be obtained using the tilt function. Firstly, an image is taken of the area of interest, the stage is tilted to 6° (*see* **Note 25**) and a second image is taken of the

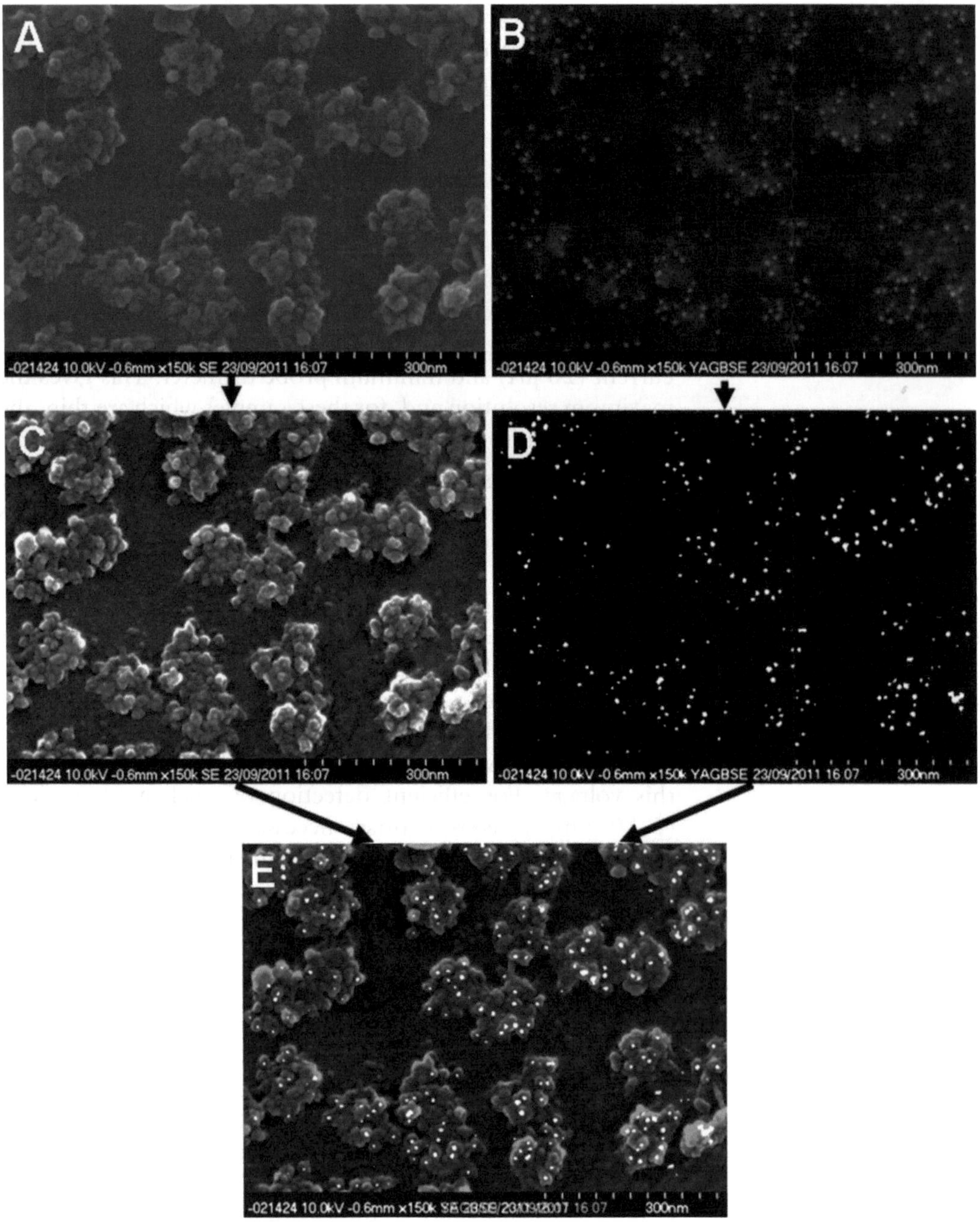

Fig. 2 Procedure for combining secondary and backscatter electron images to determine the location of immuno-gold particles on NE structures, using Image J/Fiji. (**a**) Raw secondary electron (SE) image of NE labeled with Mab414 (recognizes several nucleoporins). (**b**) Raw backscatter electron (BSE) image showing position of secondary 10 nm immunogold particles. (**c**) SE image with optimized contrast using contrast/brightness function in Image J. (**d**) BSE image filtered with median filter, contrast altered so that only gold particles are visible, then threshold function used to remove background. (**e**) SE and BSE images are combined using Overlay function in Image J as follows: (i) Open both images; (ii) Click on SE image; (iii) Click on Image tab, then Overlay, then Add Image; (iv) In the Add Image box, select the BSE image from the pull-down menu in "Image to add"; (v) Leave X location and Y location as zero; (vi) Enter Opacity at 50 % (other values may be preferable with different images); (vii) Click OK in Add Image box; (viii) Click Image, then Overlay, then Flatten to make overlay permanent; (ix) Save with a different name

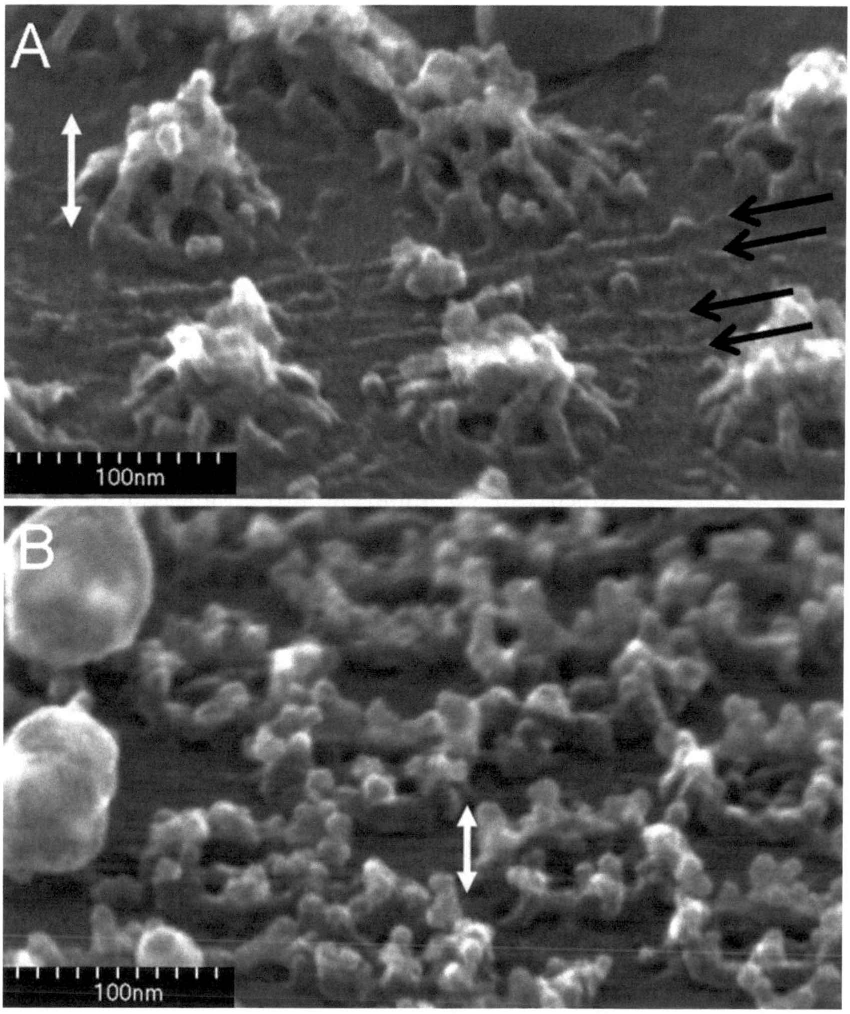

Fig. 3 Images of tilted specimens showing the nucleoplasmic (**a**) and cytoplasmic faces (**b**). *Black arrows* indicate the lamin filaments. *White arrows* indicate the height measurements that can be made

same area. One image is then colored red, the other green, and the two images are superimposed to form a red/green stereo pair, which can be viewed in 3D using red/green stereo glasses.

3. The above procedure requires a perfectly eucentric goniometer. In practice, SEMs stages are not perfect, and, particularly at the very high magnifications used to image NPCs and the lamina, there is always movement of the area of interest during tilting. This has two consequences: (1) you may lose the area of interest and not be able to obtain the second image; (2) the height of the area of interest will change, which changes the magnification when the second image is refocused. To avoid losing the area of interest, after obtaining the first image (at say, 100 kX magnification), reduce the magnification to maybe

20 kX and obtain a second image, then maybe 5 kX for a third image. You should then watch which direction the specimen moves at low magnification as the stage is tilted. Then, **without refocusing**, use the low magnification images as a guide to sequentially relocate the area of interest at progressively higher magnifications. It is important not to refocus the image with the focus controls, as this will change the magnification. Instead the image should be focussed with the stage height control, so that the image is brought into focus, bringing it back to the same height it was before tilting.

4. There are a number of ways to create a red/green anaglyph. There are commercial programs and it can be done relatively easily using Adobe Photoshop. A method is presented here using Image J. It is useful to install a plugin called Align_4, which can be found at: http://imagejdocu.tudor.lu/doku.php?id=plugin:aligning:align_4:start#align_4.
5. Open both images in Image J/Fiji.
6. Click on the "More Tools" menu (double arrow at end of menu bar).
7. Select Lookup Tables.
8. Select the left image and click on the Primary Colors Tool (red/green/blue horizontal lines) until the image is colored red.
9. Select the right image and repeat until the image is green.
10. Convert both images to RGB images (click Image→Type→RGB color).
11. Click Plugins, then Align 4. In the new box, Top Left should have one image and Top Right should have the other. The other two pull-down menus should read *None*.
12. Click OK.
13. You will see the red and green images side by side.
14. Select Image 1, click on Transp., and increase the shift value (*see* **Note 26**).
15. Click Right, Left, Up, and/or Down to overlay the images as precisely as possible (*see* **Note 27**).
16. The image should then be cropped to eliminate the black area on the left: select the Rectangular Selection Tool and select the area containing the red/green image. Click Image, then Crop.
17. Optimize the contrast/brightness (Image→Adjust→Brightness/Contrast).
18. View with red/green stereo glasses.

4 Notes

1. Generally we prefer 10 nm particles because 5 nm particles, though detectable, may be difficult to distinguish from other material, whereas 10 nm particles are usually easy to identify in the backscatter image.
2. Check by pinching with tweezers and observe any reflex. If a reflex is observed, the animal could feel pain and should be anesthetized further.
3. To remove slime and other potential contaminants and to ensure the incision area is sterile if animals are to be maintained for recovery.
4. Check they are shiny and clean with no hint of finger grease or any other dirt or smears—it is best to check this under the stereo microscope—if it is not clean nuclei will not adhere.
5. If the bore is too large it will be difficult to obtain clean nuclei, if it is too small nuclei may burst.
6. Safety Note: for both **steps 12** and **13**, although the drawn out glass cools quickly, the thicker tapered glass that was in the flame takes some time to cool down and is a burn hazard. Do not touch this part for several minutes or wear thermal protective gloves.
7. NE isolation buffer is used during nuclear isolation and spreading of NEs, giving clean NEs with well preserved NPCs and lamina. However other structures such as ribosomes and actin filaments are not well preserved and other buffers may be preferable if these are being visualized.
8. The microscope must have cool, low angle, bright illumination and a black background, in order to be able to see the nuclei.
9. The resulting hole needs to be the right size. This is determined by how sharp the needle is and how far into the oocyte it is pushed. The hole should be about the size of, or a little smaller than the diameter of the nucleus. If it is too large the nucleus may come out highly contaminated with yolk and may be difficult to see. If it is too small the nucleus will burst.
10. If the nucleus does not emerge, you can press very gently on the oocyte with the side of the needle. However this often results in bursting, in which case you may see a small puff of clear liquid coming out of the hole.
11. If there is a lot of yolk attached to the nucleus it can be cleaned off by expelling and picking up the nucleus a few times with the drawn-out pipette.
12. Ideally the nucleus should be completely free of yolk, otherwise it may not stick to the silicon chip.

13. Particularly clean nuclei, will stick readily to most surfaces including the dishes and pipette so it is necessary to not let them settle for any length of time.
14. Chips quickly become coated in proteins from the oocytes/ spread nuclei preventing further nuclei from sticking.
15. The timing here is critical. The purpose of this step is to extract material that seems to lie in between individual lamin filaments and enable resolution of lamin filaments. Prolonged incubation in this buffer results in damage to structures such as the NPC basket.
16. It is usually best not to leave the sample in this fix for extended periods (i.e., overnight) as it may affect antigenicity. Although it is best to proceed straight to Subheading 3.3, samples may be transferred to PBS overnight at 4 °C before labeling the next day.
17. Samples can be left for several hours or overnight in this fix.
18. It is important that the face with the NE attached is never allowed to air-dry.
19. The optimal concentration of primary antibody and incubation time must be determined empirically, but as a guide we usually use concentrations that are 5–10× more concentrated than you would use for immunofluorescence.
20. Secondary antibody concentration should also be determined empirically, but we generally use a 1:50 dilution as a starting point.
21. To ensure all ethanol has diffused out of sample.
22. Samples should only be coated immediately prior to imaging (because the chromium coat deteriorates over time due to oxidation) and should be stored in a vacuum to keep them dry (before and after coating).
23. Some, particularly analytical, instruments may operate at higher maximum beam currents and it may be worth experimenting with current.
24. Instruments with beam decelerators may also allow low electron landing energies to be used with similar/acceptable resolution.
25. A tilt of 6° is used because human eyes are ~6° apart, so this gives the most "real" 3D view. 3D perspective, however, can be increased by using higher tilt values.
26. This determines how much the image is shifted and should be large to start with to get the images roughly aligned, then decreased for fine adjustments.
27. It is impossible to align the images completely precisely because the tilted image will be "compressed" relative to the flat image. We usually align precisely along the midline, or with respect the part of the image that is of most interest.

Acknowledgements

Thanks to Christine Richardson, Helen Grindley, Pam Ritchie, and Sandra Rutherford for technical assistance, Terry Allen, Herbert Macgregor, Reimer Stick, and the late Rob Apkarian for help in developing these methods, and the Biotechnology and Biological Sciences Research Council, UK, grant number BB/G011818/1 for funding.

References

1. Zuleger N, Kerr AR, Schirmer EC (2012) Many mechanisms, one entrance: membrane protein translocation into the nucleus. Cell Mol Life Sci 69:2205–2216
2. Aebi U, Cohn J, Buhle L, Gerace L (1986) The nuclear lamina is a meshwork of intermediate-type filaments. Nature 323: 560–564
3. Goldberg MW, Huttenlauch I, Hutchison CJ, Stick R (2008) Filaments made from A- and B-type lamins differ in structure and organization. J Cell Sci 121:215–225
4. Peter A, Stick R (2015) Evolutionary aspects in intermediate filament proteins. Curr Opin Cell Biol 32:48–55
5. Cartwright S, Karakesisoglou I (2014) Nesprins in health and disease. Semin Cell Dev Biol 29:169–179
6. Bui KH, von Appen A, DiGuilio AL et al (2013) Integrated structural analysis of the human nuclear pore complex scaffold. Cell 155:1233–1243
7. Goldberg MW, Allen TD (1996) The nuclear pore complex and lamina: three-dimensional structures and interactions determined by field emission in-lens scanning electron microscopy. J Mol Biol 257:848–865
8. Ben-Harush K, Wiesel N, Frenkiel-Krispin D et al (2009) The supramolecular organization of the C. elegans nuclear lamin filament. J Mol Biol 386:1392–1402
9. Strelkov SV, Schumacher J, Burkhard P et al (2004) Crystal structure of the human lamin A coil 2B dimer: implications for the head-to-tail association of nuclear lamins. J Mol Biol 343:1067–1080
10. Allen TD, Rutherford SA, Bennion GR et al (1998) Three-dimensional surface structure analysis of the nucleus. Methods Cell Biol 53:125–138
11. Goldberg MW, Allen TD (1992) High resolution scanning electron microscopy of the nuclear envelope: demonstration of a new, regular, fibrous lattice attached to the baskets of the nucleoplasmic face of the nuclear pores. J Cell Biol 119:1429–1440
12. Kiseleva E, Goldberg MW, Allen TD, Akey CW (1998) Active nuclear pore complexes in Chironomus. Visualisation of transporter configuration related to mRNP export. J Cell Sci 111:223–236
13. Goldberg MW, Solovei I, Allen TD (1997) Nuclear pore complex structure in birds. J Struct Biol 119:284–294
14. Kiseleva E, Rutherford SA, Cotter LM, Allen TD, Goldberg MW (2001) Steps of nuclear pore complexes disassembly and reassembly during mitosis in early *drosophila* embryos. J Cell Sci 114:3607–3618
15. Stick R, Goldberg MW (2010) Oocytes as an experimental system to analyze the ultrastructure of endogenous and ectopically expressed nuclear envelope components by field-emission scanning electron microscopy. Methods 51:170–176
16. Kiseleva E, Drummond SP, Goldberg MW et al (2004) Actin- and protein 4.1 containing filaments link nuclear pore complexes to subnuclear organelles in Xenopus oocyte nuclei. J Cell Sci 117:2481–2490
17. Wallace RA, Jared DW, Dumont JN, Sega MW (1973) Protein incorporation by isolated amphibian oocytes. J Exp Zool 184:321–333

Chapter 28

An In Vitro Assay to Study Targeting of Membrane Proteins to the Inner Nuclear Membrane

Rosemarie Ungricht, Sumit Pawar, and Ulrike Kutay

Abstract

Newly synthesized membrane proteins are inserted into the endoplasmic reticulum (ER) from where they are constantly sorted to various cellular compartments. To analyze and visualize sorting of membrane proteins to the inner nuclear membrane (INM), we developed a trap-release system that uncouples membrane integration into the ER from transport. This assay allows the simultaneous release of a large pool of an INM-destined membrane protein from the ER and microscopy-based monitoring of targeting to the INM. The use of semi-permeabilized HeLa cells further enables the identification and characterization of essential requirements of the targeting process. This protocol provides a detailed description of reporter construction, in vitro reconstitution, and visualization of trafficking.

Key words Nuclear envelope, Inner nuclear membrane, In vitro reconstitution, Microscopy, Semi-permeabilized cell system, Membrane protein

1 Introduction

In eukaryotic cells, the nuclear envelope (NE) generates the physical barrier between the nuclear interior and the cytoplasm and plays important roles in nuclear organization and regulation of gene expression [1, 2]. The NE is composed of an inner and an outer nuclear membrane (INM and ONM) that are connected at sites where nuclear pore complexes (NPC) are situated [2]. Although the NE is continuous with the endoplasmic reticulum (ER), it is enriched in a characteristic set of membrane proteins [3]. These proteins play essential roles in maintaining various nuclear functions via associations with lamins and chromatin.

Newly synthesized membrane proteins are constantly sorted from the ER to the INM against a concentration gradient. The importance of correct sorting is underscored by the mislocalization of INM proteins in a wide spectrum of human diseases [4, 5]. Understanding how these membrane proteins are targeted to the

Sue Shackleton et al. (eds.), *The Nuclear Envelope: Methods and Protocols*, Methods in Molecular Biology, vol. 1411, DOI 10.1007/978-1-4939-3530-7_28, © Springer Science+Business Media New York 2016

INM and how their distribution is regulated may help to elucidate the mechanism of such disorders.

Translocation of proteins to the INM has been investigated by a variety of experimental approaches in recent years. Most early studies were based on steady-state enrichment of INM proteins, and did not resolve the dynamics of the transport process [6–8]. Mobility of various mammalian INM proteins has also been extensively analyzed by FRAP and photoactivation experiments [9–13]. Relocalization dynamics from the ER to the INM was for the first time visualized by exploiting the FKBP-FRB system and following the rapamycin-induced nuclear retention of a semi-artificial reporter [14]. These in vivo studies provided valuable insights into the processes of INM targeting and NE protein dynamics. In vitro reconstitution based assays, on the other hand, serve as a powerful tool to investigate the mechanism of complex cellular processes in a systematic way [15–17].

To gain mechanistic insights into targeting of INM proteins in human cells, we recently established an in vitro assay that allows us to separate the synthesis of integral membrane proteins from their translocation to the INM [18]. Therewith, our assay allows measuring transport kinetics of human INM proteins from the ER to the INM. Moreover, it enables to interfere with essential cellular processes using biochemical alterations and thereby to determine the molecular requirements of INM protein targeting. Previous studies had demonstrated that the size of the nucleoplasmic domains of INM proteins is restricted to <60 kDa [13, 14, 19–21]. We took advantage of this size limitation for NPC translocation to trap INM proteins in the ER after synthesis. We fused a 50 kDa extension (two mRFPs) to the extralumenal domain of human INM proteins, linked by a peptide containing a TEV protease cleavage site, to enable release from the ER trap by TEV digestion. Translocation of the reporter proteins from the ER to the INM could be monitored over time using a microscopy-based readout (Fig. 1).

In this chapter, we give experimental details of the in vitro INM targeting assay. We start off with a description of how plasmids encoding for INM protein reporters are generated (Subheading 3.1). Then, we explain how these reporters are stably integrated into the HeLa cell genome for tetracycline-inducible expression (Subheading 3.2). Further, we provide a protocol for the production of transport-competent HeLa cell extract (Subheading 3.3) and for recombinant NusA tagged TEV protease (Subheading 3.4) based on a published procedure [20]. Finally, we describe semi-permeabilization of HeLa cells and the reconstitution of INM trafficking including microscopic analysis (Subheading 3.5).

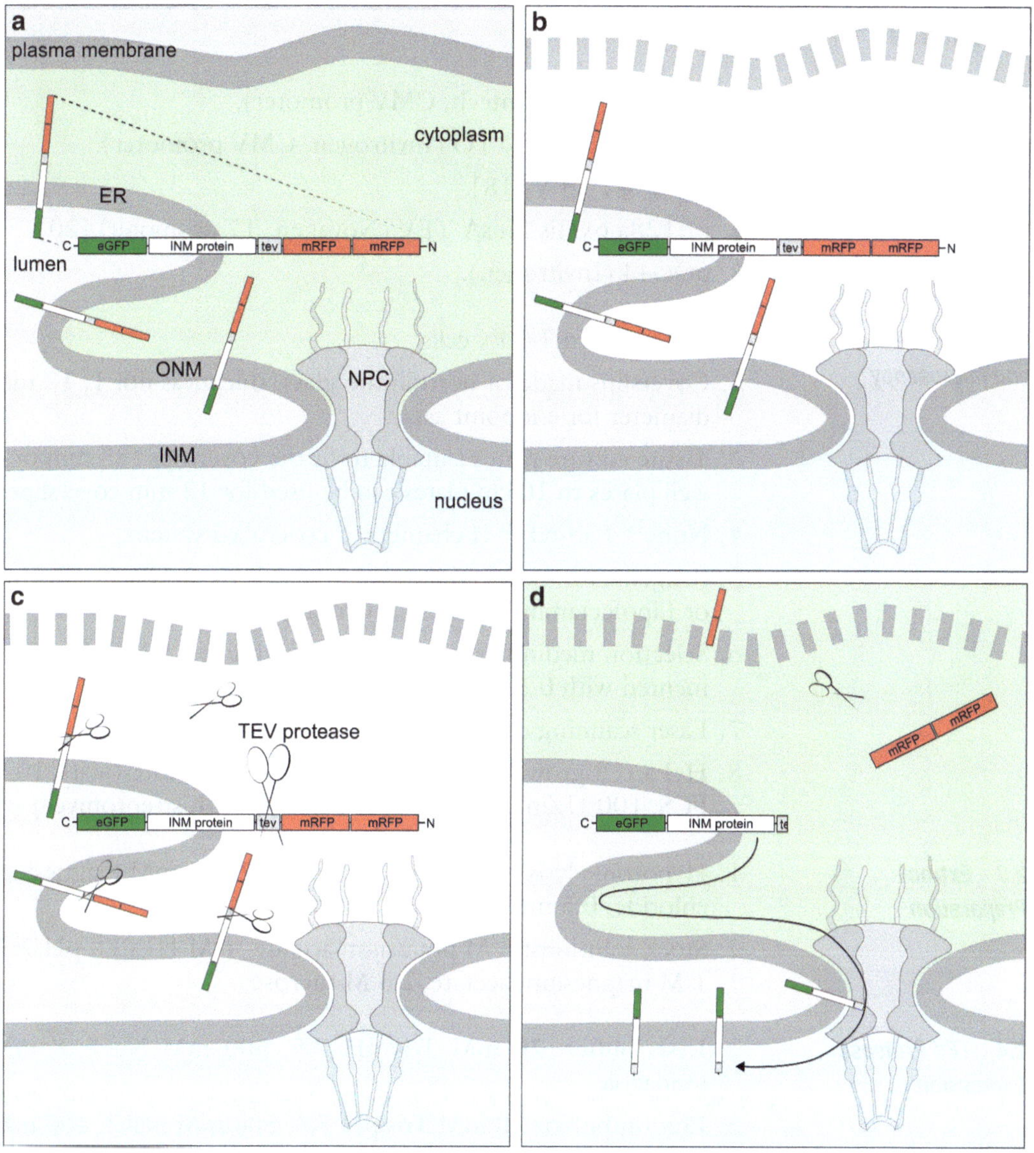

Fig. 1 Scheme illustrating the INM targeting assay. (**a**) Expression of INM protein reporters is induced by tetracycline addition in stable "tet-on" HeLa cell lines or by transient transfection of reporter constructs. Due to the extralumenal size extension (2× mRFP), the reporters are trapped in the ER after synthesis. (**b**) Cells are semi-permeabilized with a low concentration of the non-ionic detergent digitonin. Soluble cytoplasmic proteins and small components are washed out. (**c**) INM protein reporters become competent for NPC passage and thus INM targeting after cleavage by NusA-tagged TEV protease. (**d**) Semi-permeabilized cells are supplemented with HeLa lysate and an energy regenerating system. Under these conditions, reporter proteins target efficiently to the INM

2 Materials

2.1 Plasmids

1. pEGFP-N3 (Clontech, CMV promoter).
2. pcDNA™5/FRT/TO (Invitrogen, CMV promoter).
3. pCFP-C1-TEV [18].
4. pET28a 6xHis-NusA-TEV (Novagen, T7 promoter) [20].
5. pOG44 (Invitrogen).

2.2 Cell Culture and Microscopy

1. HeLa Flp-In T-REx cells.
2. Coverslips made of borosilicate glass, thickness no. 1, 12 mm diameter for endpoint assay.
3. Tissue culture plates suitable to fit the coverslips (35 mm or 6 well plates to 10 cm plates can be used for 12 mm coverslips).
4. Nunc™ Lab-tek™ II chambered coverglass system.
5. Reagents for transfection: e.g., we use X-tremeGENE (Roche) or Lipofectamine™ 2000 (Invitrogen).
6. Selection medium for generation of cell lines: DMEM supplemented with 0.3 mg/mL hygromycin.
7. Laser scanning confocal microscope.
8. HeLa cell growth medium: DMEM supplemented with 10 % FCS, 100 U/mL penicillin, and 100 μg/mL streptomycin.

2.3 Extract Preparation

1. Hypotonic lysis buffer: 5 mM Tris pH 7.5, 1 mM magnesium chloride, 0.5 mM EGTA, 0.5 mM EDTA.
2. Stock solutions: 5 M potassium acetate, 1 M HEPES pH 7.5, 1 M magnesium acetate, 2.5 M sucrose.

2.4 TEV Protease Expression

1. Lysis buffer: 20 mM Tris pH 7.5, 500 mM NaCl, 8 mM imidazole.
2. Elution buffer: 20 mM Tris pH 7.5, 500 mM NaCl, 400 mM imidazole.
3. Dialysis buffer: 20 mM HEPES pH 7.5, 50 mM potassium acetate, 2.5 mM magnesium acetate.
4. Amido Black staining solution: 0.1 % (w/v) napthol blue black, 2 % acetic acid in ddH_2O.
5. Lysogeny broth medium and selection antibiotics.
6. β-mercoptoethanol.

2.5 In Vitro INM Targeting Assay

1. Permeabilization buffer (PB): 20 mM HEPES, pH 7.5, 110 mM potassium acetate, 5 mM magnesium acetate, 0.5 mM EGTA, and 250 mM sucrose in ddH_2O. Prepare fresh PB for each experiment.

2. Digitonin is dissolved in warm ddH_2O to a 5 % stock solution. Store small aliquots (up to 50 μL) at −20 °C. Repeated freeze-thawing can lead to insolubility of digitonin and should therefore be avoided (*see* **Note 1**).
3. HeLa cell lysate (see 2.3).
4. Energy mix: An energy regenerating system is prepared as 20× solution. Mix 200 mM creatine phosphate, 10 mM GTP, 10 mM ATP, 1 mg/mL creatine kinase and 250 mM sucrose in 50 mM HEPES pH 7.5. Snap freeze small aliquots and store at −80 °C (*see* **Note 2**).
5. Fixative: 4 % paraformaldehyde in PBS. Adjust the pH to 7.5.
6. DAPI or Hoechst are dissolved at a concentration of 1 μg/mL in PBS.
7. Apyrase (NEB): 50 milliunits per reaction in PB.

3 Methods

3.1 Construction of INM Protein Reporters

In order to uncouple synthesis from transport of an INM protein of interest, a cleavable size-tag composed of two mRFP moieties is fused to the extralumenal domain of the INM-destined membrane protein. The two RFP moieties add a 50 kDa extension that leads to trapping of newly synthesized INM proteins in the ER, due to size constraints for NPC translocation. It is important to consider that ER trapping requires the size-tag to be fused extralumenally, which depending on the orientation in the membrane may be N- or C-terminally of your protein of interest. Trapping was efficient for all seven tested INM proteins (Fig. 2) (*see* **Note 3**).

For visualization of the targeting process by microscopy, we fused a lumenal eGFP to INM protein reporters. This is, however, only feasible for membrane proteins with lumenal ends. In principle, extralumenal tagging is also possible and may allow the reporter to reach an authentic steady-state accumulation level at the INM after release if the GFP does not interfere with nuclear interactions. Note, however, that an extralumenal tag will slow down translocation through NPCs [18]. INM protein reporters with varying topologies are depicted in Fig. 2a.

The following protocol describes how a type II transmembrane protein can be inserted into INM targeting reporter plasmids (Fig. 2b).

1. Reporters are assembled in a pEGFP-N3 vector backbone containing a CMV promoter. ORFs encoding for INM proteins of interest can be inserted between the *BglII* and *SalI* restriction sites (Fig. 2b).
2. Fusion proteins start with two monomeric RFPs separated by a linker (SRAQAS), followed by a peptide containing a TEV cleavage site (SNSPTTENLYFQ'GRS), the INM protein of

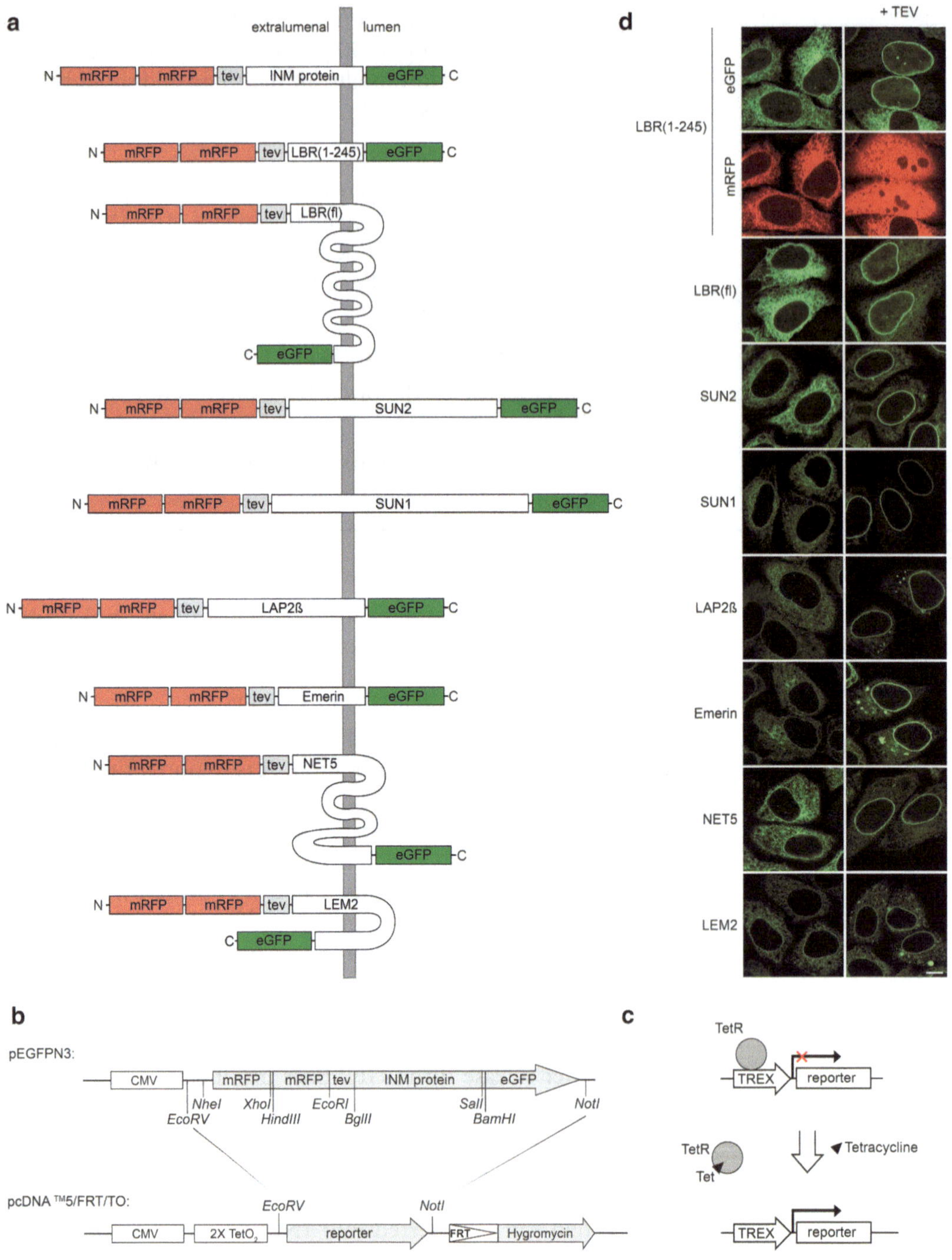

Fig. 2 (**a**) Schematic representation of membrane topology of INM protein reporters. (**b**) Example INM protein reporter ORF constructed in pEGFP-N3 and sub-cloned into pcDNA5/FRT/TO. (**c**) Principle of tetracycline-inducible expression in HeLa(FRT) T-REx cell lines. (**d**) Release of INM protein reporters from the ER to the NE upon TEV cleavage in vivo. HeLa cells expressing INM protein reporters (listed directly adjacent in **a**) were transiently transfected with CFP-TEV. After 24 h, localization was analyzed by confocal microscopy recording eGFP and mRFP fluorescence, here shown for 2xRFPtLBR(1–245)-GFP. For the other INM proteins, only eGFP fluorescence is shown

choice, and a C-terminal eGFP. After TEV cleavage, the amino acids GRS form the new start of the fusion protein followed by the natural N-terminus of the INM protein.

After insertion of an INM protein coding sequence, test whether the reporter is trapped in the ER and re-localizes to the INM after TEV cleavage.

3. Transiently transfect INM protein reporter plasmids into HeLa cells. Transfect 1 μg of plasmid DNA using a transfection reagent of your choice according to the manufacturer's instructions. After 24 h, fix cells with paraformaldehyde and analyze reporter localization by fluorescence microscopy detecting intrinsic fluorescence of eGFP and mRFP (*see* Fig. 2b) (*see* **Note 4**).
4. Analyze reporter localization after TEV protease cleavage in vivo by co-transfection of the reporter with a plasmid encoding TEV protease (pCFPC1-TEV) (Fig. 2c) (*see* **Note 5**).

3.2 Stable Tetracycline-Inducible HeLa Cell Lines

Analysis of INM targeting works best using stable cell lines that homogeneously express reporter proteins at defined levels. To control the expression level and to limit the duration of expression, we recommend the use of inducible expression, e.g., exploiting tetracycline-inducible expression in the Flp-In™ T-REx™ system (Fig. 2d). Avoid strong overexpression to guarantee excess of retention sites relative to the INM protein reporter. Limit expression time to avoid "post-mitotic" INM accumulation of the reporters during NE reformation in late anaphase or telophase (*see* **Note 4**).

The following example describes the generation of a tetracycline-inducible HeLa cell line stably expressing an INM protein reporter, integrated into a Flp Recombination Target (FRT) site by recombination (using the Flp-In™ T-REx™ system).

1. Insert the coding sequence for the INM targeting reporter via *EcoRV/NotI* into the pcDNA™5/FRT/TO vector.
2. Grow a host cell line containing an FRT site and a tetracycline repressor in HeLa cell growth medium in 2 wells of a 6-well plate to a confluency of ~60 % in a humidified incubator at 37 °C and 5 % CO_2.
3. Transiently co-transfect cells with 0.1 μg of the pcDNA™5/FRT/TO plasmid encoding the INM protein reporter and 0.9 μg of the Flp recombinase expression plasmid pOG44 per well.
4. 24 h post-transfection, transfer cells from each well into one separate 10 cm plate.
5. On the subsequent day, start antibiotic selection of positive clones by replacing the medium with DMEM containing 0.3 mg/mL hygromycin.

6. Replace the selection medium every 3–4 days until colonies are visible (takes ~2 weeks).
7. To pick single colonies, wash with PBS, add 5 mL 0.5 mM EDTA in PBS and incubate for 5 min in a humidified incubator at 37 °C and 5 % CO_2. Then, carefully pick single colonies with a 200 μL pipet under the tissue culture hood. Transfer the cells into single wells of a 24-well plate containing selection medium and resuspend gently.
8. Let cells grow until they reach confluency and then expand them into a 6-well plate. Use an aliquot of cells for the characterization of clones, by inducing expression of reporters with a standard tetracycline concentration of 200 ng/mL for 24 h. Analyze localization of the reporter by fluorescence microcopy (*see* **Note 6**).
9. Expand homogenous clones to a 10 cm plate.
10. Store selected cell lines by cryopreservation.
11. Titrate the tetracycline concentration for optimal expression of the reporter in the selected clonal cell line. For our reporter proteins, the optimal concentration ranged from 200 ng/mL to 2 ng/mL for 16 h of expression (*see* **Note** 7).

3.3 Preparation of Cytosolic Extract

Extracts were prepared using commercially available HeLa cell pellets (5.0×10^9 cells). Preparation of HeLa cell extract should be performed at 4 °C to avoid denaturation of proteins or loss of protein activity.

1. Resuspend the cell pellet in 20 mL hypotonic lysis buffer and allow cells to swell for 20 min on ice.
2. Lyse the cells by shear forces in a Dounce homogenizer with 10 strokes, followed by centrifugation at $4500 \times g$ for 25 min at 4 °C (*see* **Note 8**).
3. Keep the supernatant. Re-extract the pellet with 20 mL hypotonic lysis buffer and again allow cells to swell for 20 min. Add 220 mM potassium acetate directly to the Dounce homogenizer followed by 10 strokes. Clear the homogenate by centrifugation at $4500 \times g$ for 25 min at 4 °C.
4. Combine supernatants from the two extractions and clear by ultracentrifugation at $220{,}000 \times g$ (55,000 rpm in a Ti70 rotor) for 90 min at 4 °C.
5. Supplement the cleared lysate with stock solutions (Subheading 2.3) to a final concentration of 20 mM HEPES pH 7.5, 4 mM MgAc2 and 250 mM sucrose, aliquot and snap freeze in liquid nitrogen. Store at −80 °C. Avoid repeated freeze–thaw cycles (*see* **Note 2**).

3.4 TEV Protease Expression and Purification

A plasmid encoding His_6-tagged TEV protease fused to NusA was kindly provided by W. Antonin (FML, Tübingen, Germany). The protease was essentially purified as described [20].

1. Transform plasmid into chemically competent BL21 *E. coli* cells and inoculate in 150 mL lysogeny broth medium supplemented with 25 μg/mL kanamycin (LB + kan). Incubate on a rotary shaker at 37 °C overnight.
2. Dilute the overnight culture 1/100 in 500 mL LB + kan. Grow cells at 37 °C for around 5 h on a rotary shaker.
3. Add 500 mL LB + kan and induce expression of TEV protease with 0.5 mM IPTG, change incubator temperature to 25 °C and incubate overnight.
4. Harvest cells by centrifugation at 5900 × *g* for 15 min at 4 °C.
5. Discard the supernatant and resuspend the pellet in 35 mL lysis buffer.
6. Sonicate the lysate, placed on ice, for 6 min at 50 % power, followed by 30 s at 100 % power using a screw tip (guidelines based on using a 13 mm diameter tip, Branson Sonifier 250).
7. Clear the lysate by centrifugation at 11,000 × *g* (12,000 rpm in an SS-34 rotor) for 30 min at 4 °C.
8. Take the supernatant and add 2 mM β-mercoptoethanol.
9. Purify His_6-tagged NusA-TEV via Ni-NTA agarose using a standard protocol.
10. Elute TEV protease in 1 mL fractions with elution buffer and check for protein content by staining a 1 μL drop on nitrocellulose membrane with Amido Black.
11. Pool peak fractions and proceed to dialysis or add 250 mM sucrose and snap freeze in liquid nitrogen.
12. Prepare a dialysis tube (cutoff 15 kDa) by boiling it once in 10 mM sodium bicarbonate and twice in 10 mM EDTA. Rinse the tube in dialysis buffer.
13. Introduce protein solution into the dialysis tube and dialyze overnight in 1 L at 4 °C. Exchange dialysis buffer and dialyze for another 4 h.
14. Add 250 mM sucrose, aliquot and snap-freeze in liquid nitrogen (*see* **Note 2**).
15. Test the activity of the TEV protease: Dilute the protease in PB in a concentration series ranging from 15 to 400 μg/mL and test the activity on semi-permeabilized reporter cells (*see* **below**) for 10 min at room temperature. Choose a dilution for the subsequent experiments in which TEV cleavage is > 80 % efficient within 10 min, as judged by loss of the RFP signal.

3.5 The In Vitro INM Targeting Assay

Semi-permeabilization is performed essentially as described for in vitro transport reactions for soluble proteins [17, 22]. To prepare semi-permeabilized cells, the plasma membrane is selectively permeabilized by treatment with a low concentration of the non-ionic detergent digitonin. Under these conditions, the nuclear membranes and ER remain intact. In the subsequent washing steps, soluble cytoplasmic components are removed from cells and can be replaced by a cytosolic extract. Using semi-permeabilized cells for in vitro reconstitution of INM targeting permits biochemical manipulations that allow identifying requirements of the sorting process. These alterations range from inhibition of nucleocytoplasmic transport, modification of cytoplasmic extract or NTP levels to competition experiments (*see* Subheading 3.5.4).

3.5.1 Semi-permeabilization

Semi-permeabilization is performed on ice using cold buffers (*see* **Note 9**).

1. HeLa cells are grown on 12 mm glass coverslips or in 8-well Lab-tek™ chambers to a confluency of ~ 70 %. Cells grown on glass coverslips can be used for endpoint assays, which have the advantage that multiple conditions can be compared at one or several time-points. For kinetic measurements, we recommend live-cell imaging on cells semi-permeabilized in Lab-tek™ chambers (*see* **Note 10**).
2. Transfer 10 cm plates containing either coverslips or Lab-tek™ chambers onto a metal block, which is placed on ice.
3. Wash once with ice-cold PBS.
4. Semi-permeabilize the cells with permeabilization buffer (PB) containing 0.001 % digitonin for 10 min at 4 °C (*see* **Notes 1 and 11**).
5. Wash successively in three washing steps in PB for 2, 5, and 10 min (*see* **Note 12**).
6. Proceed with "Endpoint Assay" protocol for fixed cell analysis (Subheading 3.5.2) or "Kinetic Analysis of INM Targeting" (Subheading 3.5.3).

3.5.2 Endpoint Assay

1. Fix a first "control" coverslip for 10 min in 4 % paraformaldehyde as a reference for efficiency of subsequent TEV cleavage (perm, Fig. 3a). All further fixation steps are done alike.
2. To limit the volume used for INM targeting reactions, transfer the coverslips to the separating "walls" of a 24-well plate. Induce cleavage of the reporter by addition of 30 μL from a 75 μg/mL stock of NusA-TEV in PB. Incubate at 25 °C for 10 min (*see* **Note 13**).
3. Meanwhile prepare 30 μL of import mixture per coverslip containing the transport competent HeLa lysate (cytosolic extract) and a 1× energy regenerating system (*see* **Note 14**).

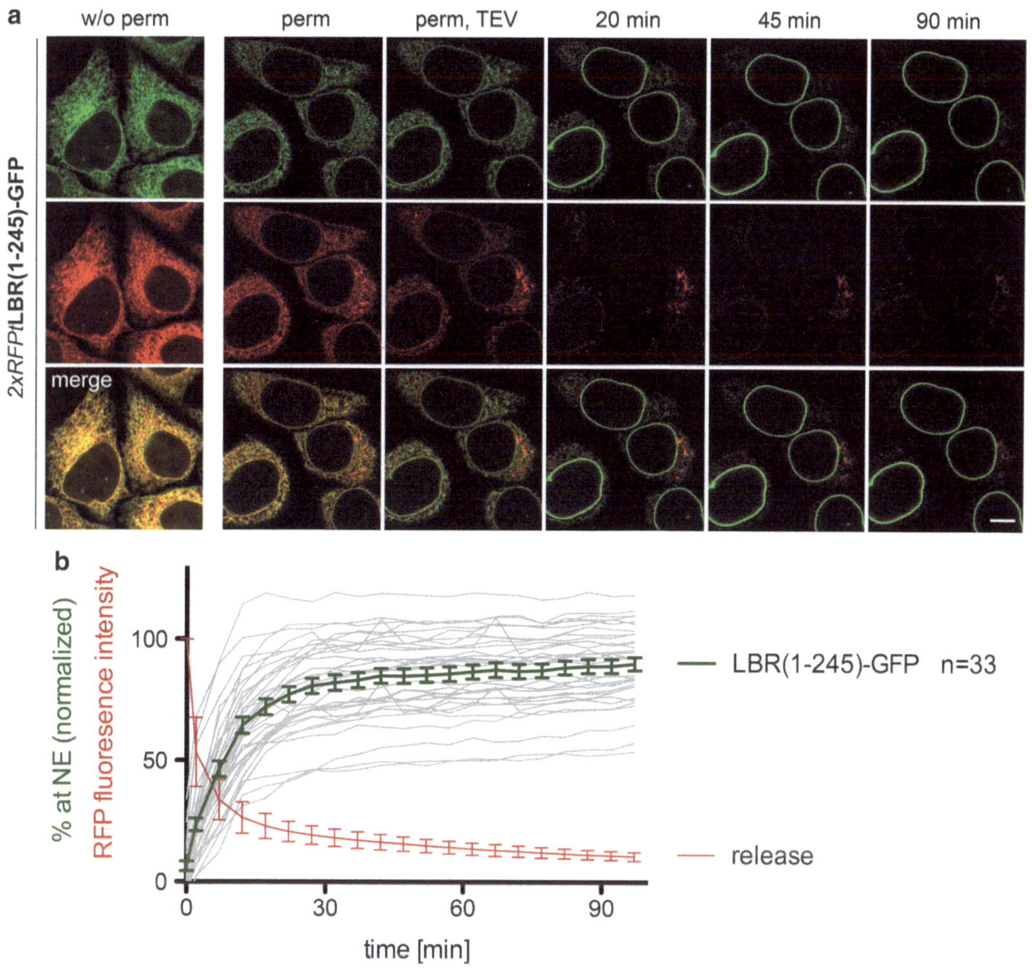

Fig. 3 Targeting kinetics of LBR(1–245)-GFP. (**a**) Representative images of reporter cells expressing 2xRFPtLBR(1–245)-GFP, before semi-permeabilization (w/o perm), after semi-permeabilization (perm), after TEV cleavage (perm, TEV) and after different times of the INM targeting reaction. (**b**) Mean NE accumulation relative to total fluorescence was quantified using a MATLAB based quantification tool [18] (*green line*). Release by TEV cleavage was quantified by the loss in RFP signal intensity (*red line*). *Gray lines* represent single cell tracks over time, *green* and *red lines* show mean +/– SD (*n* = 33)

4. Fix a second coverslip to control for TEV cleavage efficiency and as reference 0 min time point (*see* **Note 15**).

5. For the import reactions, remove liquid from coverslips onto low-lint tissue. Place 30 μL of import mixture on each coverslip. Keep coverslips on a 24-well plate in a humidified environment to avoid drying. We use a plastic box containing a wet tissue. Incubate this box at 37 °C.

6. At defined time points (e.g., 15, 30, 45, and 90 min), wash the coverslips with PB and fix the cells.

7. Stain nuclei with DAPI or Hoechst to later allow for recognition of nuclear contours during image analysis.

8. Mount coverslips on glass slides using an aqueous mounting medium. We prefer Vectashield (*see* **Note 16**).
9. Reporter signal can be imaged at any standard confocal microscope. We use a Leica TCS-SP2/AOBS microscope equipped with a HCX Pl APO lbd.Bl. 63×, N.A. 1.4 oil immersion lens. Image the mid-focal plane.

3.5.3 Kinetic Analysis of INM Targeting by Time-Lapse Imaging

1. After the last washing step in PB, add 100 μL of import mixture (HeLa lysate and 1× energy regenerating system) per well of a 8-well Nunc™ Lab-Tek™ chamber (*see* **Note 17**).
2. Mount Lab-Tek™ chambers at a Zeiss LSM 710-FCS microscope or equivalent instrument with an 63× 1.4 N.A. Oil DIC Plan-Apochromat immersion lens, and equipped with a humidified incubator equilibrated at 37 °C (*see* **Note 18**).
3. Adjust the focus and mark positions of interest. Perform multicolor imaging in two channels recording eGFP and mRFP fluorescence.
4. Acquire a pre-release z-stacked series of images of all marked positions. This image will be set as time 0. (Fig. 3a, "perm, TEV"; b quantified as time point 0) (*see* **Note 19**).
5. Carefully open the incubator and remove the lid of the Lab-tek™ chamber. Add 25 μL concentrated NusA-TEV dissolved in PB (750 μg/mL stock) directly to the import mixture to start release (*see* **Note 20**).
6. Immediately start recording with a time interval of 2–5 min for up to 2.5 h and acquiring z-stacks (*see* **Note 21**).

3.5.4 Cellular and Biochemical Alterations

Our in vitro reconstitution approach is ideally suited for a broad range of cellular and biochemical manipulations. This enables the identification of essential requirements of the transport process. Hereafter, we give a few examples of alterations that can be combined with reconstitution of INM targeting in vitro.

1. The functional involvement of cytoplasmic components can be assessed by manipulation of the HeLa cell extract. Proteins of interest can be depleted by affinity chromatography using specific antibodies coupled to protein A/G beads, or by use of other chromatographic media such as phenyl-sepharose [18] or ion exchange chromatography using Q or S sepharose.
2. The energy status of semi-permeabilized cells is normally restored by the addition of an energy regenerating system. However, reconstitution can also be performed in the absence of energy. Residual NTPs (in semi-permeabilized cells or the HeLa lysate) may be hydrolyzed by supplementing the import mixture with apyrase.

3. In vitro reconstitution can be performed in the presence of chemical inhibitors such as non-hydrolyzable nucleotide analogs (e.g., ATPγS, GTPγS) or dominant-negative protein variants (e.g., RanQ69L [23]). Semi-permeabilized cells may be pre-incubated with inhibitors or dominant-negative recombinant proteins before cleavage of the reporter by TEV protease treatment (*see* **Note 22**).
4. It is also possible to use otherwise cell-impermeable drugs/ inhibitors during in vitro reconstitution (e.g., wheat germ agglutinin (WGA) [24]).
5. To investigate the function of a nuclear or membrane-bound protein in INM targeting, cells can be subjected to RNAi-mediated knockdown before semi-permeabilization. In general, reporter cells are transfected with siRNA oligonucleotides against a gene of interest 1–3 days before reporter protein induction by tetracycline. Knockdown efficiency should always be controlled by Western blotting or immunofluorescence analysis. We used RNAi-mediated knockdown to decipher which nucleoporins contribute to the size barrier for membrane proteins [18] (*see* **Note 23**).

3.6 Quantification

For image quantification the nuclear contours are first detected, then a region extending from the contours is used to define the NE. Finally, the fluorescence (pixel) intensity within this NE region and in the ER is measured to quantify the fraction of total intensity at the NE.

3.6.1 Quantification of Endpoint Assays

1. Use an ImageJ plugin [18] to detect nuclear contours based on DAPI-stained nuclei and the ER border by thresholding on the eGFP channel.
2. From the nuclear contour, extend an area 230 nm outwards towards the ER and 920 nm inwards into the nucleus to define the NE region.
3. Measure the integrated pixel intensity at the NE.
4. Measure the integrated pixel intensity in the ER.
5. Add NE and ER pixel intensities and divide the NE pixel intensity by this number to get the percentage of total fluorescence intensity at the NE.
6. Normalize this number relative to the fraction at the NE before release by TEV cleavage (*see* **Note 24**).

3.6.2 Quantification of Time-Lapse Images

1. Open images using a MATLAB-based quantification tool [18].
2. In the selected image set, detect nuclear contours by a Laplacian of Gaussians (log) filter based edge detector and follow over time based on their spatial location.

3. Based on these contours, set the NE as an ~ 800 nm wide ring.
4. Define the ER outline as delimited by an orthogonal border in the middle of the connecting line between the barycenter of nuclei.
5. Measure the integrated pixel intensity at the NE.
6. Measure the integrated pixel intensity in the ER.
7. Add NE and ER pixel intensities and divide the NE pixel intensity by this number to get the percentage of total fluorescence intensity at the NE.
8. Normalize this number relative to the fraction at the NE before release by TEV cleavage (*see* **Note 24**).

4 Notes

1. Since the purity of digitonin may vary between different lots, the effective concentration of each lot should be tested before use.
2. Avoid repeated freeze–thaw cycles for active components like cell extract, energy regenerating system, TEV protease, and nucleotide analogs.
3. Should the extralumenal extension of your INM protein of interest be smaller then 10 kDa, it might become necessary to add a larger size-tag, e.g., 3xmRFP, to achieve efficient trapping in the ER.
4. Note that INM proteins can also be targeted during NE reformation after mitosis. This may lead to a partial INM accumulation despite of the size-tag that becomes apparent in cells that went through mitosis. Should NE accumulation after mitosis be particularly strong, you may observe a red NE signal (mRFP) that is protected from NusA-TEV cleavage. In such case, we recommend testing shorter expression times of the reporters or the addition of thymidine to arrest cells in interphase during expression.
5. For preservation of soluble 2xRFP, fix the cells in 2 % PFA, 0.25 % glutaraldehyde in PBS for 10 min. Quench residual unreacted glutaraldehyde for 10 min with 1 mg/mL $NaBH_4$. Some soluble RFP may be lost if standard formaldehyde fixation is used.
6. Choose a clonal cell line that displays homogenous tetracycline-inducible expression of the reporter. Constitutive expression of the INM protein reporter in the ER might be deleterious for cells.
7. It is important to consider that strong overexpression of INM proteins may lead to a reduced accumulation at the INM as

retention sites in the nuclear interior may become limiting. Therefore, it should be analyzed whether expression levels correlate with steady-state accumulation at the NE. This can either be done on cells transiently transfected with INM protein reporters or on cell lines induced with increasing tetracycline concentrations. Choose expression conditions that result in a similar or equal steady-state NE accumulation compared to the endogenous INM protein.

8. The Dounce homogenizer and pestle used for the preparation of extract should be prechilled on ice.
9. To ensure that the NE remains unaffected throughout the experiment, it is critical to perform semi-permeabilization at low digitonin concentrations at low temperature. Permeabilization buffer should be chilled before use and the cell culture dish/Lab-tek™ chamber should be placed on ice throughout the semi-permeabilization and subsequent washing steps.
10. An appropriate cell density is crucial for semi-permeabilization-based experiments. High cell density or growth of cells in clusters may affect the permeabilization efficiency, leaving the cells in the center of the cluster un-permeabilized. Low cell density on the other hand may lead to detachment of cells or over-permeabilization.
11. To avoid detaching of cells during semi-permeabilization, washing steps should be performed carefully by pipetting the buffer on the walls of the Lab-tek™ chamber/cell culture dishes.
12. Semi-permeabilized cells are stable for a few h when kept on ice. We, however, recommend continuing with the experiment immediately to ensure reproducibility and minimize secondary effects.
13. Release must be complete after 10 min. If not, more recombinant TEV protease can be used or a TEV protease with a higher activity should be prepared. A fast and efficient release step is crucial for measuring targeting kinetics reliably. Otherwise, targeting dynamics will be falsified and limited by the TEV cleavage rate.
14. Import mixtures can be varied according to your experimental question (*see* Subheading 3.5.4).
15. It is possible to cleave the reporters with TEV at room temperature without prematurely starting the targeting reaction, i.e., we did not observe a significant enrichment of reporter proteins at the NE after TEV cleavage at 25 °C for 10 min compared to controls.
16. Do not use hard dry mounting media. This will lead to dented nuclei when using semi-permeabilized cells.

17. Imaging on semi-permeabilized cells does not need special imaging medium or CO_2.
18. Pre-warm the microscope incubator before time-lapse experiments to avoid temperature drift while imaging.
19. A z-stack is not absolutely required if the imaging system runs stably for up to 2 h. In case the focus should drift, a z-stack centered around the focal plane on the NE covering 15 μm in 15 stacks can be recorded.
20. To minimize the dilution of HeLa lysate and energy, TEV protease should be used at high concentrations in live assays.
21. Prevent photobleaching and light-based toxicity by using low laser power and short exposure times whenever possible.
22. Whenever recombinant proteins are added to semi-permeabilized cells, it is necessary to re-buffer these proteins into PB. This can either be achieved by size-exclusion chromatography (e.g., PD-10 columns, GE healthcare) or by dialysis. Strictly avoid phosphate buffers or traces of substances like imidazole or urea when adding protein solutions to semi-permeabilized cells.
23. RNAi-mediated depletion of endogenous proteins may sometimes affect the expression of certain reporters. Ideally, conditions that allow for an efficient depletion of proteins without altering the expression of the reporter are chosen for the experiment.
24. Accumulation at the NE relative to total fluorescence can be quantified in the mid-section focal plane as this adequately reflected measurements in stacks over entire cells.

Acknowledgements

We thank Monika Mayr for comments. We are grateful to ETH Zurich and the Swiss National Science Foundation for continuous financial support. Our work on this topic is financed by an ERC Advanced Grant (NucEnv) to UK.

References

1. Mekhail K, Moazed D (2010) The nuclear envelope in genome organization, expression and stability. Nat Rev Mol Cell Biol 11:317–328
2. Hetzer MW (2010) The nuclear envelope. Cold Spring Harb Perspect Biol 2:a000539
3. Schirmer EC, Gerace L (2005) The nuclear membrane proteome: extending the envelope. Trends Biochem Sci 30:551–558
4. Katta SS, Smoyer CJ, Jaspersen SL (2014) Destination: inner nuclear membrane. Trends Cell Biol 24:221–229
5. Burke B, Stewart CL (2014) Functional architecture of the cell's nucleus in development, aging, and disease. Curr Top Dev Biol 109:1–52
6. Powell L, Burke B (1990) Internuclear exchange of an inner nuclear membrane protein (p55) in heterokaryons: in vivo evidence for the interac-

tion of p55 with the nuclear lamina. J Cell Biol 111:2225–2234
7. Smith S, Blobel G (1993) The first membrane spanning region of the lamin B receptor is sufficient for sorting to the inner nuclear membrane. J Cell Biol 120:631–637
8. Soullam B, Worman HJ (1993) The amino-terminal domain of the lamin B receptor is a nuclear envelope targeting signal. J Cell Biol 120:1093–1100
9. Ostlund C, Ellenberg J, Hallberg E et al (1999) Intracellular trafficking of emerin, the Emery-Dreifuss muscular dystrophy protein. J Cell Sci 112:1709–1719
10. Wu W, Lin F, Worman HJ (2002) Intracellular trafficking of MAN1, an integral protein of the nuclear envelope inner membrane. J Cell Sci 115:1361–1371
11. Shimi T, Koujin T, Segura-Totten M et al (2004) Dynamic interaction between BAF and emerin revealed by FRAP, FLIP, and FRET analyses in living HeLa cells. J Struct Biol 147:31–41
12. Ostlund C, Sullivan T, Stewart CL et al (2006) Dependence of diffusional mobility of integral inner nuclear membrane proteins on A-type lamins. Biochemistry 45:1374–1382
13. Zuleger N, Kelly DA, Richardson AC et al (2011) System analysis shows distinct mechanisms and common principles of nuclear envelope protein dynamics. J Cell Biol 193: 109–123
14. Ohba T, Schirmer EC, Nishimoto T et al (2004) Energy- and temperature-dependent transport of integral proteins to the inner nuclear membrane via the nuclear pore. J Cell Biol 167:1051–1062
15. Muhlhausser P, Kutay U (2007) An in vitro nuclear disassembly system reveals a role for the RanGTPase system and microtubule-dependent steps in nuclear envelope breakdown. J Cell Biol 178:595–610
16. Ribbeck K, Gorlich D (2001) Kinetic analysis of translocation through nuclear pore complexes. EMBO J 20:1320–1330
17. Adam SA, Marr RS, Gerace L (1990) Nuclear protein import in permeabilized mammalian cells requires soluble cytoplasmic factors. J Cell Biol 111:807–816
18. Ungricht R, Klann M, Horvath P et al (2015) Diffusion and Retention are Major Determinants of Protein Targeting to the Inner Nuclear Membrane. J Cell Biol 209:687–704
19. Soullam B, Worman HJ (1995) Signals and structural features involved in integral membrane protein targeting to the inner nuclear membrane. J Cell Biol 130:15–27
20. Theerthagiri G, Eisenhardt N, Schwarz H et al (2010) The nucleoporin Nup188 controls passage of membrane proteins across the nuclear pore complex. J Cell Biol 189:1129–1142
21. Turgay Y, Ungricht R, Rothballer A et al (2010) A classical NLS and the SUN domain contribute to the targeting of SUN2 to the inner nuclear membrane. EMBO J 29: 2262–2275
22. Gorlich D, Prehn S, Laskey RA et al (1994) Isolation of a protein that is essential for the first step of nuclear protein import. Cell 79:767–778
23. Klebe C, Bischoff FR, Ponstingl H et al (1995) Interaction of the nuclear GTP-binding protein Ran with its regulatory proteins RCC1 and RanGAP1. Biochemistry 34:639–647
24. Finlay DR, Newmeyer DD, Price TM et al (1987) Inhibition of in vitro nuclear transport by a lectin that binds to nuclear pores. J Cell Biol 104:189–200

Chapter 29

Nuclear Protein Transport in Digitonin Permeabilized Cells

Stephen A. Adam

Abstract

The high concentration of cholesterol in the plasma membrane relative to the endomembranes of eukaryotic cells allows the selective permeabilization of the plasma membrane with the glycoside digitonin leaving the intracellular membrane bound organelles intact. In this chapter, we describe the basic method to use digitonin permeabilized cells to reconstitute the transport of proteins containing nuclear localization signals into the nucleus. The assay requires only a target cell line that can be permeabilized with digitonin, a source of soluble transport factors, typically provided by the cytosol fraction of cultured cells, and a cargo protein of interest. No other specialized equipment is required other than a fluorescence microscope. The assay can be used to identify transport factors required to transport specific proteins, to study the regulation of protein transport, or to study nuclear protein transport under different conditions.

Key words Digitonin permeabilization, Nuclear transport, Karyopherin/importin/nuclear transport factor, Nuclear transport reconstitution

1 Introduction

The nuclear envelope in eukaryotic cells is a selectively permeable barrier separating the cytoplasmic and nuclear compartments. Macromolecular traffic between the cytoplasmic and nuclear compartments is mediated by large proteinaceous structures embedded in the nuclear envelope called nuclear pore complexes (NPCs). The NPCs form selective channels for the passive diffusion of small molecules and ions, but the movement of proteins and nucleic acids larger than ~30 kDa occurs by a regulated process of facilitated diffusion [1].

The facilitated diffusion of proteins and RNA through the NPCs is mediated by a family of transport receptors, collectively called karyopherins, and several accessory proteins [2]. The karyopherins bind to nuclear localization signals (NLSs) intrinsic to the proteins to be transported, which direct either the nuclear import or the nuclear export of the protein. Most often, these signals are

Sue Shackleton et al. (eds.), *The Nuclear Envelope: Methods and Protocols*, Methods in Molecular Biology, vol. 1411,
DOI 10.1007/978-1-4939-3530-7_29, © Springer Science+Business Media New York 2016

short stretches of amino acids, but other signals comprised of three-dimensional structures formed by protein folding or complex formation have also been described [3]. The founding member of the karyopherin family of transport receptors is known as importin β or karyopherin β. All members of this family bind to the small nuclear GTPase, Ran, which induces a large conformation change in the karyopherin, allowing it to either release cargo (as in the case of protein import) or bind cargo (as in the case of protein export). For example, a protein to be imported (a cargo) will be bound by a karyopherin (an importin) in the cytoplasm, because the concentration of Ran-GTP in the cytoplasm is very low relative to the nuclear concentration. The karyopherins have a surface chemistry that allows them to enter the channel of the NPC from one compartment and diffuse to the other compartment. Once in the nucleus, Ran-GTP binds to the karyopherin thus releasing the cargo. The opposite scheme allows karyopherins involved in export (exportins) to form a ternary complex between the exportin, Ran-GTP, and cargo. Upon entry into the cytoplasm, a Ran-GTPase activating protein stimulates Ran to hydrolyze its bound GTP to GDP triggering a conformational change in the exportin to release the cargo. Like other small GTPases, it is this cyclic conversion of Ran between the GTP and GDP-bound forms that defines the cellular compartment and allows proteins to accumulate against a concentration gradient.

Since we first described the digitonin permeabilized cell nuclear transport assay 25 years ago [4], it has been used and modified by numerous investigators to identify components of the nuclear transport system [4–11], and to study transport kinetics and regulation [12–15]. The assay is simple to perform requiring only a target cell line, a soluble source of transport factors, and a cargo protein. Any eukaryotic cell with sufficient plasma membrane cholesterol can be used in the assay. Since the transport receptors and accessory proteins are soluble in cells (with the exception of the NPCs), cell lysates can be, in theory, produced from any cell of interest that can be obtained in sufficient quantity. Cargo proteins can be endogenous proteins present in the cell lysate, tagged proteins expressed in bacteria or artificial proteins containing an inserted NLS (such as GST-NLS-EGFP), or a carrier protein such as serum albumin coupled to NLS peptides. The assay can be performed on a single sample or on multiple samples depending on the format used to grow the target cells and does not require any specialized equipment or apparatus [16]. It has also been useful for single molecule analysis of transport mechanisms [17], the identification of karyopherin substrates [14, 18], and the analysis of transport under cellular stress conditions [16].

2 Materials

2.1 Solutions

1. PBS: 1.44 g Na_2HPO_4 (anhydrous), 0.24 g KH_2PO_4, 8 g NaCl, 0.2 g KCl; adjust volume to 1 L.
2. 10× transport buffer: 23.8 g HEPES (free acid) (200 mM), 66.2 g CH_3COOK (1.35 M), 4.1 g CH_3COONa (anhydrous) (100 mM), 2.1 g $Mg(CH_3COO)_2$-$4H_2O$ (20 mM), 0.95 g EGTA (5 mM); adjust the pH to 7.3 with KOH and volume to 500 mL. Filter through a 0.2 μm filter (*see* **Note 1**).
3. Digitonin: 50 mg/mL in DMSO. The highest purity digitonin available should be used. Store at −20 °C.
4. DTT: 1 M in deionized water (DW), store in aliquots at −20 °C.
5. 20× ATP regenerating system: 20 mM ATP, 100 mM creatine phosphate, 400 U/mL creatine phosphokinase made up in transport buffer (*see* **Note 2**).
6. 1000× protease inhibitors: 1 mg/mL aprotinin in DW, 1 mg/mL leupeptin in DW, 1 mg/mL pepstatin A in DMSO. Store all in aliquots at −20 °C (*see* **Note 3**).
7. Lysis buffer: 0.595 g HEPES (free acid) (5 mM), 0.49 g CH_3COOK (10 mM), 0.21 g $Mg(CH_3COO)_2$-$4H_2O$ (2 mM); adjust pH to 7.3 and volume to 500 mL, filter through a 0.2 μm filter.

2.2 Equipment to Prepare Cell Lysates

1. Spinner culture flasks.
2. Tight fitting Dounce homogenizer, glass or stainless steel, 15 mL or less volume.
3. Dialysis tubing 12 kDa cutoff or smaller.

3 Methods

The method given below describes methods for preparing cells and lysates and performing the assay. It should be modified according to the goals of the particular experiment. For example, if one is testing the dependence of a particular cargo on a particular transport factor, one could either (1) prepare a minimal transport reaction in transport buffer alone plus individually purified proteins such as Ran and importin/karyopherin or (2) deplete the transport factor of interest from the cytosol or reticulocyte lysate.

3.1 Preparation of Rabbit Reticulocyte Lysate

1. Rabbit reticulocyte lysate can be obtained from several commercial sources.
2. Add 1 part 10× transport buffer to 9 parts reticulocyte lysate.
3. Freeze aliquots in liquid nitrogen. Store aliquots at −80 °C (*see* **Note 4**).

3.2 Preparation of Lysates from Cultured Cells

1. Culture cells in spinner culture at 37 °C in appropriate medium. Maintain the culture at a density of 3–6 × 10^5 cells/mL (*see* **Note 5**).
2. Collect the cells by centrifugation at 250 × *g* for 10 min at 4 °C.
3. Resuspend the cell pellet in <50 mL ice-cold PBS by gentle pipetting and transfer to a 50 mL conical tube.
4. Pellet the cells by centrifugation at 250 × *g* for 10 min at 4 °C.
5. Repeat **steps 3** and **4**.
6. Resuspend the cell pellet in <50 mL transport buffer containing 2 mM DTT and centrifuge at 250 × *g* for 10 min at 4 °C.
7. Gently resuspend the cells in 1.5 volumes of lysis buffer containing 2 mM DTT and protease inhibitors. Swell on ice for 10 min (*see* **Note 6**).
8. Transfer the cells to a tight fitting glass Dounce homogenizer or stainless steel tissue grinder prechilled in an ice bath.
9. Lyse the cells by 10–20 strokes. Monitor the lysis by microscopy (*see* **Note** 7).
10. Transfer the homogenate to prechilled high-speed centrifuge tubes and centrifuge at 15,000 × *g* for 20 min at 4 °C.
11. Collect the supernatant into a prechilled ultracentrifuge tube and centrifuge at 100,000 × *g* for 30 min.
12. Collect the supernatant and dialyze against transport buffer with 2 mM DTT for several h to overnight at 4 °C (*see* **Note 8**).
13. Collect the dialyzed lysate and aliquot. Freeze the aliquots in liquid nitrogen and store at −80 °C (*see* **Note 9**).

3.3 Preparation of the Reaction Mix

1. Determine the volume of reaction mix to prepare by testing different volumes of transport buffer on the coverslip, slide, or dish to be analyzed. Enough volume to cover the cells completely during the incubation step is required.
2. Mix cytosol with ATP regenerating system and 2 μM cargo protein. Adjust to desired volume with transport buffer containing 2 mM DTT and protease inhibitors. Keep on ice while permeabilizing the cells (*see* **Note 10**).

3.4 Preparation of the Digitonin Permeabilized Cells

1. 24–48 h prior to the experiment, passage the cells for the assay so they will be sub-confluent on the day of the experiment (*see* **Note 11**).
2. On the day of the assay, replace the growth medium with fresh medium 2–6 h prior to the assay (*see* **Note 12**).
3. Prepare enough transport buffer with 2 mM DTT and protease inhibitors for all the washing and permeabilization steps.

4. Remove the growth medium from the cells and rinse twice with transport buffer prepared in **step 3** (*see* **Note 13**).
5. Remove the wash buffer and add transport buffer containing digitonin, 2 mM DTT, and protease inhibitors to the cells (*see* **Note 14**).
6. Incubate the cells at the desired temperature for 2–5 min (*see* **Note 15**).
7. Remove the permeabilization buffer and rinse the cells once or twice with import buffer containing 2 mM DTT and protease inhibitors, but without digitonin.

3.5 The Nuclear Transport Reaction

1. Pipette complete transport reaction prepared in Subheading 3.3 onto a section of Parafilm (*see* **Note 16** and Fig. 1b).
2. Lift coverslip from dish and remove excess buffer by touching the edge to a Kimwipes or paper towel (*see* **Note 17**).
3. Invert the coverslip onto the drop of transport reaction so the cells are in contact with the liquid (*see* **Note 18** and Fig. 1c).

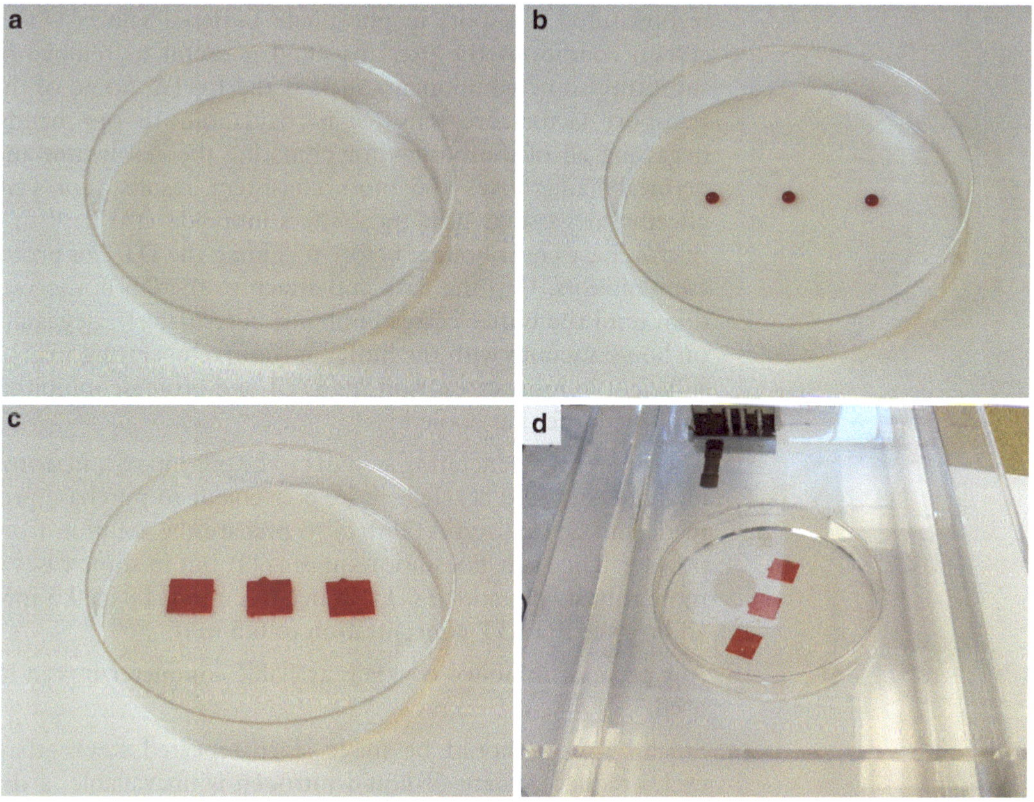

Fig. 1 Setting up the nuclear transport assay. (**a**) Place a piece of blotting paper on the bottom of a dish and wet with water or buffer. Place a piece of Parafilm large enough to accommodate the reactions on the wet blotting paper. (**b**) Pipette each reaction mix onto the Parafilm. (**c**) Invert coverslips with permeabilized cells onto the reaction droplets. (**d**) Cover the dish and float in a water bath set to desired temperature

4. Cover the humidified chamber and float on the surface of a water bath set at 30 °C for 15–30 min (Fig. 1d).
5. Lift the coverslip carefully and return to a dish containing ice cold transport buffer to wash the cells (*see* **Note 19**).
6. Remove the wash buffer and fix the cells by adding 2–4 % formaldehyde in transport buffer at room temperature for 5–10 min.
7. (Optional) Stain the DNA with DAPI or Hoechst.
8. Rinse the coverslip with transport buffer and mount on a microscope slide in a small quantity of transport buffer (*see* **Note 20**).
9. After imaging, the amount of the fluorescent cargo in the nucleus can be quantified with individual lab software.

4 Notes

1. Other buffers and salts likely will work equally well as long as the solution is close to isotonic. We have even successfully reconstituted transport in phosphate buffered saline. Under certain conditions the user may find it useful to remove or substitute some components such as the EGTA. Some of the transport factors are sensitive to oxidation. In our hands, degassing all of the buffers for preparing the cell lysates and permeabilizing gives the most consistent results. For very effective degassing, heat the buffers in a side arm flask in a microwave to near boiling before to adding the DTT or protease inhibitors. Cap the flask and attach to the lab house vacuum until the buffer ceases bubbling. Alternatively, degassing on house vacuum with the buffer at room temperature may be sufficient in most cases. Add the DTT and protease inhibitors after the buffer has cooled.
2. The 20× ATP regenerating system can be pre-mixed and stored in aliquots at −80 °C. It is more convenient to purchase premade and neutralized ATP than to prepare the solution from powder. The assay does not require GTP, but in some cases is more robust with added GTP. If needed, add GTP to 0.5 mM and reduce the ATP concentration to 0.5 mM.
3. Any protease inhibitor cocktails available commercially can be used.
4. Small aliquots should be made since repeated freeze–thaw cycles reduce activity. If liquid nitrogen is unavailable, a dry ice–methanol bath is sufficient.
5. Cells grown in spinner flasks have several advantages over cells grown attached in dishes. In spinner culture, most of the cells

are actively cycling, unlike cells grown on solid substrates. It is also easier to obtain the large mass of cells needed to prepare lysate and to collect the cells into a tight pellet without using enzymatic dissociation. Lysates can be made from cells grown on dishes, but they usually have much lower activity than lysates prepared from spinner cells.

6. It is important to keep all buffers ice cold during the swelling and lysis steps and to include DTT. The buffer should all be degassed prior to use.
7. In the phase contrast microscope, lysed cells will no longer be refractile. If more than ~10 % of the cells remain intact after the first homogenization, repeat with five additional strokes until >90 % of the cells are broken.
8. Dialysis membrane with a molecular weight cut-off of 12 kDa or smaller should be used. To maintain the dialysis in the cold, place the vessel containing the dialysis buffer into an insulated container and surround with ice, replenishing the ice as necessary.
9. Small aliquots should be made since repeated freeze–thaw cycles reduce activity. If liquid nitrogen is unavailable, a dry ice–methanol bath is sufficient.
10. The optimal concentration must be determined for each batch of cytosol. 10–80 % of the incubation volume can be provided by the cytosol fraction. For lysates prepared from cultured cells, add cytochalasin B to 20 μM to prevent actin polymerization during the subsequent incubation with the permeabilized cells.
11. Actively dividing cells yield the best results, although in some specialized applications serum starved, contact inhibited, or senescent cells may be used. Cells may be grown on glass coverslips or multiwell slides, directly on plastic, or in suspension.
12. This step can be omitted, but for some cells gives more uniform results.
13. The temperature of the buffer in this step does not seem to matter. If you are planning to permeabilize the cells in an ice bath, chill the buffer for the wash. For suspension cells, pellet the cells in a low speed centrifuge.
14. Use an adequate amount of buffer here. For a 35 mm dish, 2 mL of buffer would be appropriate. Permeabilization is a function of digitonin-to lipid ratio. Therefore, there is a limit as to how many cells can be effectively permeabilized in a fixed amount of digitonin solution. 50 μg/mL digitonin is a good starting concentration, although some cells may require lower or higher concentrations of digitonin. Some cells can be permeabilized with up to 100 μg/mL digitonin without loss of

nuclear envelope integrity. Each lot of digitonin must be titrated to test its effectiveness for permeabilizing cells. The cells may be permeabilized directly in the dish or moved to a separate container such as a coplin jar to permeabilize several coverslips together. For suspension cells, first resuspend the cell pellet in one half the final volume of import buffer, and then add one half volume of import buffer containing twice the final concentration of digitonin. This will avoid exposing the cells to a high concentration of detergent.

15. The temperature does not seem to matter. We routinely permeabilize on ice, but have also had success permeabilizing at room temperature.
16. Prepare a humidified chamber by placing a piece of blotting paper in the bottom of a plastic dish and wet the paper with water. Place a piece of Parafilm, slightly smaller than the blotting paper on the paper (Fig. 1a).
17. Do not allow the cells to dry out.
18. The volume of liquid will depend on the size of the coverslip. For an 18 × 18 mm coverslip, 50 μL of liquid should be sufficient. If doing the assay in a plate or on a multiwell slide, simply aspirate the wash buffer and add the appropriate amount of transport reaction mix.
19. If a small volume of transport reaction is used, the coverslip can be floated off the Parafilm by pipetting 200 μL of transport buffer under the edge of the coverslip.
20. If using a nonfluorescent transport substrate, the fixed cells can be processed for indirect immunofluorescence using standard techniques.

References

1. Wente SR, Rout MP (2010) The nuclear pore complex and nuclear transport. Cold Spring Harb Perspect Biol 2:a000562
2. Kimura M, Imamoto N (2014) Biological significance of the importin-beta family-dependent nucleocytoplasmic transport pathways. Traffic 15:727–748
3. Marfori M, Mynott A, Ellis JJ, Mehdi AM, Saunders NF, Curmi PM, Forwood JK, Boden M, Kobe B (2011) Molecular basis for specificity of nuclear import and prediction of nuclear localization. Biochim Biophys Acta 1813:1562–1577
4. Adam SA, Marr RS, Gerace L (1990) Nuclear protein import in permeabilized mammalian cells requires soluble cytoplasmic factors. J Cell Biol 111:807–816
5. Adam EJ, Adam SA (1994) Identification of cytosolic factors required for nuclear location sequence-mediated binding to the nuclear envelope. J Cell Biol 125:547–555
6. Chi NC, Adam EJ, Adam SA (1995) Sequence and characterization of cytoplasmic nuclear protein import factor p97. J Cell Biol 130:265–274
7. Gorlich D, Kostka S, Kraft R, Dingwall C, Laskey RA, Hartmann E, Prehn S (1995) Two different subunits of importin cooperate to recognize nuclear localization signals and bind them to the nuclear envelope. Curr Biol 5:383–392
8. Imamoto N, Shimamoto T, Takao T, Tachibana T, Kose S, Matsubae M, Sekimoto T, Shimonishi Y, Yoneda Y (1995) In vivo evi-

dence for involvement of a 58 kDa component of nuclear pore-targeting complex in nuclear protein import. EMBO J 14:3617–3626
9. Moore MS, Blobel G (1993) The GTP-binding protein Ran/TC4 is required for protein import into the nucleus. Nature 365:661–663
10. Moore MS, Blobel G (1994) Purification of a Ran-interacting protein that is required for protein import into the nucleus. Proc Natl Acad Sci U S A 91:10212–10216
11. Paschal BM, Gerace L (1995) Identification of NTF2, a cytosolic factor for nuclear import that interacts with nuclear pore complex protein p62. J Cell Biol 129:925–937
12. Hu W, Jans DA (1999) Efficiency of importin alpha/beta-mediated nuclear localization sequence recognition and nuclear import. Differential role of NTF2. J Biol Chem 274:15820–15827
13. Kehlenbach RH, Dickmanns A, Kehlenbach A, Guan T, Gerace L (1999) A role for RanBP1 in the release of CRM1 from the nuclear pore complex in a terminal step of nuclear export. J Cell Biol 145:645–657
14. Kohler M, Speck C, Christiansen M, Bischoff FR, Prehn S, Haller H, Gorlich D, Hartmann E (1999) Evidence for distinct substrate specificities of importin alpha family members in nuclear protein import. Mol Cell Biol 19:7782–7791
15. Smith A, Brownawell A, Macara IG (1998) Nuclear import of Ran is mediated by the transport factor NTF2. Curr Biol 8:1403–1406
16. Furuta M, Kose S, Kehlenbach RH, Imamoto N (2014) Analysis of nucleocytoplasmic transport in digitonin-permeabilized cells under different cellular conditions. Methods Cell Biol 122:331–352
17. Tu LC, Musser SM (2011) Single molecule studies of nucleocytoplasmic transport. Biochim Biophys Acta 1813:1607–1618
18. Kimura M, Kose S, Okumura N, Imai K, Furuta M, Sakiyama N, Tomii K, Horton P, Takao T, Imamoto N (2013) Identification of cargo proteins specific for the nucleocytoplasmic transport carrier transportin by combination of an in vitro transport system and stable isotope labeling by amino acids in cell culture (SILAC)-based quantitative proteomics. Mol Cell Proteomics 12:145–157

Chapter 30

Analysis of CRM1-Dependent Nuclear Export in Permeabilized Cells

Ralph H. Kehlenbach and Sarah A. Port

Abstract

Nuclear protein import and export assays in permeabilized cells have been instrumental for the identification of transport factors and for the molecular characterization of nucleocytoplasmic transport pathways. Our original assay to quantitatively analyze CRM1-dependent export was based on stably transfected cells expressing GFP-NFAT. We now present a simplified version of the assay using transiently transfected cells expressing GFP-NFAT or GFP-snurportin1 as a fluorescent export cargo and mCherry-emerin as a marker protein for transfected cells. CRM1- and Ran-dependent export is recapitulated in digitonin-permeabilized cells and quantified by flow cytometry. The assay should be applicable to other combinations of cargo and marker proteins.

Key words Nucleus, Nuclear transport, Export, CRM1, Ran, NFAT, Snurportin 1, Digitonin, Flow cytometry

1 Introduction

Twenty years after the identification of nuclear export sequences (NESs) in the HIV-1 Rev protein [1] and in PKI, the heat stable inhibitor of the catalytic subunit of cAMP-dependent protein kinase [2], the number of proteins carrying characteristic hydrophobic NESs is still growing. CRM1, a member of the importin β-family of nuclear transport receptors, has been recognized as the major factor that mediates transport of proteins out of the nucleus, although other family members are involved in export of selected cargoes as well (for reviews *see* refs. 3–5). More recently, CRM1 was identified as a potential target in anti-cancer therapy (for review *see* ref. 6), based on the observation that it is upregulated in many tumor cells. This probably relates to the fact that many tumor suppressors like p53 [7] or p27 [8] are bona fide CRM1 cargoes.

We initially designed an assay to analyze nuclear export in digitonin-permeabilized cells that was based on similar principles as the classic nuclear import assay developed by Adam and Gerace [9].

Sue Shackleton et al. (eds.), *The Nuclear Envelope: Methods and Protocols*, Methods in Molecular Biology, vol. 1411, DOI 10.1007/978-1-4939-3530-7_30, © Springer Science+Business Media New York 2016

The assay used GFP-NFAT (nuclear factor of activated T cells) as a CRM1-dependent reporter protein [10]. In vivo, nuclear trafficking of GFP-NFAT can easily be controlled by adding the calcium ionophore ionomycin to cells, a treatment that leads to rapid nuclear import of the protein [11, 12]. Nuclear export, on the other hand, occurs upon medium exchange and can be recapitulated in digitonin-permeabilized cells in vitro. After such an export reaction in permeabilized suspension cells, residual nuclear fluorescence can be quantified by flow cytometry, allowing the analysis of a large number of samples in a short period of time. Again, the method for quantification was based on a previous approach [13]. Although this assay proved very powerful for the identification of CRM1 [10], as well as for the characterization of co-factors like Ran [14], RanBP1 [15], RanGAP [16], and Nup214 [17], it has not widely been used by other researchers, because it requires a cell line stably transfected with the GFP-NFAT plasmid. We therefore set out to develop a similar assay that now uses transiently transfected cells expressing GFP-NFAT or GFP-snurportin1 (GFP-SPN1) as CRM1-dependent export cargoes and mCherry-emerin as a marker that allows gating for transfected cells in the subsequent flow cytometry analysis. Other combinations of cargoes and markers are possible. Nuclear accumulation of the export cargoes is achieved by treating the cells with the selective CRM1-inhibitor leptomycin B (LMB). For GFP-NFAT, ionomycin is used as an additional import-inducing agent (*see* **Note 1**). The marker protein should remain associated with the cells upon digitonin permeabilization. Emerin is a protein that localizes to the inner nuclear membrane. In our transfection assays, it is also found in stable association with other membranes, e.g., the endoplasmic reticulum. Alternatively, other membrane proteins can be used that should, of course, not interfere with nucleocytoplasmic transport (*see* **Note 2**).

Using the same principles described here, it should be possible to monitor nuclear export in vitro of virtually any CRM1-dependent export cargo (*see* **Note 3**). Several aspects of the methods are similar to those described in a previous chapter in *Methods in Molecular Biology* [18], where we presented the method using cells that stably express GFP-NFAT.

2 Materials

2.1 Plasmids and Cells

1. NFAT reporter: for the specific construct we use, the coding sequence of human NFAT was amplified from pSH107c (kindly provided by G.R. Crabtree, Stanford, CA) and cloned via *Hind*III and *Bam*HI into a pS65T-C1 (Clontech)-based eukaryotic expression vector, behind the coding sequence of a modified version of GFP (S65T; [10]).

2. SPN1 reporter: for the specific construct we use, the coding sequence of human SPN1 was amplified from GST-SPN1 [19] and cloned via *Xho*I and *Eco*RI into the pEGFP-C1 vector [20].
3. Emerin reporter: for the specific construct we use, the coding sequence of human emerin was amplified from emerin-GFP (kindly provided by Eric Schirmer, Edinburgh) and cloned via *Xho*I and *Bam*HI into the pmCherry-C1 vector (Clontech), yielding mCherry-emerin.
4. Ran expression plasmid: we use human wild-type Ran in pET11d.
5. CRM1 expression plasmid: we use a His-tagged human CRM1 in the pQE60 vector (kindly provided by I. Mattaj, EMBL, Heidelberg, Germany).
6. Cells for transport assay: HeLa p4 cells (P4.R5 MAGI) [21] were obtained through the NIH AIDS Reagent Program (Division of AIDS, NIAID, NIH, Dr. Nathaniel Landau; *see* **Note 4**).
7. Growth medium for assay cells: Dulbecco's Modified Eagle's medium (DMEM), containing 10 % fetal bovine serum (FBS), penicillin (100 U/mL), and streptomycin (100 μg/mL).
8. Cells for cytosol preparation: HeLa suspension cells (CSH HeLa strain, obtained from Dr. Frauke Melchior, Heidelberg).
9. Growth medium for cytosol preparation: Joklik's modified S-MEM containing 10 % fetal bovine serum (FBS), 100 U/mL and 100 μg/mL penicillin streptomycin.

2.2 Buffers for Assay and Preparation of Cytosol

1. Lysis buffer: 5 mM HEPES-KOH, pH 7.3, 10 mM KOAc, 2 mM $Mg(OAc)_2$, 2 mM dithiothreitol (DTT), 0.1 mM phenylmethyl sulfonyl fluoride (PMSF), 1 μg/mL of each leupeptin, pepstatin, and aprotinin.
2. Transport buffer: 20 mM HEPES-KOH, pH 7.3, 110 mM KOAc, 2 mM $Mg(OAc)_2$, 1 mM EGTA, DTT, and protease inhibitors as above (*see* **Note 5**).

2.3 Buffers and Reagents for Preparation of Recombinant Transport Factors

1. Transport buffer: as above.
2. Ran buffer: 50 mM HEPES-NaOH, pH 8.0, 500 mM NaCl, 5 mM $MgCl_2$, 0.1 mM PMSF, and 1 μg/mL of each leupeptin, pepstatin, and aprotinin.
3. CRM1 buffer: 50 mM HEPES NaOH, pH 7.4, 500 mM NaCl, 2 mM $MgCl_2$, 20 mM β-mercaptoethanol, 0.1 mM PMSF, and 1 μg/mL of each leupeptin, pepstatin, and aprotinin.

   CRM1 low salt buffer: 50 mM HEPES-NaOH, pH 7.4, 50 mM NaCl, 2 mM $Mg(OAc)_2$, 2 mM DTT.

4. CRM1 high salt buffer: 50 mM HEPES-NaOH, pH 7.4, 500 mM NaCl, 2 mM $Mg(OAc)_2$, 2 mM DTT.
5. PBS: 137 mM NaCl, 2.7 mM KCl, 10 mM Na_2HPO_4, 1.8 mM KH_2PO_4.
6. Imidazole 5 M stock.
7. Standard bacterial culture reagents.
8. DEAE Sepharose.
9. Preparative S75 column (see GE Healthcare for specifications to match if another is used).
10. PD-10 Desalting Columns or a HiPrep 26/10 Desalting (Sephadex G-25) type column (e.g., GE Healthcare, though any similar desalting column will do).
11. MonoQ anion exchange column (see GE Healthcare for specifications to match if another is used).
12. Isopropylthio-β-Δ-galactoside (IPTG) 1 M stock.
13. Ammonium sulfate.
14. GDP 100 mM stock.

2.4 Additional Reagents for Transport Assays

1. ATP-regenerating system: (A) 100 mM ATP in 20 mM HEPES, 100 mM $Mg(OAc)_2$ (*see* **Note 6**). Adjust pH to 7.4 with NaOH. (B) 80 mg/mL creatine phosphate in H_2O. (C) 2000 U/mL creatine phosphokinase in 50 % glycerol, 20 mM HEPES, pH 7.4. Store all components at −20 °C and mix 2A:2B:1C (ATP:creatine phosphate:creatine phosphokinase) before use.
2. Leptomycin B (LMB; Enzo Life Sciences): prepare 370 μM stock (for preincubation of permeabilized cells) and 10 μM working dilution in ethanol (for treatment of intact cells); store at −20 °C.
3. Ionomycin: prepare stock (1 mM in DMSO) and store at −80 °C.
4. Digitonin: prepare stock (1 % in DMSO) and store at −20 °C.
5. Trypan blue: use 1:1 with cell suspension.
6. Oligonucleotide (*see* **Note 7**): dissolve desalted oligonucleotides (5′AGAG GAAAATTTGTTTCATA and 5′TATGAAACAAA TTTTCCTCT), each at 200 μM, in 40 mM Tris–HCl, pH 7.4, 20 mM $MgCl_2$, 50 mM NaCl, combine equal volumes and anneal by heating to 65 °C for 5 min and slow cooling to room temperature. Freeze in aliquots and store at −20 °C.
7. Wheat germ agglutinin (WGA): WGA, which inhibits most nucleocytoplasmic transport pathways, is dissolved in transport buffer at 2 mg/mL and stored at −80 °C.

3 Methods

3.1 Preparation of Cytosol

1. Grow HeLa suspension cells in spinner flasks to mid-log phase and collect them by centrifugation at 300 × *g* for 15 min (3–10 L can easily be handled). All of the following steps are performed on ice or at 4 °C.
2. Wash twice with PBS and once with transport buffer.
3. Resuspend in 1 volume of lysis buffer and swell for 10 min on ice.
4. Lyse the cells in a dounce homogenizer (*see* **Note 8**). After centrifugation at 1500 × *g* for 15 min, the supernatant is cleared by centrifugation at 120,000 × *g* for 1 h.
5. Dialyze overnight against transport buffer (10,000 K_d cutoff), freeze the resulting cytosol (~10 mg/mL) in aliquots in liquid nitrogen, and store at −80 °C.

3.2 Preparation of Recombinant Transport Factors

3.2.1 Ran

Various methods for the preparation of Ran have been described in the literature. We typically use a modified version of the one described by Melchior et al. [22]. Human wild-type Ran is expressed from pET11d in BL21-(DE3) cells.

1. Grow 2 L of a bacterial culture to an OD_{600} of 0.6 and induce expression with 0.5 mM isopropylthio-β-Δ-galactoside (IPTG) for 3 h at 37 °C. All subsequent steps are performed on ice or at 4 °C.
2. Harvest cells and resuspend them in 100 mL of Ran buffer, and lyse them. We find that the use of an EmulsiFlex-C3 (Avestin) for lysing produces reproducibly good protein, but other lysing methods should be able to be applied.
3. Clear lysate by centrifugation at 100,000 × *g* for 30 min.
4. Incubate the supernatant in batch with 70 mL DEAE Sepharose, equilibrated in Ran buffer, for 1 h with gentle agitation. Pour the slurry into a column and collect the flowthrough and approximately 100 mL of the subsequent wash (Ran buffer; wash until no protein can be detected).
5. Precipitate Ran from the DEAE flowthrough with ammonium sulfate: add salt gradually to 55 % saturation, stir for 2 h, and centrifuge for 20 min at 100,000 × *g*. Resuspend precipitate in 6–8 mL transport buffer containing 250 μM GDP, and clarify by centrifugation at 14,000 × *g* for 15 min.
6. Load onto a preparative S75 column, equilibrated in transport buffer. Collect 5-mL fractions and analyze them by SDS-PAGE.
7. Pool Ran-containing fractions, freeze in aliquots in liquid nitrogen, and store at −80 °C.

The final concentration of Ran (ca. 95 % pure) should be 2–3 mg/mL.

3.2.2 CRM1

The preparation of His-tagged CRM1 from pQE60-CRM1 (kindly provided by I. Mattaj, EMBL, Heidelberg, Germany) was originally described in [23].

1. Express the protein at 37 °C overnight in TG1 cells without induction by IPTG.
2. Harvest bacteria, wash them in PBS, and store cell pellets at −80 °C until purification.
3. Thaw and resuspend bacteria in cold CRM1 buffer, and lyse them. As for Ran we recommend an EmulsiFlex-C3 (Avestin). All following steps are performed on ice or at 4 °C.
4. Clear the lysate by centrifugation at 100,000 × *g* for 30 min.
5. Incubate the supernatant with Ni-NTA sepharose equilibrated in CRM1 buffer at 4 °C for 1 h.
6. Wash beads four times with 50 mL CRM1 buffer, each supplemented with 30 mM imidazole.
7. Transfer beads to an empty column and elute with CRM1 buffer containing 300 mM imidazole.
8. Change to CRM1 low salt buffer using a desalting column attached to an FPLC system.
9. Load the protein onto a MonoQ anion exchanger column equilibrated in CRM1 low salt buffer and elute with an increasing concentration of CRM1 high salt buffer.
10. Pool the peak fractions containing CRM1, change into transport buffer with desalting columns, and freeze in aliquots in liquid nitrogen. Store at −80 °C.

3.3 Transfections and Treatment of Cells

1. Grow HeLa p4 cells in DMEM medium in 24-well dishes.
2. Transfect cells with Lipofectamine (Life Technologies) or the calcium phosphate method [24], using 0.8 μg plasmid per ~10^5 cells per 24-well, with or without coverslips (*see* **Note 9**).
3. Incubate overnight and induce nuclear import of the reporter protein with 5 nM LMB for 90 min. For GFP-NFAT, also add 1 μM ionomycin for the last 30 min. The localization of GFP-NFAT and mCherry-emerin before and after treatment of cells with ionomycin/LMB is shown in Fig. 1. Continue as described in Subheading 3.4 or 3.5.

3.4 Analysis of Transport by Fluorescence Microscopy

Although the major advantage of our approach is the quantitative analysis by flow cytometry (*see* Subheading 3.5), we briefly outline the method for microscopic analysis as well.

1. Wash cells grown on coverslips with cold transport buffer and permeabilize with digitonin (final concentration ~70 μg/mL in transport buffer) on ice (*see* Fig. 1). The required amount of the detergent should be tested on a separate slide for each cell type, using trypan blue (*see* **Note 8**).

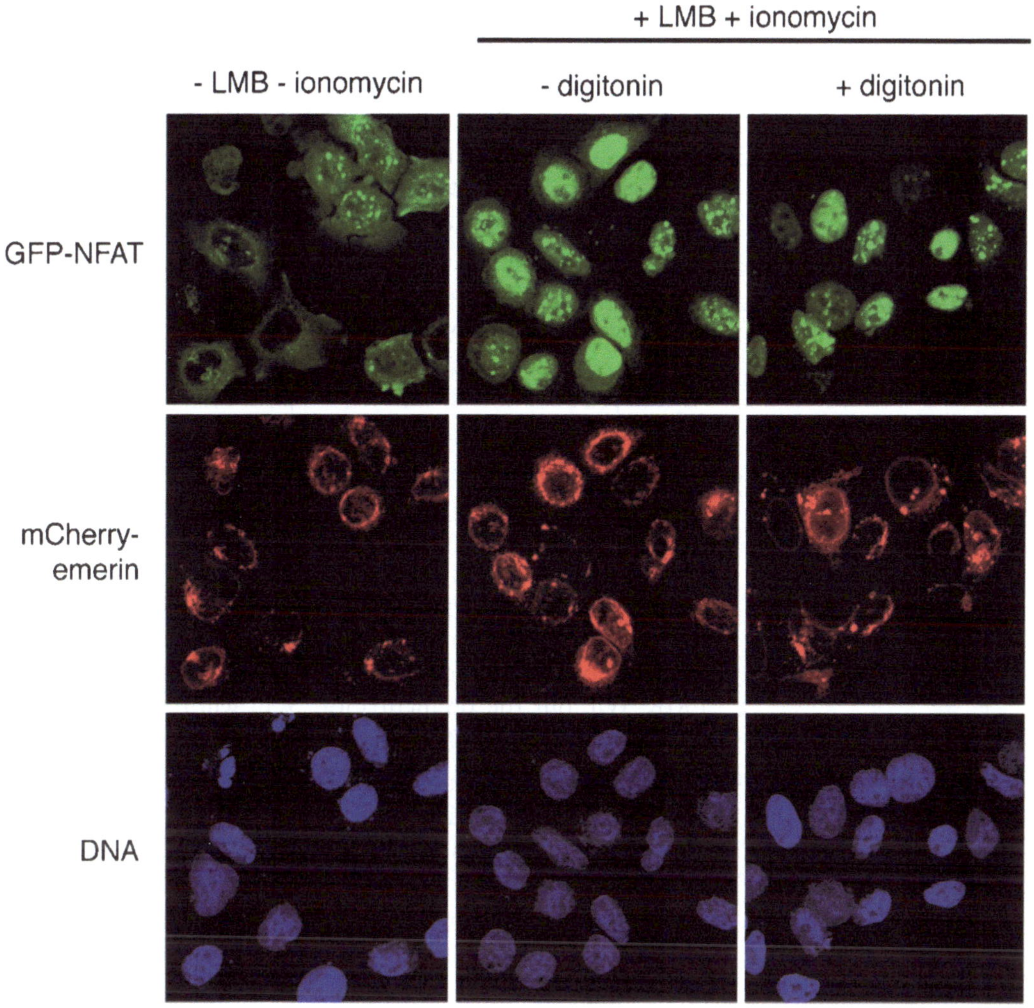

Fig. 1 Subcellular localization of GFP-NFAT and mCherry-emerin. HeLa p4 cells were transfected with plasmids coding for GFP-NFAT (*green*) and mCherry-emerin (*red*) and analyzed by confocal microscopy. GFP-NFAT localized to the cytoplasm or was homogenously distributed in the cells. Nuclear import of GFP-NFAT was induced by LMB and ionomycin. The nuclear localization of GFP-NFAT did not change upon permeabilization of the cells with digitonin. Note that the membrane association of mCherry-emerin did not vary in any of the tested conditions, confirming its suitability as a marker protein

2. Subject cells to preincubation and transport reaction, using reagents as outlined in Subheading 3.5, **steps 3** and **4**. Use ~40 μL solution per coverslip and incubate at 30 °C in a moisturized chamber.
3. Wash with cold transport buffer. Fix cells with 3.7 % formaldehyde in PBS.
4. Analyze cells by fluorescence microscopy and quantify images using local software.

3.5 Analysis of Transport by Flow Cytometry

1. Trypsinize cells: wash with PBS, add trypsin. Remove most of the trypsin solution immediately to minimize carryover of the protease. Incubate at 37 °C until cells detach. Collect cells in cold transport buffer containing 10 % fetal bovine serum (FBS) to inactivate the trypsin. The following steps are performed on ice or at 4 °C unless otherwise indicated. Centrifuge for 5 min at 300 × *g* and wash once in transport buffer.
2. Resuspend cells in transport buffer at 10^7/mL. Add digitonin to 70 μg/mL (0.7 μL of a 1 % stock per 10^6 cells). Leave on ice for 3 min and check permeabilization with trypan blue, using 5 μL of cells. Add more digitonin if less than 90–95 % of the nuclei appear blue. Wash with transport buffer and centrifuge as in **step 1**.
3. Preincubation: Resuspend cells at 10^7/mL in transport buffer and add 100 nM LMB and 25 μL ATP-regenerating system per mL of cell suspension. Incubate for 15 min in a 30 °C water bath. Wash cells twice with transport buffer. The preincubation results in depletion of transport factors like CRM1 that initially remain associated with the permeabilized cells. Furthermore, endogenous CRM1 will be modified by LMB. Without the preincubation, Ran will be the only rate-limiting factor for CRM1-dependent export [10].
4. Resuspend cells in transport buffer at 2×10^7/mL. The transport reaction (40 μL in a 5-mL flow cytometry tube) should contain: 10 μL of cells (200,000), 1 μL ATP-regenerating system, 0.5 μL annealed oligonucleotide for cells transfected with GFP-NFAT (final concentration: 1.25 μM). Add cytosol (~1 mg/mL usually yields optimal transport) and/or recombinant transport factors.
5. Incubate on ice or in a 30 °C water bath for 30 min.
6. Add 4 mL cold transport buffer. Centrifuge for 10 min at 300 × *g* and 4 °C. Remove most of supernatant by aspiration, leaving ~200 μL behind.
7. Analyze export of the GFP-cargo by flow cytometry. We use a FACSCanto II Cell Analyzer (Becton Dickinson), and typically count 5000–10,000 cells. For measuring GFP-fluorescence, gate the cells such that only mCherry-positive ones are analyzed. In most cases, the median fluorescence is best suited for statistical analysis of transport. Export reactions may be standardized by assigning a GFP-cargo fluorescence value of 100 to a 4 °C control (*see* **Note 10**).

Typical results for nuclear transport reactions as analyzed by flow cytometry are shown in Figs. 2 and 3. With respect to the efficiency of nuclear export of GFP-NFAT, they are very similar to those obtained using stably transfected cells [10, 18]. Cytosol, which should contain all relevant transport factors, strongly

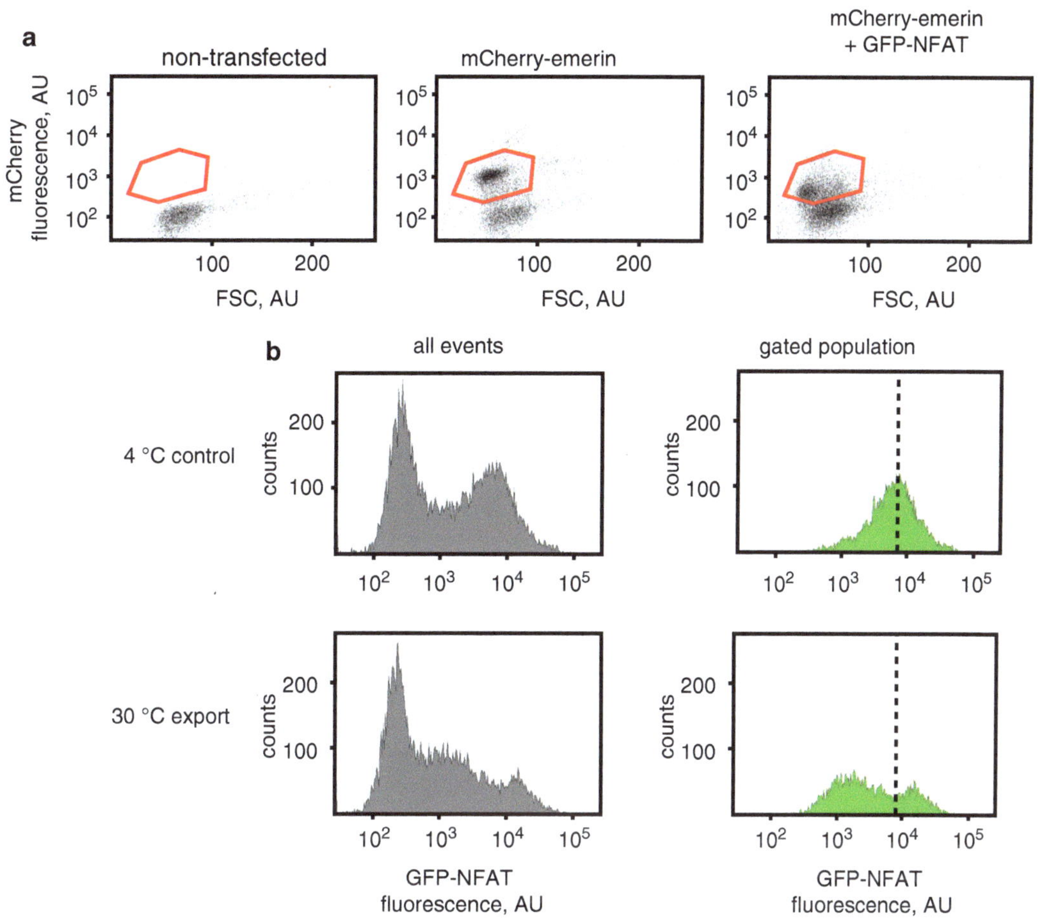

Fig. 2 Analysis of nuclear export by flow cytometry. HeLa p4 cells were transfected with plasmids coding for GFP-NFAT and/or mCherry-emerin. (**a**) Cell populations negative or positive for mCherry-emerin could be clearly distinguished by analyzing non-transfected cells and cells transfected with mCherry-emerin, alone or in combination with GFP-NFAT. A gate (*red area*) was set in a way that only mCherry-emerin-positive cells were included in further analysis. *FSC* forward scatter, *AU* arbitrary units. (**b**) Comparison of GFP-NFAT fluorescence in the total and the gated population. A prominent peak for non-transfected cells can be observed on the *left side* of the fluorescence intensity profile in the total population (*left panel*). Gating for the transfection marker mCherry-emerin (as described above) led to a simplified fluorescence intensity profile of GFP-NFAT, as most non-transfected cells were excluded from the analysis (*right panel*). The fluorescence intensity of nuclei was normalized to a 4 °C control sample (indicated by the *dashed line*) and was clearly reduced upon export of GFP-NFAT. Note the small remaining peak with a high GFP-NFAT fluorescence in the export reaction, which could result from cells that had not been sufficiently permeabilized and were therefore not accessible for export factors

promoted nuclear export (i.e., resulted in low levels of fluorescence). CRM1 and Ran were only active when added together, indicating that both factors are rate-limiting under our experimental conditions. The physiological significance of the results was validated using several control conditions. First, we included wheat germ agglutinin (WGA) in our reaction. WGA is a lectin that binds

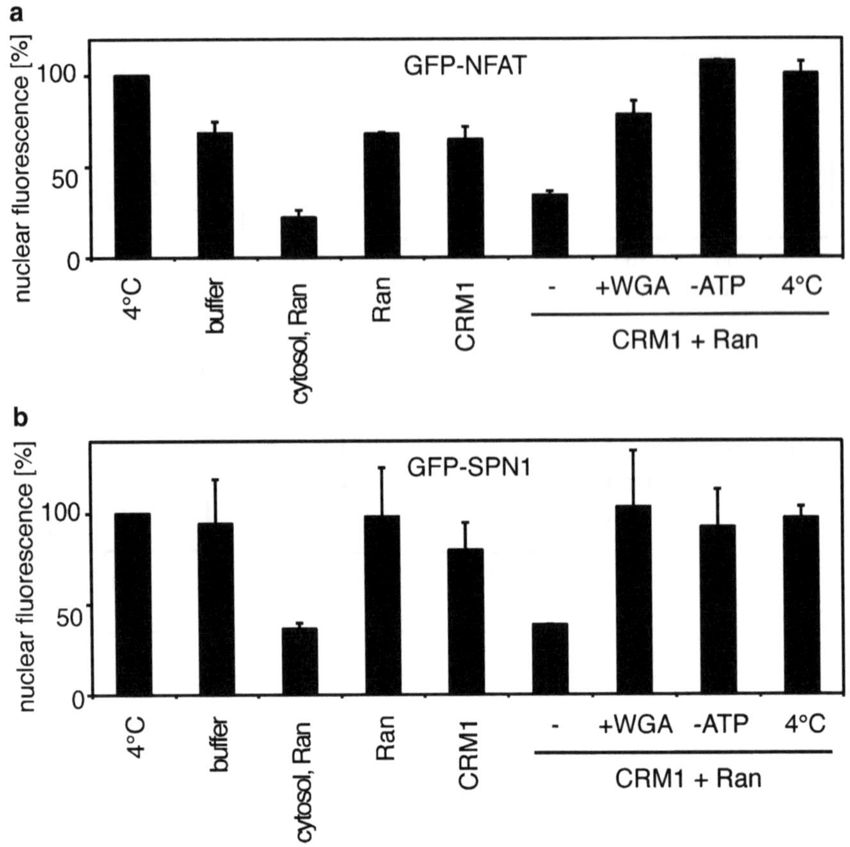

Fig. 3 CRM1 and Ran promote nuclear export of GFP-NFAT and GFP-SPN1. HeLa p4 cells were transfected with plasmids coding for GFP-NFAT or GFP-SPN1 and mCherry-emerin and subjected to nuclear export reactions in vitro. Nuclear fluorescence of GFP-NFAT (**a**) and GFP-SPN1 (**b**) was analyzed by flow cytometry. Total populations were gated for mCherry-emerin-positive cells and the median fluorescence intensities of samples incubated at 4 °C were set as 100 %. Reactions contained 0.7 mg/mL cytosol, 2 μM Ran, 300 nM CRM1, and/or 150 μg/mL WGA, as indicated. The average and standard deviation of two independent experiments are shown

to various *O*-glycosylated nucleoporins [25] and is known to inhibit the majority of nuclear transport pathways [26, 27]. Here, it clearly inhibited CRM1-dependent export of GFP-NFAT and GFP-SPN1. Next, reactions were performed in the absence of an ATP-regenerating system (–ATP) or at 4 °C. Under both conditions, the cells retained a high level of fluorescence. Taken together, these controls validate our in vitro assay as a transport system that faithfully reconstitutes nuclear export in transiently transfected, permeabilized cells.

4 Notes

1. For GFP-NFAT, ionomycin alone is sufficient to induce nuclear import. If LMB is used, cells must be washed thoroughly to prevent the drug from modifying CRM1 that is added to the reaction.
2. In our experiments, we did not observe an obvious influence of mCherry-emerin on CRM1-dependent export. New combinations of cargoes and transfection markers should be analyzed by fluorescence microscopy to verify efficient nuclear import of the cargo and unaltered localization of the marker upon LMB treatment as well as their unchanged localization upon permeabilization of the plasma membrane by digitonin.
3. Depending on the size of the CRM1-cargo, the protein might leak out of the nucleus in digitonin-permeabilized cells. This is not a problem for GFP-NFAT, which is >100 kDa in size. Smaller proteins can be fused to larger tags (e.g., GST or double-GFP) to minimize passive diffusion.
4. We typically use HeLa p4, because the transfection rate in these cells is very high.
5. EGTA inhibits the cellular phosphatase calcineurin, which dephosphorylates NFAT, resulting in nuclear import of the protein [12]. Thus, re-import of the reporter protein, which would complicate the analysis, is prevented under our conditions.
6. Use sodium-ATP. Lithium-ATP may interfere with export of GFP-NFAT [10].
7. This double-stranded oligonucleotide mimics the DNA-binding site of NFAT. It can stimulate export of GFP-NFAT about twofold, probably by promoting the release of GFP-NFAT from chromatin.
8. Check permeabilization with trypan blue, which stains the nuclei. As an alternative to dounce homogenization, the plasma membrane can be permeabilized with digitonin. The nuclear membrane remains intact, because of its lower level of cholesterol, as compared to the plasma membrane. Add 0.5–1 μL of a 10 % solution of digitonin in DMSO for 10^7 cells.
9. Transfection conditions are chosen such that essentially all mCherry-positive cells are also positive for our GFP-export cargo. In the final analysis by flow cytometry, this allows gating on mCherry-positive cells and measuring the residual GFP-fluorescence, assuming that even GFP-negative cells did contain the export cargo at the beginning of the reaction. Depending on the efficiency of the transfection, we obtain 15–30 % of total cells that express both, the red and the green

fluorescent protein (*see* Fig. 1). The calcium phosphate method should not be used for proteins like NFAT, whose intracellular localization is controlled by the cellular calcium level.

10. It is also possible to simultaneously monitor nuclear import of fluorescently labeled import substrates (e.g., Cy5-BSA-NLS; [10]). Furthermore, one may use propidium iodide to determine the DNA content of the cells. This can be done for nuclear import or export, allowing analysis of nuclear transport with respect to the cell cycle [28].

References

1. Fischer U, Huber J, Boelens WC, Mattaj IW, Lührmann R (1995) The HIV-1 Rev activation domain is a nuclear export signal that accesses an export pathway used by specific cellular RNAs. Cell 82:475–483
2. Wen W, Meinkoth JL, Tsien RY, Taylor SS (1995) Identification of a signal for rapid export of proteins from the nucleus. Cell 82:463–473
3. Görlich D, Kutay U (1999) Transport between the cell nucleus and the cytoplasm. Annu Rev Cell Dev Biol 15:607–660
4. Hutten S, Kehlenbach RH (2007) CRM1-mediated nuclear export: to the pore and beyond. Trends Cell Biol 17:193–201
5. Pemberton LF, Paschal BM (2005) Mechanisms of receptor-mediated nuclear import and nuclear export. Traffic 6:187–198
6. Turner JG, Dawson J, Cubitt CL, Baz R, Sullivan DM (2014) Inhibition of CRM1-dependent nuclear export sensitizes malignant cells to cytotoxic and targeted agents. Semin Cancer Biol 27:62–73
7. Santiago A, Li D, Zhao LY, Godsey A, Liao D (2013) p53 SUMOylation promotes its nuclear export by facilitating its release from the nuclear export receptor CRM1. Mol Biol Cell 24:2739–2752
8. Ishida N, Hara T, Kamura T, Yoshida M, Nakayama K, Nakayama KI (2002) Phosphorylation of p27Kip1 on serine 10 is required for its binding to CRM1 and nuclear export. J Biol Chem 277:14355–14358
9. Adam SA, Marr RS, Gerace L (1990) Nuclear protein import in permeabilized mammalian cells requires soluble cytoplasmic factors. J Cell Biol 111:807–816
10. Kehlenbach RH, Dickmanns A, Gerace L (1998) Nucleocytoplasmic shuttling factors including Ran and CRM1 mediate nuclear export of NFAT in vitro. J Cell Biol 141:863–874
11. Flanagan WM, Corthesy B, Bram RJ, Crabtree GR (1991) Nuclear association of a T-cell transcription factor blocked by FK-506 and cyclosporin A. Nature 352:803–807
12. Shibasaki F, Price ER, Milan D, McKeon F (1996) Role of kinases and the phosphatase calcineurin in the nuclear shuttling of transcription factor NF-AT4. Nature 382: 370–373
13. Paschal BM, Gerace L (1995) Identification of NTF2, a cytosolic factor for nuclear import that interacts with nuclear pore complex protein p62. J Cell Biol 129:925–937
14. Kehlenbach RH, Assheuer R, Kehlenbach A, Becker J, Gerace L (2001) Stimulation of nuclear export and inhibition of nuclear import by a Ran mutant deficient in binding to Ran-binding protein 1. J Biol Chem 276: 14524–14531
15. Kehlenbach RH, Dickmanns A, Kehlenbach A, Guan T, Gerace L (1999) A role for RanBP1 in the release of CRM1 from the nuclear pore complex in a terminal step of nuclear export. J Cell Biol 145:645–657
16. Hutten S, Flotho A, Melchior F, Kehlenbach RH (2008) The Nup358-RanGAP complex is required for efficient importin alpha/beta-dependent nuclear import. Mol Biol Cell 19:2300–2310
17. Roloff S, Spillner C, Kehlenbach RH (2013) Several phenylalanine-glycine motives in the nucleoporin Nup214 are essential for binding of the nuclear export receptor CRM1. J Biol Chem 288:3952–3963
18. Kehlenbach RH, Gerace L (2002) Analysis of nuclear protein import and export in vitro using fluorescent cargoes. Methods Mol Biol 189:231–245
19. Strasser A, Dickmanns A, Schmidt U, Penka E, Urlaub H, Sekine M, Lührmann R, Ficner R (2004) Purification, crystallization and preliminary crystallographic data of the m3G cap-

binding domain of human snRNP import factor snurportin 1. Acta Crystallogr D Biol Crystallogr 60:1628–1631

20. Waldmann I, Spillner C, Kehlenbach RH (2012) The nucleoporin-like protein NLP1 (hCG1) promotes CRM1-dependent nuclear protein export. J Cell Sci 125:144–154
21. Charneau P, Mirambeau G, Roux P, Paulous S, Buc H, Clavel F (1994) HIV-1 reverse transcription. A termination step at the center of the genome. J Mol Biol 241:651–662
22. Melchior F, Sweet DJ, Gerace L (1995) Analysis of Ran/TC4 function in nuclear protein import. Methods Enzymol 257:279–291
23. Guan T, Kehlenbach RH, Schirmer EC, Kehlenbach A, Fan F, Clurman BE, Arnheim N, Gerace L (2000) Nup50, a nucleoplasmically oriented nucleoporin with a role in nuclear protein export. Mol Cell Biol 20:5619–5630
24. Ausubel FM, Brent R, Kingston RE, Moore DD, Seidman JG, Smith JA, Struhl K (1994) Current protocols in molecular biology. Greene Publishing Associates and Wiley-Interscience, New York
25. Hanover JA, Cohen CK, Willingham MC, Park MK (1987) O-linked N-acetylglucosamine is attached to proteins of the nuclear pore. Evidence for cytoplasmic and nucleoplasmic glycoproteins. J Biol Chem 262:9887–9894
26. Dargemont C, Kuhn LC (1992) Export of mRNA from microinjected nuclei of Xenopus laevis oocytes. J Cell Biol 118:1–9
27. Yoneda Y, Imamoto-Sonobe N, Yamaizumi M, Uchida T (1987) Reversible inhibition of protein import into the nucleus by wheat germ agglutinin injected into cultured cells. Exp Cell Res 173:586–595
28. Kehlenbach RH, Gerace L (2000) Phosphorylation of the nuclear transport machinery down-regulates nuclear protein import in vitro. J Biol Chem 275: 17848–17856

Chapter 31

SPEED Microscopy and Its Application in Nucleocytoplasmic Transport

Jiong Ma, Joseph M. Kelich, and Weidong Yang

Abstract

In eukaryotic cells, the nuclear pore complexes (NPCs) selectively mediate the bidirectional trafficking of macromolecules between the cytoplasm and the nucleus. The selective barrier is formed by intrinsically disordered phenylalanine–glycine (FG) nucleoporins anchored on the wall of the submicrometer NPC, which allows for passive diffusion and facilitated translocation through the nuclear pore. Dysfunction of nucleocytoplasmic transport has been associated with many human diseases. However, due to the technical challenge of imaging the native tomography of the FG-nucleoporin barrier and its interactions with transiting molecules in the native NPC, the precise nucleocytoplasmic transport mechanism remains unresolved. To refine the transport mechanism, single-molecule fluorescence microscopy methods have been employed to obtain the transport kinetics and the spatial transport route of individual fluorescent molecules through the NPC. In this method paper, we particularly highlight a newly developed high-speed super-resolution three-dimensional microscopy approach, termed as SPEED (single-point edge-excitation subdiffraction) microscopy, and its application in characterizing nucleocytoplasmic transport.

Key words Nuclear pore complex (NPC), Transport receptor, Nucleocytoplasmic transport, Single-molecule tracking, Super-resolution microscopy, Single-point edge-excitation subdiffraction microscopy (SPEED)

1 Introduction

The bidirectional trafficking of proteins and genetic material across the double-membrane nuclear envelope is mediated by the nuclear pore complex (NPC) in eukaryotic cells. A highly selective permeable barrier formed by phenylalanine–glycine (FG) nucleoporins (FG-Nup) in the NPC allows for two transport modes: passive diffusion of signal-independent small molecules (<40 kDa) and transport receptor-facilitated translocation of cargos carrying a nuclear localization sequence (NLS) or nuclear export signal (NES) [1–3]. Strict regulation of nucleocytoplasmic transport is crucial for cell survival, differentiation, growth, and other cellular activities essential for propagation and health [4, 5]. However, due to limited knowledge regarding the native configuration and conformation

Sue Shackleton et al. (eds.), *The Nuclear Envelope: Methods and Protocols*, Methods in Molecular Biology, vol. 1411,
DOI 10.1007/978-1-4939-3530-7_31, © Springer Science+Business Media New York 2016

of the FG-Nup barrier as well as the transient interactions undergone between the transiting molecules and the FG-Nup barrier in the NPC, the precise nucleocytoplasmic transport mechanism remains unresolved [6–11].

To further understand the mechanism of nucleocytoplasmic transport, single-molecule fluorescence microscopy has been developed and applied into the studies of nucleocytoplasmic transport. Back in 2004, Yang and Musser developed narrow-field epifluorescence microscopy and utilized it to determine the transport time and efficiencies of the passive and facilitated transport thought the NPC in a digitonin-permeabilized cell system [12–14]. Soon after, several other single-molecule fluorescence approaches have been developed to study nucleocytoplasmic transport as well [15–17]. These pioneering studies have provided great insight into the dwell times, the diffusion patterns, and the transport efficiencies of macromolecules with different sizes as they transport through the NPCs, though the spatial locations of individual transiting molecules were obtained mostly in one dimension along the nucleocytoplasmic transport axis.

Here, we provide a very detailed introduction of a newly developed high-speed super-resolution microscopy method, named single-point edge-excitation subdiffraction (SPEED) microscopy, and also the experimental details of preparing permeabilized cell systems and fluorescent candidates for single-molecule tracking through the native NPC. The technical advances of SPEED microscopy include: first, the optical setup of SPEED microscopy endures a light-diffraction-limit illumination volume and a higher signal-to-noise ratio for single-molecule detection in cells, resulting in a spatiotemporal resolution of 8–12 nm and 0.4–2 ms for imaging transiting molecules through individual NPCs; second, the 2D to 3D deconvolution algorithm enables a final demonstration of 3D pathways converted from the 2D super-resolution spatial locations of various transiting molecules with different sizes [18, 19]. Finally, with the new method, 3D distinct transport routes of passive and facilitated transport have been successfully obtained in native NPCs [18–21].

2 Materials

2.1 Preparation of Protein Cargos and Transport Receptors for Nucleocytoplasmic Transport

1. Bacterial cells for protein expression. Typically *E. coli* strains JM109 or BL21(DE3).
2. Plasmids encoding the protein of interest and typically also containing a T7 promoter that can be activated by IPTG (Isopropyl-D-1-thiogalactopyranoside). Plasmids also typically encode a 6× Histidine tag at either the N or C terminus of the

protein construct for purification. The histidine tag is preferred for transport receptors as longer tags sometimes interfere with function (see other methods on individual transport factors for specifics). Antibiotic resistance (typically ampicillin) is additionally incorporated into the plasmid.

3. LB (Luria Broth) media.
4. Appropriate antibiotic: e.g., 100 mg/mL stock of ampicillin.
5. 1 M IPTG.
6. Cell lysis buffer: we use CelLytic B (Sigma-Aldrich), though other buffers should work similarly. Adjust buffer composition according to requirements for different tags.
7. Protease inhibitors: 100× stocks at 200 μg/mL pepstatin A, 200 μg/mL leupeptin, 2 mg/mL trypsin inhibitor.
8. Ni-NTA resin (for histidine tags).
9. Chromatography columns plugged with glass wool.
10. Nickel pre-wash buffer: 50 mM Na-phosphate, 300 mM NaCl, 20 mM imidazole, pH 8.0.
11. Nickel wash buffer: 10 mM imidazole, 100 mM NaCl, pH 8.0.
12. Nickel elution buffer: 250 mM imidazole, 100 mM NaCl, pH 8.0.
13. 5× sample buffer (SB) for SDS-PAGE: 300 mM Tris 6.8, 25 % BME, 10 % SDS, 50 % glycerol.
14. SDS-PAGE polyacrylamide gels.
15. 1× PBS (137 mM NaCl, 2.7 mM KCl, 4.3 mM Na_2HPO_4, 1.47 mM KH_2PO_4).
16. MonoQ or Superdex 200 columns (e.g., Amersham Pharmacia).
17. Protein filter concentrators: e.g., we use Nanosep filters with a pore size of 10 kDa.
18. Reagents for determining protein concentration.
19. Cysteine-reactive organic dyes: e.g., Alexa fluor 555 or Alexa fluor 647 maleimide (Molecular Probes).
20. TCEP tris (2-carboxyethyl) to be used at a concentration of 0.5 % v/v.
21. Coupling buffer: 50 mM sodium phosphate, 150 mM NaCl.
22. 2-mercaptoethanol at a final concentration of 0.5 % for 10 min.
23. Dialysis system to remove free unbound dye.
24. Spectrophotometer (such as Nanodrop Thermo Scientific) to determine the concentration and labeling ratio of dye to protein in the final labeled protein mixture.

2.2 Preparation of the Permeabilized Cell System

1. HeLa cells stably expressing GFP fused to the NPC scaffold protein POM 121.
2. HeLa growth medium: DMEM, 10 % FBS, 1 % penicillin/streptomycin.
3. Standard tissue culture supplies.
4. Autoclaved glass slides.
5. Cover slips.
6. Silicone grease.
7. Diamond knife.
8. Filter paper.
9. 10 mL plastic syringe with a pipette tip.
10. Aluminum frame—temper, which serves to hold the slide in the microscope sample holder. This can be made by your local machine shop as described in Fig. 1.
11. Pointed tweezers.
12. Cotton-tipped swab.
13. 100 % ethanol.
14. Transport buffer: 20 mM Hepes, 110 mM KOAc, 5 mM NaOAc, 2 mM MgOAc, 1 mM EGTA, pH 7.3 (adjust with NaOH).
15. 40 mg/mL digitonin stock solution.
16. Polyvinylpyrrolidone (PVP; 360 kDa).

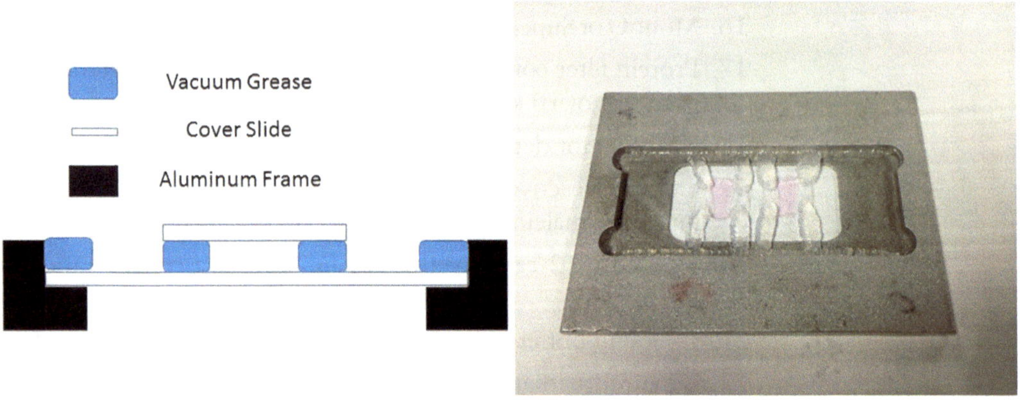

Fig. 1 Chamber setup for single-molecule experiments. The dimensions of the aluminum frame for the flow chamber are 7.5 cm by 5.5 cm. It is designed to hold a 2.5 cm × 7.5 cm glass slide. Each cover slip is cut to be approximately 6 mm × 6 mm and set upon grease two separated grease lines along the edges of the cover slip. This forms a flow chamber allowing liquid to move between the glass slide and cover slip. Ideally, two flow chambers fit well on one glass slide as shown in the figure. Each constructed flow chamber allows for a small volume of 25 μL of sample to flow through. In this way we can conserve on the amount of sample used and set up multiple conditions per slide

2.3 SPEED Microscopy Tracking of Single Fluorescent Molecules

1. SPEED microscope setup: Olympus IX81 equipped with a 1.4 NA 100× oil-immersion apochromatic objective (UPLSAPO 100×, Olympus), a 35 mW 633 nm He-Ne laser (our specific setup from Melles Griot), a 120 mW ArKr tunable ion laser (our specific setup from Melles Griot), an on-chip multiplication gain charge-coupled device camera (our specific setup from Cascade 128+, Roper Scientific).
2. Slidebook software package (Intelligent Imaging Innovations) for data acquisition and processing.
3. An optical chopper (Newport) that can be used to generate an on-off mode of laser excitation.
4. Optical filter (FFF555/646 Di01, Semrock).
5. Dichroic filter (Di01-R405/488/561/635-25x36, Semrock).
6. Emission filter (NF01-405/488/561/635-25X5.0, Semrock).
7. The overall setup is described in Fig. 2.
8. Cargo protein prepared as described in Subheading 3.1.

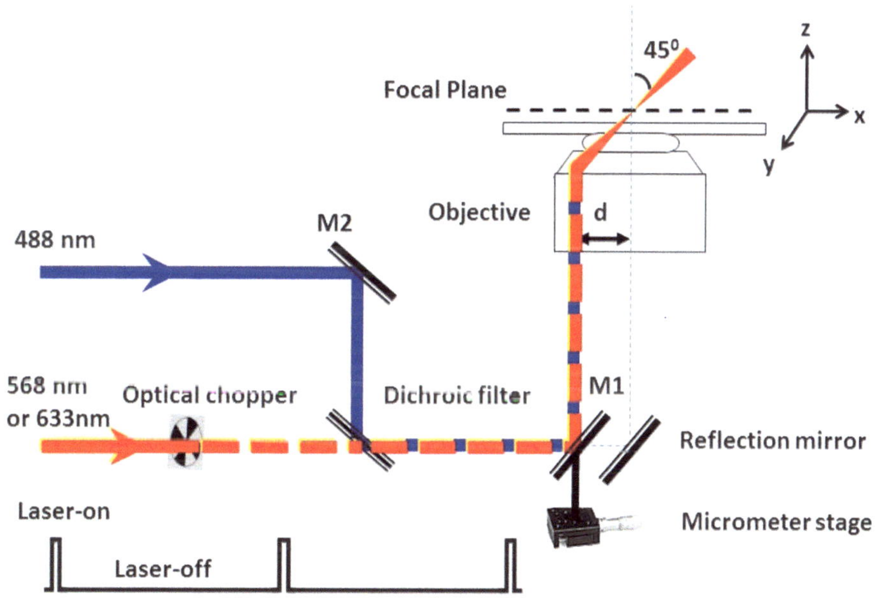

Fig. 2 Diagram of SPEED microscopy setup [18]. Diagram of the single-point edge excitation in SPEED microscopy. Equivalent to the illumination pattern of a 488 nm laser, the 568 nm or 633 nm laser is shifted about 237 μm from the center of the objective (*d*) by a micrometer stage to generate an inclined illumination volume at a 45° angle to the perpendicular direction. An optical chopper on the path of the 568 or 633 nm laser beam provides an on-off laser mode with a shorter laser-on time than the laser-off time. The longer laser-off time is sufficient for the transiting molecules in the NPC to escape from the illumination volume and for fresh cargo molecules with intact fluorescence to diffuse from the cytoplasm or the nucleus into the NPC. To ensure the imaging of complete transport events through the NPC, the photobleaching time of the single fluorescent molecules is longer than their nuclear transport time

9. 1 M GTP stock solution.
10. Purified transport factors for facilitated transport: Importin-α, Importin-β1, Ran, and NTF2. These can be prepared using the same strategy as for the cargo.
11. 10 kDa Dextran for passive transport.

2.4 2D to 3D Transform Process

1. Matrix calculator software.
2. Photoshop or similar image manipulation software.
3. Amira software.

3 Methods

3.1 Preparation of Protein Cargos and Transport Receptors for Nucleocytoplasmic Transport

1. Generate bacterial cells to express needed proteins: By means of a conventional heat-shock bacterial transformation, strains of *E. coli* (JM109 or BL21(DE3)) are to be transformed in order to overexpress the genes responsible for production of proteins of interest and are kept in stock at −80 °C. This protocol is specific for 6× Histidine tags under a promoter inducible with IPTG. When ready to grow cells for expression, inoculate a loopful or ~50 μL of a frozen stock of transformed *E. coli* cells into 5 mL of LB media with addition of 100 μg/mL ampicillin (or the appropriate measure if a different antibiotic resistance marker is used) and grow aerobically with shaking for 12–14 h at 37 °C at 225 rpm.
2. Amplify and induce the bacterial culture: Transfer the 5 mL saturated starter culture into 1 L of LB media containing 100 μg/mL ampicillin (or appropriate measure if a different antibiotic is used) and shake in a 37 °C incubator for roughly 5–6 h until an OD of ~0.6 is reached. Add 1 mM IPTG to activate protein production and continue incubation for another 5–6 h at 30 °C (*see* **Note 1**).
3. Pellet the culture in a cooled high-speed centrifuge at 20,000 × *g* for 15 min at 4 °C and discard the supernatant. Resuspend cells in 40 mL of CelLytic B (Sigma-Aldrich). The volume should be based on 10 mL per 1 g of pellet.
4. Add protease inhibitors fresh at 1:100 (400 μL each) to final concentrations 2 μg/mL pepstatin A, 2 μg/mL leupeptin, and 20 μg/mL trypsin inhibitor to inhibit protein degradation from the *E. coli* lysate.
5. For histidine tagged plasmids prepare 1 mL of Ni-NTA resin by removing the ethanol it is stored in via centrifugation at 10,000 × *g* for 1 min followed by aspiration of the supernatant. Then resuspend the resin with Cell lytic B.

6. Repeat the pelleting and resuspension from **step 5** to further wash the resin three times.
7. Add the equilibrated Ni-NTA resin-Cell Lytic B solution to the protein lysate.
8. Agitate the mixture in the dark for 30 min in a cold room (~4 °C) to allow an adequate time for the binding reaction between the histidine tag and Ni-NTA beads.
9. Centrifuge at 20,000 × *g* for 10–20 min in the cooling centrifuge at 4 °C.
10. Discard the supernatant and prepare a chromatography column plugged with glass wool.
11. Resuspend the pellet from **step 9** containing the Ni-NTA beads and bound protein in 20 mL of pre-wash buffer and pour into the column.
12. Perform column chromatography as follows: the first fraction of 12 fractions is to be the resuspended protein in 20 mL prewash buffer, the second is 10 mL of wash buffer, and finally the fourth and the remaining fractions are each 0.5 mL of elution buffer.
13. Prepare an SDS-PAGE polyacrylamide gel and prepare the samples by mixing 4 μL of a fraction and 1 μL of 5× SB. Before loading, boil samples for 5 min. After loading, run the gel at 120 V for 60 min. The fractions that contain the highest concentration of protein and yielding the purest product can be identified. Depending on the protein of your interest, the bands should be located respectively to the size of a protein using a protein standard as a reference.
14. Further purify the purest fraction with MonoQ and/or Superdex 200 (e.g., Amersham Pharmacia) to ensure purity above 90 % (*see* **Note 2**).
15. Concentrate proteins by spinning the most concentrated fractions through a filter concentrator (e.g., Nanosep filter) with a pore size of 10 kDa at 10,000 rpm (14,000 × *g*) in a microfuge for 5 min, washing it with 1× PBS buffer every time.
16. Determine the final protein concentration utilizing a protein concentration assay (e.g., the BSA assay) for the subsequent protein labeling procedure (**steps 16** and **17**) if cargo and to adjust concentrations for use in the assay if transport factors.
17. For protein cargos containing surface cysteine residues labeling can be accomplished with an excess of a cysteine-reactive organic dye, e.g., Alexa fluor 555 or Alexa fluor 647 maleimide (Molecular Probes). First reduce the protein with pure TCEP (tenfold molar excess to protein) for 20 min to prepare surface cysteines for binding.

18. Next add the maleimide dye in 20-fold molar excess for 2 h at room temperature in coupling buffer.
19. Quench reactions with 0.5 % 2-mercaptoethanol for 10 min.
20. Dialyze the proteins to remove the free unbound dye (*see* **Notes 3** and **4**).
21. Determine the concentration and labeling ratio of final labeled protein by spectrometry utilizing absorption at 280 nm and the corresponding absorbance for the dye used (*see* **Note 5**).

3.2 Preparation of the Permeabilized Cell System

1. HeLa cells stably expressing GFP fused to the NPC scaffold protein POM 121 are used for single-molecule nuclear transport assays. Start a fresh culture of HeLa cell line from a stock by thawing at 37 °C and inoculating into a 25 cm^2 culture flask with 5 mL HeLa cell medium. Incubate the cells at 37 °C within a 5 % CO_2 incubator for 24 h and split the cells at least three times to ensure good health.
2. One day before the experiment, plate cells intended for a single-molecule experiment on an autoclaved glass slide placed in a sterile Petri dish containing HeLa cell medium. Place these Petri dishes with the glass slides into the incubator and keep for 24 h prior to the experiment.
3. On the day of the experiment, construct flow chambers by adding a top cover slip together with two lines of silicone grease as spacers. The top cover slips are constructed as small slivers of glass cover slip cut with a diamond knife to be approximately 6 mm^2 and are placed upside down on filter paper. Place two lines of silicone grease finely along the edges of the intact glass slides with a 10 mL plastic syringe that has a pipette tip tightly lodged (to avoid leakage) in the opening to carefully control the size of the grease bed (*see* **Note 6**).
4. Drip-dry the HeLa cell-coated slide and mount with grease as glue in a home-machined aluminum frame—temper, which serves to hold the slide in the microscope sample holder (Fig. 1).
5. Pick up the top cover slips with the two lines of grease using pointed tweezers and then invert them over the cell-coated slide, forming a flow chamber. Make a relatively firm attachment by pressing gently over the greased edges of the cover slip (Fig. 1).
6. Add HeLa cell medium to the flow chamber to prevent the cells from drying out. Solution flow is promoted by a small piece filter paper to absorb the fluid out of the exit side of the flow chambers. With steady hands, the solution can be added simultaneously by micropipette to one side of the flow chamber with one hand, while being absorbed on the other side by filter paper (Fig. 1).

7. After the construction, wash the bottom of the slide with a wet cotton-tipped swab, twice with water, and then twice with 100 % ethanol. Then allow to air-dry.
8. Once the slide is mounted on the microscope, wash the cells two times with 25 μL Transport buffer (*see* **Note 7**).
9. Permeabilize the cells for ~2 min with 25 μL 40 μg/mL digitonin in Transport buffer and wash two times again with 25 μL Transport buffer supplemented with 1.5 % polyvinylpyrrolidone (PVP; 360 kDa) (*see* **Note 8**).

3.3 SPEED Microscopy Tracking of Single Fluorescent Molecules

1. The SPEED microscope setup is described in Fig. 2. GFP and Alexa Fluor 647 fluorescence are excited by 488 nm and 633 nm lasers respectively. The two lasers are combined by an optical filter, collimated and focused into an overlapped illumination volume in the focal plane. The green and red fluorescence emissions are collected by the same objective, filtered by a dichroic filter and an emission filter, and imaged by identical CCD cameras. As shown in Fig. 2, equivalent to the illumination pattern of a 488-nm laser, the 633- or 568 nm laser is shifted about 237 μm (*d*) by a micrometer stage off the center of the objective to generate an inclined illumination point spread function at 45° to the perpendicular direction. An optical chopper on the path of the laser beam provides an on–off laser mode (*see* **Note 9**).
2. Add cargo and transport proteins. For import experiments in permeabilized cells, the cargo protein and essential transport cofactors must be added together with transport buffer and 1.5 % PVP. Typical import cargo reactions contain 0.5 nM Alexa labeled cargo, 1 mM GTP, 0.5 μM Importin-α, 0.5 μM Importin-β1, 2 μM Ran, and 1 μM NTF2 for facilitated transport and 0.5 nM of dextran or other small molecules for passive transport (*see* **Note 10**).
3. Focus the microscope to the equator of the NE. Target one NPC at the far end of the equator.
4. First image a single NPC with a 10 μW 488 nm laser for 1 s. The fluorescent spot of NPC can then be fitted to a 2D elliptical Gaussian function to determine the centroid position of the NPC (*see* **Note 11**).
5. Photobleach the autofluorescence utilizing a 633 nm laser for 60 s until completely dark.
6. Set the 633 nm laser to 2–5 mW and the chopper with 2 Hz and 1:10 on-off ratio.
7. Set the intensification of the CCD to maximum and reduce the detection region to 128 pixels × 20 pixels. Use the "Stream" mode and set the exposure time to 0. With this setup, the CCD camera works on the speed of 400 μs/frame.

8. The prepared sample from **step 2** is then added to the flow chamber at a volume of 25 μL and pulled through the chamber by applying a wick of filter paper on the opposite side (*see* **Note 12**). Be careful not to absorb too much of the mixture leaving the flow chamber dry.
9. As soon as the cargo molecules are added, take a series of videos of single-molecule translocations for 5 min per cell (*see* **Note 13**).
10. An example typical trajectory is seen in Fig. 3a. The fluorescent spot is fitted to a 2D symmetrical Gaussian function with the peak position representing the spatial location of the labeled cargo molecule. The data can then be plotted onto a 2-dimensional Cartesian coordinate system (Fig. 3b) (*see* **Note 15**).

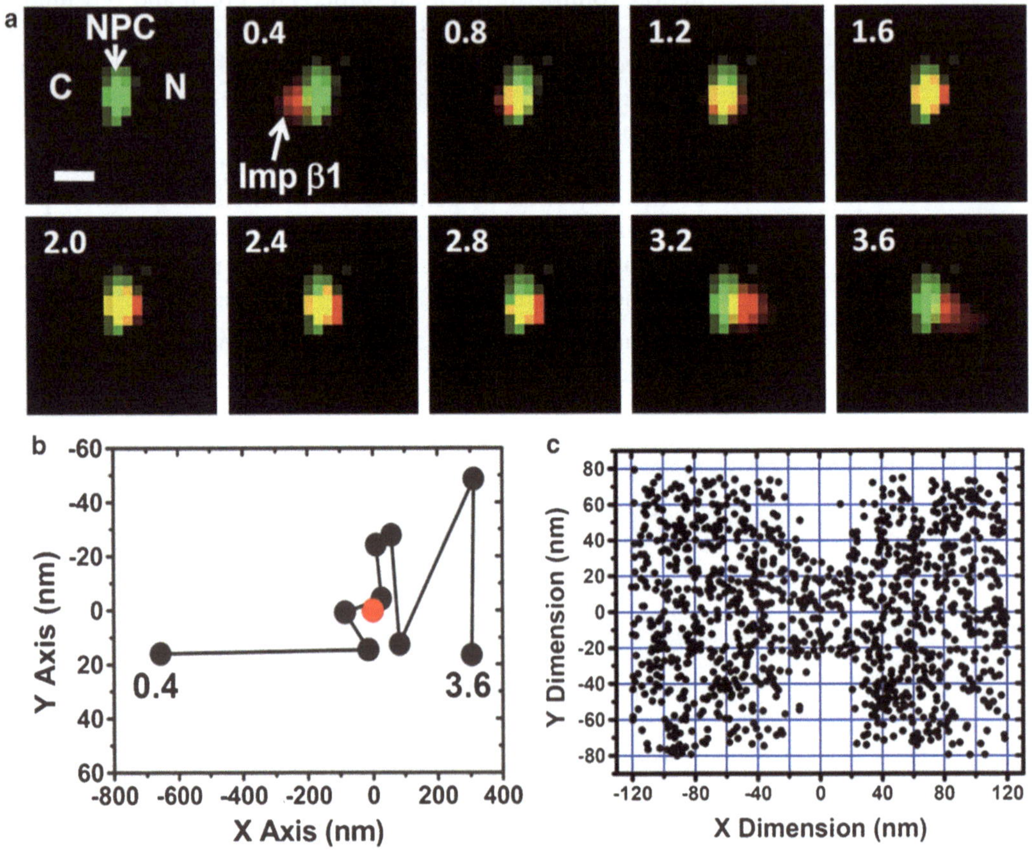

Fig. 3 Single-molecule trajectories and 2D spatial locations of Importin-β1 (Imp β1) in single NPCs [18]. (**a**) A typical nuclear transport event of Imp β1 molecules from the cytoplasm to the nucleus. First, a single GFP-NPC (*green spot*) was visualized in the illumination volume. Then, a single fluorescent Imp β1 molecule (*red spot*) entered the illumination volume, starting in the cytoplasm (C), interacting with the NPC, and entering the nucleus (N). *Numbers* denote time in milliseconds. (Scale bar, 1 μm.) (**b**) Single-molecule trajectories of the transport event in (**a**). Based on the centroid (*red dot*) and the dimensions of the NPC, the Imp β1 molecule was determined to interact with the NPC from 0.8 to 2.8 ms. (**c**) Superimposed plots of 1093 spatial localizations of single Imp β1 molecules located primarily within a rectangular area of 240 × 160 nm

11. Combine all collected trajectories and generate the 2D localization distributions via histograms (Fig. 3c) (*see* **Note 14**).
12. Repeat **steps 3–11** for a different NPC in another cell until a suitable number of locations are obtained (*see* **Note 15**).

3.4 2D to 3D Transform Process

1. When over 1000 localizations are obtained, select one region along the *x*-dimension, for example from −20 to 30 nm relative to the obtained NPC centroid (Fig. 4b). Histograms indicating the number of the locations obtained in 5 nm increments along the *y*-dimension must be generated for the chosen *x*-dimensional region (Fig. 4c) (*see* **Note 16** for determining *x*-dimensional regions).
2. In order to obtain the *r*-dimensional data corresponding to the radial dimension of the NPCs cylindrical shape, the *y*-dimensional data is then placed into the vector $F(yj)$. Use the Matrix calculator (Fig. 6) to calculate the vector $f(ri)$, which corresponds with the relative density distributions in the *r*-dimension seen in Fig. 4d. $F(yj)$ is the total interaction sites at the position yj; $f(ri)$ is the density of the interaction sites at ri; Δr is the bin size.

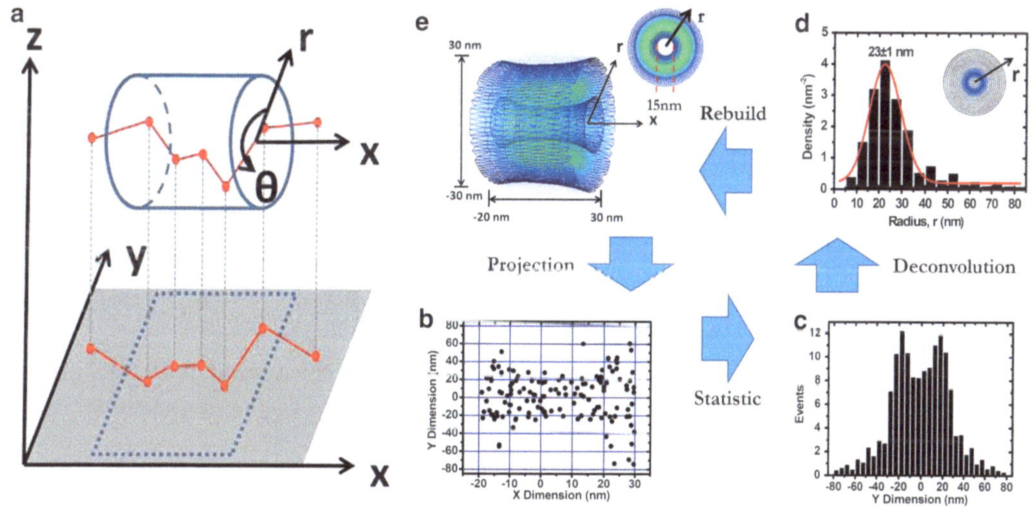

Fig. 4 Diagram of the 2D to 3D transform process [18]. (**a**) A diagram to show that the obtained 2D spatial locations of Imp β1 are a projection effect of the actual 3D spatial locations of Imp β1 in the *xy* plane. (**b**) Two-dimensional spatial locations of Imp β1 in the central scaffold region of the NPC. (**c**) Histogram of Imp β1 locations in the central scaffold region in the *y*-dimension. (Bin size, 5 nm.) (**d**) The obtained histogram of spatial densities along the radius (*r*) at the cross section of NPC in the central scaffold. The radii were used to plot concentric rings. The darker the ring, the higher the density. (Bin size, 5 nm.) (**e**) The obtained 3D spatial densities of Imp β1 (*blue shaded region* and *isolated surface lines*, *brighter green color* indicates higher density) in the central pore

$$\begin{pmatrix} F(r_1) \\ F(r_2) \\ F(r_3) \\ F(r_4) \\ F(r_5) \\ F(r_6) \\ F(r_7) \\ F(r_8) \\ F(r_9) \\ F(r_{10}) \\ F(r_{11}) \\ F(r_{12}) \\ F(r_{13}) \\ F(r_{14}) \\ F(r_{15}) \\ F(r_{16}) \\ F(r_\infty) \end{pmatrix} = \Delta r^2 \cdot \left[\begin{array}{ccccccccccccccccc} 0.8660 & 1.0705 & 1.0216 & 1.0106 & 1.0063 & 1.0042 & 1.0030 & 1.0022 & 1.0017 & 1.0014 & 1.0011 & 1.0010 & 1.0008 & 1.0007 & 1.0006 & 1.0005 & 20.0078 \\ 0 & 1.3229 & 1.2752 & 1.1100 & 1.0616 & 1.0398 & 1.0279 & 1.0207 & 1.0616 & 1.0127 & 1.0104 & 1.0086 & 1.0073 & 1.0062 & 1.0054 & 1.0047 & 20.0705 \\ 0 & 0 & 1.6583 & 1.4642 & 1.2076 & 1.1242 & 1.0840 & 1.0610 & 1.0465 & 1.0367 & 1.0297 & 1.0246 & 1.0207 & 1.0176 & 1.0152 & 1.0133 & 20.1965 \\ 0 & 0 & 0 & 1.9365 & 1.6342 & 1.3027 & 1.1888 & 1.1316 & 1.0978 & 1.0759 & 1.0608 & 1.0499 & 1.0417 & 1.0355 & 1.0305 & 1.0265 & 20.3875 \\ 0 & 0 & 0 & 0 & 2.1796 & 1.7892 & 1.3933 & 1.2525 & 1.1799 & 1.1361 & 1.1071 & 1.0869 & 1.0720 & 1.0608 & 1.0520 & 1.0451 & 20.6459 \\ 0 & 0 & 0 & 0 & 0 & 2.3979 & 1.9322 & 1.4794 & 1.3144 & 1.2277 & 1.1746 & 1.1391 & 1.1139 & 1.0952 & 1.0809 & 1.0697 & 20.9750 \\ 0 & 0 & 0 & 0 & 0 & 0 & 2.5981 & 2.0656 & 1.5613 & 1.3744 & 1.2748 & 1.2130 & 1.1712 & 1.1413 & 1.1189 & 1.1017 & 21.3798 \\ 0 & 0 & 0 & 0 & 0 & 0 & 0 & 2.7839 & 2.1911 & 1.6394 & 1.4324 & 1.3208 & 1.2509 & 1.2032 & 1.1688 & 1.1429 & 21.8667 \\ 0 & 0 & 0 & 0 & 0 & 0 & 0 & 0 & 2.9580 & 2.3098 & 1.7143 & 1.4884 & 1.3656 & 1.2881 & 1.2349 & 1.1962 & 22.4446 \\ 0 & 0 & 0 & 0 & 0 & 0 & 0 & 0 & 0 & 3.1225 & 2.5308 & 1.7862 & 1.5427 & 1.4094 & 1.3247 & 1.2662 & 23.1256 \\ 0 & 0 & 0 & 0 & 0 & 0 & 0 & 0 & 0 & 0 & 3.4278 & 2.5308 & 1.8554 & 1.5953 & 1.4520 & 1.3606 & 23.9273 \\ 0 & 0 & 0 & 0 & 0 & 0 & 0 & 0 & 0 & 0 & 0 & 3.4278 & 2.6344 & 1.9222 & 1.6463 & 1.4936 & 24.8757 \\ 0 & 0 & 0 & 0 & 0 & 0 & 0 & 0 & 0 & 0 & 0 & 0 & 3.5707 & 2.7341 & 1.9868 & 1.6959 & 26.0125 \\ 0 & 0 & 0 & 0 & 0 & 0 & 0 & 0 & 0 & 0 & 0 & 0 & 0 & 3.7081 & 2.8303 & 2.0494 & 27.4122 \\ 0 & 0 & 0 & 0 & 0 & 0 & 0 & 0 & 0 & 0 & 0 & 0 & 0 & 0 & 3.8406 & 2.9233 & 29.2361 \\ 0 & 0 & 0 & 0 & 0 & 0 & 0 & 0 & 0 & 0 & 0 & 0 & 0 & 0 & 0 & 3.9686 & 32.0314 \\ 0 & 0 & 0 & 0 & 0 & 0 & 0 & 0 & 0 & 0 & 0 & 0 & 0 & 0 & 0 & 0 & 36 \end{array}\right] \begin{pmatrix} f(r_1) \\ f(r_2) \\ f(r_3) \\ f(r_4) \\ f(r_5) \\ f(r_6) \\ f(r_7) \\ f(r_8) \\ f(r_9) \\ f(r_{10}) \\ f(r_{11}) \\ f(r_{12}) \\ f(r_{13}) \\ f(r_{14}) \\ f(r_{15}) \\ f(r_{16}) \\ f(r_\infty) \end{pmatrix}$$

Fig. 5 Transform calculation Function 1 (adapted from ref. 18)

3. Normalize the relative density data with the thickness of the selected *x*-dimensional region (divide by 50 nm for the region -20 to 30 nm).
4. Draw concentric rings to represent circular slices of the NPC along the axis and fill the rings with grayscale color based on the relative density for *r*-dimension by Photoshop, like the inset picture in Fig. 4d highlighted in blue. The highest density area should be blue (100 %), while the areas lacking density will be completely white (0 %).
5. Repeat **steps 1–4** for other *x*-dimensional regions (*see* **Note 16**).
6. Exclude any region that demonstrates a random distribution, such as region I and VII in Fig. 5.
7. Assemble the 3D structure by combining and stacking the generated ring picture obtained for each *x*-dimensional region selected by the program Amira (Fig. 4e).

4 Notes

1. IPTG is a lactose metabolite that triggers transcription of the lac operon, where the lacZ gene is replaced with the gene of interest and IPTG is then used to induce gene expression. This is used for protein constructs under the regulation of T7 promoters.
2. The purification of the target protein should be more than 90 % for single-molecule experiments.

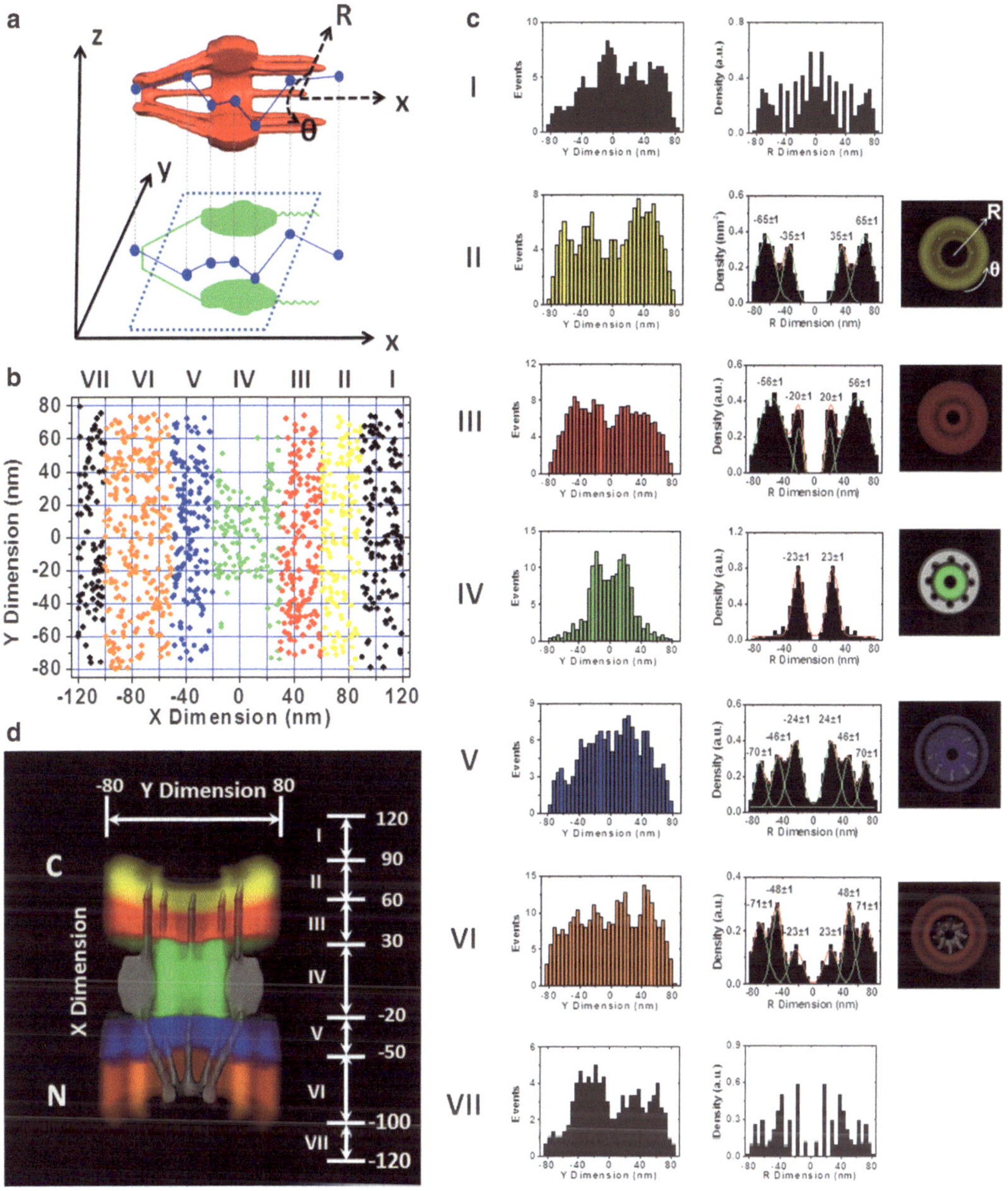

Fig. 6 3D spatial density map of the interaction sites of Imp β1 in the NPCs [21]. (**a**) The diagram demonstrates that the 2D spatial locations of Imp β1 obtained in the focal plane of the objective (the corresponding *xy* plane) are projections of the actual 3D spatial locations in the NPC. The 3D spatial locations can be described by a cylindrical coordinate system (*R*, *y*, *x*). Coordinate *x* represents the location along the NPC axis, whereas (*R*, *y*) refers to the positions at the cross section of the NPC. (**b**) The 2D spatial locations of Imp β1 within the NPC are spatially divided into ranges I–VII and purposely plotted in seven different colors. The separated regions are based on the different clusters of the interaction sites between Imp β1 and the FG repeats in the NPC as shown in (**c** and **d**). (**c**) Histograms of the spatial densities of the Imp β1 molecules along the radii (*R*) at the cross section of the NPC in ranges I–VII. The first column contains the histograms of the spatial locations of Imp β1 in ranges I–VII along the *y*-dimension in (**b**), the second column contains the corresponding histograms in theFig. 6 (continued) *R* dimension after the deconvolution process (the major peaks are obtained by Gaussian fittings), and the third column is the cross-sectional image of each region. Because the Imp β1 spatial locations are randomly distributed outside the NPC, there is no interaction group in ranges I or VII. Bin size: 5 nm. (**d**) Cutaway view of the 3D spatial density map of Imp β1 superimposed on the NPC architecture (*gray*). Different colored regions illustrate different spatial configurations driven by the interactions between Imp β1 and the FG repeats in the NPC. The *numbers* in nanometers denote the distance from the centroid of the NPC

3. Check the protein information to make sure there are surface cysteines available for labeling. For proteins without cysteines available, it could be labeled with another strategy.
4. The concentration of the free dye should be determined and be controlled to be less than 1 % of the target dye.
5. For a labeling ratio >1, the activity of the target molecule should be checked and the final concentration for the experiment should be calculated with the labeled molecule. For a labeling ratio <1, the final concentration for the experiment should be calculated by the real concentration of the labeled protein, which equals the concentration of the protein multiplied by the labeling ratio. The effect of unlabeled protein should be considered.
6. We usually prefer to have two chambers made on one cover slip to allow two different conditions for the experiment. Leakage under the grease should be checked to avoid contamination between the two channels.
7. All the DMEM media should be removed prior to experimentation and replaced with transport buffer.
8. Monitor the permeabilization in real time using bright field imaging. The intact nucleus, together with a controlled cytoplasmic environment, provides a greatly simplified transport system that allows elucidation of the complicated nuclear transport step by step.
9. Before single-molecule experiments, the localization precision and photobleaching time of the single cargo molecule should be determined. Plate 10 pM cargo molecules on a glass cover slip, put the cover slip in 4 °C for 30 min in the dark, and wash away the moving molecules after taking it back to microscope for imaging. Thus, only the immobile fraction of cargo molecules is left after the above process due to nonspecific absorption. The intensity and time course of the fluorescence can then be measured. The intensity could be transformed to figure out the number of emitted photons to determine the localization precision for stationary single molecules. For example, the localization precision of Importin-β1 bearing four copies of Alexa Fluor 647 is about 9 nm, emitting over 1100 photons per frame (0.4 ms). By adjusting laser power, the photobleaching time should be more than tenfold the time needed for nuclear transport of the targeted cargo.
10. Fluorescent cargo concentrations were >100 nM in bulk experiments, and ~100 pM for single-molecule assays. At these concentrations, >99 % of the cargo is expected to form complexes with Importin-α and Importin-β1 at 0.5 μM and lone single molecules are able to be imaged.

11. Examine the ratio of Gaussian widths in the long and short axis of the chosen GFP-NPC fluorescence spot. The ratio needs to fall in between 1.74 and 1.82. Within this range, an illuminated NPC only has a free angle of 1.4° to the perpendicular direction to the NE.
12. Be careful not to touch the sample. Take an image of the NE before and after the experiment to help to determine if the cell was shifted.
13. The drift of the optical system should be determined before the experiment. The total detection time should be less than the time by which a large drift in precision can occur.
14. The position of the NPC could be rechecked with the NE and with the symmetry of the *y*-dimension distribution of single-molecule spatial locations.
15. The single-molecule locations should be filtered with high signal-to-noise ratio and by the width and the intensity of the fluorescent spot in order to exclude the points with bad precision or outside the image plane.
16. The separate *x*-dimensional subregions seen in Fig. 5 (represented by different colors (I–VII)) are determined by pooling together the positional data demonstrating similar *r*-dimensional distributions for each 10-nm bin of *r*-dimensional data along the *x*-dimension. A new subregion is started once the *r*-dimensional distribution pattern differs from the previous 10-nm selection. A computer program is used to determine the best fit for each subregion.

Acknowledgements

The work was supported by grants from the National Institutes of Health (NIH GM094041 and GM097037 to W.Y.).

References

1. Tran EJ, Wente SR (2006) Dynamic nuclear pore complexes: life on the edge. Cell 125: 1041–1053
2. Beck M et al (2004) Nuclear pore complex structure and dynamics revealed by cryoelectron tomography. Science 306:1387–1390
3. Terry LJ, Wente SR (2009) Flexible gates: dynamic topologies and functions for FG nucleoporins in nucleocytoplasmic transport. Eukaryot Cell 8:1814–1827
4. Xu SL, Powers MA (2009) Nuclear pore proteins and cancer. Semin Cell Dev Biol 20:620–630
5. Capelson M, Hetzer MW (2009) The role of nuclear pores in gene regulation, development and disease. EMBO Rep 10:697–705
6. Macara IG (2001) Transport into and out of the nucleus. Microbiol Mol Biol Rev 65: 570–594
7. Peters R (2005) Translocation through the nuclear pore complex: selectivity and speed by reduction-of-dimensionality. Traffic 6:421–427
8. Lim RYH et al (2007) Nanomechanical basis of selective gating by the nuclear pore complex. Science 318:640–643

9. Ribbeck K, Gorlich D (2001) Kinetic analysis of translocation through nuclear pore complexes. EMBO J 20:1320–1330
10. Frey S, Gorlich D (2007) A saturated FG-repeat hydrogel can reproduce the permeability properties of nuclear pore complexes. Cell 130: 512–523
11. Yamada J et al (2010) A bimodal distribution of two distinct categories of intrinsically disordered structures with separate functions in FG nucleoporins. Mol Cell Proteomics 9: 2205–2224
12. Yang WD, Gelles J, Musser SM (2004) Imaging of single-molecule translocation through nuclear pore complexes. Proc Natl Acad Sci U S A 101:12887–12892
13. Yang WD, Musser SM (2006) Nuclear import time and transport efficiency depend on importin beta concentration. J Cell Biol 174:951–961
14. Sun C, Yang W, Tu LC, Musser SM (2008) Single-molecule measurements of importin alpha/cargo complex dissociation at the nuclear pore. Proc Natl Acad Sci U S A 105:8613–8618
15. Dange T et al (2008) Autonomy and robustness of translocation through the nuclear pore complex: a single-molecule study. J Cell Biol 183:77–86
16. Kubitscheck U et al (2005) Nuclear transport of single molecules: dwell times at the nuclear pore complex. J Cell Biol 168:233–243
17. Lowe AR et al (2010) Selectivity mechanism of the nuclear pore complex characterized by single cargo tracking. Nature 467:600–603
18. Ma J, Yang WD (2010) Three-dimensional distribution of transient interactions in the nuclear pore complex obtained from single-molecule snapshots. Proc Natl Acad Sci U S A 107:7305–7310
19. Ma J et al (2012) Self-regulated viscous channel in the nuclear pore complex. Proc Natl Acad Sci U S A 109:7326–7331
20. Ma J et al (2013) High-resolution three-dimensional mapping of mRNA export through the nuclear pore. Nat Commun 4:2414
21. Goryaynov A, Ma J, Yang W (2012) Single-molecule studies of nuclear transport: from one dimension to three dimensions. Integr Biol 4:10–21

Index

Sue Shackleton et al. (eds.), *The Nuclear Envelope: Methods and Protocols*, Methods in Molecular Biology, vol. 1411, DOI 10.1007/978-1-4939-3530-7, © Springer Science+Business Media New York 2016

E

F

G

H

I

MIX
Papier aus verantwortungsvollen Quellen
Paper from responsible sources
FSC® C105338

If you have any concerns about our products,
you can contact us on
ProductSafety@springernature.com

In case Publisher is established outside the EU,
the EU authorized representative is:
Springer Nature Customer Service Center GmbH
Europaplatz 3, 69115 Heidelberg, Germany

Printed by Libri Plureos GmbH
in Hamburg, Germany